Environmental Pollution: Evaluation, Assessment and Mitigation

Environmental Pollution: Evaluation, Assessment and Mitigation

Editor: Nathan Payne

CALLISTO REFERENCE

www.callistoreference.com

Callisto Reference,
118-35 Queens Blvd., Suite 400,
Forest Hills, NY 11375, USA

Visit us on the World Wide Web at:
www.callistoreference.com

ISBN: 978-1-63239-940-3 (Hardback)

Cataloging-in-Publication Data

Environmental pollution : evaluation, assessment and mitigation / edited by Nathan Payne.
    p. cm.
Includes bibliographical references and index.
ISBN 978-1-63239-940-3
1. Pollution. 2. Environmental impact analysis. 3. Pollution--Risk assessment.
4. Environmental protection. I. Payne, Nathan.
TD174 .E58 2018
363.73--dc23

Table of Contents

Preface.. VII

Chapter 1   **Estimating Soil Organic Carbon Stocks and Spatial Patterns with Statistical and GIS-Based Methods**.. 1
Junjun Zhi, Changwei Jing, Shengpan Lin, Cao Zhang, Qiankun Liu,
Stephen D. DeGloria, Jiaping Wu

Chapter 2   **Dynamics of Soil Organic Carbon and Microbial Biomass Carbon in Relation to Water Erosion and Tillage Erosion**..9
Nie Xiaojun, Zhang Jianhui, Su Zhengan

Chapter 3   **Latitudinal Gradients in Degradation of Marine Dissolved Organic Carbon**............................... 16
Carol Arnosti, Andrew D. Steen, Kai Ziervogel, Sherif Ghobrial, Wade H. Jeffrey

Chapter 4   **Environmental Impact of the Production of Mealworms as a Protein Source for Humans – A Life Cycle Assessment**...22
Dennis G. A. B. Oonincx, Imke J. M. de Boer

Chapter 5   **Phosphoglycerate Mutases Function as Reverse Regulated Isoenzymes in *Synechococcus elongatus* PCC 7942**... 27
Jiri Jablonsky, Martin Hagemann, Doreen Schwarz, Olaf Wolkenhauer

Chapter 6   **Identifying Like-Minded Audiences for Global Warming Public Engagement Campaigns: An Audience Segmentation Analysis and Tool Development**... 34
Edward W. Maibach, Anthony Leiserowitz, Connie Roser-Renouf, C. K. Mertz

Chapter 7   **The Environmental Price Tag on a Ton of Mountaintop Removal Coal**....................................... 43
Brian D. Lutz, Emily S. Bernhardt, William H. Schlesinger

Chapter 8   **The Role of Stream Water Carbon Dynamics and Export in the Carbon Balance of a Tropical Seasonal Rainforest, Southwest China**... 48
Wen-Jun Zhou, Yi-Ping Zhang, Douglas A. Schaefer, Li-Qing Sha, Yun Deng,
Xiao- Bao Deng, Kai-Jie Dai

Chapter 9   **Biomass Partitioning and its Relationship with the Environmental Factors at the Alpine Steppe in Northern Tibet**...57
Jianbo Wu, Jiangtao Hong, Xiaodan Wang, Jian Sun, Xuyang Lu, Jihui Fan,
Yanjiang Cai

Chapter 10   **Comparisons of Three Methods for Organic and Inorganic Carbon in Calcareous Soils of Northwestern China**... 65
Xiujun Wang, Jiaping Wang, Juan Zhang

Chapter 11   **Life Cycle Assessment of Metals: A Scientific Synthesis**... 71
Philip Nuss, Matthew J. Eckelman

Chapter 12   **Environmental Management in Small and Medium-Sized Companies: An Analysis from the Perspective of the Theory of Planned Behavior**................................. 83
Agustín J. Sánchez-Medina, Leonardo Romero-Quintero, Silvia Sosa-Cabrera

Chapter 13   **Multi-Scale Measures of Rugosity, Slope and Aspect from Benthic Stereo Image Reconstructions**.. 95
Ariell Friedman, Oscar Pizarro, Stefan B. Williams, Matthew Johnson-Roberson

Chapter 14   **Analysis of Changes in Traumatic Symptoms and Daily Life Activity of Children Affected by the 2011 Japan Earthquake and Tsunami over Time**......................... 109
Masahide Usami, Yoshitaka Iwadare, Kyota Watanabe, Masaki Kodaira, Hirokage Ushijima, Tetsuya Tanaka, Maiko Harada, Hiromi Tanaka, Yoshinori Sasaki, Kazuhiko Saito

Chapter 15   **Potential of Best Practice to Reduce Impacts from Oil and Gas Projects in the Amazon**... 117
Matt Finer, Clinton N. Jenkins, Bill Powers

Chapter 16   **Regulatory Response to Carbon Starvation in *Caulobacter crescentus**......................... 131
Leticia Britos, Eduardo Abeliuk, Thomas Taverner, Mary Lipton, Harley McAdams, Lucy Shapiro

Chapter 17   **Spatial Distribution of Soil Organic Carbon and its Influencing Factors in Desert Grasslands of the Hexi Corridor, Northwest China**.................................... 150
Min Wang, Yongzhong Su, Xiao Yang

Chapter 18   **Rapid Response of Hydrological Loss of DOC to Water Table Drawdown and Warming in Zoige Peatland: Results from a Mesocosm Experiment**........................ 158
Xue-Dong Lou, Sheng-Qiang Zhai, Bing Kang, Ya-Lin Hu, Li-Le Hu

Chapter 19   **Carbon Footprint of Telemedicine Solutions - Unexplored Opportunity for Reducing Carbon Emissions in the Health Sector**.. 167
Åsa Holmner, Kristie L. Ebi, Lutfan Lazuardi, Maria Nilsson

Chapter 20   **Combining XCO$_2$ Measurements Derived from SCIAMACHY and GOSAT for Potentially Generating Global CO$_2$ Maps with High Spatiotemporal Resolution**.............................. 177
Tianxing Wang, Jiancheng Shi, Yingying Jing, Tianjie Zhao, Dabin Ji, Chuan Xiong

Permissions

List of Contributors

Index

Preface

Environmental pollution is defined as the disposal of contaminants and unwanted material that can cause damage to the environment. Pollution can be classified as point source and non-point source pollution. Air dispersion models, soil sample assessments, radiation dose monitoring, oxygen reduction potential are some of the measurement techniques related to the various types of pollution. This book is compiled in such a manner, that it will provide in-depth knowledge about the theory and practice of environmental pollution assessment and mitigation. It is a resource guide for experts as well as students.

This book is a result of research of several months to collate the most relevant data in the field.

When I was approached with the idea of this book and the proposal to edit it, I was overwhelmed. It gave me an opportunity to reach out to all those who share a common interest with me in this field. I had 3 main parameters for editing this text:

1. Accuracy – The data and information provided in this book should be up-to-date and valuable to the readers.

2. Structure – The data must be presented in a structured format for easy understanding and better grasping of the readers.

3. Universal Approach – This book not only targets students but also experts and innovators in the field, thus my aim was to present topics which are of use to all.

Thus, it took me a couple of months to finish the editing of this book.

I would like to make a special mention of my publisher who considered me worthy of this opportunity and also supported me throughout the editing process. I would also like to thank the editing team at the back-end who extended their help whenever required.

Editor

Estimating Soil Organic Carbon Stocks and Spatial Patterns with Statistical and GIS-Based Methods

Junjun Zhi[1], Changwei Jing[2], Shengpan Lin[1], Cao Zhang[1], Qiankun Liu[1], Stephen D. DeGloria[3], Jiaping Wu[2]*

1 College of Environmental and Resource Sciences, Zhejiang University, Hangzhou, China, 2 Ocean College, Zhejiang University, Hangzhou, China, 3 Department of Crop and Soil Sciences, Cornell University, Ithaca, New York, United States of America

Abstract

Accurately quantifying soil organic carbon (SOC) is considered fundamental to studying soil quality, modeling the global carbon cycle, and assessing global climate change. This study evaluated the uncertainties caused by up-scaling of soil properties from the county scale to the provincial scale and from lower-level classification of Soil Species to Soil Group, using four methods: the mean, median, Soil Profile Statistics (SPS), and pedological professional knowledge based (PKB) methods. For the SPS method, SOC stock is calculated at the county scale by multiplying the mean SOC density value of each soil type in a county by its corresponding area. For the mean or median method, SOC density value of each soil type is calculated using provincial arithmetic mean or median. For the PKB method, SOC density value of each soil type is calculated at the county scale considering soil parent materials and spatial locations of all soil profiles. A newly constructed 1:50,000 soil survey geographic database of Zhejiang Province, China, was used for evaluation. Results indicated that with soil classification levels up-scaling from Soil Species to Soil Group, the variation of estimated SOC stocks among different soil classification levels was obviously lower than that among different methods. The difference in the estimated SOC stocks among the four methods was lowest at the Soil Species level. The differences in SOC stocks among the mean, median, and PKB methods for different Soil Groups resulted from the differences in the procedure of aggregating soil profile properties to represent the attributes of one soil type. Compared with the other three estimation methods (i.e., the SPS, mean and median methods), the PKB method holds significant promise for characterizing spatial differences in SOC distribution because spatial locations of all soil profiles are considered during the aggregation procedure.

Editor: Ben Bond-Lamberty, DOE Pacific Northwest National Laboratory, United States of America

Funding: This study was supported by the National Natural Science Foundation of China (No. 30771253) and the Key Project of the Science and Technology Department of Zhejiang Province (No. 2006C22026). The funders had no role in study design, data collection and analysis, decision to publish, or preparation of the manuscript.

Competing Interests: The authors have declared that no competing interests exist.

* E-mail: jw67@zju.edu.cn

Introduction

Soil organic carbon (SOC), which plays a critical role in the global carbon cycle, comprises a major part of the terrestrial carbon reservoir [1–3]. In terrestrial ecosystems, SOC stock is almost three times the size of carbon storage in the vegetation of terrestrial ecosystems [4], and approximately twice as large as carbon storage in the atmosphere [5]. Because of the important role of SOC and its large quantity stored in terrestrial ecosystems, a slight change in SOC stock may influence global climate [6–8]. Accurately quantifying SOC stock has become a focus of present research on global climate change, and is considered essential for studying soil quality, modeling the global carbon cycle, and assessing global climate change [8–12].

Different SOC stock estimates, however, can vary greatly at both global and regional scales [13–17]. For global scales, Bohn [18] estimated the total SOC stock was 3,000 Pg (1 Pg = 10^{15} g), whereas Bolin [19] estimated only 710 Pg, over a four-fold difference. In China, estimated SOC stocks for terrestrial ecosystems range from 50 Pg [20] to 185.7 Pg [21], also approximately a four-fold difference. The methodology for estimating SOC stocks, such as soil profile-based [22–24] and model-based (e.g., the CENTURY [25] and DeNitrification-DeComposition models [26]), is one of the main factors attributed to the wide range of differences in SOC stock estimation from different studies. The soil profile-based methodology calculates SOC stock using soil profiles and their corresponding areas obtained from soil survey products (the soil type method), vegetation type maps (the vegetation type method), or life zone maps (the life zone method), among which the soil type method is the most widely used [9,18–19,27–28]. According to the sources of soil area, the soil type method contains the Soil Profile Statistics (SPS) method and the GIS-based Soil Type (GST) method [2]. The SPS method calculates SOC stock by multiplying the SOC density value of a soil type by its corresponding field survey area recorded in soil survey reports (e.g., Soil Species of China [29]). The GIS-based Soil Type method calculates areas of various soil types accurately based on digital soil map and can provide information on the spatial distribution of SOC stocks.

In China, different scale (i.e., county, provincial, and national) soil survey products (e.g., soil survey reports, soil maps) of the Second National Soil Survey of China are the most important data sources for SOC stock estimations [15,20–21,30–33]. For the

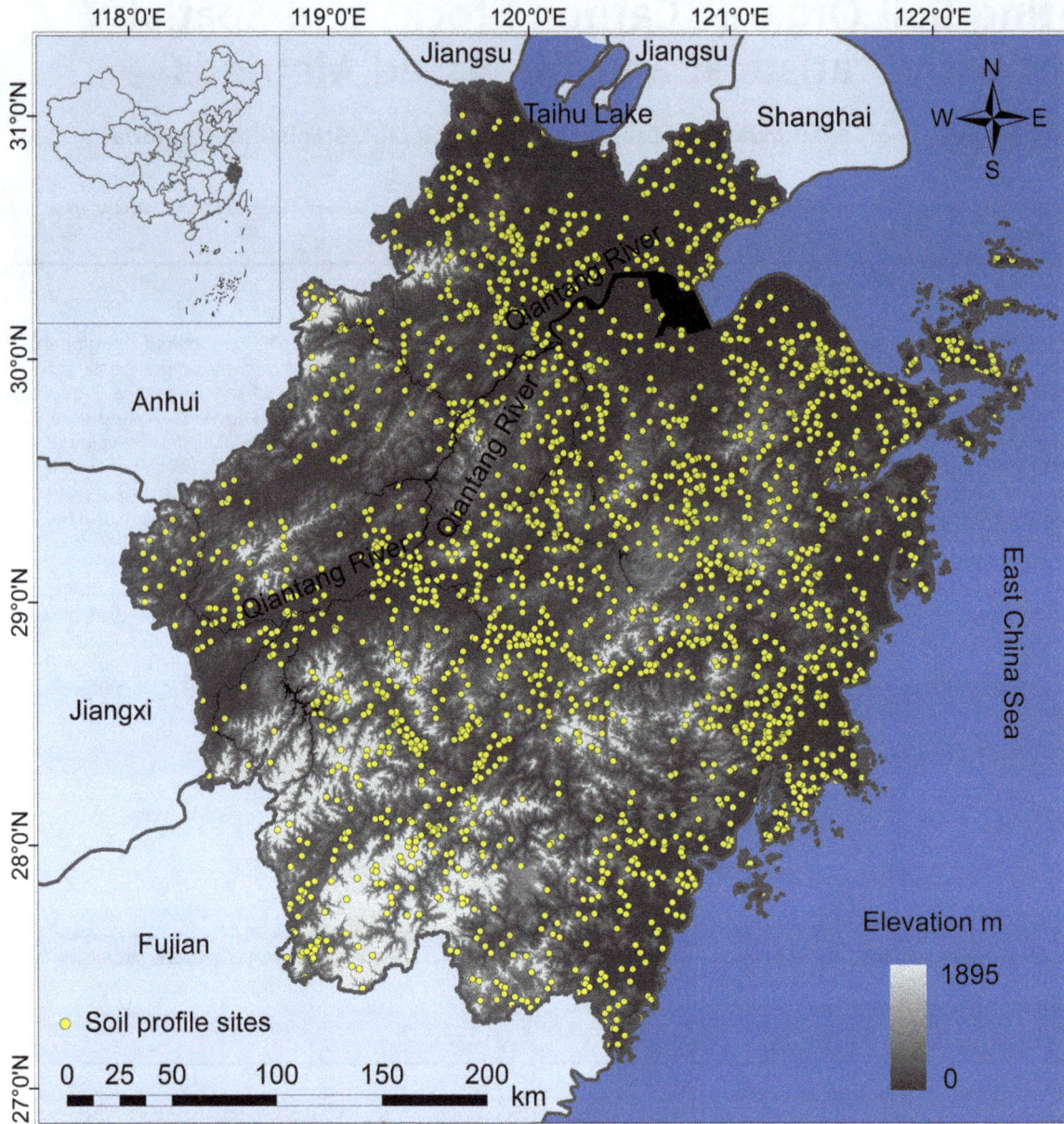

Figure 1. Study area and distribution of soil profile sites.

Second National Soil Survey of China, soil profile properties were aggregated sequentially in the county, provincial, and national scales. Different number of sampling profiles can be used to represent the attributes of one soil type at different scales and can cause uncertainties in SOC stock estimation. At the provincial scale, calculating the mean and median of SOC density values of multiple soil profiles of the same soil type name to represent the SOC density value of that soil type are the two most commonly used GIS-based Soil Type methods [32–34]. At the county scale, the SPS method [20,30–32] and the pedological professional knowledge based (PKB) method [2,15,34–36] are the two most commonly used methods. Compared with the median or mean method, the PKB method aggregates soil profile properties downscaling from the provincial scale to more detail soil map units, which links a SOC density value of each soil profile to a digital soil map according to the identity or similarity in soil parent

materials and spatial locations of all soil profiles at the county scale. However, the uncertainties of soil profile properties caused by up-scaling from the county scale to the provincial scale among the four methods remain unknown.

Soil map up-scaling includes soil classification level up-scaling and resolution up-scaling. The former aggregates Soil Species to higher soil classification levels (e.g., Soil Group, Subgroup, Soil Family); the latter aggregates higher resolution soil maps to lower resolution ones, such as from 1:50,000-county-scale map to 1:1,000,000-country-scale map. Both the soil classification level up-scaling and resolution up-scaling can cause uncertainties in SOC stock estimation. Zhao et al. [34] tested the effects of soil map resolution up-scaling from 1:500,000 to 1:10,000,000 using the three GIS-based Soil Type methods (i.e., the mean, median, and PKB methods) on SOC stock estimation for Hebei Province, China. However, the effects of these three GIS-based Soil Type

Table 1. Methods used to estimate SOC (soil organic carbon) stocks in Zhejiang Province, China.

Method	SOC density value	Scale	Note
SPS[a]	Mean	County; soil species	(1) One soil species has multiple areas, which were surveyed county by county. (2) Mean SOC density value was calculated from one or multiple profiles within the county. (3) 2154 soil profiles were used.
Mean or median	Mean or median	Province; soil species	(1) One soil species has one area calculated from the digital soil map. (2) Mean or median SOC density value was calculated from one or multiple profiles within the Province. (3) 2154 soil profiles were used.
PKB[a]	Mean	County; soil map unit	(1) Soil map units were derived from the digital soil map county by county. (2) One soil map unit in one county may have one or multiple areas calculated from the digital soil map. (3) Mean SOC density value was calculated from one or multiple profiles located within one polygon; polygons belong to one soil map unit in one county may assigned different SOC density values. (4) 2154 soil profiles were used.

[a]SPS, Soil Profile Statistics; PKB, pedological professional knowledge based.

methods on SOC stock estimation at soil map resolution larger than 1:500,000 remain unknown. Moreover, the uncertainties in SOC stock estimation caused by soil classification level up-scaling using the SPS, mean, median, and PKB methods remain unknown.

The goal of this study is to evaluate the uncertainties of changes in scale among the four estimation methods (i.e., the SPS, mean, median, and PKB methods) by using a newly completed 1:50,000 soil survey geographic database of Zhejiang Province, China. Specifically, the main objectives of this study were to: (1) evaluate the uncertainties caused by up-scaling of soil profile properties from the county scale to the provincial scale and soil classification levels from Soil Species to Soil Group among the four methods; and (2) quantify spatial differences in SOC stock estimation among the three GIS-based Soil Type methods.

Materials and Methods

Study Area

Zhejiang Province is located between 27°06′ and 31°03′ N latitude and 118°01′ and 123°10′ E longitude in southeastern China (Figure 1). The Province covers a land area of $1.02*10^5$

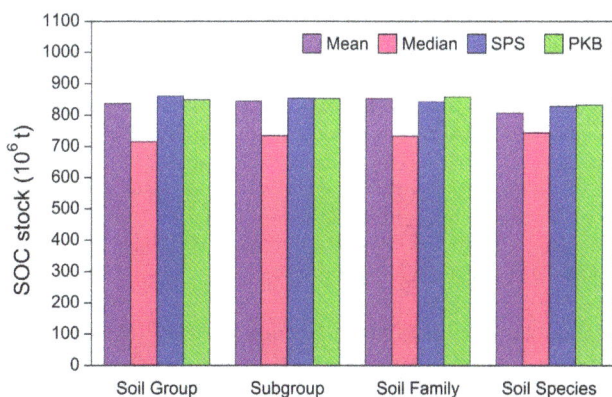

Figure 2. The estimates of SOC (soil organic carbon) stocks for soil classification levels up-scaling from Soil Species to Soil Group using the mean, median, SPS (soil profile statistic), and PKB (pedological professional knowledge based) methods.

km^2. Hills and mountains are dominant terrain, occupying 70.4% of the total land area. The plains and basins account for 23.2%, and the remaining 6.4% is comprised of lakes and rivers. Provincial topography is characterized by high mountains in the southwest and low plains in the northeast. The elevation ranges from 0 to 1895 m with an average of 296 m above sea level.

With a subtropical monsoon climate, Zhejiang Province has an annual average temperature from 15 to 18°C and an annual precipitation of 1200 to 1800 mm. As substantial differences exist in climate, geomorphology, geohydrology, land use, and parent material throughout the Province, its vegetation and soils and their spatial distribution patterns vary greatly. According to Soils of Zhejiang Province [37], Red soils are the dominant Soil Group, accounting for approximately 39% of the total area of Zhejiang soils.

Data Sources

A recently completed 1:50,000 soil survey geographic database of Zhejiang Province [38] was used in the study, which is the most detailed soil survey data at the provincial scale in China to date. The database includes Soil Spatial Data and Soil Attributes Data. The Soil Spatial Data include a 1:50,000 digital soil map of the Province and a digital map of 2154 geo-referenced soil sampling profile sites, both of which were derived by digitizing and re-compiling field soil survey maps at 1:50,000 from 76 counties in Zhejiang. The Soil Attributes Data contain the properties of the 2154 geo-referenced soil sampling profiles, which were taken from original county soil survey reports. For each profile, there are one to seven soil layers; for each layer, there are up to 104 soil physical and chemical properties, including geographic location, depth, bulk density, organic matter content, and gravel content, etc.

The soil survey geographic database was used to estimate SOC stocks with the mean, median, and PKB methods. This information from the Second National Soil Survey of China conducted in early 1980's was the most comprehensive and detailed inventory of soil characteristics in Zhejiang Province. The Genetic Soil Classification System of Zhejiang Province was used during the field soil survey. The soils were classified using a six-level hierarchical scheme: Order, Suborder, Group, Subgroup, Family, and Species. The Soil Species is the basic classification level, and the Soil Group is the most commonly used classification level in China [2]. According to the Genetic Soil Classification

Table 2. Descriptive statistics of profile soil organic carbon densities by Soil Group.

Soil Group	Soil Order of U.S. Taxonomy[a]	N[b]	Range	Min[b]	Max[b]	Mean	Median	SD[b]	CV/%[b]	Skew[b]	Kurt[b]
Red soils	Alfisols, ultisols, inceptisols	372	32.98	0.87	33.85	6.97	6.15	3.93	15.5	2.29	9.48
Yellow soils	Alfisols, inceptisols	126	91.26	1.93	93.19	16.56	13.39	14.61	213.3	3.25	13.16
Purple soils	Inceptisols, entisols	84	12.10	0.29	12.38	5.19	4.84	2.36	5.6	0.70	0.28
Limestone soils	Mollisols, inceptisols	22	20.52	1.95	22.47	9.14	9.68	5.04	25.4	0.51	0.92
Skel soils	Inceptisols, entisols	114	36.71	0.10	36.81	4.89	3.30	5.27	27.8	3.56	16.24
Red clay soils	Inceptisols, alfisols	4	2.22	4.80	7.02	5.48	5.04	1.04	1.1	1.92	3.75
Mountain meadow soils	Histosols, inceptisols	4	241.54	37.98	279.52	104.32	49.89	116.98	13680.0	1.98	3.94
Fluvio-aquic soils	Inceptisols, entisols	189	20.34	0.94	21.28	6.38	5.78	3.48	12.1	0.96	1.66
Coastal saline soils	Inceptisols	64	14.25	0.92	15.17	7.37	7.50	3.19	10.2	−0.05	−0.55
Paddy soils	Anthrosols	1175	143.90	1.97	145.87	9.82	8.54	7.13	50.9	8.10	124.50
All profiles		2154	279.42	0.10	279.52	9.06	7.62	9.41	88.6	14.17	346.41

[a]Reference conversion between Soil Group of the Genetic Soil Classification System of Zhejiang Province and Soil Order of the U.S. Taxonomy.
[b]N, Min, Max, SD, CV, Skew, Kurt are the abbreviations of the number of soil profiles occurring in a Soil Group, minimum, maximum, standard deviation, coefficient of variation, skewness and kurtosis, respectively.

System of Zhejiang Province, the collected 2154 soil profiles originated from 10 Soil Groups, 21 Subgroups, 99 Soil Families and 277 Soil Species.

Estimation of Organic Carbon Density for Soil Profiles

For each of the 2 154 soil profiles, SOC density was calculated with the formula [34,39]:

$$\text{SOCD}_D = \sum_{i=1}^{n} [(1 - \theta_i\%) \times \rho_i \times C_i \times T_i / 100] \qquad (1)$$

where SOCD_D (kg m^{-2}) is the SOC density of a soil profile within a depth of D (cm), n is the number of soil layers in the soil survey, $\theta_i\%$ represents the volumetric percentage of gravel (>2 mm) content, ρ_i is the soil bulk density (g/cm^3), C_i is the organic carbon content (C g/kg), and T_i represents the thickness (cm) of the layer i. Density of SOC was estimated to a maximum depth of 100 cm to facilitate comparison among data sets. For profiles with actual depths greater than or equal to 100 cm, but less than 100 cm was observed, data for the unobserved profile sections were derived from the mean values of all the corresponding soil profiles of the same Soil Species or Soil Family [15,33,40]. Organic carbon content is calculated by multiplying soil organic matter content by 0.58 (the Bemmelen index), which is based on the assumption that soil organic matter contains approximately 58% organic carbon [41].

Estimation of Regional Soil Organic Carbon Stocks

The SOC stocks of the Province were estimated by using the SPS, mean, median, and PKB methods (Table 1).

For the SPS method, the mean SOC density value of each soil type in one county was calculated from one or multiple profiles belonging to that soil type; then the SOC stock value for that soil type was calculated by multiplying its SOC density value by its corresponding area; finally, the SOC stock values of all soil types were summed up as the total SOC stock for the Province. The area calculated from the 1:50,000 digital soil map was used in this study to facilitate comparison among different methods.

For the mean or median method, the arithmetic mean or median of SOC density values of all soil profiles of the same soil type name in the Province was calculated first as the SOC density value of that soil type; secondly, SOC stocks for polygons on the 1:50,000 digital soil map were calculated by multiplying the mean or median values by corresponding areas of the soil types and then were summed up as the total stock for the Province.

For the PKB method, the SOC density value of each of the 2154 soil profiles was linked to corresponding polygons on the 1:50,000 digital soil map according to the same soil type name, the soil parent materials, and the spatial locations by county. When two or more soil profiles of the same soil type name were located in one polygon, the mean SOC density value of these soil profiles was calculated and used for the linkage. The SOC stock values for all polygons were calculated and then were summed up as the total stock for the Province.

The estimation of SOC stocks using the three GIS-based Soil Type methods were performed using ESRI software ArcGIS 10.0 (Redlands, CA) with water and urban areas excluded from the calculation.

Table 3. Estimates of SOC (soil organic carbon) stocks and SOC density values of various Soil Groups using the mean, median, SPS (soil profile statistic), and PKB (pedological professional knowledge based) methods.

Soil Group	Area[a] (km²)	PKB SOC density (kg m⁻²)	PKB SOC stock (10⁶ t)	Mean SOC density (kg m⁻²)	Mean SOC stock (10⁶ t)	Mean %[b]	Median SOC density (kg m⁻²)	Median SOC stock (10⁶ t)	Median %[b]	SPS SOC density (kg m⁻²)	SPS SOC stock (10⁶ t)	SPS %[b]
Red soils	39681.3	6.53	259.10	6.73	266.96	3.0	6.24	247.71	−4.4	6.50	258.11	−0.4
Yellow soils	10013.5	16.92	169.45	14.00	140.15	−17.3	12.24	122.54	−27.7	16.72	167.44	−1.2
Purple soils	3597.7	5.73	20.60	5.00	18.01	−12.6	4.77	17.15	−16.7	5.65	20.34	−1.2
Limestone soils	1571.0	10.18	16.00	8.56	13.45	−15.9	9.43	14.82	−7.4	10.47	16.45	2.8
Skel soils	13736.8	5.11	70.20	5.26	72.30	3.0	4.29	58.99	−16.0	5.09	69.95	−0.4
Red clay soils	29.1	5.45	0.16	5.45	0.16	0.0	5.45	0.16	0.0	5.45	0.16	0.0
Mountain meadow soils	3.3	45.30	0.15	104.32	0.34	130.3	49.89	0.16	10.1	45.30	0.15	0.0
Fluvio-aquic soils	4318.3	7.43	32.10	7.33	31.66	−1.4	7.17	30.98	−3.5	7.44	32.13	0.1
Coastal saline soils	2793.3	7.79	21.75	7.58	21.19	−2.6	7.50	20.95	−3.7	7.80	21.78	0.1
Paddy soils	24995.8	9.68	241.98	9.67	241.66	−0.1	9.20	229.86	−5.0	9.64	241.06	−0.4
Total soils	100740.1	8.25	831.49	8.00	805.88	−3.1	7.38	743.32	−10.6	8.21	827.57	−0.5

[a]Area used for the four methods in this study was from the 1:50,000 digital soil map.
[b]Percentage of difference of estimated SOC stocks between the mean, median, or SPS method and the PKB method.

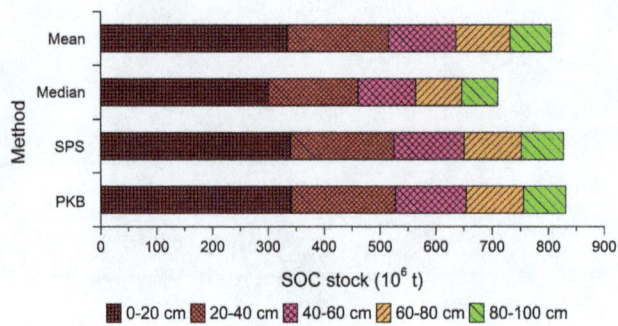

Figure 3. Estimated SOC stocks at different soil depths using the mean, median, SPS (soil profile statistic), and PKB (pedological professional knowledge based) methods.

Results

Effect of Soil Classification Levels Up-Scaling from Soil Species to Soil Group on SOC Estimation Using Different Methods

The SOC stocks for soil classification levels up-scaling from Soil Species to Soil Group using the mean, median, SPS, and PKB methods were calculated (Figure 2). Among the four methods, the estimated SOC stocks for the study area using the median method were always the lowest regardless of soil classification level. Because the SOC densities of the 2154 soil profiles showed a positively skewed distribution with skewness of 14.17 (Table 2), the estimated SOC stocks using the mean method were much higher than those using the median method at all four soil classification levels.

With soil classification levels up-scaling from Soil Species to Soil Group, the estimated SOC stocks presented small variations with coefficient of variations of 2.1%, 1.4%, 1.4%, and 1.2% for the mean, median, SPS, and PKB methods, respectively. The

coefficient of variations of estimated SOC stocks among different methods presented an increasing trend, with 4.4%, 6.2%, 6.3%, and 7.1% for soil classification levels of Soil Species, Soil Family, Subgroup, and Soil Group, respectively. The estimated SOC stocks using the mean, median, and SPS methods at the Soil Species level were 3.1%, 10.6%, and 0.5% lower than that obtained by the PKB method, respectively. The variation of estimated SOC stocks among different methods was obviously larger than that among different soil classification levels.

Effect of Different Methods on the Estimates of SOC Stocks for Various Soil Groups

Soil Group is the most stable and consistent soil classification level commonly used in China [2]. In this study, the SOC stocks and SOC densities for various Soil Groups estimated by the mean, median, SPS, and PKB methods are presented in Table 3. For the Soil Groups with SOC density values of soil profiles presented positively skewed distribution, the estimated SOC stocks using the median method were usually lower than that using the mean and PKB methods (Table 2, Table 3). The estimated SOC stocks for various Soil Groups using the SPS method were most similar to that obtained by the PKB method because both methods aggregate soil profile properties at the county scale. The estimated SOC stocks for various Soil Groups using the mean and median methods showed large differences compared with the PKB method because both the mean and median methods aggregating soil profile properties at the provincial scale that the SOC density values of all soil profiles belonging to a soil type in the Province were aggregated to one value.

The most pronounced differences of the estimated SOC stocks for the ten Soil Groups (i.e., Red soils, Yellow soils, Purple soils, Limestone soils, Skel soils, Red clay soils, Mountain meadow soils, Fluvio-aquic soils, Coastal saline soils and Paddy soils) among the three GIS-based Soil Type methods occurred on Yellow soils and Mountain meadow soils, which presented high SOC density values with high coefficient of variances. Thus, the process of linking the

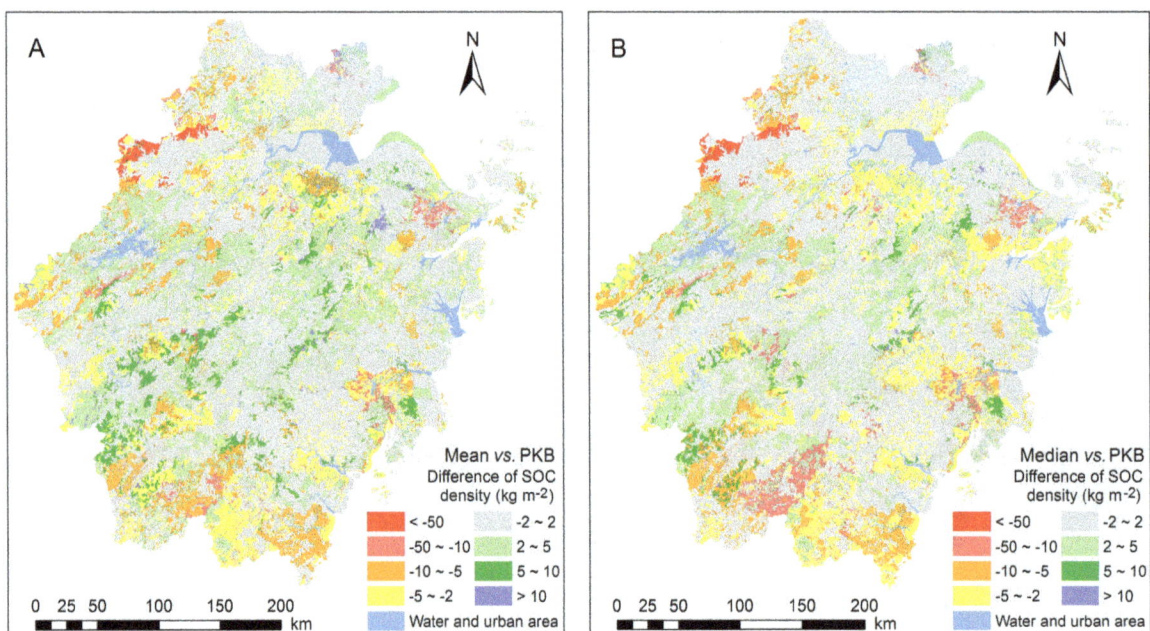

Figure 4. Spatial differences of estimated SOC (soil organic carbon) densities between the mean and PKB (pedological professional knowledge based) methods (A), and between the median and PKB methods (B).

aggregated SOC density values of soil profiles belonging to these two Soil Groups to corresponding polygons on the 1:50,000 digital soil map would significantly influence the estimates of SOC stocks. One important factor leading to the significantly high variation of SOC in Mountain meadow soils was the limited number of soil profiles (n = 4) in the soil survey area.

Effect of Different Methods on the SOC Stock Estimation at Different Soil Depths

For the four methods, the trend of estimated SOC stocks at each of the five soil depths (0–20, 20–40, 40–60, 60–80, and 80–100 cm) is consistent with the soil depth of 0–100 cm, followed the order: PKB>SPS>mean>median (Figure 3). At each of the five soil depths, the estimated SOC stock using the SPS method presented the smallest difference while the median method presented the largest difference compared with the PKB method. Decreasing SOC stock with increasing soil depth was evident independent of estimation method, and these decreasing trends are consistent among all four methods. At the depth of 0–20 cm, SOC stocks estimated by the mean, median, SPS, and PKB methods account for 41.5%, 42.3%, 41.1%, and 41.1%, of total SOC stocks, respectively, while the SOC stocks at the depth of 80–100 cm estimated by the four methods account for only 9.2%, 8.9%, 9.1%, and 9.1%, of the total SOC stocks, respectively.

Comparison of the Differences in the Spatial Patterns of SOC Distribution among the Three GIS-based Soil Type Methods

Estimated SOC densities using three GIS-based Soil Type methods showed clearly spatial differences (Figure 4). The largest differences were as high as -229.63 kg m^{-2} between the median and PKB methods and -175.20 kg m^{-2} between the mean and PKB methods. Polygons mostly belonging to the Soil Groups of Mountain meadow soils and Yellow soils showed substantial differences (<-50 kg m^{-2}) in SOC densities among the three GIS-based Soil Type methods. For the polygons with large spatial differences in SOC densities ranging from -50 kg m^{-2} to -5 kg m^{-2} and >5 kg m^{-2}, the locations were usually in mountain and hilly areas with dominate Soil Groups of Yellow soils, Red soils, and Skew soils, with the remainder in plain areas with dominate Soil Group of Paddy soils. The total areas for spatial differences in SOC densities with absolute value larger than 5 kg m^{-2} between the mean and PKB methods and the median and PKB methods were estimated to be 12883.4 km^2 and 11428.0 km^2, accounting for 12.8% and 11.3% of the total area of Zhejiang soils, respectively. The total areas for spatial differences in SOC densities with absolute value lower than 2 kg m^{-2} between the mean and PKB methods and the median and PKB methods were estimated to be 55316.1 km^2 and 59147.7 km^2, accounting for 54.9% and 58.7% of the total area of Zhejiang soils, respectively.

Discussion

With soil classification levels up-scaling from Soil Species to Soil Group, the estimated SOC stock values using the SPS method were most similar to that obtained by the PKB method. The difference of the estimated SOC stocks between the SPS method and the PKB method occurred in the cases where two or more soil profiles belonging to one soil species existed in one county. However, for most cases, there was only one soil profile belonging

to one soil species in one county, the SOC density value of that soil profile was used to represent the SOC density value of that soil species in that county, which resulted in the same estimated SOC stock value of that soil species in that county using the SPS and PKB methods. Therefore, more number of soil profiles is needed to further compare the difference between these two methods.

The estimated SOC stock values using the mean method were similar to that using the PKB method. When two or more soil profiles located in the same polygon within a county, the mean SOC density value of these soil profiles was calculated and used to link with that polygon, so the PKB method approaches the mean method in linking SOC density values with polygons to some extent [34], especially for higher soil classification levels (e.g., Soil Group, Subgroup, Soil Family).

With soil profile properties up-scaling from the county scale (the SPS and PKB methods) to the provincial scale (the mean and median methods), the SOC density values of all soil profiles belonging to a soil type in the Province were aggregated to one value. The aggregating procedure for the mean or median method could cause large uncertainties because information about how much area one profile can represent was missing, thus different polygons belonging to the same map unit on the digital soil map were linked with the same SOC density value. For the PKB method, different polygons belonging to the same map unit may be assigned different SOC density values depending on their locations, thus has an obvious advantage in the demonstration of spatial differences in SOC distribution. The difference in aggregating procedure led to Soil Groups (e.g., Mountain meadow soils, Yellow soils) with soil profiles of high variations in SOC density values presented significant spatial differences among the three GIS-based Soil Type methods.

Conclusions

The up-scaling of soil classification levels from Soil Species to Soil Group has small effect on SOC stock estimation. However, obvious differences occurred with different estimation methods, especially for Soil Groups with high variations in SOC densities and spatial distribution. The PKB method, which links soil profile properties to spatial databases by county, produced more stable results for soil types, soil parent materials, and spatial locations of all soil profiles under consideration. Thus this method has an obvious advantage in the demonstration of differences in spatial patterns of SOC distribution than both the mean and median methods. We recommend the PKB method as a prior option rather than the mean, median, and SPS methods for SOC stock estimation in China, especially when 1:50,000 soil survey geographic database (soil map) is available. It potentially reduces uncertainties related to up-scaling soil profile properties.

Acknowledgments

We are very grateful to the Academic Editor and two reviewers for their useful comments and suggestions on the original manuscript.

Author Contributions

Conceived and designed the experiments: JJZ JPW. Performed the experiments: JJZ CWJ CZ QKL. Analyzed the data: JJZ CZ QKL. Contributed reagents/materials/analysis tools: JJZ. Wrote the paper: JJZ. Revised the manuscript: SPL JPW SDD.

References

1. Davidson EA, Janssens IA (2006) Temperature sensitivity of soil carbon decomposition and feedbacks to climate change. Nature 440: 165–173.
2. Zhang Y, Zhao YC, Shi XZ, Lu XX, Yu DS, et al. (2008) Variation of soil organic carbon estimates in mountain regions: a case study from Southwest China. Geoderma 146: 449–456.
3. Zhou Y, Pei Z, Su J, Zhang J, Zheng Y, et al. (2012) Comparing Soil Organic Carbon Dynamics in Perennial Grasses and Shrubs in a Saline-Alkaline Arid Region, Northwestern China. PLoS ONE 7 (8): e42927. doi:10.1371/journal.pone.0042927
4. Schlesinger WH (1990) Evidence from chronosequence studies for a low carbon-storage potential of soils. Nature 348: 232–234.
5. Eswaran H, Van Den Berg E, Reich PF (1993) Organic carbon in soils of the world. Soil Science Society of America Journal 57: 192–194.
6. Lal R, Kimble JM, Follett RF, Cole CV (1998) The Potential of U.S. Cropland to Sequester Carbon and Mitigate the Greenhouse Effect. Chelsea: Ann Arbor Press.
7. Uri ND (2000) Conservation practices in U.S. agriculture and their implication for global climate change. Science of the Total Environment 256: 23–38.
8. Li MM, Zhang XC, Pang GW, Han FF (2013) The estimation of soil organic carbon distribution and storage in a small catchment area of the Loess Plateau. Catena 101: 11–16.
9. Batjes NH (1996) Total carbon and nitrogen in the soils of the world. European Journal of Soil Science 47: 151–163.
10. Tan G, Shibasaki R (2003) Global estimation of crop productivity and the impacts of global warming by GIS and EPIC integration. Ecological Modelling 168: 357–370.
11. Janzen HH (2004) Carbon cycling in earth systems-a soil science perspective. Agriculture, Ecosystems & Environment 104: 399–417.
12. Morisada K, Ono K, Kanomata H (2004) Organic carbon stock in forest soils in Japan. Geoderma 119: 21–32.
13. Wang SQ, Huang M, Shao XM, Robert AM, Li KR, et al. (2004) Vertical distribution of soil organic carbon in China. Environmental Management 33: 200–209.
14. Leifeld J, Bassin S, Fuhrer J (2005) Carbon stocks in Swiss agricultural soils predicted by land-use, soil characteristics, and altitude. Agriculture, Ecosystems & Environment 105: 255–266.
15. Yu DS, Shi XZ, Wang HJ, Sun WX, Chen JM, et al. (2007) Regional patterns of soil organic carbon stocks in China. Journal of Environmental Management 85: 680–689.
16. Arnold RW (1995) Role of soil survey in obtaining a global carbon budget. In: Lal R, Kimble J, Levine E, Stewart BA, editors. Soils and global change. Boca Raton: CRC Press. 257–263.
17. Rapalee G, Trumbore SE, Davidson EA, Harden JW, Veldhuis H (1998) Soil carbon stocks and their rates of accumulation and loss in a boreal forest landscape. Global Biogeochemical Cycles 12: 687–701.
18. Bohn HL (1976) Estimate of organic carbon in world soils. Soil Science Society of America Journal 40: 468–470.
19. Bolin B (1977) Change of land biota and their importance for the carbon cycle. Science 196: 613–615.
20. Pan GX (1999) Study on carbon reservoir in soils of China. Bulletin of Science and Technology 15: 330–332.
21. Fang JY, Liu GH, Xu SL (1996) Carbon sinks of terrestrial ecosystems in China. In: Wang RS, editor. Study on Key Issues of Modern Ecology. Beijing: Press of Chinese Sciences and Technologies, 251–267 p.
22. Homann PS, Sollins P, Fiorella M, Thorson M, Kern JS (1998) Regional soil organic carbon storage estimates for western Oregon. Soil Science Society of America Journal 62: 789–796.
23. Batjes NH (2000) Effects of mapped variation in soil conditions on estimates of soil carbon and nitrogen stocks for South America. Geoderma 97: 135–144.
24. Wu J, Ransom MD, Kluitenberg GJ, Nellis MD, Seyler HL (2001) Land-use management using a soil survey geographic database for Finney County, Kansas. Soil Science Society of America Journal 65: 169–177.
25. Alvaro-Fuentes J, Plaza-Bonilla D, Arrue JL, Lampurlanes J, Cantero-Martinez C (2014) Soil organic carbon storage in a no-tillage chronosequence under Mediterranean conditions. Plant and Soil 376: 31–41.
26. Xu SX, Zhao YC, Shi XZ, Yu DS, Li CS, et al. (2013) Map scale effects of soil databases on modeling organic carbon dynamics for paddy soils of China. Catena 104: 67–76.
27. Batjes NH (2006) Soil carbon stocks of Jordan and projected changes upon improved management of croplands. Geoderma 132: 361–371.
28. Liebens J, VanMolle M (2003) Influence of estimation procedure on soil organic carbon stock assessment in Flanders, Belgium. Soil Use and Management 19: 364–371.
29. National Soil Survey Office (1993) Soil Species of China, Volumes 1–6. Beijing: China Agriculture Press.
30. Wang SQ, Zhou CH (1999) Estimating soil carbon reservoir of terrestrial ecosystem in China. Geographical Research 18: 349–356.
31. Jin F, Yang H, Cai ZC, Zhao QG (2001) Calculation of density and reserve of organic carbon in soils. Acta Pedologica Sinica 38: 522–528.
32. Yu JJ, Yang F, Wu KN, Li L, Lü QL (2008) Soil organic carbon storage and its spatial distribution in Henan Province. Chinese Journal of Applied Ecology 19: 1058–1063.
33. Zhao YC, Shi XZ, Yu DS, Pagella TF, Sun WX, et al. (2005) Soil organic carbon density in Hebei Province, China: estimates and uncertainty. Pedosphere 15: 293–300.
34. Zhao YC, Shi XZ, Yu DS, Sun WX (2006) Map scale effect on soil organic carbon stock estimation in north China. Soil Science Society of America Journal 70: 1377–1386.
35. Shi XZ, Yu DS, Warner ED, Pan XZ, Petersen GW, et al. (2004) Soil database of 1:1000000 digital soil survey and reference system of the Chinese Genetic Soil Classification System. Soil Survey Horizons 45: 111–148.
36. Liu QH, Shi XZ, Weindorf DC, Yu DS, Zhao YC, et al (2006) Soil organic carbon storage of paddy soils in China using the 1:1000000 soil database and their implications for C sequestration. Global Biogeochemical Cycles 20: GB3024.
37. Zhejiang Soil Survey Office (1994) Soils of Zhejiang Province. Hangzhou: Zhejiang Science and Technology Press.
38. Wu JP, Hu YL, Zhi JJ, Jing CW, Chen HJ, et al. (2013) A 1:50000 soil database of Zhejiang Province, China. Acta Pedologica Sinica 50: 30–40.
39. Kazuhito M, Ono K, Kanomata H (2004) Organic carbon stock in forest soils in Japan. Geoderma 119: 21–32.
40. Sun WX, Shi XZ, Yu DS (2003) Distribution pattern and density calculation of soil organic carbon in profile. Soils 35: 236–241.
41. Nelson DW, Sommers LE (1982) Total carbon, organic carbon and organic matter. In: Page AL, Miller RH, Keeney DR, editors. Methods of Soil Analysis. Part 2: Chemical and Microbiological Properties. Wisconsin: American Society of Agronomy. 539–579.

2

Dynamics of Soil Organic Carbon and Microbial Biomass Carbon in Relation to Water Erosion and Tillage Erosion

Nie Xiaojun[1,2], Zhang Jianhui[1]*, Su Zhengan[1]

1 Key Laboratory of Mountain Surface Processes and Ecological Regulation, Chinese Academy of Sciences, Institute of Mountain Hazards and Environment, Chinese Academy of Sciences and Ministry of Water Conservancy, Chengdu, China, **2** School of Surveying and Land Information Engineering, Henan Polytechnic University, Jiaozuo, China

Abstract

Dynamics of soil organic carbon (SOC) are associated with soil erosion, yet there is a shortage of research concerning the relationship between soil erosion, SOC, and especially microbial biomass carbon (MBC). In this paper, we selected two typical slope landscapes including gentle and steep slopes from the Sichuan Basin, China, and used the ^{137}Cs technique to determine the effects of water erosion and tillage erosion on the dynamics of SOC and MBC. Soil samples for the determination of ^{137}Cs, SOC, MBC and soil particle-size fractions were collected on two types of contrasting hillslopes. [137]Cs data revealed that soil loss occurred at upper slope positions of the two landscapes and soil accumulation at the lower slope positions. Soil erosion rates as well as distribution patterns of the <0.002-mm clay shows that water erosion is the major process of soil redistribution in the gentle slope landscape, while tillage erosion acts as the dominant process of soil redistribution in the steep slope landscape. In gentle slope landscapes, both SOC and MBC contents increased downslope and these distribution patterns were closely linked to soil redistribution rates. In steep slope landscapes, only SOC contents increased downslope, dependent on soil redistribution. It is noticeable that MBC/SOC ratios were significantly lower in gentle slope landscapes than in steep slope landscapes, implying that water erosion has a negative effect on the microbial biomass compared with tillage erosion. It is suggested that MBC dynamics are closely associated with soil redistribution by water erosion but independent of that by tillage erosion, while SOC dynamics are influenced by soil redistribution by both water erosion and tillage erosion.

Editor: David L. Kirchman, University of Delaware, United States of America

Funding: The funding for our study: the National Natural Science Foundation of China (41001157, 41271242), the 135 Strategic Program of the Institute of Mountain Hazards and Environment, CAS (SDS-135-1206), and the Doctoral Program Foundation of Henan Polytechnic University (B2010-47). The funders had no role in study design, data collection and analysis, decision to publish, or preparation of the manuscript.

Competing Interests: The authors have declared that no competing interests exist.

* E-mail: zjh@imde.ac.cn

Introduction

Soil erosion has a severe impact on soil organic carbon (SOC) pools, and as a consequence affects the soil carbon cycle. Historically, water erosion was assumed to be a major soil disturbance in agricultural slope landscapes. Numerous studies have emphasized the influences of water erosion on SOC removal and carbon sequestration in agricultural slope landscapes [1–6]. Since the widespread application of [137]Cs (i.e., Caesium-137) technique to assess soil erosion in the 1990s, field evidence from different regions of the world has documented that tillage erosion (i.e. the net movement of soil downslope in response to the action of farm implements) is another major soil disturbance in agricultural slope landscapes with intensive tillage [5,7–11]. The impacts of tillage erosion on within-field variability of SOC have recently been addressed [12–14]. For example, on steep hillslopes of the Sichuan Basin, China, there is a complete depletion of SOC at summit positions and remarkable accumulation at bottom positions after 15-yr tillage erosion [14].

Although the physical process of SOC transfer on the slope under water and tillage erosion has been well known, the SOC biochemical process (e.g. carbon mineralization), mostly linked with the carbon cycle, remains unclear in eroding landscapes.

Polyakov and Lal (2004) reported that no significant differences in SOC mineralization exist between the eroding and control soils, but there is a 26% higher CO_2-efflux in the deposited soils compared with those without deposition [15]. In contrast, a recent study conducted by Hemelryck et al. (2011) shows that SOC mineralization in soils with deposition does not apparently differ from soils without deposition [16]. Van Oost et al. (2007) concluded that SOC mineralization associated with transport are relatively minor and that most deposited carbon is effectively preserved [17].

Indeed, it is difficult to make clear the relationship between erosion and SOC mineralization without considering the type of erosion event and soil microbial conditions. In fact, soil microbes participate in all of the biochemical process of SOC, thus microbial biomass affects the emission of CO_2 under soil erosion. Soil microbial biomass carbon (MBC) generally making up 1–5% of total SOC [18], acts as the most active component in the biochemical process of SOC turnover [19] and responds more rapidly to soil disturbance than does SOC [20]. Relatively, few studies have dealt with relationships between MBC dynamics and soil erosion [21]. In the process of water erosion, selective entrainment of fine soil particles is controlled by the transport capacity of the overland flow [22], and the labile SOC fraction adhered to fine soil particles is easily mineralized [23]. In contrast, tillage erosion unselectively moves the soil downslope [14,24], with no transport-related mineralization of SOC in the process of tillage erosion [25]. Given different mechanisms of soil transport, it would be expected that the two different processes of soil erosion may exert different impacts on MBC and the resultant SOC mineralization.

Caesium-137 (^{137}Cs) is a well-established tracer of soil erosion [26]. By comparing ^{137}Cs inventories of the study sites with those of reference sites, one can determine whether erosion (less ^{137}Cs in study sites than in reference sites) or deposition (more ^{137}Cs in study sites than in reference sites) has occurred. The relationships among patterns of soil erosion, SOC and some physio-chemical properties have been established for a number of fields and small agricultural watersheds in different climate zones [5,27–29], showing that this method is a valuable tool to investigate the erosion-carbon relationship. In this study, we use the same methodology to evaluate the relationship between erosion and carbon dynamics on cultivated slopes of the Sichuan Basin, China, where both water erosion and tillage erosion have occurred.

This study aims to (1) compare the difference in soil erosion between the two different slope landscapes (i.e., gentle and steep slopes), (2) examine spatial patterns of SOC and MBC distribution over eroded slopes, and (3) elucidate effects of water erosion and tillage erosion on the dynamics of SOC and MBC.

Materials and Methods

Study Area

The study area was located in Jianyang County of the Sichuan Basin, southwestern China (30°04′28″–30°39′00″N, 104°11′34″–104°53′36″E). This region is typical of the hilly areas of Sichuan (400–587 m above sea level) and has a humid subtropical climate (mean annual temperature 17°C; mean annual rainfall of 872 mm with 90% of precipitation between May and October). Most of the hillslopes have been dissected into slope segments with different slope gradients during previous long-term agricultural practices to minimize soil erosion and optimize farming operations. With the hoe, a predominant tillage implement in this area, one major tillage operation per year always starts at the bottom of the slope and moves upslope step by step, but at every step the tillage direction is always downslope (i.e., always pulling down). Due to long-term intensive hoeing, those in the upper parts of the slope are characterized by a thin soil layer, underlain by bedrock, whereas deeper soils are found in the lower parts. The soils in the study area, derived from purple mudstone and sandstone of Jurassic Age, are classified as Orthic Regosols in the FAO soil taxonomy. Soil texture is clay loam (27% clay, 29% silt, 44% sand), containing 1.0–2.5% organic matter and with a pH of 7.9–8.2. Crop rotation involves [wheat (*Triticum aestivum* L.), maize (*Zea mays* L.)/sweet potato (*Ipomoea batatas*)] and fertilizers (N: 330 kg ha^{-1} year^{-1} as urea and ammonium bicarbonate, P$_2$O$_5$:166 kg ha^{-1} year^{-1} as superphosphate) are uniformly applied on the cultivated slopes. Residue management is also uniformly on the slopes in the study area. For wheat and maize, the residues are cut by hand as low as possible and removed, retained residues with about 10-cm height above the ground are mixed into the tillage layer by manual hoeing. For sweet potato, all the residues are removed from the field.

Soil Sampling

Soil sampling was conducted on two types of contrasting hillslopes. Two neighboring cultivated slopes with a length of 44 and 53 m and a corresponding gradient of 7% and 4%,

respectively, were considered as two replicates for gentle slope landscapes (Table 1). Another two neighboring slopes with a length of 17 and 22 m and a corresponding slope gradient of 26% and 19%, respectively, were considered as two replicates for steep slope landscapes (Table 1). Samples were collected using a soil corer (6.8-cm diameter) along the downslope transect at 10-m intervals on the two gentle slopes and at 5-m intervals on the two steep slopes. For steep slope landscapes, there were 4, 1 and 4 samples, respectively, in the upper, middle and lower positions. For gentle slope landscapes, 4, 3 and 4 samples were collected in the upper, middle and lower positions, respectively. The coordinates of each sampling point as well as elevation were measured using a survey-grade Differential Global Positioning System (DGPS).

For each sampling point, five soil cores were collected. Two cores for the determination of ^{137}Cs inventory were taken to the bedrock (depths range from 20 to 40 cm depending on thickness of total soil layer at different slope positions) and were then bulked to make a composite sample. In the context of this study, soil depth reached a maximum of 40 cm, and therefore sampling to this depth has all ^{137}Cs in soil profiles contained in samples. Another three core samples for the determination of SOC, MBC, and soil particle-size fractions were collected from the till layer (0–20 cm). One of major considerations for this is that soil redistribution by water and tillage happened mainly in the till layer. Another is that soil microbes are the most active in the till layer where a large portion of organic matter such as crop residues are concentrated, thus the effects of soil erosion on the biochemical process of SOC can be expected to be dominant. After the three soil cores were taken, they were mixed into a composite sample and then divided into two subsamples. One was sieved (2-mm) and immediately stored at 4°C in plastic bags loosely tied to ensure sufficient aeration and to prevent moisture loss until MBC analysis. The other was air-dried and sieved (2-mm) for analyses of SOC and particle-size fractions.

Laboratory Analysis

Soil samples for ^{137}Cs determination were air-dried, crushed and passed through a 2-mm mesh sieve to remove gravels. Samples of the <2 mm particle-size fraction were packed into plastic (PVC) beakers with a 320-cm^3 volume, and ^{137}Cs activity was measured using a hyperpure lithium-drifted germanium detector (HpC –40% efficiency) coupled with a Nuclear Data 6700 multichanel γ-ray spectrophotometer. Caesium-137 was detected at 662 KeV, and count time for each sample ranged from 40,000 to 60,000 s, providing a measurement precision of ±5%. The contents of ^{137}Cs were originally expressed as a unit mass basis (Bq kg^{-1}) and were then converted into an area basis (Bq m^{-2}) using the total weight of the bulked core sample and the cross-sectional area of the sampling device. Soil bulk densities (kg m^{-3}) were determined using oven-dried weight and sample volume.

SOC was determined using wet oxidation with K$_2$Cr$_2$O$_7$ and soil particle-size fractions were determined by pipette method following H$_2$O$_2$ treatment to destroy organic matter and dispersion of soil suspensions in Na-hexametaphosphate [30]. MBC was determined by fumigation with ethanol-free CHCl$_3$ and extraction with K$_2$SO$_4$ [31]. Measurement precision for SOC, MBC, and soil particle-size fractions is ±2%, ±4%, and ±2%, respectively.

Calculation of Soil Erosion Rates

The application of the ^{137}Cs technique and tillage erosion models provides a new perspective to assess the contribution of

Table 1. Total soil erosion rates, tillage erosion rates and water erosion rates.

	Gentle slope landscape		Steep slope landscape	
	4%-slope	7%-slope	19%-slope	26%-slope
Total length (m)	53	44	22	17
Sample number	6	5	5	4
Total soil erosion rate (t ha^{-1} yr^{-1})	24.41	28.01	39.77	44.83
Tillage erosion rate (t ha^{-1} yr^{-1})	6.96	9.29	26.16	39.67
Percentage of total erosion (%)	29	33	66	88
Water erosion rate (t ha^{-1} yr^{-1})	17.45	18.72	13.61	5.16
Percentage of total erosion (%)	71	67	34	12

water and tillage to total soil erosion [5,9–11]. For an eroding site where a total ^{137}Cs inventory A (Bq m^{-2}) is less than the reference inventory A_0 (Bq m^{-2}) at year t (date), soil erosion rates can be expressed as [32]:

$$Y = \frac{10dB}{P}\left[1 - \left(1 - \frac{A_0 - A}{A_0}\right)^{1/(t-1963)}\right] \qquad (1)$$

where Y is the soil erosion rate (t ha^{-1} yr^{-1}), d the sampling depth (m), B the bulk density of soil (kg m^{-3}) and P the particle size correction factor (assumed to be $P = 1.0$). According to the previous study conducted in the study area [5], a local ^{137}Cs reference inventory was calculated as 1318 Bq m^{-2} with radioactivity decay corrected to the year 2008.

Based on a large number of experimental observations conducted previously in the study area, an empirical model of tillage erosion has been well established by Zhang et al. (2004) [33]. We adopted the model in this study because of similar terrain attributes such as the shape (linear), ranges of gradient and slope length, and even similar soil properties. In the empirical model, tillage erosion rates increases with increasing slope gradient and is inversely proportional to the slope length, expressed as:

$$R = (k_3 + k_4 S)10/L_d \qquad (2)$$

where R is the tillage erosion rate (t ha^{-1} yr^{-1}), k_3 and k_4 are the soil transport coefficient (kg m^{-1} yr^{-1}), S is the slope gradient (m m^{-1}) and L_d is the downslope parcel length (m). Tillage transport coefficients in the study area reach 30.72 and 141.28 kg m^{-1} yr^{-1}, respectively, for k_3 and k_4 [33]. Water erosion rates were obtained by the differences between total soil erosion rates derived from ^{137}Cs data and tillage erosion rates [5,11].

Data Analysis

Statistical analyses were carried out using SPSS 11.0 for Windows version (SPSS Inc., US, 2002). Linear regression analysis was used to test the correlations between ^{137}Cs, SOC and MBC. One-way analysis of variance was used to test the significance of differences in MBC/SOC ratios between gentle and steep slope landscapes, and between upper and lower positions of the two slope landscapes.

Ethics Statement

In this study, soil sampling and sample determinations conducted were permitted by the local government (i.e. Jianyang County

People's Government). We also obtained a permission from the local government for reporting research results to the public.

Results

Soil Redistribution

For steep slope landscapes, tillage erosion rates were 26.16 and 39.67 t ha^{-1} yr^{-1} and accounted for 66% and 88% of total erosion rates for the 19%- and 26%-slopes, respectively (Table 1). For gentle slope landscapes, tillage erosion rates were 9.29 and 6.96 t ha^{-1} yr^{-1} and contributed only 33% and 29% to total erosion rates for the 7%- and 4%-slopes, respectively (Table 1). The results indicate that tillage erosion dominates the process of soil redistribution in steep slope landscapes, whereas water erosion is a major process of soil redistribution in gentle slope landscapes. Differences in dominant processes of soil erosion between the two slope landscapes were also supported by the distribution of the <0.002-mm particle-size fraction on the eroded slopes. The fine particle fraction gradually increased downslope on each gentle slope, but such a trend of the fine particle was not observed on the two steep slopes (Figure 1). This result suggests that soil fine particles are preferentially transported downslope mainly by water in gentle slope landscapes, while the soils are unselectively or

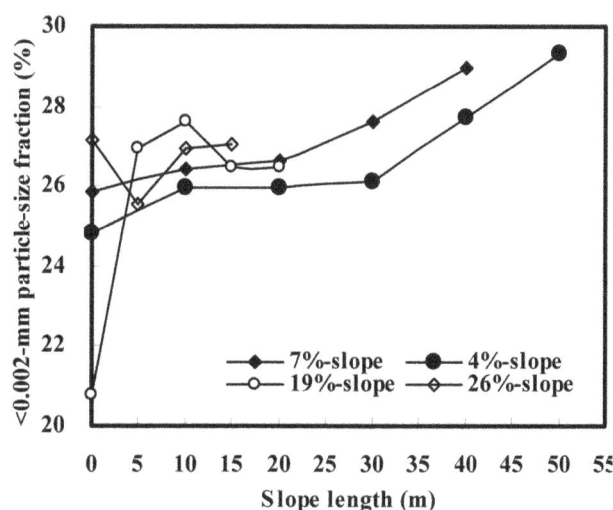

Figure 1. Distribution of the <0.002-mm particle-size fraction on eroded slopes.

entirely transported downslope due to the dominant process of tillage erosion in steep slope landscapes.

Figure 2 shows a similar ^{137}Cs distribution pattern in gentle and steep slope landscapes, i.e., ^{137}Cs inventory increased along the downslope transect. For gentle slope landscapes, ^{137}Cs inventory increased from 861 to 1336 Bq m^{-2} and 786 to 1289 Bq m^{-2} from the upper to lower positions of the 7%- and 4%-slopes, respectively (Table 2). For steep slope landscapes, ^{137}Cs inventory ranged from lows of 402 and 668 Bq m^{-2} at upper slope positions to highs of 1179 and 1002 Bq m^{-2} at lower slope positions of the 19%- and 26%-slopes, respectively (Table 2). The results suggest that soil loss occurs at upper slope positions and soil accumulation at lower slope positions.

SOC and MBC Dynamics

In gentle slope landscapes, both SOC and MBC contents increased downslope in a roughly consecutive increment (Figure 3a). SOC contents averaged 12.99 and 12.42 g kg^{-1} at lower slope positions of the 7%- and 4%-slopes with an increase of 44% and 31%, respectively, compared with those at respective upper slope positions (Table 2). From the upper to lower slope positions, MBC contents changed from 182.13 to 217.80 mg kg^{-1} with an increase of 20% on the 7%-slope, and from 168.78 to 221.13 mg kg^{-1} with an increase of 31% on the 4%-slope (Table 2). These patterns are in agreement with the distribution pattern in ^{137}Cs inventory that mirrors soil redistribution rates on the two gentle slopes. Correlation analysis also showed that inventories of SOC (kg m^{-2}) and MBC (g m^{-2}) were significantly

correlated with ^{137}Cs inventory (Bq m^{-2}) in the landscape ($r = 0.82$, $P < 0.01$ and $r = 0.88$, $P < 0.01$, respectively, for SOC and MBC; see Figures 4, 5). As a consequence, it is clear that the dynamics of SOC and MBC are closely associated with soil redistribution in gentle slope landscapes.

Different distribution patterns between SOC and MBC contents were observed in steep slope landscapes, showing a rough increase in SOC content but a discrete pattern in MBC content along the downslope transect (Figure 3b). From the upper to lower slope positions, SOC contents changed from 7.33 to 10.15 g kg^{-1} with an increase of 38% on the 19%-slope, and 9.33 to 11.82 g kg^{-1} with an increase of 27% on the 26%-slope (Table 2). An increase in MBC content was observed only at middle positions of the 19%-slope, but little change in MBC between the upper and lower positions of each steep slope (Table 2). The distribution pattern in SOC content is consistent with ^{137}Cs inventory on each steep slope. A significant correlation ($r = 0.84$, $P < 0.05$) between SOC (kg m^{-2}) and ^{137}Cs (Bq m^{-2}) was also found in steep slope landscapes (Figure 4). However, MBC distribution could not be explained by ^{137}Cs data on each steep slope, as a result of no correlation ($r = 0.25$, $P > 0.05$) between MBC (g m^{-2}) and ^{137}Cs (Bq m^{-2}) in steep slope landscapes (Figure 5). Thus it is suggested that SOC dynamics are remarkably impacted by soil redistribution in steep slope landscapes, whereas MBC dynamics are independent of soil redistribution in the landscape. The close relationship between the <0.002-mm fine particle and MBC contents in gentle slope landscapes dominated by water erosion ($r = 0.86-0.93$, $P = 0.02-0.03$; data not shown) also showed that the microbial biomass decreased with decreasing soil fine particles under water erosion. However, no significant correlations between the two parameters ($r = -0.22-0.79$, $P = 0.11-0.78$; data not shown) were found in steep slope landscapes, implying that the microbial biomass is scarcely impacted by tillage erosion which produces unselective soil particle movement on the slope.

Relationship between MBC and SOC

A close correlation ($r = 0.91$, $P < 0.01$) between MBC and SOC contents (mass basis) was found in gentle slope landscapes, whereas the two were not correlated in steep slope landscapes ($r = 0.18$, $P > 0.05$) (Figure 6a). When the contents of SOC and MBC were expressed on a unit area basis (kg m^{-2} and g m^{-2}, respectively), a similar correlation ($r = 0.95$, $P < 0.01$) between the two variables was also found in gentle slope landscapes, but there was still no correlation ($r = 0.18$, $P > 0.05$) in steep slope landscapes (Figure 6b). In this study, additional information on the relationship between MBC and SOC was also obtained from changes in the ratio of MBC (g kg^{-1}) to SOC (g kg^{-1}). The MBC/SOC ratio was slightly greater at upper positions than at lower positions of each slope (Table 2). However, there were no statistically significant differences in MBC/SOC ratios between upper and lower positions of the two slope landscapes ($P > 0.05$). The results indicate the difference in the MBC fraction per unit of SOC between the upper and lower positions was insignificant in the landscape. In terms of the whole slope, the MBC/SOC ratio in steep slope landscapes averaged 2.77×10^{-2}, and was significantly greater than that (mean 1.87×10^{-2}) in gentle slope landscapes ($P < 0.05$). This showed a smaller MBC fraction per unit of SOC in gentle slope landscapes than in steep slope landscapes, suggesting that MBC is influenced by different processes of soil redistribution in the landscape.

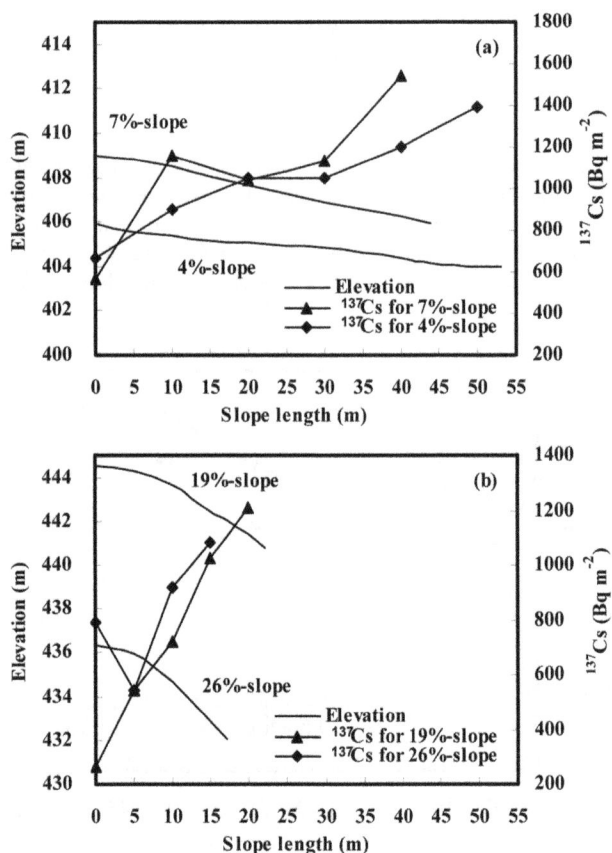

Figure 2. Distribution of ^{137}Cs inventories over eroded slopes.
(a) gentle slope landscape; (b) steep slope landscape.

Table 2. Descriptive statistics of ^{137}Cs inventories, SOC and MBC contents in different positions for gentle and steep slope landscapes.

Gentle slope landscape	4%-slope			7%-slope		
Slope position	Upper	Middle	Lower	Upper	Middle[♀]	Lower
Slope length	0–10	20–30	30–53	0–10	20	30–44
Sample number	2	2	2	2	1	2
Mean [137]Cs (Bq m^{-2})	786	1051	1296	861	1040	1336
Mean SOC (g kg^{-1})	9.33	9.84	12.42	9.20	9.39	12.99
Mean MBC (mg kg^{-1})	168.78	202.67	221.13	182.13	180.95	217.80
Mean MBC/SOC ratio	1.81×10^{-2}	2.06×10^{-2}	1.78×10^{-2}	1.99×10^{-2}	1.93×10^{-2}	1.69×10^{-2}
Steep slope landscape	19%-slope			26%-slope		
Slope position	Upper	Middle[♀]	Lower	Upper	Lower	
Slope length	0–5	10	10–22	0–10	10–17	
Sample number	2	1	2	2	2	
Mean [137]Cs (Bq m^{-2})	402	723	1117	668	1002	
Mean SOC (g kg^{-1})	7.33	8.49	10.15	9.33	11.82	
Mean MBC (mg kg^{-1})	242.45	308.44	234.41	257.09	273.62	
Mean MBC/SOC ratio	3.28×10^{-2}	3.63×10^{-2}	2.32×10^{-2}	2.83×10^{-2}	2.32×10^{-2}	

[♀]single value in the position.

Discussion

The distribution pattern of SOC content was consistent with redistribution patterns of ^{137}Cs inventory in the two slope landscapes. The result agrees with previous studies [5,11,28], further confirming a close relationship between SOC translocation and soil redistribution by water and tillage. The MBC distribution pattern was in agreement with soil redistribution in gentle slope landscapes but independent of soil redistribution in steep slope landscapes. This is attributed to impacts of water-induced soil redistribution on SOC and MBC in gentle slope landscapes, and impacts of tillage-induced soil redistribution in steep slope landscapes. The difference in the relationship between MBC and SOC under the disturbances of water and tillage erosion

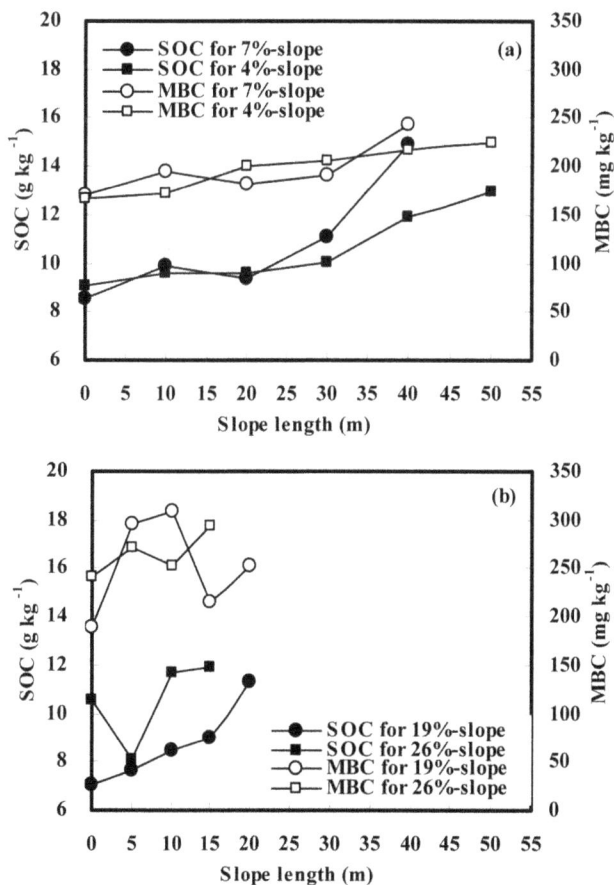

Figure 3. Distribution of SOC and MBC contents over eroded slopes. (a) gentle slope landscape; (b) steep slope landscape.

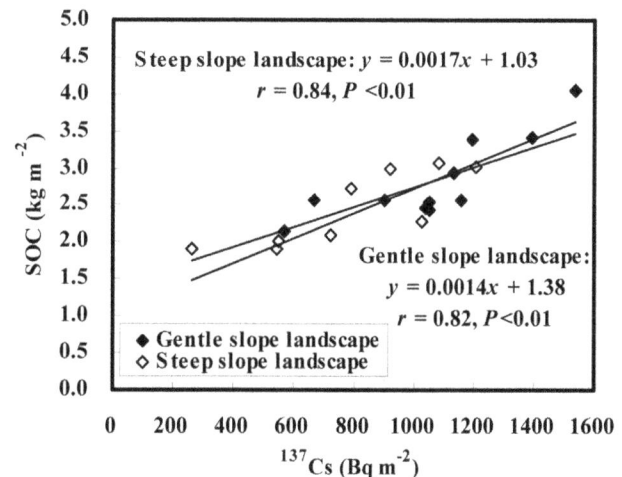

Figure 4. SOC contents vs. [137]Cs inventories in gentle and steep slope landscapes.

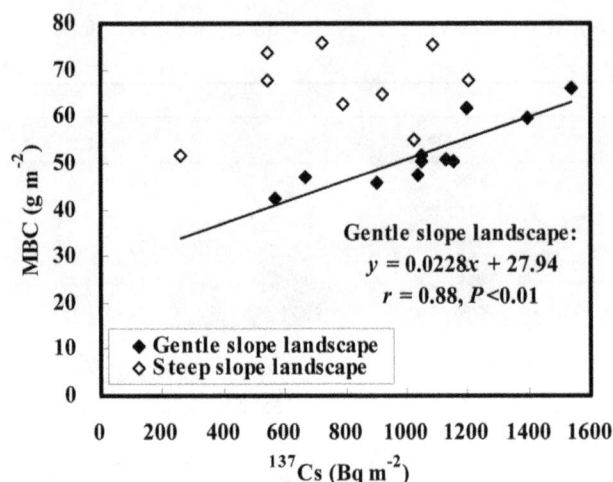

Figure 5. MBC contents vs. ^{137}Cs inventories in gentle and steep slope landscapes.

differed from the results of previous studies. A close correlation between MBC and SOC has been reported under various anthropic disturbances such as plant cover, fertilization, organic amendments, and tillage [34–37]. It is, therefore, that the dominant erosion process should be first considered when one assesses the dynamics of MBC and SOC in eroded slope landscapes.

In the two slope landscapes, lower SOC contents were found at upper slope positions with soil loss than at lower slope positions with soil accumulation. The result indicates the consistent physical process of SOC depletion in upper slope positions and SOC accumulation in lower slope positions regardless of water- or tillage-induced soil transport. However, changes in MBC contents were inconsistent between upper and lower positions in the two slope landscapes. On the two gentle slopes, MBC contents were apparently 20% and 31% higher in lower slope positions than upper slope positions, thus indicating that the physical transfer of MBC is related to soil redistribution. On the two steep slopes, however, there was little difference in MBC content between upper and lower positions, suggesting a negligible impact of soil redistribution on the physical dynamics of MBC. This similarity of MBC content between the two positions in steep slope landscapes could be attributed to unselective removal and the dilution effect due to the incorporation of the soil from deeper layers (e.g. subsoils) into the till layer by tillage (i.e. MBC contents in the newly formed surface soil decrease compared to the original surface soil, due to poor MBC in the subsoils) [14,17]. In the study area, the farmers have to till into subsoils, parent material, or even bedrocks to offset soil loss in the till layer. The soil translocated from the upslope moves entirely downslope and subsequently deposits at lower slope positions under consecutive tillage operations. Consequently, little impact on soil microbial biomass was present in the till layer in the line of slope, as a result of the redistribution of low fertility soils. No statistically significant difference ($P>0.05$) in MBC/SOC ratios between upper and lower positions in the two slope landscapes showed that the erosion-induced SOC depletion did not apparently decrease the ratio of MBC to SOC, and the deposition-induced SOC accumulation did not obviously increase this ratio either.

The MBC/SOC ratio represents the contribution of microbial biomass to organic carbon in soil [38], and is a more useful assessment index for soil health than either MBC or SOC [39].

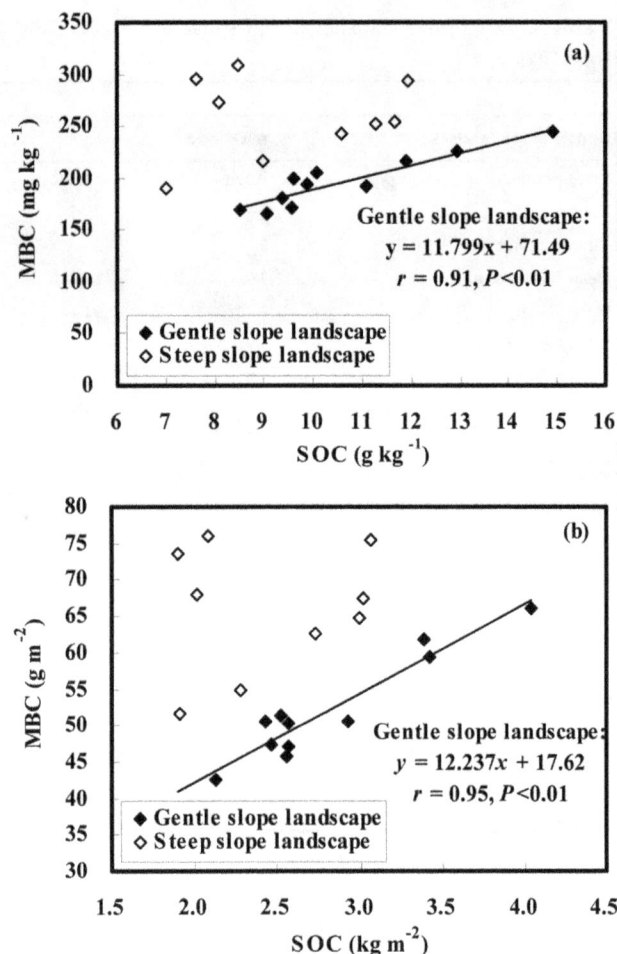

Figure 6. Relationship between MBC and SOC in gentle and steep slope landscapes. (a) mass basis; (b) area basis.

The lower mean of MBC/SOC ratios in gentle slope landscapes than in steep slope landscapes ($P<0.05$) shows a smaller MBC contribution to SOC pools in gentle slope landscapes. This also implies that water erosion with selective removal of soil fine particles has a more severe disturbance to soil microbial biomass than tillage erosion with unselective soil movement. The difference in MBC/SOC between the two slope landscapes is probably associated with different responses of microbial biomass to distinctive changes in soil fine particles. Insam et al. (1989) reported that the clay content of soil has an important influence on the variance in MBC/SOC [40]. It is well known that soil clay (i.e. the <0.002-mm fine particle) has a strong adhesion to organic matter due to its large surface area, and thus is regarded as the mineral binding agents to form soil aggregates. In the process of water erosion, soil clay is preferentially removed by runoff, thus disrupting soil aggregates [5,41]. Accordingly, microbes incorporated into initial soil aggregates die due to their exposure to an unprotected circumstance [23], which results in a decline in soil microbial biomass. It should be noted that the same tillage operation including tillage intensity (one time per year) and tillage tool (hoe) is performed in the two slope landscapes, although tillage reduces aggregate stability by exposing the encapsulated SOC binding agents to mineralization [42]. As a result, impacts of aggregate breakdown by tillage would be similar between the two slope landscapes. In this sense, the difference in soil microbial biomass between the two landscapes resulted from different processes of soil redistribution (dominated by water erosion or by tillage erosion), excluding direct impacts of aggregate breakdown due to tillage operations.

Conclusion

In water-eroded slope landscapes, both SOC and MBC contents showed a significant correlation with soil redistribution rates. In tillage-eroded slope landscapes, SOC contents also exhibited a similar relationship, whereas MBC contents did not reveal any clear pattern. The MBC/SOC ratio was lower on the water-eroded slope than on the tillage-eroded slope, while the contribution of MBC to SOC pools was similar between the erosion-induced SOC depletion and SOC accumulation sites irrespective of landscape types. It is clear that water erosion has a severe disturbance to soil microbial biomass compared with tillage erosion. We conclude that SOC dynamics are associated closely with soil redistribution by water erosion and by tillage erosion as well, while MBC dynamics are significantly influenced by water-induced soil redistribution, but independent of tillage-induced soil redistribution.

Author Contributions

Conceived and designed the experiments: ZJH. Performed the experiments: NXJ SZA. Analyzed the data: NXJ ZJH. Contributed reagents/materials/analysis tools: NXJ. Wrote the paper: NXJ ZJH.

References

1. Verity GE, Anderson DW (1990) Soil erosion effects on soil quality and yield. Can J Soil Sci 70: 471–484.
2. Mabit L, Bernard C (1998) Relationship between soil ^{137}Cs inventories and chemical properties in a small intensively cropped watershed. Earth Planet Sci 327: 527–532.
3. Ritchie JC, McCarty GW (2003) [137]Caesium and soil carbon in a small agricultural watershed. Soil Till Res 69: 45–51.
4. Shukla MK, Lal R (2005) Erosional effects on soil organic carbon stock in an on-farm study on Alfisols in west central, Ohio. Soil Till Res 81: 173–181.
5. Zhang JH, Quine TA, Ni SJ, Ge FL (2006) Stocks and dynamics of SOC in relation to soil redistribution by water and tillage erosion. Global Change Biol 12: 1834–1841.
6. Nie XJ, Wang XD, Liu SZ, Liu HJ, Gu SX (2010) ^{137}Cs tracing dynamics of soil erosion, organic carbon and nitrogen in sloping farmland converted from original grassland in Tibetan plateau. Appl Radiat Isot 68: 1650–1655.
7. Lindstrom MJ, Nelson WW, Schumacher TE (1992) Quantifying tillage erosion rates due to moldboard plowing. Soil Till Res 24: 243–255.
8. Lobb DA, Kachanoski RG, Miller MH (1995) Tillage translocation and tillage erosion on shoulder slope landscape positions measured using ^{137}Cs as a tracer. Can J Soil Sci 75: 211–218.
9. Govers G, Quine TA, Desmet PJJ, Walling DE (1996) The relative contribution of soil tillage and overland flow erosion to soil redistribution on agricultural land. Earth Surf Proc Land 21: 929–946.
10. Quine TA, Walling DE, Chakela OK, Mandiringana OT, Zhang X (1999) Rates and patterns of tillage and water erosion on terraces and contour strips: Evidence from caesium-137 measurements. Catena 36: 115–142.
11. Li Y, Lindstrom MJ (2001) Evaluating soil quality–soil redistribution relationship on terraces and steep hillslope. Soil Sci Soc Am J 65: 1500–1508.
12. Heckrath G, Djurhuus J, Quine TA, Van Oost K, Govers G, et al. (2005) Tillage erosion and its effect on soil properties and crop yield in Denmark. J Environ Qual 34: 312–324.
13. Papiernik SK, Lindstrom MJ, Schumacher JA, Farenhorst A, Stephans KD, et al. (2005) Variation in soil properties and crop yield across an eroded prairie landscape. J Soil Water Conserv 60: 388–395.
14. Zhang JH, Nie XJ, Su ZA (2008) Soil profile properties in relation to soil redistribution by intense tillage on a steep hillslope. Soil Sci Soc Am J 72: 1767–1773.
15. Polyakov VO, Lal R (2004) Soil erosion and carbon dynamics under simulated rainfall. Soil Sci 169: 590–599.
16. Hemelryck HV, Govers G, Van Oost K, Merckx R (2011) Evaluating the impact of soil redistribution on the in situ mineralization of soil organic carbon. Earth Surf. Process. Landforms 36: 427–438.
17. Van Oost K, Quine TA, Govers G, Gryze SD, Six J, et al. (2007) The impact of agricultural soil erosion on the global carbon cycle. Science 318: 626–629.
18. Jenkinson DS, Ladd JN (1981) Microbial biomass in soil, measurement and turn over. In: Paul EA, Ladd JN, editor. Soil Biochemistry. New York: Marcel Dekker press, 415–471.
19. Marinari S, Mancinelli R, Campiglia E, Grego S (2006) Chemical and biological indicators of soil quality in organic and conventional farming systems in Central Italy. Ecol Indic 6: 701–711.
20. Bergstrom DW, Monreal CM, King DJ (1998) Sensitivity of soil enzyme activities to conservation practices. Soil Sci Soc Am J 62: 1286–1295.
21. Mabuhay JA, Nakagoshi N, Isagi Y (2004) Influence of erosion on soil microbial biomass, abundance and community diversity. Land Degrad Dev 15: 183–195.
22. Schiettecatte W, Gabriels D, Cornelis WM, Hofinan G (2008) Impact of deposition on the enrichment of organic carbon in eroded sediment. Catena 72: 340–347.
23. Schlesinger WH (1995) Soil respiration and changes in soil carbon stocks. In: Biotic Woodwell GM, Mackenzie FT, editor. Feedback in the Global Climatic System: Will the Warming Feed the Warming? New York: Oxford University Press, 159–168.
24. De Alba S, Lindstrom M, Schumacher TE, Malo DD (2007) Soil landscape evolution due to soil redistribution: a new conceptual model of soil catena evolution in agricultural landscapes. Catena 58: 77–100.
25. Van Oost K, Govers G, Quine TA, Heckrath G (2004) Comment on 'Managing soil carbon'. Science 305: 1567b.
26. Walling DE, Quine TA (1991) The use of caesium-137 measurements to investigate soil erosion on arable fields in the UK: potential applications and limitations. J Soil Sci 42: 147–165.
27. Quine TA, Zhang Y (2002) An investigation of spatial variation in soil erosion, soil properties, and crop production within an agricultural field in Devon, United Kingdom. J Soil Water Conserv 57: 55–65.
28. Ritchie JC, McCarty GW (2003) [137]Caesium and soil carbon in a small agricultural watershed. Soil Till Res 69: 45–51.
29. Mabit L, Bernard C, Makhlouf M, Laverdière MR (2008) Spatial variability of erosion and soil organic matter content estimated from ^{137}Cs measurements and geostatistics. Geoderma 145: 245–251.
30. Liu GS (1996) Soil Physical and Chemical Analysis and Description of Soil Profiles. Beijing: Chinese Standard Press, 123–166 (in Chinese).
31. Vance ED, Brookes PC, Jenkinson DS (1987) An extraction method for measuring soil microbial biomass-C. Soil Biol Biochem 19: 703–707.
32. Zhang XB, Higgitt DL, Walling DE (1990) A preliminary assessment of the potential for using caesium-137 to estimate rates of soil erosion in the Loess Plateau of China. Hydrolog Sci J 35: 267–276.
33. Zhang JH, Lobb DA, Li Y, Liu GC (2004) Assessment for tillage translocation and tillage erosion by hoeing on the steep land in hilly areas of Sichuan, China. Soil Till Res 75: 99–107.
34. García C, Hernández T, Roldán A, Martin A (2002) Effect of plant cover decline on chemical and microbiological parameters under Mediterranean climate. Soil Biol Biochem 34: 635–642.
35. Melero S, Porras JCR, Herencia JF, Madejon E (2006) Chemical and biochemical properties in a silty loam soil under conventional and organic management. Soil Till Res 90: 162–170.
36. Madejón E, Moreno F, Murillo JM, Pelegrín F (2007) Soil biochemical response to long-term conservation tillage under semi-arid Mediterranean conditions. Soil Till Res 94: 346–352.
37. Vineela C, Wani SP, Srinivasarao CH, Padmaja B, Vittal KPR (2008) Microbial properties of soils as affected by cropping and nutrient management practices in several long-term manurial experiments in the semi-arid tropics of India. Appl Soil Ecol 40: 165–173.
38. Sparling GP (1992) Ratio of microbial biomass to soil organic carbon as a sensitive indicator of changes in soil organic matter. Aust J Soil Res 30: 195–207.
39. Anderson TH, Domsch KH (1989) Ratios of microbial biomass carbon to total organic carbon in arable soils. Soil Bio Biochem 21: 471–479.
40. Insam H, Parkinson D, Domsch K (1989) Influence of macroclimate on the soil microbial biomass. Soil Bio Biochem 21: 211–221.
41. Lal R (2003) Soil erosion and the global carbon budget. Environ Int 29: 437–450.
42. Hajabbasi MA, Hemmat A (2000) Tillage impacts on aggregate stability and crop productivity in a clay-loam soil in central Iran. Soil Till Res 56: 205–212.

3

Latitudinal Gradients in Degradation of Marine Dissolved Organic Carbon

Carol Arnosti[1]*, Andrew D. Steen[1]¤, Kai Ziervogel[1], Sherif Ghobrial[1], Wade H. Jeffrey[2]

1 Department of Marine Sciences, University of North Carolina, Chapel Hill, North Carolina, United States of America, **2** Center for Environmental Diagnostics and Bioremediation, University of West Florida, Pensacola, Florida, United States of America

Abstract

Heterotrophic microbial communities cycle nearly half of net primary productivity in the ocean, and play a particularly important role in transformations of dissolved organic carbon (DOC). The specific means by which these communities mediate the transformations of organic carbon are largely unknown, since the vast majority of marine bacteria have not been isolated in culture, and most measurements of DOC degradation rates have focused on uptake and metabolism of either bulk DOC or of simple model compounds (e.g. specific amino acids or sugars). Genomic investigations provide information about the potential capabilities of organisms and communities but not the extent to which such potential is expressed. We tested directly the capabilities of heterotrophic microbial communities in surface ocean waters at 32 stations spanning latitudes from 76°S to 79°N to hydrolyze a range of high molecular weight organic substrates and thereby initiate organic matter degradation. These data demonstrate the existence of a latitudinal gradient in the range of complex substrates available to heterotrophic microbial communities, paralleling the global gradient in bacterial species richness. As changing climate increasingly affects the marine environment, changes in the spectrum of substrates accessible by microbial communities may lead to shifts in the location and rate at which marine DOC is respired. Since the inventory of DOC in the ocean is comparable in magnitude to the atmospheric CO_2 reservoir, such a change could profoundly affect the global carbon cycle.

Editor: Meni Wanunu, Northeastern University, United States of America

Funding: This work was supported by the National Science Foundation (Dr. Arnosti, Dr. Jeffrey), EPA Graduate Fellowship (Dr. Steen), and the Deutsche Forschungsgemeinschaft Postdoctoral Fellowship (Dr. Ziervogel). The authors are also grateful to the Max-Planck Institute for Marine Microbiology (Bremen) for supporting the work on Svalbard. The funders had no role in study design, data collection and analysis, decision to publish, or preparation of the manuscript.

Competing Interests: The authors have declared that no competing interests exist.

* E-mail: arnosti@email.unc.edu

¤ Current address: Center for Geomicrobiology, Aarhus University, Herning, Denmark

Introduction

Marine DOC (dissolved organic carbon) is one of the largest actively cycling reservoirs of organic carbon on earth, comparable in magnitude to the atmospheric reservoir of CO_2 [1]; heterotrophic microbial communities play a key role in driving the DOC cycle [2–4]. DOC consists of many thousands of different compounds, and is operationally divided into labile, semi-labile, and recalcitrant fractions that are defined based on timescales of removal in bioassays or by direct measurement in the ocean [5]. Structural or mechanistic explanations for varying timescales of DOC degradation, however, are lacking [6]. Since heterotrophic bacteria are unable to transport directly into the cell most substrates with a molecular weight greater than 600 Da [7], hydrolysis via extracellular enzymes is required prior to substrate uptake. The requirement for enzymatic hydrolysis is therefore a promising starting point to search for mechanistic explanations of variations in the abilities of marine bacteria to utilize specific fractions of DOC as a substrate. The activities and structural specificities of polysaccharide-hydrolyzing enzymes are of particular importance in this respect, since carbohydrates constitute a large proportion of marine high molecular weight DOC: 54% of surface water DOC and 25% of DOC in the deep ocean [8]. Most of the rest of DOC is classified as 'uncharacterized' on a molecular

basis, since lipids, amino acids, and amino sugars together constitute less than ca. 5% of the total [9].

Assessing the enzymatic capabilities of marine heterotrophic microbial communities can best be done directly in seawater, since cultured microbial isolates constitute only a small, unrepresentative fraction of extent marine microbes [10]. Most measurements of enzyme activity in seawater are based on small chromogenic or fluorogenic substrates (e.g. [11]), which provide very little information about enzymatic substrate specificities [12]. Genomic investigations can yield valuable insights into community and organism potential [13–15], but provide no information about the extent to which such potential might be expressed. Metatranscriptomic profiling is a promising route to investigate the extent to which genetic potential is realized, but assignment of sequences from the environment to functions such as specific enzyme activities is still problematic due to limitations in database coverage [16] and the vast structural and functional diversity of polysaccharide-hydrolyzing enzymes [17].

To gain insight into the capabilities of natural microbial communities to access polysaccharides, we measured extracellular enzyme activities in surface waters at 32 stations in the Atlantic, Pacific, Arctic, and Southern Oceans, as well as the Gulf of Mexico, spanning latitudes from 76°S to 79°N and a temperature range from −1.8°C to 29°C (Table 1; Fig. S1). We focused on

Table 1. Station locations and water temperatures.

Station	Latitude	Longitude	Water temp. (°C)
J	79.4N	11.1E	4
AB	77.4N	15.1E	4
P2	66.5N	168.1W	8.7
DO	38.4N	74.6W	13.5
CO	36.4N	74.8W	22
P10	35.5N	164.2W	24.2
GOM1	30.2N	87.4W	28
GOM11	29.6N	87W	28
GOM072	28.5N	89.4W	28.7
GOM073	28.3N	89.4W	28.9
P15	15.5N	161.4W	27.4
BOT12	15.1N	105.5W	29.1
BOT10	10.2N	99.4W	28.6
BOT8	5.4N	92.4W	27.4
T33	1.1N	83.5W	26
BOT7	0.005S	86W	24.1
P21	7.2S	168.4W	29.2
BOT5	8.2S	83.5W	19.8
BOT4	12.2S	81.4W	18
BOT3	17S	79.1W	17.3
T15	23.1S	79.2W	17.5
BOT1	26.3S	75W	15.2
P27	26.5S	173W	21.1
T3	39.2S	77.6W	11
G1	49.3S	174.4W	8.5
R1	56.3S	176.2E	6
R3	62.3S	178.4W	−1.5
M9	65S	176W	−1.7
G11B	74.3S	173.3W	0
R10C	76.1S	170.3E	−1.8
G9A	76.3S	179W	0
R13f	76.5S	177.5E	−1.8

direct detection of the hydrolysis of specific polysaccharides, rather than on investigations of genetic potential, in order to measure the abilities of microbial communities to access substrates irrespective of the multiplicity of enzyme(s) [18] —of known and unknown sequence—that might hydrolyze a specific substrate. Since determining the specific structures of marine dissolved carbohydrates is not possible using currently available analytical techniques [19], we used as substrates a suite of polysaccharides that are components of marine algae [20], and/or whose hydrolytic enzymes have been identified in the genomes of recently-sequenced marine bacteria [13,15,21]. These polysaccharides—laminarin, xylan, fucoidan, arabinogalactan, pullulan, and chondroitin sulfate—are structurally diverse, vary in monomer composition and linkage position, and because they are constituents of marine plankton, many are present in considerable quantities in the ocean. The production of laminarin by diatoms and *Phaeocystis* in the ocean, for example, has been estimated at 5 to15 billion metric tons annually [22]. A previous investigation at a

few locations had shown evidence of spatial variations in microbial extracellular enzyme activities in surface ocean waters [23]. The present study, the culmination of a series of investigations carried out over the course of a decade, demonstrates that there are recognizable patterns in microbial potential to access DOC on large-scale gradients, mirroring emerging patterns of microbial biogeography in the ocean [24–28].

Results and Discussion

Differing abilities to hydrolyze a diverse range of substrates, as evident in our survey (Fig. 1), demonstrate functional differences among pelagic microbial communities. All substrates were hydrolyzed at only 4 of the 32 stations, all in the Gulf of Mexico. At the other 28 sites, the spectrum of hydrolysis ranged from one to five substrates. Only laminarin was hydrolyzed at every station; chondroitin and xylan were hydrolyzed at the majority (81% and 78%, respectively) of the stations, while pullulan, arabinogalactan, and fucoidan were hydrolyzed at 63%, 47%, and 34% of the stations, respectively. Summed hydrolysis rates were maximal at tropical/subtropical stations, and these summed rates as well as the spectrum of detected enzyme activities, decreased towards the poles (Figs. 1 and 2). The relative contribution of each enzyme activity to the sum at a given station (evenness of hydrolysis rates; Fig. 3) showed a similar pattern, with increasing evenness at higher temperatures (and lower latitudes).

The broad correlation between latitude and summed hydrolysis rates points to the relationship between summed hydrolysis rates and water temperature (n = 35, including 4 visits to a single station (Station J; Table 1), $r^2 = 0.64$, $p<0.005$), a relationship that could be due to the kinetic effect of temperature. This correlation, however, is driven primarily by the correlation with laminarin ($r^2 = 0.79$; $p<0.005$) and by the fact that a broader spectrum of enzyme activities is detected across the range of stations in lower-latitude waters (Figs. 1 and 2). Temperature was poorly correlated with hydrolysis of chondroitin ($r^2 = 0.15$; $p<0.05$), arabinogalactan and pullulan ($r^2 = 0.20$ and 0.21, respectively; $p<0.01$), and fucoidan ($r^2 = 0.23$, p<0.005). The correlation with xylan was stronger ($r^2 = 0.49$, $p<0.005$), but hydrolysis rates of the substrates at stations with similar temperatures varied greatly. At temperatures close to 28°C, for example, the variation in hydrolysis of xylan, fucoidan, arabinogalactan, and chondroitin was an order of magnitude or more (Table 1; Table S1; Fig. 1). Moreover, the differences in hydrolysis rate evenness (Fig. 3) are not explained solely by temperature, since a purely kinetic effect of temperature on hydrolysis rates would be expected to cause a generally proportionate change in the activities of all enzymes, and greatly extended incubation times do not markedly broaden the spectrum of substrates hydrolyzed at high latitude [29].

To the extent that these differences in enzymatic capabilities cannot be explained by temperature, they may derive from variations in microbial community composition that cascade into these functional differences. Arctic microbial communities differ in composition from their temperate counterparts [24,30–31]. Recent investigations have demonstrated latitudinal gradients in microbial community richness, with markedly reduced diversity at high latitudes [26–28]. The functional consequences of these variations are unknown, since microbial phylogeny and function are not well correlated. Our results demonstrate that microbial community function varies systematically across latitudinal gradients. The lower summed hydrolysis rates and the more limited spectrum of substrates enzymatically accessible to microbial communities at high latitudes coincides with a reduction in community richness (Figs. 1 and 2).

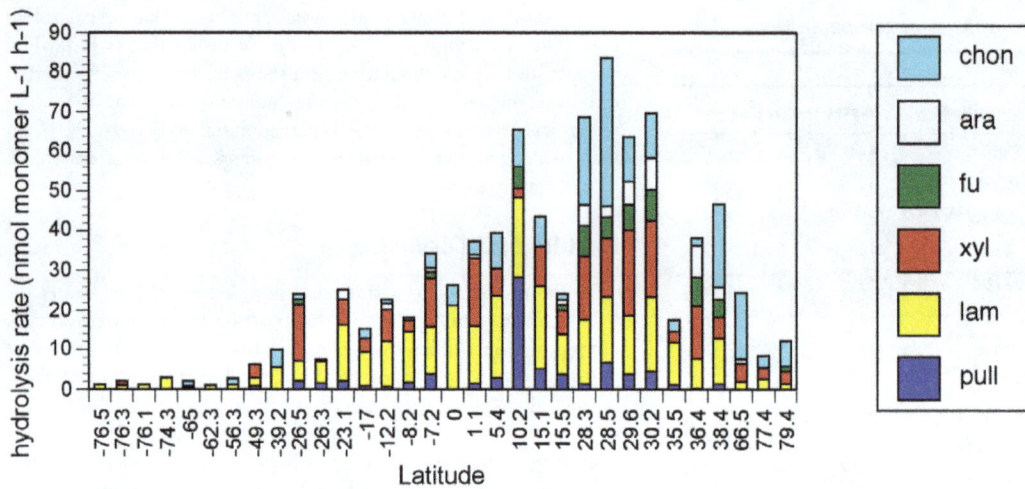

Figure 1. Summed enzymatic hydrolysis rates in surface water at each station plotted against latitude (south latitudes shown with negative numbers). Bar height shows the sum of the maximum enzymatic hydrolysis rate of each substrate at each station. All stations were visited once, with the exception of Station J (79°N); values shown for Station J are averages from 4 visits. (See Fig. 4 and Table S1 for data from each sampling time at Station J.) Pullulan hydrolysis is shown in blue, laminarin in yellow, xylan in red, fucoidan in green, arabinoglactan in white, and chondroitin sulfate in aqua. Hydrolysis rates and standard deviations for all substrates and stations are in Table S1.

Differences in the extent to which genes for polysaccharide hydrolases are expressed may also contribute to differences in enzyme activities among microbial communities, but the genetic diversity of polysaccharide hydrolases even among well-studied organisms is barely beginning to be explored [32], complicating efforts for a concerted genetic investigation. One study suggests that a high diversity of hydrolases is potentially available to microbial communities in the North Atlantic [14], but the extent and the conditions under which this genetic potential might be expressed are still matters of speculation. A search of the CAMERA database (http://camera.calit2.net/), for example, yielded numerous gene sequences related to an alpha-L-fucosidase of *Pseudoalteromonas atlantica* T6c (http://img.jgi.doe.gov/cgi-bin/pub/main.cgi). These sequences were found at 60 different Global Ocean Survey (GOS) sites (available through CAMERA), spanning latitudes of 32°S to 45°N, despite the fact that this

enzyme activity was not detected in many of our samples from the same range of latitudes (Fig. 1). Likewise, a search (via CAMERA) of the two marine sites in the Antarctic Aquatic Metagenome produced sequences from Newcomb Bay (66°S) closely matching pullulanases from fully-sequenced marine bacteria (http://blast.ncbi.nlm.nih.gov/Blast.cgi), although pullulanase activity was not detectable at any of our sites at latitudes higher than 49°S or 38°N (Fig. 1). Limited geographical overlap between the current CAMERA database and our samples presently preclude a more detailed comparison among sites that would provide insight into the extent to which microbial communities vary in their genetic response to environmental parameters.

Patterns of microbial community composition at a given location are temporally repeatable [33–37]. Results from our investigation suggest that patterns of hydrolytic activities are also consistent over multi-year timescales. One high-latitude station

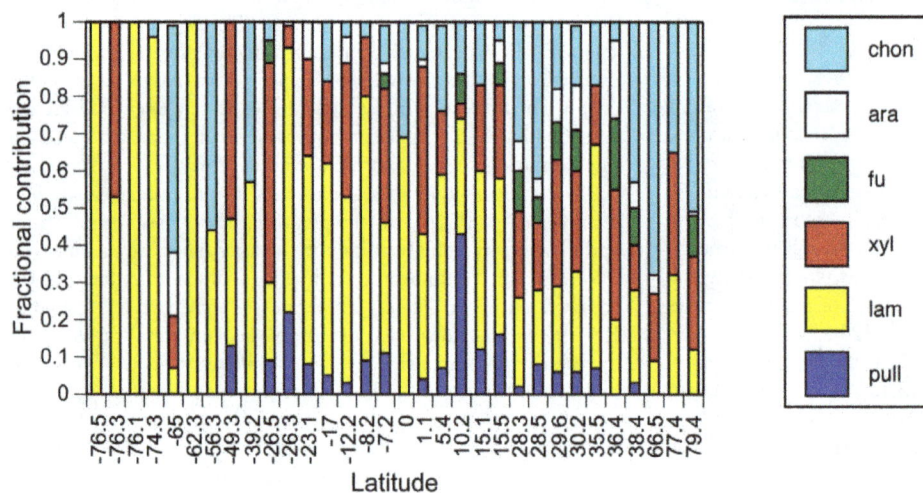

Figure 2. Proportionate contribution of each enzyme activity to summed hydrolysis rates, normalized to 100%. Station latitude and color key for enzyme identity are as in Fig. 2.

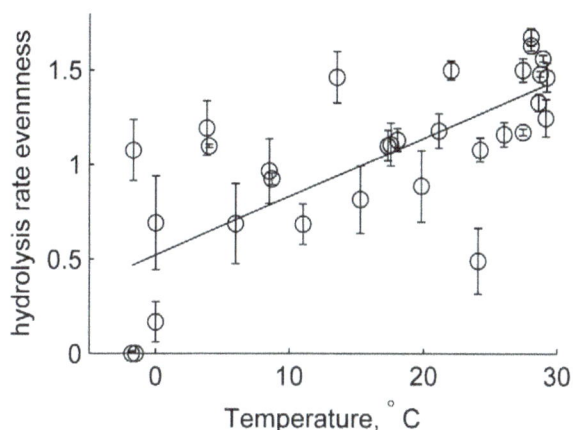

Figure 3. Evenness of enzymatic hydrolysis rates as a function of temperature. Average values (as shown in Figs. 1 and 2) were used for Station J, the only station that was sampled more than once. Line fit: $r^2 = 0.6154$, $n = 26$, $p < 0.001$.

(79°N, Stn. J; Table 1) sampled four times over the course of 10 years consistently showed hydrolytic activity dominated by chondroitinase (>50% of total activity), with similar patterns and levels of total activity (Table S1; Fig. 4). Likewise, two pairs of stations close to one another in the Gulf of Mexico (GOM1 and GOM11, 28°N; also GOM072 and GOM073, 28.2–30.2°N; Table 1) sampled 6 years apart showed the same broad substrate spectrum (nearly unique among sample locations; Fig. 2) and high levels of total activity (Fig. 1).

The patterns in enzymatic activities observed here provide insight into the potential of microbial communities to access components of the DOC pool, rather than a snapshot of enzyme activities expressed at the time of sampling, since our enzyme incubations lasted several days to two weeks. These experiments demonstrate the potential of a microbial community to access specific substrates over timescales sufficiently long to allow for cellular growth as well as for enzyme induction. The observation that many substrates remained unhydrolyzed over these timescales suggests that some microbial communities collectively lack the

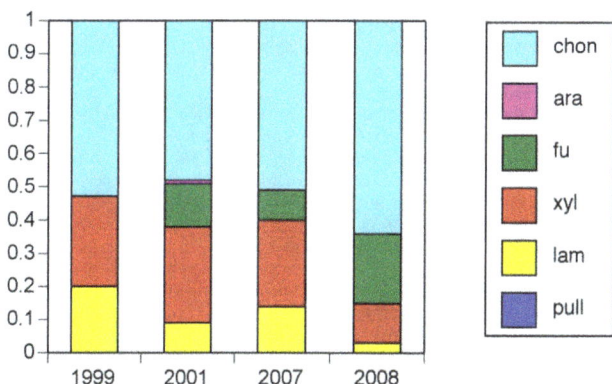

Figure 4. Relative contributions to summed hydrolysis rates for four visits to Station J, Svalbard. Each enzyme activity is shown in a different color; summed rates are normalized to 1.0. Pullulan hydrolysis is shown in blue, laminarin in yellow, xylan in red, fucoidan in green, arabinoglactan in white, and chondroitin sulfate in aqua. All rates plus standard deviations are listed in Table S1.

capabilities (enzymes, inducers, or organisms) to access specific substrates [38].

This information is unique, since no other currently available analytical method tests the abilities of microbial communities to access specific polysaccharide structures. Likewise, the concentrations of specific polysaccharides in the ocean cannot be measured with currently-available techniques [19], precluding direct measurement of polysaccharide production or concentration. Comparable data cannot be obtained by measuring surface water DOC concentrations at diverse locations, because DOC concentrations are a function of DOC production, DOC degradation, and water mass history [39] (the radiocarbon age of bulk DOC is ca. 6000 years [40].) Measurements of DOC concentrations in the surface ocean therefore cannot differentiate unambiguously between changes in the metabolic capabilities of heterotrophic microbes, changes in DOC production, and differences in water mass history. Moreover, attempts to constrain production and consumption terms for DOC across broad latitudinal gradients are greatly complicated by the multitude of sources and varying bioavailabilities of different components of the DOC pool [6], as well as the paucity of data from high latitude environments [9,41]. A measure of the spectrum of enzyme activities capable of hydrolyzing a class of biomolecules constituting the largest identified component of the DOC pool thus yields insight into processes otherwise not amenable to quantification.

The broader spectrum of enzyme activities observed at temperate and tropical sites indicates that those microbial communities can access a wider range of substrates than their high-latitude counterparts. The fact that this trend was discernable in our global data set despite variations in season, levels of productivity, oceanic province, and a host of other factors, is remarkable. The functional factors controlling the breadth of microbial community metabolic capabilities remain to be determined. However, the pattern observed here matches the decrease in bacterial species richness observed at high latitudes [26–27]. The reason for that trend is not well understood [42]; bacterial richness correlates to a similar extent with water temperature and latitude [26]. A recent model investigation of global patterns of phytoplankton diversity, however, points at the magnitude of seasonal variability in environments and at relative rates of organism dispersal as key factors controlling latitudinal diversity gradients for phytoplankton [43]. Similar factors may control diversity gradients of heterotrophic bacteria.

Projected changes in ocean environments over the coming decades driven by global warming [44–45] may increase the rates and widen the spectrum of enzyme activities due to changes in microbial community composition that are facilitated by changes in ocean temperature, circulation, or biogeochemical parameters [46]. Particularly in the Arctic, where rapid temperature increases are projected for the near future, the input of terrestrially-derived DOC into the Arctic basin may be greatly accelerated due to melting of permafrost and increased runoff [47]. Currently, much of this organic matter is buried or exported to the North Atlantic [41,48–49]. In the future, more organic matter may reach the Arctic Ocean, and if the range of complex substrates available to heterotrophic microbial communities is broader in a future Arctic Ocean, a larger fraction of it may be remineralized there.

Kirchman et al. [46] predict that climate change in polar waters may also lead to changes in food web structure resulting in greater transfer of carbon from phytoplankton to DOC to bacteria, and thus to the respiration of a larger fraction of marine primary production to CO_2. These processes would be greatly facilitated if future polar microbial communities can use a wider range of enzymatic tools, as their temperate to tropical counterparts currently do. Since DOC reactivity is a function of the metabolic capabilities of microbial communities as well as intrinsic chemical characteristics, predictive understanding of the global carbon cycle will require further investigation of the ways in which these metabolic capabilities are likely to change in response to changing climate.

Materials and Methods

Sample collection, substrate incubation, and sample analysis

Surface water was collected at each site (Table 1; Fig. S1), by Niskin or Go-Flo bottles; surface water at Stns. J and AB was collected by bucket. Water samples were dispensed into replicate vials. Each vial received a single fluorescently labeled polysaccharide [50] as a substrate, at a concentration of 3.5 µmol monomer equivalent L^{-1}. Triplicate vials were incubated at *in situ* temperature. Controls were autoclaved or poisoned with mercuric chloride before substrate addition. Subsamples from each vial were filtered through 0.2 µm pore-sized filters, and were stored frozen prior to analysis. Changes in substrate molecular weight with time were quantified using gel permeation chromatography, with Sephadex G-50 and G-75 columns linked in series, and fluorescence detection (excitation and emission wavelengths of 490 and 530 nm, respectively), as previously described [51]. Hydrolysis rates were calculated from changes in molecular weight distribution of the polysaccharides, as described previously [50–51]. No specific permits were required for the collection of water from the ocean as described above. The field studies did not involve any endangered or protected species.

Statistical analyses

Statistical comparisons were carried out using two-tailed Student's t-test. The evenness of hydrolysis rates (the contribution of each activity to summed activities) was calculated using Shannon's entropy [52], $H = -\sum_{i=1}^{6} f_i \ln f_i$, where f_i represents the hydrolysis rate of the ith substrate expressed relative to the sum of all hydrolysis rates for that station. As previously discussed [53], H is maximized ($H \approx 1.79$) when enzymatic hydrolysis rates of all substrates are equal, and minimized ($H = 0$) when only one hydrolysis rate was measurable. Error bars in Fig. 3 represent the standard deviation of the ensemble results from a Monte Carlo error simulation, as described in [53].

Acknowledgments

We thank the captains, crews, and scientific parties of R/V *Farm*, R/V *Cape Hatteras*, R/V *Cape Henlopen*, R/V *N.B. Palmer*, R/V *Hugh Sharp*, and R/V *Pelican* for help with sampling, and A.J. Baldwin for her assistance with shipboard experiments.

Author Contributions

Conceived and designed the experiments: CA WHJ. Performed the experiments: CA ADS KZ SG WHJ. Analyzed the data: CA ADS KZ SG. Wrote the paper: CA. Contributed to writing the manuscript: ADS KZ SG WHJ. Performed statistical analyses: ADS KZ.

References

1. Hedges JI (2002) Why dissolved organics matter. In: Hansell DA, Carlson CA, eds. Biogeochemistry of Marine Dissolved Organic Matter. San Diego: Academic Press. pp 1–33.

2. Azam F (1998) Microbial control of oceanic carbon flux: The plot thickens. Science 280: 694–696.

3. Hansell DA, Carlson C (2002) Biogeochemistry of Dissolved Organic Matter. San Diego: Academic Press.

4. McCarren J, Becker JW, Repeta DJ, Shi Y, Young CR, et al. (2010) Microbial community transcriptomes reveal microbes and metabolic pathways associated with dissolved organic matter turnover in the sea. PNAS.

5. Carlson CA (2002) Production and removal processes. In: Hansell DA, Carlson CA, eds. Biogeochemistry of Dissolved Organic Matter. San Diego: Academic Press. pp 91–139.

6. Jiao N, Herndl GJ, Hansell DA, Benner R, Kattner G, et al. (2010) Microbial production of recalcitrant dissolved organic matter: long-term carbon storage in the global ocean. Nature Rev Microbiol 8: 593–599.

7. Weiss MS, Abele U, Weckesser J, Welte W, Schiltz E, et al. (1991) Molecular architecture and electrostatic properties of a bacterial porin. Science 254: 1627–1630.

8. Benner R, Pakulski JD, McCarthy M, Hedges JI, Hatcher PG (1992) Bulk chemical characteristics of dissolved organic matter in the ocean. Science 255: 1561–1564.

9. Benner R (2002) Chemical composition and reactivity. In: Hansell DA, Carlson CA, eds. Biogeochemistry of Marine Dissolved Organic Matter. New York: Academic Press. pp 59–90.

10. DeLong EF, Pace N (2001) Environmental diversity of bacteria and archea. Syst Biol 50: 470–478.

11. Hoppe H-G (1983) Significance of exoenzymatic activities in the ecology of brackish water: measurements by means of methylumbelliferyl-substrates. Mar Ecol Prog Ser 11: 299–308.

12. Arnosti C (2011) Microbial extracellular enzymes and the marine carbon cycle, In: Carlson C, Giovannoni S, eds. (2011) Annual Review of Marine Science, vol 3. Palo Alto: Annual Reviews. pp 401–425.

13. Bauer M, Kube M, Telling H, Richter M, Lombardot T, et al. (2006) Whole genome analysis of the marine Bacteroidetes 'Gramella forsetii' reveals adaptations to degradation of polymeric organic matter. Environmental Microbiology 8: 2201–2213.

14. Elifantz H, Waidner LA, Michelou VK, Cottrell MT, Kirchman DL (2008) Diversity and abundance of glycosyl hydrolase family 5 in the North Atlantic Ocean. FEMS Microbiol Ecol 63: 316–327.

15. Weiner RM, Taylor LE, II, Henrissat B, Hauser L, Land M, et al. (2008) Complete genome sequence of the complex carbohydrate-degrading marine bacterium, Saccharophagus degradans Strain 2-40T. PLOS Genetics 5(5): e1000087.

16. Gilbert J, Field D, Swift P, Simon T, Cummings D, et al. (2010) Taxonomic and functional diversity of microbes at a temperate coastal site: A 'multi-omic' study of seasonal and diel temporal variation. PLoS ONE 5(11): e15545.

17. Cantarel BL, Coutinho PM, Rancurel C, Bernard T, Lombard V, et al. (2008) The carbohydrate-active enzyme database (CAZy): an expert resource for glycogenomics. Nucleic Acids Res 37: 233–38.

18. Arrieta JM, Herndl GJ (2002) Changes in bacterial β-glucosidase diversity during a coastal phytoplankton bloom. Limnol Oceanogr 47: 594–599.

19. Hedges JI, Eglinton G, Hatcher PG, Kirchman DL, Arnosti C, et al. (2000) The Molecularly-uncharacterized component (MUC) of nonliving organic matter in natural environments. Organic Geochem 31: 945–958.

20. Painter TJ (1983) Algal Polysaccharides. In: Aspinall GO, ed. The Polysaccharides. New York: Academic Press. pp 195–285.

21. Glockner FO, Kube M, Bauer M, Teeling H, Lombardot T, et al. (2003) Complete genome sequence of the marine planctomycete Pirellula sp. strain 1. Proc Nat Acad Sci 100: 8298–8303.

22. Alderkamp A-C, van Rijssel M, Bolhuis H (2007) Characterization of marine bacteria and the activity of their enzyme systems involved in degradation of the algal storage glucan laminarin. FEMS Microbiol Ecol 59: 108–117.

23. Arnosti C, Durkin S, Jeffrey WH (2005) Patterns of extracellular enzyme activities among pelagic marine microbial communities: implications for cycling of dissolved organic carbon. Aq Microb Ecol 38: 135–145.

24. Baldwin AJ, Moss JA, Pakulski JD, Catala P, Joux F, et al. (2005) Microbial diversity in a Pacific Ocean transect from the Arctic to Antarctic circles. Aq Microb Ecol 41: 91–102.

25. Martiny JBH, Bohannan BJM, Brown JH, Colwell RK, Fuhrman JA, et al. (2006) Microbial biogeography: putting microorganisms on the map. Nature Microb Rev 6: 102–112.

26. Fuhrman JA, Steele JA, Hewson I, Schwalbach MS, Brown MV, et al. (2008) A latitudinal diversity gradient in planktonic marine bacteria. Proc Nat Acad Sci 105(22): 7774–7778.

27. Pommier T, Canback B, Riemann L, Bostrom KH, Simu K, et al. (2007) Global patterns of diversity and community structure in marine bacterioplankton. Molecular Ecol 16: 867–880.

28. Wietz M, Gram L, Jorgensen B, Schramm A (2010) Latitudinal patterns in the abundance of major marine bacterioplankton groups. Aq Microb Ecol 61: 179–189.

29. Arnosti C (2008) Functional differences between Arctic sedimentary and seawater microbial communities: Contrasts in microbial hydrolysis of complex substrates. FEMS Microb Ecol 66: 343–351.

30. Bano N, Hollibaugh JT (2002) Phylogenetic composition of bacterioplankton assemblages from the Arctic Ocean. Appl Environ Microbiol 68: 505–518.

31. Malmstrom RR, Straza TRA, Cottrell MT, Kirchman DL (2007) Diversity, abundance, and biomass production of bacterial groups in the western Arctic Ocean. Aq Microb Ecol 47: 45–55.

32. Hess M, Sczyrba A, Egan R, Kim T-W, Chokhawala H, et al. (2011) Metagenomic discovery of biomass-degading genes and genomes from cow rumen. Science 331: 463–467.

33. Murray AE, Preston CM, Massana R, Taylor LT, Blakis A, et al. (1998) Seasonal and spatial variability of bacteria and archael assemblages in the coastal waters near Anvers Island, Antarctica. Appl Environ Microbiol 64: 2585–2595.

34. Morris RM, Vergin KL, Cho J-C, Rappe M, Carlson CA, et al. (2005) Temporal and spatial response of bacterioplankton lineages to annual convective overturn at the Bermuda Atlantic Time-series Study site. Limnol Oceanogr 50: 1687–1696.

35. Kan J, Crump BC, Wang K, Chen F (2006) Bacterioplankton communities in Cheasapeake Bay: Predictable or random assemblages. Limnol Oceanogr 51: 2157–2169.

36. Fuhrman JA, Steele JA (2008) Community structure of marine bacterioplankton: patterns, networks, and relationships to function. Aq Microb Ecol 53: 69–81.

37. Fuhrman JA, Hewson I, Schwalbach MS, Steele JA, Brown MV, et al. (2006) Annually reoccurring bacterial communities are predictable from ocean conditions. Proc Nat Acad Sci USA 103: 13104–13109.

38. Arnosti C (2004) Speed bumps and barricades in the carbon cycle: Substrate structural effects on carbon cycling. Mar Chem 92: 263–273.

39. Goldberg SJ, Carlson CA, Bock B, Nelson NB, Siegel DA (2010) Meridional variability in dissolved organic matter stocks and diagenetic state within the euphotic and mesopelagic zone of the North Atlantic subtropical gyre. Mar Chem 119: 9–21.

40. Bauer JE, Williams PM, Druffel ERM (1992) ^{14}C activity of dissolved organic carbon fractions in the north-central Pacific and Sargasso Sea. Nature 357: 667–670.

41. Amon RMW (2004) The role of dissolved organic matter for the organic carbon cycle in the Arctic Ocean. In: Stein R, Macdonald RW, eds. The organic carbon cycle in the Arctic Ocean. Berlin: Springer. pp 83–100.

42. Fuhrman JA (2009) Microbial community structure and its functional implications. Nature 459: 193–199.

43. Barton AD, Dutkiewicz S, Flierl G, Bragg J, Follows MJ (2010) Patterns of diversity in marine phytoplankton. Science 327: 1509–1511.

44. Behrenfeld MJ, O'Malley RT, Siegel DA, McClain CR, Sarmiento JL, et al. (2006) Climate-driven trends in contemporary ocean productivity. Nature 444: 752–755.

45. Vezina AF, Hoegh-Guldberg O (2008) Effects of ocean acidification on marine ecosystems: Introduction. Mar Ecol Prog Ser 373: 199–201.

46. Kirchman DL, Moran XAG, Ducklow H (2009) Microbial growth in the polar oceans—role of temperature and potential impact of climate change. Nature Rev Microbiol 7: 451–459.

47. Stocker TF, Raible CC (2005) Water cycle shifts gear. Nature 434: 830–832.

48. Stein R, Macdonald RW (2004) Organic carbon budget: Arctic Ocean vs. Global Ocean. In: Stein R, Macdonald RW, eds. The organic carbon cycle in the Arctic Ocean. Berlin: Springer. pp 315–323.

49. Benner R, Louchouarn P, Amon RMW (2005) Terrigenous dissolved organic matter in the Arctic Ocean and its transport to surface and deep waters of the North Atlantic. Global Biogeochem Cycles 19: GB2025. doi:2010.1029/2004GB002398.

50. Arnosti C (1996) A new method for measuring polysaccharide hydrolysis rates in marine environments. Org Geochem 25: 105–115.

51. Arnosti C (2003) Fluorescent derivatization of polysaccharides and carbohydrate-containing biopolymers for measurement of enzyme activities in complex media. J Chromatog B 793: 181–191.

52. Legendre P, Legendre L (1998) Numerical Ecology. New York: Elsevier. 853 p.

53. Steen AD, Ziervogel K, Arnosti C (2010) Comparison of multivariate microbial datasets with the Shannon Index: an example using enzyme activities from diverse marine environments. Org Geochem 41: 1019–1021.

4

Environmental Impact of the Production of Mealworms as a Protein Source for Humans – A Life Cycle Assessment

Dennis G. A. B. Oonincx[1]*, **Imke J. M. de Boer**[2]

1 Department of Plant Sciences, Wageningen University, Wageningen, The Netherlands, 2 Animal Department of Animal Sciences, Wageningen University, Wageningen, The Netherlands

Abstract

The demand for animal protein is expected to rise by 70–80% between 2012 and 2050, while the current animal production sector already causes major environmental degradation. Edible insects are suggested as a more sustainable source of animal protein. However, few experimental data regarding environmental impact of insect production are available. Therefore, a lifecycle assessment for mealworm production was conducted, in which greenhouse gas production, energy use and land use were quantified and compared to conventional sources of animal protein. Production of one kg of edible protein from milk, chicken, pork or beef result in higher greenhouse gas emissions, require similar amounts of energy and require much more land. This study demonstrates that mealworms should be considered a more sustainable source of edible protein.

Editor: Gregory A. Sword, Texas A&M University, United States of America

Funding: This study was directly funded by Wageningen University, Wageningen, The Netherlands (www.wur.nl) as part of a PhD program. Wageningen University had no other role in study design, data collection and analysis, decision to publish, or preparation of the manuscript, than can be expected with the academic supervision of a PhD candidate.

Competing Interests: The authors have declared that no competing interests exist.

* E-mail: dennis.oonincx@wur.nl

Introduction

The demand for food of animal origin is rising globally and is expected to increase by 70–80% between 2012 and 2050 [1,2,3]. Currently, the livestock sector uses about 70% of all agricultural land [2,4] and is responsible for about 15% of the total emission of anthropogenic greenhouse gas (GHG) [2,3]. Expansion of agricultural acreage by land clearing is a major source of GHG emissions [2,5] and one of the largest contributors to global warming [6]. People's choices for certain diets influence GHG emissions and other environmental parameters [6,7]. A suggested mitigation measure is a shift towards protein from lower impact animal species [1,4,8]. Various authors have suggested insects as an environmentally more friendly alternative to conventional livestock [9,10,11,12,13]. However, little data are available on the environmental impact associated with insect production. Husbandry contributions to GHG emissions is much lower for insects (2–122 g/kg mass gain) than for beef cattle (2850 g/kg mass gain), and in the lower range when compared to pigs (80–1130 g/kg mass gain) [14]. However, this is only a part of the total GHG emissions in animal production chains. To choose among different sources of animal protein, GHG emissions, and other environmental parameters, such as land or fossil energy use, need to be assessed. Life cycle assessment (LCA) is a widely accepted method to quantify these parameters [15] and has been used for various animal products [16]. Within an LCA preselected parameters are quantified along the entire life cycle of a product. For mealworms, for instance, not only direct GHG emissions through respiration, but also GHG emissions related to feed production, and distribution as well as emissions due to the heating of the climate-controlled-rearing facility are quantified and attributed to a product. Although claims that insects are a more sustainable protein source than conventional livestock are widespread, to our knowledge, an LCA of any insect species used as a protein source has never been published. The objective of this paper, therefore, was to quantify the environmental impact attributed to the production of two tenebrionid species, viz. the mealworm (*Tenebrio molitor*) and the super worm (*Zophobas morio*). This impact was then compared to conventional sources of animal protein, such as milk, chicken, pork and beef.

Materials and Methods

Through collaboration with a commercial mealworm producer in The Netherlands (van de Ven Insectenkwekerij, Deurne, The Netherlands), insight in the production process of mealworms was acquired. This farm produces two mealworm species in the same way and at equal quantity (kg/year). Therefore, we conducted a combined LCA for both species. First, the system boundary was defined (Figure 1). A cradle-to-farm-gate approach was chosen, which means that the environmental impact was assessed up to the moment that the fresh product leaves the farm gate.

An LCA relates the environmental impact of a product to a functional unit (FU). Comparing LCA results among animal products demands identical FU's. Mealworms, like other animal products, can nutritionally be seen as a source of protein. Therefore, we defined two FUs in our study: 1) kg of fresh product, and 2) kg of edible protein. To compute the amount of edible protein, we first multiplied the kg of fresh product with the average reported dry matter (DM) content (*T. molitor* 38%; *Z. morio*

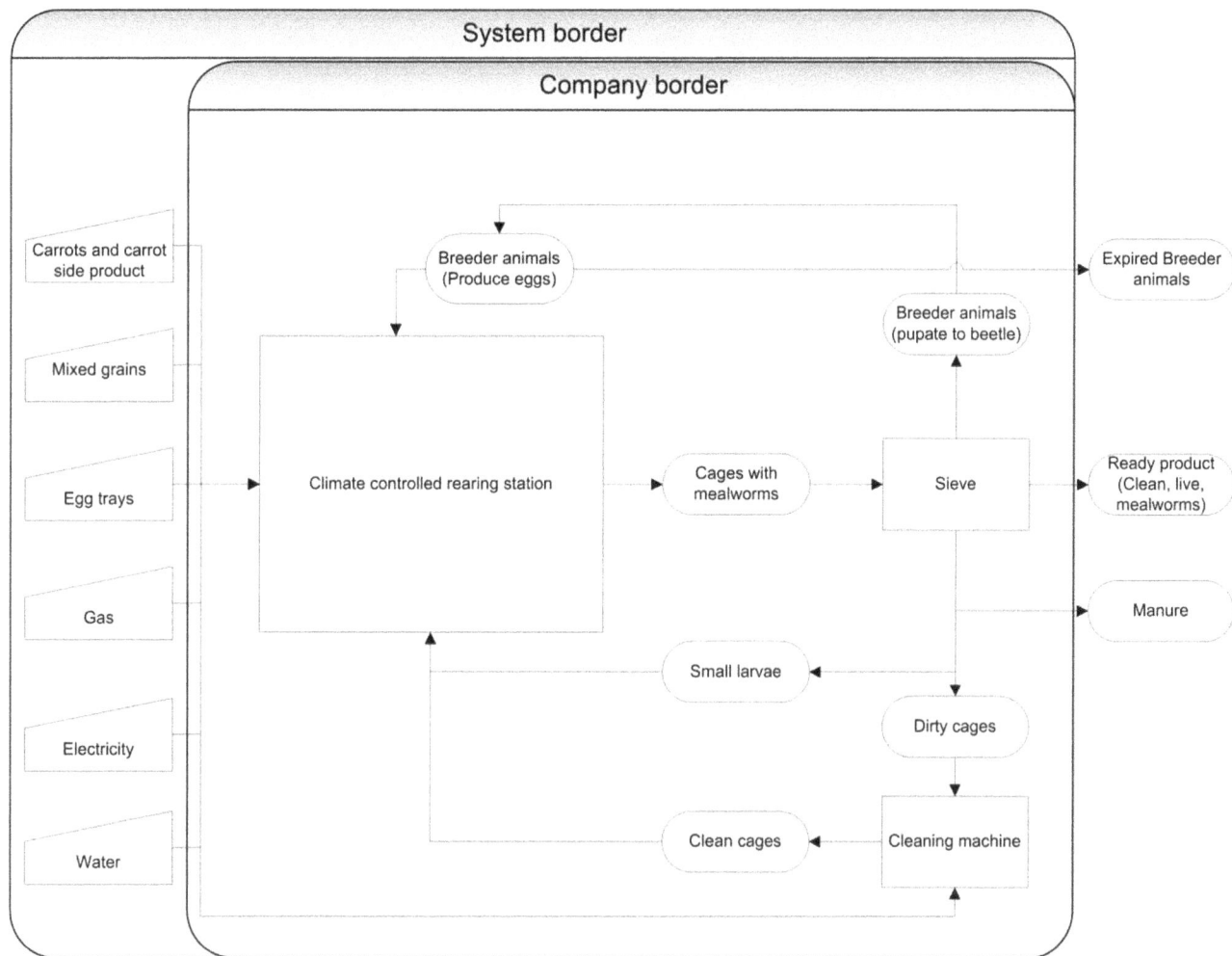

Figure 1. The mealworm production system. Flows entering the company are on the left, centrally the production steps are shown and flows exiting the system are on the right. For flow quantities see Table 1.

43%) and the average percentage of reported crude protein in the dry matter (*T. molitor* 53%; *Z. morio* 45%) [17,18,19,20]. Subsequently, we multiplied with the edible portion, which we consider to be 100% for mealworms since they are consumed by humans as the whole animal. Protein content [16,21,22] and edible portion [16,22] of common production animals vary depending on breed, country of production and other factors. In this study we used the data reported by De Vries & De Boer [16].

We quantified three environmental indicators: 1) global warming potential (GWP), 2) fossil energy use (EU), and, 3) land use (LU). Global warming potential was expressed in CO_2-equivalents (CO_2-eq); the sum of CO_2, CH_4, and N_2O emissions. The conversion factor to CO_2-eq is 1 for CO_2, 25 for CH_4 and 298 for N_2O [23]. Land use was expressed in m^2 per year, and fossil energy use in mega joules (MJ). When several products stem from one production process, such as grain and straw, it is called a multifunctional production process. Its environmental impact is then allocated to the various outputs. Multifunctional processes included in our system were: (1) production of feed ingredients and their co-products, and (2) production of mealworms and its co-product manure. We allocated the impact of feed production to its outputs based on their relative economic value, whereas the impact of mealworm production was fully allocated to mealworms.

All inputs of the mealworm production system were quantified in this assessment. The production system, including the diet, is identical for both mealworm species (Figure 1). The diet consisted of fresh carrots and a mixed grain feed (i.e. wheat bran, oats, soy, rye and corn supplemented with beer yeast). For industrial competitive protection the exact composition of the diet is not disclosed. The feed conversion ratio (FCR) for concentrates was calculated by dividing the amount of concentrates used by the amount of live mealworms produced. Egg cartons are used to increase the surface area for the adult mealworms. For the environmental impact of the egg cartons, data for recycled cardboard were used. A batch of twenty egg cartons was dried at 70°C until a stable weight was reached and the average dry weight was assumed representative.

In order to create an optimal rearing environment, the climate-controlled-rearing facility is heated and ventilated by the usage of natural gas and electricity from the Dutch power grid. In order to minimize seasonal influences on energy usage, data for a complete year were used.

The quantities of all inputs and the output of the production system were disclosed by the mealworm producing company. Quantitative data for each input regarding GWP, EU, and LU (Table 1) were based on Ecoinvent [24] and new data from the

Dutch animal feed industry [25]. This data includes: the production, processing, and transportation of carrots, feed ingredients, and egg trays as well as the production, transportation and use of natural gas, electricity and water. The LU of the farm and the direct GHG emissions from the mealworms were added to this data. Direct GHG emission for $Z.$ $morio$ and $T.$ $molitor$ were assumed equal and data from Oonincx et al. (2010b), in which the same diet was used, were assumed representative. Finally, the total GWP, EU, and LU were divided by kg of fresh mealworm, or the kg of edible protein produced, resulting in an expression for both FUs.

Results

The Feed Conversion Ratio (FCR) for concentrates was 2.2 kg/ kg of live weight in this study. The absolute and relative GWP, EU, and LU for the production of one kilogram of fresh mealworms based on economic allocation are provided in Table 2. The GWP of one kg of fresh mealworms was 2.7 kg of CO_2-eq, of which 42% results from the production and transport of feed grains, 14% from the production and transport of carrots, 26% from gas used for heating, and 17% from the use of electricity. The EU of one kg of fresh mealworms was 34 MJ, of which 31% results from the emission of production and transport of feed grains, 13% from the production and transport of carrots, 35% from gas used for heating, and 21% from the use of electricity. The LU of one kg of fresh mealworms was 3.6 m^2 per year, of which 85% was required to cultivate feed grains, and 14% to produce carrots. When expressed per kg of edible protein from mealworms, the GWP was 14 kg of CO_2-eq, the EU was 173 MJ and the LU was 18 m^2. The relative contributions remain the same.

Discussion

Both mass and economic allocation can be used when describing the environmental impact of a product. Since the latter is more commonly used, values in this discussion are based on economic allocation allowing comparison with other food sources of animal origin.

Differences in environmental impact of products from pork, chicken, and beef are caused by three main factors; 1) enteric CH_4

Table 2. Environmental impact of inputs in a mealworm production system.

	GWP (kg CO_2-eq)		EU (MJ)		LU (m^2)	
Carrots (kg)	0.38	14.27%	4.31	12.80%	0.51	14.39%
Mixed grains (kg)	1.11	41.98%	10.47	31.09%	3.03	85.14%
Gas (MJ)	0.70	26.26%	11.71	34.77%	0.00	0.02%
Egg trays (kg)	0.00	0.12%	0.04	0.13%	0.00	0.01%
Electricity (MJ)	0.45	17.06%	7.13	21.17%	0.01	0.24%
Water (M³)	0.00	0.03%	0.01	0.04%	0.00	0.00%
Animal (kg)	0.01	0.29%	0.00	0.00%	0.00	0.00%
Farm	0.00	0.00%	0.00	0.00%	0.01	0.20%
Total	2.65	100.00%	33.68	100.00%	3.56	100.00%

Absolute and relative contribution global warming potential (GWP), energy use (EU) and land use (LU) for the production of one kg of fresh mealworms based on economic allocation.

production, 2) reproduction rate and 3) feed conversion efficiency [16]. Based on these three factors one would expect a lower environmental impact from mealworm production. Firstly, mealworms do not produce CH_4 [14], secondly they have a high reproduction rate; one female of $T.$ $molitor$ produces 160 eggs in her life (3 months) and a $Z.$ $morio$ female produces 1500 eggs in her life (1 year). Furthermore, the maturation period is short; $T.$ $molitor$ reaches adulthood in 10 weeks and $Z.$ $morio$ in 3.5 months [27]. Thirdly, feed conversion efficiency, depends amongst other things, on the diet provided. The feed conversion ratio (FCR) for concentrates (kg/kg of fresh weight) for the mealworms in this study (2.2) was similar to values reported for chicken (2.3) but lower than for pigs (4.0) and beef cattle (2.7–8.8) [28]. The large spread reported for beef cattle is due to variation in the proportion of concentrates, relative to roughage (for instance grass), used in the diet.

We assessed three indicators to provide insight in the sustainability of mealworm production and compared these to literature values (Figure 2, 3 and 4). The GWP of mealworms per kg of edible protein is low compared to milk (1.77–2.80× as high), chicken (1.32–2.67× as high), pork (1.51–3.87× as high) or beef (5.52–12.51× as high) [16]. The EU of mealworm production per kg of edible protein is higher than for milk (21–83% of the value for mealworm) or chicken (46–88% of the value for mealworm), similar to pork (55–137% of the value for mealworm) and lower than for beef (1.02–1.58× as high). Mealworms, being poikilothermic, depend on suitable ambient temperatures for growth and development. When ambient temperatures are low, heating is required, increasing energy use. Mitigation measures are being investigated: larger larvae in this system produce a surplus of metabolic heat, which could be used to heat the heat-demanding small larvae. The LU of the described production system was very low compared to milk (1.81–3.23× as high), chicken (2.30–2.85× as high), pork (2.57–3.49× as high higher) and beef (7.89–14.12× as high).

Over the last two decades productivity of chickens and pigs has increased annually by 2.3%, due to the application of science and new technologies [29]. Further improvement of the mealworm production system by, for instance, automation, feed optimization or genetic strain selection is expected to increase productivity and

Table 1. Resource use per year and environmental impact.

Resource	Turnover/Year	GWP	EU	LU
		kg CO_2-eq	MJ	m^2
Carrots (kg)	260000	0.12	1.38	0.16
Mixed grains (kg)	182000	0.51	4.79	1.39
Egg trays (kg)	262	0.98	13.70	0.10
Gas (MJ)	811200	0.07	1.20	0.00
Electricity (MJ)	187200	0.20	3.17	0.00
Water (M³)	211	0.32	5.55	0.04
Animal (kg)	83200	0.01	0.00	0.00
Farm	1	0	0.00	588

Global warming potential (GWP), energy use (EU) and land use (LU) are expressed per unit of input based on economic allocation.

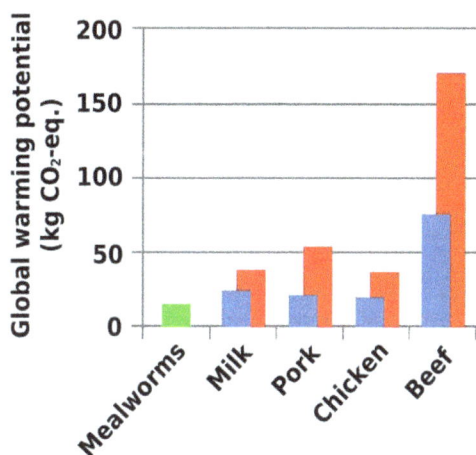

Figure 2. Environmental impact of mealworms compared to other animal products. Global warming potential due to the production of one kg of edible protein. Results from this study depicted in green. Minimum (blue) and maximum (red) literature data is adapted from de Vries & de Boer (2010).

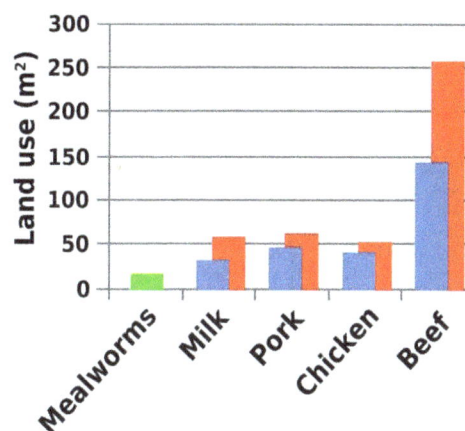

Figure 4. Environmental impact of mealworms compared to other animal products. Land use due to the production of one kg of edible protein. Results from this study depicted in green. Minimum (blue) and maximum (red) literature data is adapted from de Vries & de Boer (2010).

decrease the environmental impact. Since these aspects are currently underdeveloped, the potential rate of productivity improvement is expected to be higher for mealworms compared to the more common production animals.

Two further aspects can influence the environmental impact of mealworm production: off-farm differences in storage and processing, and the economic value of by-products. Processing and storage has a large effect on the environmental impact of a food product: emissions during slaughtering, transportation and storage of pork and chicken contribute 17–25% to the total GWP [7,16]. For mealworms, used for human consumption, there is currently no standard method for processing and storing. If the environmental impact due to processing and storage is similar to that of other animal derived food products, the results of this study are also representative beyond the farm gate.

Besides food, there are also non-food by-products from common production animals, such as leather, blood meal or feathers. The

relative value of these products is not taken into account for the expression of the selected indicators, expressed per kg of edible protein. A decrease of the environmental impact due to the economic value of these by-products would be relatively small (3% for pork, 5% for chicken, and 3–5% for beef [26]).

We consider the low LU of mealworms to be particularly important; effects of GHG emissions can be countered by carbon fixation [30], and forest regrowth and afforestation [5,31], depletion of fossil fuels can be countered by usage of alternative sources of energy [32,33], but land availability is fixed and limited. Expansion of agricultural land is a major source of GHG production especially in tropical regions [4,5,6]. Slowing down the expansion of agricultural land is a critical step towards sustainable agriculture [4]. The increasing world population will therefore need to be fed using the same area of land that is available now [6]. Mealworms require only 43% of the amount of land used for the production of one kg of edible animal protein as milk, and only 10% of the land used for production of beef.

Conclusions

The EU of mealworm production is higher than for milk or chicken and similar to pork and beef. However, mealworms, when considered as a human protein source, produce much less GHG's and require much less land, than chickens, pigs and cattle. With land availability being the most stringent limitation in sustainably feeding the world's population, this study clearly shows that mealworm should be considered as a more sustainable alternative to milk, chicken, pork and beef.

Acknowledgments

The authors kindly acknowledge Roland and Michel van de Ven for providing company data and insight in their production system. The authors are grateful to Theo Viets for his support with data handling and to Merel Schouten for assisting in depicting the results.

Author Contributions

Conceived and designed the experiments: DO IB. Performed the experiments: DO IB. Analyzed the data: DO. Contributed reagents/materials/analysis tools: IB. Wrote the paper: DO IB.

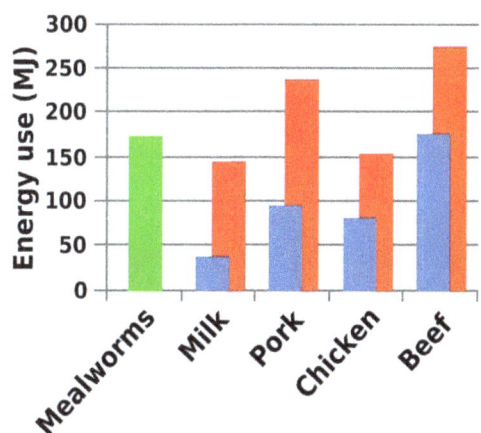

Figure 3. Environmental impact of mealworms compared to other animal products. Energy use due to the production of one kg of edible protein. Results from this study depicted in green. Minimum (blue) and maximum (red) literature data is adapted from de Vries & de Boer (2010).

References

1. Pelletier N, Tyedmers P (2010) Forecasting potential global environmental costs of livestock production 2000–2050. Proceedings of the National Academy of Sciences of the United States of America 107: 18371–18374.
2. Steinfeld H, Gerber P, Wassenaar T, Castel V, Rosales M, et al. (2006) Livestock's Long Shadow. Environmental issues and options.: 319 pp.
3. Steinfeld H (2012) Sustainability Issues in Livestock Production. Exploratory Workshop Sustainable Protein Supply. Amsterdam. 26pp.
4. Foley JA, Ramankutty N, Brauman KA, Cassidy ES, Gerber JS, et al. (2011) Solutions for a cultivated planet. Nature 478: 337–342.
5. Pan Y, Birdsey RA, Fang J, Houghton R, Kauppi PE, et al. (2011) A Large and Persistent Carbon Sink in the World's Forests. Science 333: 988–993.
6. Godfray HCJ, Pretty J, Thomas SM, Warham EJ, Beddington JR (2011) Linking Policy on Climate and Food. Science 331: 1013–1014.
7. Carlsson-Kanyama A (1998) Climate change and dietary choices – how can emissions of greenhouse gases from food consumption be reduced? Food Policy 23: 277–293.
8. McMichael AJ, Powles JW, Butler CD, Uauy R (2007) Food, livestock production, energy, climate change, and health. Lancet 370: 1253–1263.
9. Mercer CWL (1997) Sustainable production of insects for food and income by New Guinea villagers. Ecology of Food and Nutrition 36: 151–157.
10. RamosElorduy J (2005) Insects: a hopeful food source. Ecological Implications of minilivestock Potential of insects, rodents, and frogs (Ed: MG Paoletti) Science Publishers, Inc Enfield, New Hampshire, USA ISBN 1-57808-339-7 Chapter 14: 263–291.
11. van Huis A (2003) Insects as food in sub-Saharan Africa: Insect Science and its Application. 2003 July–September; 23(3):163–185.
12. DeFoliart G (1997) An overview of the role of edible insects in preserving biodiversity. Ecology of Food and Nutrition 36: 109–132.
13. Meijer-Rochow VB (1975) Can insects help to ease the problem of world food shortage? Search 6: 261–262.
14. Oonincx DGAB, van Itterbeeck J, Heetkamp MJW, van den Brand H, van Loon JJA, et al. (2010) An Exploration on Greenhouse Gas and Ammonia Production by Insect Species Suitable for Animal or Human Consumption. Plos One 5: 12.
15. Guinee J (2001) Handbook on life cycle assessment - Operational guide to the ISO standards. International Journal of Life Cycle Assessment 6: 255–255.
16. de Vries M, de Boer IJM (2010) Comparing environmental impacts for livestock products: A review of life cycle assessments. Livestock Science 128: 1–11.
17. Barker D, Fitzpatrick MP, Dierenfeld ES (1998) Nutrient composition of selected whole invertebrates. Zoo biology 17: 123–134.
18. Bernard JB, Allen ME (1997) Feeding captive insectivorous animals: Nutrional aspects of insects as food. Nutrition Advisory Group handbook 1997.
19. Finke MD (2002) Complete nutrient composition of commercially raised invertebrates used as food for insectivores. Zoo Biology 21: 269–285.
20. Oonincx DGAB, Stevens Y, van den Borne JJGC, van Leeuwen JPTM, Hendriks WH (2010) Effects of vitamin D-3 supplementation and UVb exposure on the growth and plasma concentration of vitamin D-3 metabolites in juvenile bearded dragons (Pogona vitticeps). Comparative Biochemistry and Physiology B-Biochemistry & Molecular Biology 156: 122–128.
21. FAO (2007) Meat processing technology for small- to medium-scale producers. Bangkok: FAO. 456 p.
22. Smil V (2002) Worldwide transformation of diets, burdens of meat production and opportunities for novel food proteins. Enzyme and Microbial Technology 30: 305–311.
23. IPCC (2007) Summary for Policymakers. In: Climate Change 2007: The Physical Science Basis. Contribution of Working Group I to the Fourth Assessment Report of the Intergovernmental Panel on Climate Change. Cambridge, United Kingdom and New York, NY, USA.: IPCC. 18pp.
24. Ecoinvent (2004) Final Reports Ecoinvent 2000. In: Inventories SCfLC, editor. Ecoinvent Data v1.1 ed. Dubendorf, Switzerland: Swiss Centre for Life Cycle Inventories. 67pp.
25. WUR-Lifestock Research (2012) LCI data for the calculation tool Feedprint for greenhouse gas emissions of feed production and utilization; other products. Gouda/Lelystad, The Netherlands: Blonk Milieu Advies and WUR-Lifestock Research. 51 p.
26. Luske B, Blonk H (2009) Milieueffecten van dierlijke bijproducten (in Dutch). Gouda: Blonk Milieu advies. 70 p.
27. Friederich U, Volland W (2004) Breeding Food Animals: Live Food for Vivarium Animals; Friederich U, Volland W, editors. Malabar, Florida: Krieger publishing company. 178pp.
28. Wilkinson JM (2011) Re-defining efficiency of feed use by livestock. Animal 5: 1014–1022.
29. Steinfeld H, Gerber P (2010) Livestock production and the global environment: Consume less or produce better? Proceedings of the National Academy of Sciences of the United States of America 107: 18237–18238.
30. Piao S, Fang J, Ciais P, Peylin P, Huang Y, et al. (2009) The carbon balance of terrestrial ecosystems in China. 458: 1009–1013.
31. Wise M, Calvin K, Thomson A, Clarke L, Bond-Lamberty B, et al. (2009) Implications of Limiting CO2 Concentrations for Land Use and Energy. Science 324: 1183–1186.
32. Graham-Rowe D (2011) Agriculture: Beyond food versus fuel. 474: S6–S8.
33. Tilman D, Socolow R, Foley JA, Hill J, Larson E, et al. (2009) Beneficial Biofuels—The Food, Energy, and Environment Trilemma. Science 325: 270–271.

5

Phosphoglycerate Mutases Function as Reverse Regulated Isoenzymes in *Synechococcus elongatus* PCC 7942

Jiri Jablonsky[1]*, Martin Hagemann[2], Doreen Schwarz[2], Olaf Wolkenhauer[1]

1 Department of Systems Biology and Bioinformatics, University of Rostock, Rostock, Germany, **2** Department of Plant Physiology, University of Rostock, Rostock, Germany

Abstract

Phosphoglycerate-mutase (PGM) is an ubiquitous glycolytic enzyme, which in eukaryotic cells can be found in different compartments. In prokaryotic cells, several PGMs are annotated/localized in one compartment. The identification and functional characterization of PGMs in prokaryotes is therefore important for better understanding of metabolic regulation. Here we introduce a method, based on a multi-level kinetic model of the primary carbon metabolism in cyanobacterium *Synechococcus elongatus* PCC 7942, that allows the identification of a specific function for a particular PGM. The strategy employs multiple parameter estimation runs in high CO_2, combined with simulations testing a broad range of kinetic parameters against the changes in transcript levels of annotated PGMs. Simulations are evaluated for a match in metabolic level in low CO_2, to reveal trends that can be linked to the function of a particular PGM. A one-isoenzyme scenario shows that PGM2 is a major regulator of glycolysis, while PGM1 and PGM4 make the system robust against environmental changes. Strikingly, combining two PGMs with reverse transcriptional regulation allows both features. A conclusion arising from our analysis is that a two-enzyme PGM system is required to regulate the flux between glycolysis and the Calvin-Benson cycle, while an additional PGM increases the robustness of the system.

Editor: Nestor V. Torres, Universidad de La Laguna, Spain

Funding: The work presented here has been supported by the German Research Foundation (DFG) as part of PROMICS research group 1186. The funders had no role in study design, data collection and analysis, decision to publish, or preparation of the manuscript.

Competing Interests: The authors have declared that no competing interests exist.

* E-mail: jiri.jablonsky@gmail.com

Introduction

Phosphoglycerate-mutases (PGMs) are a group of non-homologous glycolytic enzymes [1], having independent evolutionary origins, catalyzing the reversible conversion of 3-phosphoglycerate (3PGA) to 2-phosphoglycerate (2PGA). PGMs can be found in different compartments in the eukaryotic cells fulfilling different tasks [2]. PGMs are also present in prokaryotes [3] such as cyanobacteria, in which the different PGM isoenzymes are localized in one compartment. PGMs can be divided into two analogous subgroups, cofactor-independent and cofactor (2,3-bisphosphoglycerate)-dependent enzymes [3]. The occurrence of these enzyme types seems to be scattered among prokaryotes. Based on our own analysis, the presence of PGM proteins from these two groups is unpredictable among cyanobacteria, as has been reported before for other bacteria [1]. This scattered occurrence seems to imply that the type of PGM has no significant role in the regulation of the system. However, the occurrence of more than one PGM in one compartment certainly needs some regulation or functional specialization, unknown so far.

The preferred substrate of PGM is 3-phosphoglycerate (3PGA). In cyanobacteria, 3PGA is made by photosynthetic CO_2 fixation, in which 3PGA represents the first stable carbon-fixation product of the Calvin-Benson cycle. The majority of 3PGA is used for the regeneration of the Calvin-Benson cycle, CO_2 acceptor molecule ribulose-bisphophate. Excess carbon is taken out of the cycle and stored as glycogen or shuttled into the primary carbon metabolism

via glycolysis (Fig. 1). The flux of 3PGA in the compartmented cell of land plants is regulated by phosphate translocator (chloroplast ↔ cytosol), which is not present in cyanobacteria. In the non-compartmented cyanobacterial cell, PGMs could be the key regulator of the flux out of the Calvin-Benson cycle. Marked changes in the relative flux of carbon through the Calvin-Benson cycle and its export into the glycolytic path have been observed in cyanobacteria exposed to excess or limited amounts of inorganic carbon [4,5].

Steady state metabolomic and transcriptomic analysis with the model cyanobacterium *Synechococcus elongatus* PCC 7942 (hereafter *Synechococcus*) indicated that cells shifted to low CO_2 availability reduce Calvin-Benson cycle activity, stop glycogen storage, and increase the export of organic carbon through the glycolysis [6]. In the *Synechococcus* genome (http://genome.kazusa.or.jp/cyanobase/SYNPCC7942) four PGM isoenzymes are annotated (Fig. 2). These annotations are mainly based on sequence similarities determined with the BLAST algorithm [7]. However, automated BLAST-based annotation can provide false functional assignments [8]. In the case of *Synechococcus*, the occurrence of isoenzymes in one compartment suggests a regulatory/specialized role of these isoenzymes in central metabolism. Moreover, in the cyanobacterial carbon metabolism, PGMs have a cardinal position at the crossroads of Calvin-Benson cycle and associated carbon metabolism via glycolysis (Fig. 1). Therefore, it is important to

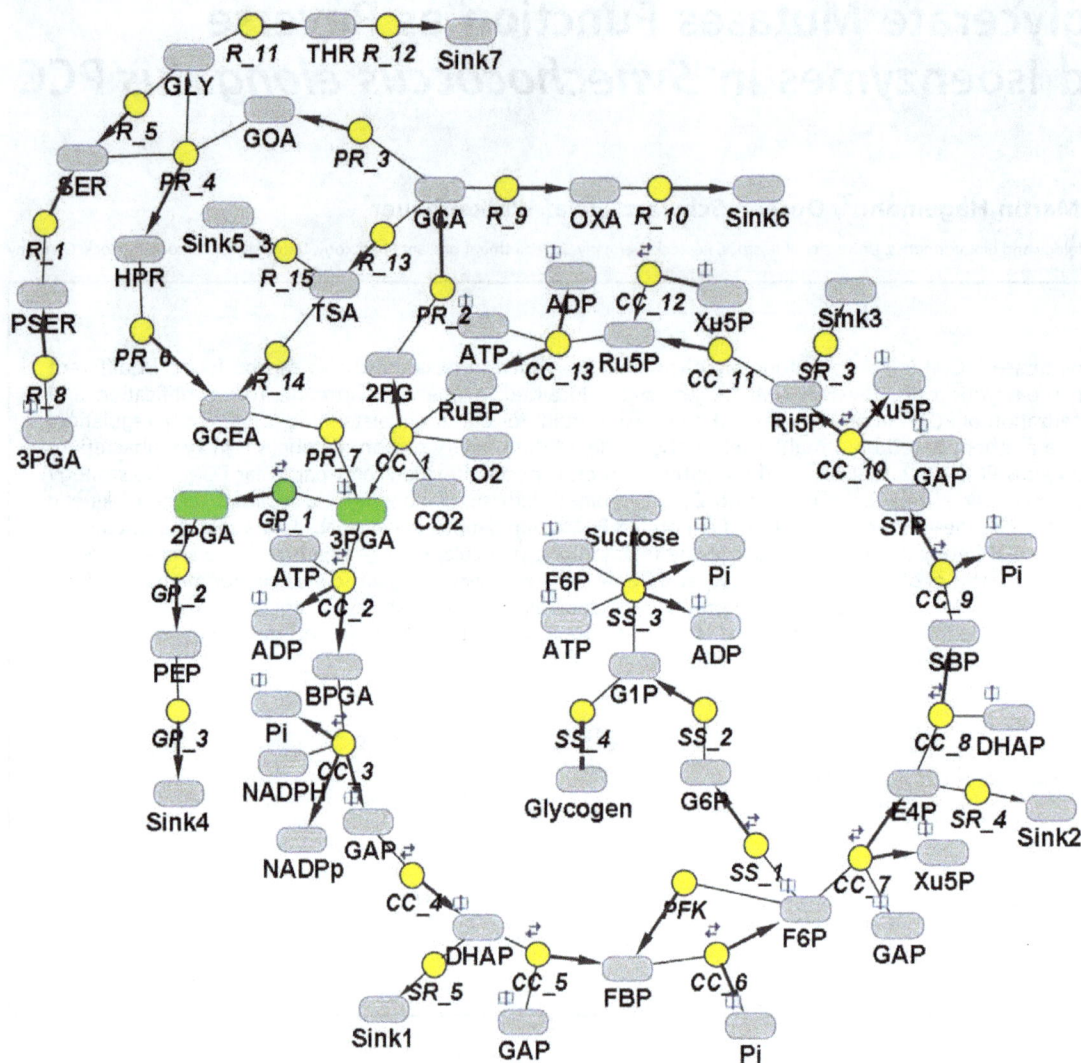

Figure 1. Scheme of the primary carbon metabolism, encoded as a kinetic model of *Synechococcus elongatus* **PCC 7942.** The model includes the Calvin-Benson cycle, sucrose and glycogen synthesis, photorespiratory pathways, glycolysis and sink reactions, representing the adjacent pathways. Green color represents the reaction catalyzed by phosphoglycerate mutase (PGM) and indicates its cardinal position in the crossroads of metabolic pathways (need for complex model). Note: The reactions are described in the model by reversible and irreversible Michaelis-Menten kinetics; reversibility of particular reaction is indicated by two little arrows.

understand the regulation of PGMs and thereby to validate their annotation.

We here propose the use of kinetic modeling for such analysis. Kinetic modeling is a standard method for predicting the behavior of biological systems [9]. However, reactions catalyzed by isoenzymes are commonly described by single enzymatic kinetics, which would not explain hidden regulatory mechanisms. In order to address this challenge, we have designed a multi-level kinetic model of carbon core metabolism of *Synechococcus*. The model not only helps in the validation of the annotation but also in the

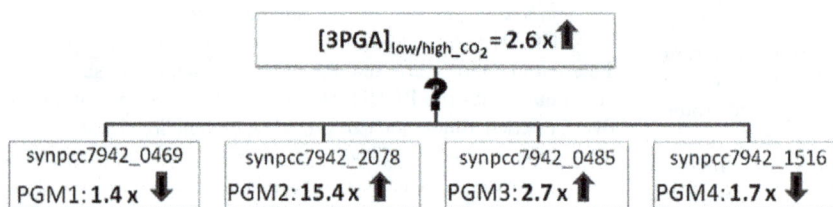

Figure 2. Comparison of fold changes in concentration for 3PGA and in expression levels of the four annotated PGM isoenzymes in cells of *Synechococcus elongatus* **PCC 7942 after shifting from high to low CO₂ level.** Note: synpcc7942_0485 probably represents a gene encoding a phosphoserine phosphatase but we cannot exclude if it functions as PGM.

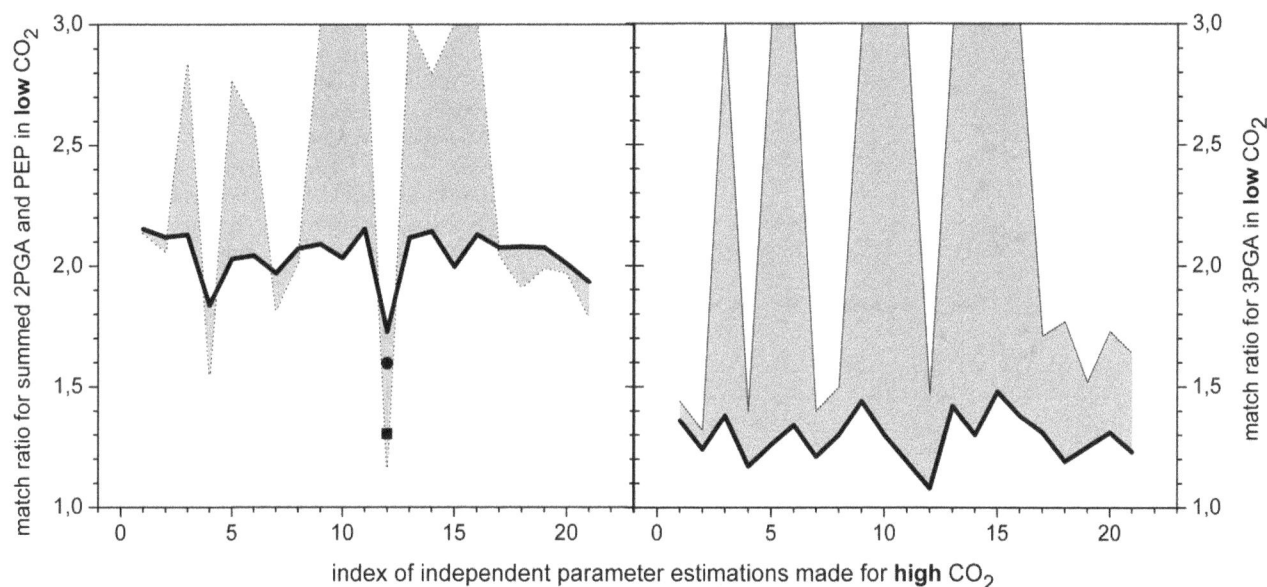

Figure 3. Quality for the match of simulated and measured data in low CO$_2$ for single PGM scenario in dependence on (i) varying PGM activities fitted in high CO$_2$ and (ii) regulation by transcript amounts. Figures represent the match between simulated and measured data in low CO$_2$ in dependence of estimated kinetic parameters (V$_{max}$ and k$_M$ values for preferred substrate and product) in high CO$_2$ for a single PGM scenario. The V$_{max}$, fitted to steady state in high CO$_2$, was modified by the amount of PGM isoforms taken from the changes in mRNA values (one by one) after shift from high to low CO$_2$. Results are shown for randomly chosen set of twenty parameters estimation runs for PGM and the best fit (Nr. 12). The **left figure** shows the match for preferred product of PGM, 2-phosphoglycerate (2PGA). The black solid line shows the impact of 1.7-fold down-regulated enzyme, corresponding to PGM4; gray contours indicate the difference in matching the data if PGM is 15.4-fold up-regulated (corresponding to PGM2). In order to illustrate the impact of transcriptomic changes for the other two annotated PGMs, the results for 1.4-fold down-regulated (circle) and 2.7-fold up-regulated (square) isoenzymes are presented in the case of the best fit. The **right figure** shows the match for preferred substrate of PGM, 3-phosphoglycerate (3PGA); colors/lines have the same meaning as for the figure on the left. Notes: 1) the top boundary of axis y shows results equal or worse than match ratio equals to 3, 2) The match ratio is calculated as X/Y where X(Y) is a higher(lower) number from a pair of simulated and experimental values for a particular data point, 3) Fit from high CO$_2$ was included (saved as a result) if the difference between simulated and measured data was smaller than 15%, 4) each point represents an independent simulation run compared to experimental data - the lines improve the perception for the differences in match ratios and have no other meaning.

explanation of regulatory mechanism for PGM isoforms in one compartment.

Materials and Methods

The model was developed and simulations were executed using the SimBiology toolbox of MATLAB (Mathworks Inc.). The routine employed for parameter estimation was a hybrid genetic algorithm (ga_hybrid, Mathworks Inc.). The model is available in the supplement (Model S1).

Systems Biology Workflow

The kinetic model, a successor of the corrected Zhu model [10,11] of the Calvin-Benson cycle, was redesigned for cyanobacteria, extended (photorespiratory pathways, glycolysis) and validated on available metabolic data from *Synechococcus el.* PCC 7942 [6]. The constraints of the model are: 1) The ATP · (ADP +ATP)$^{-1}$ ratio was maintained in the physiological range 0.74–0.76 [12], both in high and low CO$_2$ steady states and 2) the biomass production in high CO$_2$ was calculated from growth rate data. On average, *Synechococcus* shows a 3.4-fold higher growth at 5% CO$_2$ (defined as high CO$_2$, [6]) compared to ambient air CO$_2$ (defined as low CO$_2$, [6]).

The parameter estimation for up to four annotated isoenzymes of PGM was run in high CO$_2$ steady state and the result (kinetic parameters) stored if the difference between the simulated and experimental metabolic data was lower than 15%. We also

assumed that, in the case of isoenzymes in non-compartmented prokaryotic cell, a change in gene expression in the steady state is equal to changes in protein activity, i.e., minimal or no post-translation modifications. For the unidentified part of carbon metabolism - reaction catalyzed by PGM - four transcriptomic profiles (four PGM isoforms) were tested for combinations of one to four isoforms and the quality of match with metabolic data in low CO$_2$ was stored for each profile.

Experimental Data

Relative transcriptomic and metabolomics data of high CO$_2$ as well as low CO$_2$ *Synechococcus* cells were taken from [6]. A considering of two environmental conditions was necessary to known and implement the changes in transcriptomic level of PGMs in the model and helpful for constraining the model by doubling the amount of metabolic data. For 3PGA, cellular concentrations were calculated using the data from [13]. These authors reported that cyanobacteria contain 1300 nmol 3PGA/g fresh weight. According to our calibration of cyanobacterial fresh weight and total cellular volume, this amount corresponds to 5.39 mM 3PGA in the total cell volume in low CO$_2$ grown cells. The concentration of 2-phosphoglycerate was recalculated on the basis of known ratio to 3PGA [6].

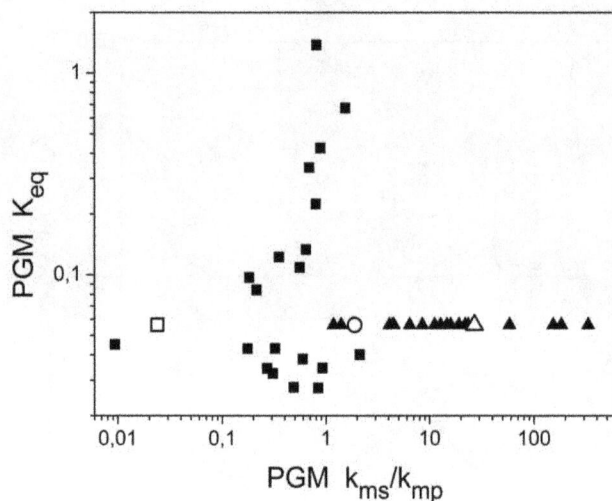

Figure 4. Parameter space of the equilibrium constant and ratio of k_M constants for single PGM and dual PGMs scenarios. The values k_{ms} and k_{mp} indicate the k_M values for the preferred substrate and product (reversibility), respectively. Keq indicates the equilibrium constant. **Solid squares** denote independent runs of parameter estimation, presented in Fig. 3, for single PGM scenario. **The open square** denotes the best fit for single PGM scenario. **Solid triangles** denote independent runs of parameter estimation, presented in Fig. 5, for dual reverse regulated PGMs scenario for PGM beta. **The open triangle** denotes the best fit for dual reverse regulated PGMs scenario for PGM beta. **The open circle** denotes the best fit for triple PGMs scenario for PGM gamma. This analysis shows how multiple sets of kinetic parameters match experimental data in one steady state for unconstrainted parameter estimation.

Results and Discussion

When One Enzyme is not Enough

The implementation of isoenzymes within a kinetic model, when localized in one compartment, deals with two extremes. First, for only one enzyme and the model not being constrained by a lack of data, there are still non-identifiability problems due to multiple sets of parameter value combinations that are equally able to fit the data related to a particular state. Second, considering two or more enzymes, the combinational explosion in the parameter space leads to computational requirements that make it virtually impossible to fit the model to sets of experimental data for different states of the system.

Various studies have focused on the response of cyanobacteria to changes in the environment, by considering a range of omics data [6,14,15], which allow constraining the model. On the basis of these results, we were aiming to develop a method that would permit the identification and analysis of isoenzymes in a one-compartment system and within a reasonable amount of time (days to weeks). We here particularly focus on PGMs because of their key position in regulating the flux of carbon through the Calvin-Benson cycle and its export into glycolysis in cyanobacteria. This position of PGMs in the metabolic network makes it necessary to establish a complex model (Fig. 1).

We started with the simplest scenario, encoding only one PGM, which is a common approach to un-constrained kinetic modeling. Although this approach cannot reveal anything about the regulation of catalyzed reactions in the un-constrained model, it is yet unknown if the outcome based on the constrained model can reveal some additional information about the system regulation or not. We therefore run a set of independent parameter estimations

in high CO_2 for a single PGM and used these values in numerical simulations. The quality of the estimates is evaluated by comparing the quality of match of simulations with metabolic data, separately for preferred substrate, 3-phosphoglycerate (3PGA) and product, 2-phosphoglycerate (2PGA), in low CO_2 (for details see Materials and Methods).

The results are striking. The analysis reveals two trends, showing that one PGM provides a match either for preferred substrate or preferred product (Fig. 3) but never for both. Moreover, these trends indicate an impact of transcriptomic regulation, which affects V_{max} in the enzymatic kinetics that describes the PGM. The first trend, exhibiting robustness in the system behavior (Fig. 3, black solid line), is associated with down-regulated PGM4. The other one (Fig. 3, gray contours), connected with up-regulated PGM2, shows a high sensitivity towards variations in the kinetic parameters (K_{eq} and k_M). This pattern was tested for a broad range of K_{eq} and k_M values (Fig. 4, solid squares). Furthermore, we can observe both, a very good and bad match, with experimental data (Fig. 3, gray contours), implying a major role in the kinetic regulation of the system. Taken together, single PGM cannot regulate the interconversion between 3PGA and 2PGA, however, the single isoenzyme scenario has provided valuable information about the requirements for homeostasis of this metabolic crossroad.

Reverse Regulated Versus Co-regulated Isoenzymes

We have shown that the one-isoenzyme scenario cannot keep the expected balance between 3PGA and 2PGA (Fig. 3). We can either get a robust system, which is however out of the physiological range, or a system that is very sensitive to changes in the parameter values. The natural next step in the analysis is to test for two PGMs in one compartment. This approach requires metabolic data from two steady states and transcriptomic data describing the shift between these two states, in our case, from high to low CO_2. However, parameter estimation for two isoenzymes, running for two steady states of metabolism, has enormous computational demands. At this point, data collected from the single PGM scenario, based on the constrained model, proved to be very useful.

The single PGM scenario suggests that it is the up-regulated PGM2 which is likely to provide the kinetic regulation for the reaction due to high sensitivity with respect to kinetic parameter values. We therefore tested the dual PGM scenario with two PGMs, denoted as alpha and beta. We employed the parameter values from the best fit provided by the single PGM scenario for alpha PGM (Fig. 4, open square) and varied the parameters for beta PGM, see Fig. 4 (solid triangles); the equilibrium constant is the same for both PGMs. Moreover, for the sake of comparison, we have added the robust case from the single PGM scenario (1.7-fold down-regulated PGM4) and another dual PGM scenario in which both PGMs were co-down-regulated (PGM1 and PGM4).

The comparison with other scenarios clearly shows a large improvement, both in robustness against varying parameter values for the beta PGM and in the ability to describe the experimental data if we proceed from single PGM scenario (Fig. 5, dashed line), over dual co-regulated PGMs (Fig. 5, dotted line) to dual reverse regulated PGMs (Fig. 5, solid line). It is therefore the combination of two PGMs with reverse transcriptomic regulation, PGM2 and PGM4, giving the closest match with experimental data. Moreover, if we have a look at the estimated parameters for the best fit for this scenario (Fig. 4, open square and open triangle), the Michaelis constants for preferred substrate and product for alpha and beta PGMs in Table 1 have nearly interchanged values. This is an expected outcome from a evolutionary point of view for two

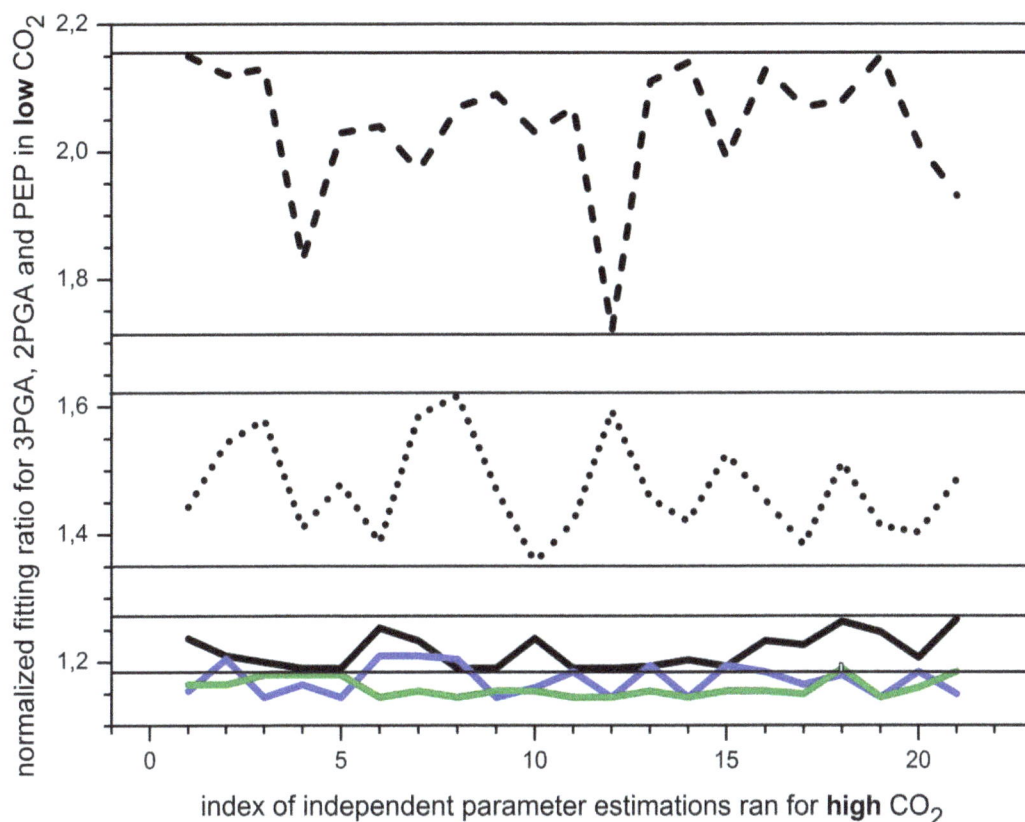

Figure 5. Impact of reverse and similar transcriptional regulation of PGM isoenzyme genes in one compartment. For the purpose of comparison, the dashed line represents the single enzyme scenario from Fig. 3, 1.7-fold down-regulated PGM4 (merged solid lines from Figure 3). Further, two dual PGMs scenarios are presented. K_{eq} and k_M values for alpha PGM are from the best fit for single PGM scenario (Fig. 4, open square) and V_{max} for alpha PGM and all kinetic parameters for PGM beta were estimated to fit the experimental data in high CO_2 steady state. Alpha and beta stand for two PGMs in dual PGMs scenario. The dotted line shows the cooperation of co-regulated PGM4 and PGM1 (1.7- and 1.4-fold down-regulation, respectively). The solid line indicates the case in which reverse regulation of PGM2 and PGM4 takes a place (15.4-fold up-regulation and 1.7-fold down-regulation). The blue line indicates the scenario of three PGMs in which PGM1 (1.4-fold down-regulation) is considered as gamma PGM. The k_M values for beta PGM are taken from the best fit of dual reverse PGMs scenario (Fig. 4, open triangle). The green line indicates another triple PGMs scenario where k_M values, both for beta and gamma PGMs, were varied. Note: each point represents an independent simulation run compared to experimental data - the lines improve the perception for the differences in match ratios and have no other meaning.

enzymes catalyzing the same reaction in one compartment, thus supporting the applied approach. In summary, our in silico experiment led us to the conclusion that reverse gene regulation for two isoenzymes in one compartment, together with the

Table 1. Estimated kinetic parameters for the best fit describing the reverse regulated dual PGMs scenario.

kinetic parameter	value	unit
Keq	0.056	dimensionless
Vmax_alpha	0.050	$mM \cdot s^{-1}$
kms_alpha	0.042	mM
kmp_alpha	1.772	mM
Vmax_beta	0.027	$mM \cdot s^{-1}$
kms_beta	1.257	mM
kmp_beta	0.047	mM

Activity of PGMs is in the model described by reversible Michaelis-Menten kinetics. Note: V_{max} values are normalized to the activity of RuBisCO. The routine employed for parameter estimation was a hybrid genetic algorithm.

opposite affinity for the substrate and product, provides very good explanation of the measured metabolic data.

The results clearly show high robustness of the system (Fig. 5, solid line) against varying the kinetic parameters (Fig. 4, solid triangles), which is equivalent to noise from fluctuating metabolite concentrations in a single cell. Hence the results support our initial assumption that system of isoenzymes does not require an additional regulatory mechanism, for instance the post-translational modifications. This was implicit in the assumed 1:1 ratio between the transcript level and protein activity for isoenzymes in the non-compartmented cell.

The Case of Two other Annotated PGMs

We have shown that two reverse regulated PGMs are able to explain the experimental data (Fig. 5), i.e., to regulate the interconversion between 3PGA and 2PGA and thus the flux between the Calvin-Benson cycle and glycolysis. This raises the questions whether there is any benefit of having more than two PGMs in one compartment. In order to test the triple PGMs scenario, we have taken the best fits for alpha and beta PGMs (Fig. 4, open square and open triangle) and varied the k_M values for gamma PGM. Our analysis of the triple PGMs scenario shows that there is a negligible improvement in robustness and only

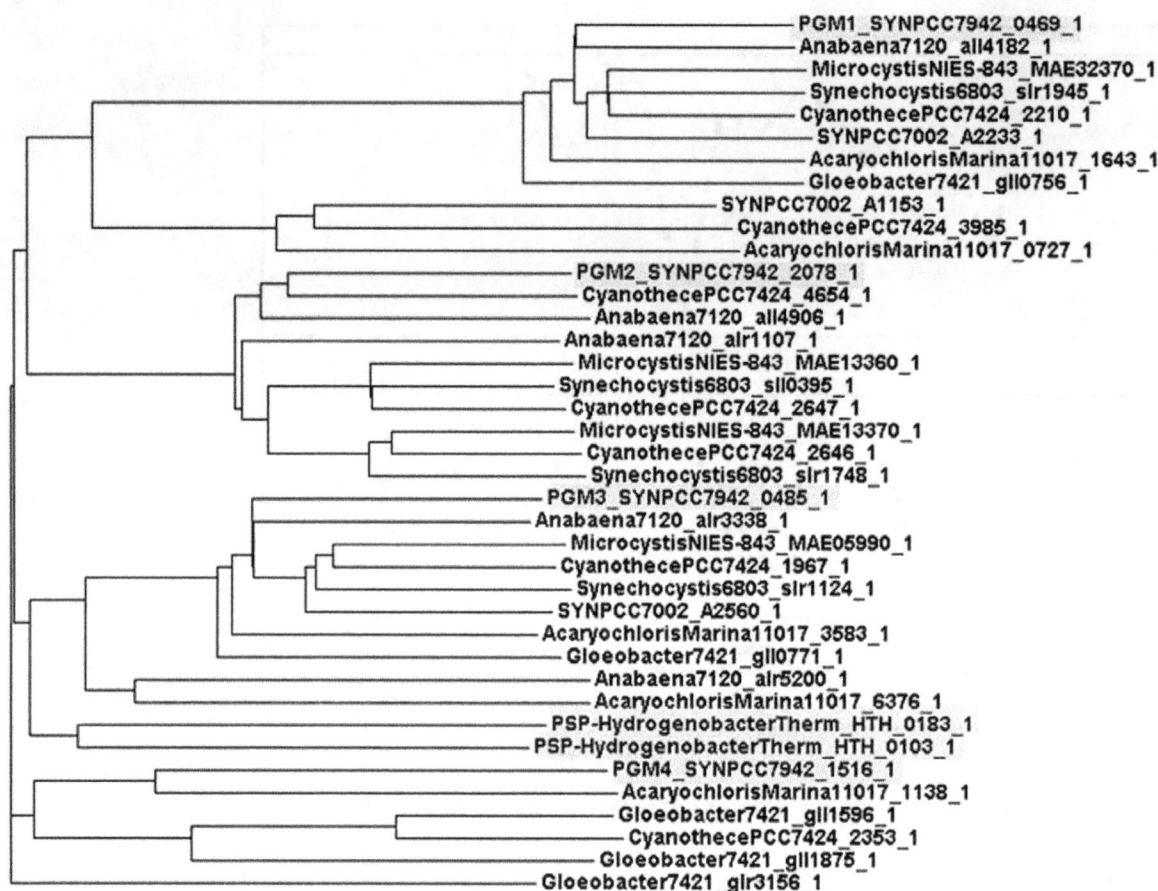

Figure 6. Grouping of PGMs and PSPs by using cluster analysis. ClustalW2 2.1 (http://www.ebi.ac.uk/Tools/) was employed as a tool for protein alignment analysis, codon table for bacteria was selected. PGMs 1–4 (*Synechococcus*) and two PSPs (*Hydrogenobacter thermophilus*) are highlighted. PGM3 is clustered with two PSPs.

a small improvement in matching experimental data (Fig. 5, blue line). The k_M values for gamma PGM, in the case of the best fit, is in between the k_M values of PGMs alpha and beta (Fig. 4, open circle). There was however no improvement in the biomass production (our estimation of growth speed), which implies that only two PGMs are necessary to control the flux between the Calvin-Benson cycle and glycolysis in the controlled environment. Since two-component response regulators have been proposed for the glycolytic pathway [16], this further supports our conclusion that only two PGMs are essential for the regulation of the system.

Recently, two assumed members of PGM family, cofactor dependent PGMs annotated in *Hydrogenobacter thermophilus*, were identified in a wet lab experiment as a phosphoserine phosphatases [17]. It is known that phosphoserine phosphatase (PSP) is essential for serine and glycine metabolism [18,19] but no PSP is annotated for *Synechococcus*. The alternative pathways for serine and glycine synthesis in cyanobacteria were proposed [20,21], however, we tested if one or more of the PGMs annotated in *Synechococcus* can be identified as PSP. Cluster analysis of PGM-like proteins shows that each PGM belongs to different protein family (Fig. 6) but, interestingly, PSPs in *Hydrogenobacter thermophilus* are in the same cluster as PGM3 (Fig. 6). Moreover, we have detected conservative regions between that PSPs and PGM3. It is therefore likely that we can identify PGM3 as PSP, however, we cannot entirely exclude that this enzymes functions as PGM as well. Finally, PGM3 is up-regulated in low CO_2 (Fig. 3) which lead to speculation on

contribution of photorespiration for serine biosynthesis as suggested [20].

As for the PGM1 and PGM4, due to almost the same expression in low CO_2 (Fig. 2), we do not have sufficient data to clearly distinguish between these two down-regulated enzymes. We however showed that stronger down-regulation gives higher robustness to the system, as indicated in Fig. 3. There might also be other reasons why there are three, or even four, PGMs in *Synechococcus*. For instance, it is known that the majority of proteins have more than one function [22].

The simultaneous parameter estimation for two PGMs for more steady states is very difficult and nearly impossible for three PGMs. One can therefore assume a rather small likelihood for the right combination of substrate-product affinities for every PGM to occur during evolution. Moreover, our analysis shows that the system in not very sensitive to varying separately the affinity of beta or gamma PGM although the best fit fulfills the expectation regarding the affinities for the product/substrate for two isoenzymes (Tab. 1). Therefore, an occurrence of several PGMs could be explained, from the evolutionary point of view, as the simplest means to achieve the robust response of the regulated system. As we have shown, the number of isoenzymes has a big impact on the system robustness (Fig. 5). Indirect support for such speculation is our analysis for triple PGMs scenario where k_M values, both for beta and gamma PGMs, were varied. The results demonstrate even higher robustness of the system (Fig. 5, green line) than the case of fixed beta PGM and varied gamma PGM (Fig. 5, blue line). Furthermore, the mean quality of match with experimental data improved by 5.6% in comparison to dual reverse PGMs scenario. Taken together, these results might imply that in the case of essential metabolic crossroads, the occurrence of more than two isoenzymes catalyzing the same reaction, especially in non-controlled environment, work as a buffer keeping the homeostasis of the system.

Despite the fact that our analysis cannot provide a clear cut answer why there are more PGMs, the presented approach contributes to an understanding of enzymatic regulation and provides a ration approach to identify the roles of particular, even non-homologous, isoenzymes.

Supporting Information

Model S1 The model of primary carbon metabolism in cyanobacterium *Synechococcus elongatus* PCC 7942 is provided and encoded in SBML L2V4 and set up for triple PGMs scenario, based on the best fits of alpha, beta and gamma PGM (Fig. 4).

Author Contributions

Jointly discussed ideas and concepts: JJ MH DS OW. The modeling was done by: JJ. Contributed to the writing of the final manuscript: OW MH. Provided the experimental data: DS MH. Read and approved the final manuscript: JJ MH DS OW. Conceived and designed the experiments: MH. Performed the experiments: MH DS. Analyzed the data: JJ MH. Contributed reagents/materials/analysis tools: MH OW. Wrote the paper: JJ.

References

1. Foster JM, Davis PJ, Raverdy S, Sibley MH, Raleigh EA, et al. (2010) Evolution of Bacterial Phosphoglycerate Mutases: Non-Homologous Isofunctional Enzymes Undergoing Gene Losses, Gains and Lateral Transfers. PLoS ONE 5.
2. Carreras J, Mezquita J, Bosch J, Bartrons R, Pons G (1982) Phylogeny and ontogeny of the phosphoglycerate mutases–IV. Distribution of glycerate-2,3-P2 dependent and independent phosphoglycerate mutases in algae, fungi, plants and animals. Comp Biochem Physiol B 71: 591–597.
3. Fraser HI, Kvaratskhelia M, White MF (1999) The two analogous phosphoglycerate mutases of *Escherichia coli*. FEBS Lett 455: 344–348.
4. Huege J, Goetze J, Schwarz D, Bauwe H, Hagemann M, et al. (2011) Modulation of the Major Paths of Carbon in Photorespiratory Mutants of *Synechocystis*. PLoS ONE 6.
5. Young JD, Shastri AA, Gregory S, Morgan JA (2011) Mapping photoautotrophic metabolism with isotopically nonstationary C-13 flux analysis. Metabolic Engineering 13: 656–665.
6. Schwarz D, Nodop A, Hüge J, Purfürst S, Forchhammer K, et al. (2011) Metabolic and transcriptomic phenotyping of inorganic carbon acclimation in the cyanobacterium *Synechococcus elongatus* PCC 7942. Plant Physiology 155: 1640–1655.
7. Zhang Z, Schaffer AA, Miller W, Madden TL, Lipman DJ, et al. (1998) Protein sequence similarity searches using patterns as seeds. Nucleic Acids Research 26: 3986–3990.
8. Ashkenazi S, Snir R and Ofran Y (2012) Assessing the Relationship between Conservation of Function and Conservation of Sequence Using Photosynthetic Proteins. Bioinformatics 28: 3203–3210.
9. Heinrich R, Schuster (1996) The Regulation of Cellular Systems, Springer, Berlin, Germany.
10. Zhu XG, de Sturler E, Long SP (2007) Optimizing the distribution of resources between enzymes of carbon metabolism can dramatically increase photosynthetic rate: a numerical simulation using an evolutionary algorithm. Plant Physiol 145: 513–526.
11. Jablonsky J, Bauwe H, Wolkenhauer H (2011) Modelling the Calvin-Benson cycle. BMC Systems Biology 5: 185.
12. Kallas T, Castenholz RW (1982) Internal pH and ATP-ADP pools in the cyanobacterium *Synechococcus* sp. During exposure to growth inhibiting low pH. Journal of Bacteriology 149: 229–236.
13. Takahashi H, Uchimiya H, Hihara Y (2008) Difference in metabolite levels between photoautotrophic and photomixotrophic cultures of *Synechocystis* sp. PCC 6803 examined by capillary electrophoresis electrospray ionization mass spectrometry. J Exp Bot 59: 3009–3018.
14. Eisenhut M, von Wobeser EA, Jonas L, Schubert H, Ibelings BW, et al. (2007) Long-term response toward inorganic carbon limitation in wild type and glycolate turnover mutants of the cyanobacterium *Synechocystis* sp. strain PCC 6803. Plant Physiol 144: 1946–1959.
15. Hackenberg C, Hüge J, Engelhardt A, Wittink F, Laue M, et al. (2012) Low-carbon acclimation in carboxysome-less and photorespiratory mutants of the cyanobacterium *Synechocystis* sp. strain PCC 6803. Microbiology 158: 398–413.
16. Tabei Y, Okada K, Tsuzuki M (2007) Sll1330 controls the expression of glycolytic genes in *Synechocystis* sp. PCC 6803. BBRC 355: 1045–1050.
17. Chiba Y, Oshima K, Arai H, Ishii M, Igarashi Y (2012) Discovery and Analysis of Cofactor-dependent Phosphoglycerate Mutase Homologs as Novel Phosphoserine Phosphatases in *Hydrogenobacter thermophilus*. Journal of Biological Chemistry 287: 11934–11941.
18. Borkenhagen LF, Kennedy EP (1959) The enzymatic exchange of L-serine with O-phospho-L-serine catalyzed by a specific phosphatase. J Biol Chem 234: 849–53.
19. Ponce-de-Leon MM, Pizer LI (1972) Serine biosynthesis and its regulation in Bacillus subtilis. J Bacteriol 110: 895–904.
20. Knoop H, Zilliges Y, Lockau W, Steuer R (2010) The Metabolic Network of *Synechocystis* sp. PCC 6803: Systemic Properties of Autotrophic Growth. Plant Physiology 154: 410–422.
21. Steuer R, Knoop H, Machne R (2012) Modelling cyanobacteria: from metabolism to integrative models of phototrophic growth. J Exp Bot 63: 2259–2274.
22. Zaretsky JZ, Wreschner DH (2008) Protein multifunctionality: principles and mechanisms. Translational oncogenomics 3: 99–136.

6

Identifying Like-Minded Audiences for Global Warming Public Engagement Campaigns: An Audience Segmentation Analysis and Tool Development

Edward W. Maibach[1]*, Anthony Leiserowitz[2], Connie Roser-Renouf[1], C. K. Mertz[3]

1 Center for Climate Change Communication, George Mason University, Fairfax, Virginia, United States of America, 2 Yale Project on Climate Change, Yale University, New Haven, Connecticut, United States of America, 3 Decision Research, Eugene, Oregon, United States of America

Abstract

Background: Achieving national reductions in greenhouse gas emissions will require public support for climate and energy policies and changes in population behaviors. Audience segmentation – a process of identifying coherent groups within a population – can be used to improve the effectiveness of public engagement campaigns.

Methodology/Principal Findings: In Fall 2008, we conducted a nationally representative survey of American adults ($n = 2,164$) to identify audience segments for global warming public engagement campaigns. By subjecting multiple measures of global warming beliefs, behaviors, policy preferences, and issue engagement to latent class analysis, we identified six distinct segments ranging in size from 7 to 33% of the population. These six segments formed a continuum, from a segment of people who were highly worried, involved and supportive of policy responses (18%), to a segment of people who were completely unconcerned and strongly opposed to policy responses (7%). Three of the segments (totaling 70%) were to varying degrees concerned about global warming and supportive of policy responses, two (totaling 18%) were unsupportive, and one was largely disengaged (12%), having paid little attention to the issue. Certain behaviors and policy preferences varied greatly across these audiences, while others did not. Using discriminant analysis, we subsequently developed 36-item and 15-item instruments that can be used to categorize respondents with 91% and 84% accuracy, respectively.

Conclusions/Significance: In late 2008, Americans supported a broad range of policies and personal actions to reduce global warming, although there was wide variation among the six identified audiences. To enhance the impact of campaigns, government agencies, non-profit organizations, and businesses seeking to engage the public can selectively target one or more of these audiences rather than address an undifferentiated general population. Our screening instruments are available to assist in that process.

Editor: Jon Moen, Umea University, Sweden

Funding: This work was funded by the Yale Center for Environmental Law and Policy; the Betsy and Jessie Fink Foundation; the 11th Hour Project; the Pacific Foundation; and by a Robert Wood Johnson Foundation Investigator Award in Health Policy. The funders had no role in study design, data collection and analysis, decision to publish, or preparation of the manuscript.

Competing Interests: The authors have declared that no competing interests exist.

* E-mail: emaibach@gmu.edu

Introduction

Global warming is a classic "wicked problem." [1] Wicked problems have no easy solutions in that they are beyond the capacity of any one organization to solve, and there is disagreement among organizations about both the causes and the best means by which to solve the problem [2]. Managing wicked problems requires working successfully within and across organizational boundaries, engaging citizens and other stakeholders in policy-making and implementation of those policies, and ultimately changing the behavior of groups of citizens or all citizens [2,3].

Successfully mitigating and adapting to global warming will require significant modifications in public policy and population behavior [4]. Public engagement campaigns are an important strategy to encourage population behavior change and build support for appropriate public policies [5–8]. Many factors limit the success of engagement campaigns, however, some of them inherent (e.g., the myriad influences on human behavior that are largely beyond the reach of a communication campaign) [5,7,8] and others situational (e.g., the tendency of governments to prematurely terminate public engagement campaigns) [7,9].

Although the research literature on global warming communication campaigns is relatively new and not yet well developed [5,7], other fields including commercial marketing [10], social marketing [11], public health [12] and political science [13] offer considerable research on the attributes of effective public engagement campaigns. Audience segmentation is one of the methods widely supported in all of these diverse research literatures.

Audience segmentation is a process of identifying groups of people within a larger population who are homogeneous with regard to critical attributes (e.g., beliefs, behaviors, political ideology) that are most relevant to the objectives of a public engagement campaign (e.g., product sales, consumer boycotts, political participation) [14]. Audience segmentation research – conducted insightfully – provides organizations with an important strategic planning asset: empirical information about how best to focus the organization's limited resources, both human and financial, to advance its objectives [15]. For example, a smaller audience segment whose members are willing to behave in ways sought by the organization may be a more productive target than a larger, less predisposed audience segment.

The principal aim of our current research was to identify audience segments within the American adult population that could be considered as potential targets for global warming public engagement campaigns. The nature of the global warming public engagement challenge – i.e., the need to build public understanding and support for appropriate public policies, and to change the behavior of large numbers of people – necessitated that we adapt and extend previously used segmentation methods.

Specifically, there is strong precedent in the research literature for segmenting audiences based on what people are doing (i.e., behaviors) and why (i.e., motivations) [16–19]. That method is well suited to population behavior change campaigns (e.g., smoking cessation campaigns), but it largely ignores a second potential focus for global warming public engagement campaigns: building public understanding of and support for appropriate public policies. Here, we extend the method of segmenting audiences based on what people are doing and why to also include people's policy preferences as an additional dimension in the analysis.

The other aim of our research was to develop an easily implemented, survey-based identification tool that can be used to identify the audience segments in independent population samples with acceptable levels of accuracy. Such a tool will enable social science researchers and public engagement campaign planners to further study the audience segments identified in our research, and to test public engagement methods with them. We believe that both aims of our research were achieved.

Results

We conducted a nationally representative survey of adults (n = 2,164) and used three major categories of variables as inputs into a segmentation analysis: global warming motivations, behaviors, and policy preferences. The global warming motivations category included two distinct sub-categories: beliefs about global warming and degree of involvement in the issue. We measured a total of 36 variables across these four categories (Tables 1, 2, 3 and 4). To maximize the practical value of the segmentation findings, we limited the analysis to five, six and seven segment solutions. As described in the Methods section below, we determined that the six-segment solution was optimal.

The six identified segments – each of which was given a concise name to summarize its essential qualities – differ dramatically with regard to what they believe about global warming, how engaged they are with the issue, what they are doing about it, and what they would like to see American government officials, businesses, and citizens do about it. The six segments also differ dramatically with regard to size: the largest represents 33% of the U.S. adult population, and the smallest only 7% (Figure 1). These six audience segments represent a spectrum of concern and action about global warming, ranging from the Alarmed (18% of the population), to the Concerned (33%), Cautious (19%), Disengaged (12%), Doubtful (11%) and Dismissive (7%).

Mean values for (or in the case of three variables, percent agreement with) each of the variables used in the segmentation analysis, by segment, are presented in Tables 1, 2, 3 and 4. The between-segment differences on all of these variables, as ascertained by ANOVA or chi-square tests, were significant at $p < .001$. Additional profiling information about the audience segments – i.e., how the six segments differ with regard to a range of additional relevant beliefs, behaviors (including media use), values, and demographics – is available at: http://environment.yale.edu/climate/publications/global-warmings-six-americas-2009/.

In brief, the *Alarmed* are the segment most engaged in the issue of global warming. They are very convinced it is happening, human-caused, and a serious and urgent threat. The Alarmed are already making changes in their own lives and support an aggressive national response.

The *Concerned* are also convinced that global warming is a serious problem, but while they support a vigorous national response, they are distinctly less involved in the issue, and less likely than the Alarmed to be taking personal action.

The *Cautious* also believe that global warming is a problem, although they are less certain that it is happening than the Alarmed or the Concerned. They don't view it as a personal threat, and don't feel a sense of urgency to deal with it through personal or societal actions.

The *Disengaged* haven't thought much about the issue. They are the segment most likely to say that they could easily change their minds about global warming, and they are the most likely to select the "don't know" option in response to every survey question about global warming where "don't know" was presented as an option.

The *Doubtful* are evenly split among those who think global warming is happening, those who think it isn't, and those who don't know. Many within this group believe that if global warming is happening, it is caused by natural changes in the environment, that it won't harm people for many decades into the future, if at all, and that America is already doing enough to respond to the threat.

Finally, the *Dismissive*, like the Alarmed, are actively engaged in the issue, but on the opposite end of the spectrum. The large majority of the people in this segment believe that global warming is not happening, is not a threat to either people or non-human nature, and is not a problem that warrants a personal or societal response.

To validate the predictive utility of these audience segments, we conducted four regression analyses using demographics (i.e., age, household income, gender, marital status, employment status, and race/ethnicity), political ideology, and segment membership as predictors of an outcome measure. A scale measuring support for nine specific potential federal greenhouse gas emission reduction policies was used as the outcome measure; these specific policy support measures are distinct from the preferred societal response measures used in the segmentation analysis, which are more general in nature (see Table 5). As shown in Table 6, demographics (Model 1, $F = 2.8$; $p < .01$), political ideology (Model 2, $F = 267$; $p < .001$) and segment status (Model 3, $F = 1,411$; $p < .001$) are each significant predictors of policy support when assessed in isolation of each other. Conversely, when assessed simultaneously (Model 4), demographic variables are not significant predictors, political ideology is a significant predictor with a moderately sized beta coefficient ($B = .10$; $p < .001$) and audience segment status is a significant predictor with a large beta coefficient ($B = .60$; $p < .001$). Audience segment alone explains as much variance in policy preferences (41%), as do demographics, political ideology and audience segment combined. We interpret these findings as validation of the predictive validity of the audience segmentation.

Table 1. Global Warming Beliefs by Audience Segment.

Survey Questions	Audience Segment						Scale Points
	Alarmed	Con-cerned	Cautious	Dis-engaged	Doubtful	Dis-missive	
1. & 1a. Certainty global warming is occurring	8.70	7.92	6.54	5.91	5.06	3.06	9
2. Human causation (% agree)	88	79	49	39	8	1	---
3. Scientific consensus (% agree)	80	64	37	23	11	8	---
4. Personal risk	3.09	2.59	1.90	2.75	1.29	1.02	4
5. Risk to future generations	3.98	3.78	2.96	4.00	1.89	1.04	4
6. Risk to plant & animal species	3.97	3.78	3.00	3.40	1.94	1.12	4
7. Timing of harm to Americans	5.46	4.83	3.53	3.85	1.77	1.01	6
8. Ability of humans to successfully mitigate warming	3.90	3.74	3.45	3.38	2.33	1.57	5
9. Actions of individual can make a difference	3.36	3.07	2.69	2.76	2.35	1.86	4
10. Technological optimism	1.70	2.05	2.32	2.03	2.38	2.33	4
11. Perceived impact of own mitigation actions	2.94	2.72	2.31	2.41	1.53	1.02	4
12. Impact of own actions if widely adopted in United States	3.69	3.48	3.01	2.90	1.94	1.10	4
13. Impact of own actions if widely adopted in modern industrialized countries	3.84	3.76	3.34	3.24	2.27	1.18	4

(p<.001 for all differences).

To enable identification of segment status with new, independent samples, we created an identification tool based on a linear discriminant function of all 36 variables used in the segmentation analysis. This identification tool – termed the "full discriminant model tool" – correctly classified 90.6% of the sample (ranging from 79 to 99% in the six segments; see Table 7). We also developed a shorter, more practical 15-item identification tool by eliminating the 20 least predictive variables from the discriminant function. This short identification tool – termed the "reduced discriminant model tool" – when applied to our dataset, correctly classified 83.8% of the sample (ranging from 60 to 97% in the six segments).

Discussion

With this research, we set out to identify and validate an audience segmentation system that can be used to inform global warming public engagement campaigns, and to develop easy-to-

use survey-based identification tools that can be used to replicate our results with acceptable levels of accuracy. Both aims were achieved with a large representative sample.

To be useful in supporting public engagement campaigns, a market segmentation scheme must demonstrate five attributes: (1) segments must be distinct from one another, and members of each segment must be sufficiently similar to be effectively targeted by the same marketing strategy; (2) segments must have direct relevance to the campaign objectives being pursued; (3) segments must be large enough to justify the time and effort required to target them; (4) the segment status of individuals in the market must be identifiable; (5) the campaign organization – or organizations – must be capable of targeting one or more of the identified segments (which may involve making the necessary changes to its structure, information and decision-making systems) [19].

Table 2. Global Warming Issue Involvement by Audience Segment.

Survey Questions	Audience Segment						Scale Points
	Alarmed	Con-cerned	Cautious	Dis-engaged	Doubtful	Dis-missive	
14. Rating of global warming (good = 1 to bad = 6)	5.72	5.31	4.35	4.04	3.66	3.19	6
15. Worry about global warming	3.65	3.08	2.44	2.31	1.56	1.12	4
16. Thought given to global warming	3.65	2.75	2.22	1.71	2.19	2.82	4
17. Need for information (4 = low need)	2.74	2.16	1.89	1.60	2.50	3.58	4
18. Personal importance of issue	4.44	3.39	2.59	2.54	1.81	1.38	4
19. Unwilling to change opinion	3.77	2.95	2.41	2.16	3.02	3.69	5
20. Personally experienced global warming	2.92	2.26	1.95	1.96	1.52	1.19	4
21. Global warming discussion frequency	3.02	2.36	1.86	1.29	1.88	2.05	4
22. Friends share views on global warming	3.59	2.71	2.21	1.65	2.85	3.61	5

(p<.001 for all differences).

Table 3. Global Warming and Energy Use Behaviors by Audience Segment.

Survey Questions	Audience Segment						Scale Points
	Alarmed	Con-cerned	Cautious	Dis-engaged	Doubtful	Dis-missive	
14. Contacted govt. officials re mitigation	1.53	1.11	1.07	1.07	1.06	1.00	5
15. Rewarded companies that reduced emissions	3.34	2.18	1.50	1.38	1.31	1.19	5
16. Intend to reward companies that reduce emissions	2.76	2.51	2.17	2.14	2.06	1.92	3
17. Punished companies that are not reducing emissions	3.14	1.92	1.32	1.28	1.18	1.08	5
18. Intend to punish companies that are not reducing emissions	2.73	2.51	2.13	2.18	2.03	1.79	3
19. Stage of change for lowering thermostat in winter	7.02	6.50	5.99	5.74	6.21	6.18	10
20. Stage of change for using public transportation or car pool	3.92	3.06	2.74	3.14	2.11	2.27	10
21. Stage of change for walking/biking instead of driving	4.73	3.49	3.14	2.59	2.68	2.72	10
22. Stage of change for CFL use	3.49	3.26	2.86	2.97	2.71	2.40	4

($p<.001$ for all differences).

The audience segments we identified possess the first four of these five attributes. The six segments – all of which are substantial in size, and whose members can be identified with the tools we developed – are distinct from one another in ways that have direct bearing on efforts to promote global warming mitigation and adaptation. The last of these five attributes, ultimately, is demonstrated by whether or not campaign organizations find value in making campaign decisions using the segmentation system. In the following paragraphs, we briefly elaborate on how global warming campaign organizations might select among the six audiences identified.

Members of the *Alarmed* segment are a highly engaged and active audience, at least in their capacity as consumers (with the exception of their travel behavior, which is more-or-less similar to that of other segments). They have a strong demonstrated tendency to use their consumer purchasing power to reward businesses they believe are contributing to solutions, and punish businesses they believe are not. They are markedly less active in

their role as citizens, however; only about one in four had contacted an elected official in the past year to urge them to take action to reduce global warming. Organizations seeking to promote policy advocacy – and possibly those seeking to modify people's travel behavior -- should consider targeting this audience.

Members of the *Concerned* segment are moderately engaged in the issue, but they are less active than are the *Alarmed*. As a result of their high prevalence in the population (1 out of every 3 adults), and their high stated intention to use their consumer purchasing power more frequently in the future to reward businesses they believe are contributing to solutions, organizations seeking to promote change through markets – rather than, or in addition to, change through public policy – should consider targeting this audience.

Members of the *Cautious* segment are only modestly engaged in the issue, and they don't appear ready to take action either as consumers or citizens. Organizations that are interested in expanding the number of Americans who are actively considering

Table 4. Preferred Societal Responses by Audience Segment.

Survey Questions	Audience Segment						Scale Points
	Alarmed	Con-cerned	Cautious	Dis-engaged	Doubtful	Dis-missive	
23. Priority of global warming for president & Congress	3.54	2.89	2.29	2.57	1.54	1.11	4
24. Corporations should do more/less to reduce warming	4.81	4.37	3.93	3.62	3.07	2.01	4
25. Citizens should do more/less to reduce warming	4.75	4.23	3.74	3.58	3.03	1.97	4
26. Desired US effort to reduce warming, given associated costs	3.78	3.33	2.89	2.83	2.01	1.37	4
27. Contingent int'l conditions for US mitigation action (% regardless of actions of other countries)	98	93	74	84	59	40	--

($p<.001$ for all differences).

Proportion represented by area

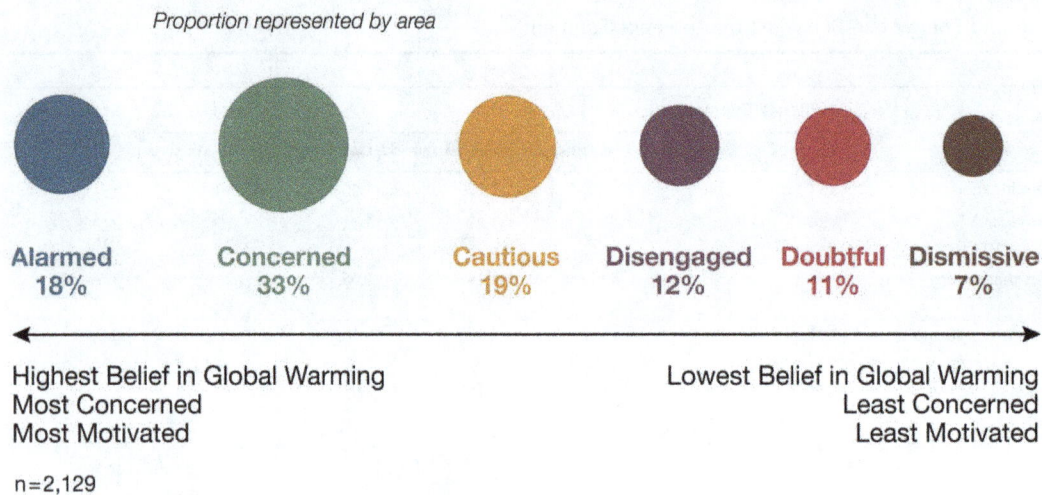

Alarmed	Concerned	Cautious	Disengaged	Doubtful	Dismissive
18%	33%	19%	12%	11%	7%

Highest Belief in Global Warming   Lowest Belief in Global Warming
Most Concerned   Least Concerned
Most Motivated   Least Motivated

n=2,129

Figure 1. Proportion of the U.S. adult population in the Six Americas.

the issue of climate change (rather than attempting to change people's behavior, or develop support for policy responses) should consider targeting members of this audience. Narrative-based communication [20], and reframing the issue in terms of human health may be productive approaches [21].

Members of the *Disengaged* segment currently have no involvement in the issue. The *Disengaged* stand apart from other segments in that they are less educated and have lower household incomes, both of which place them at higher than average risk of being harmed by global warming [22]. This is a difficult segment to reach using news media and other traditional science communication channels, both due to their current lack of interest and their financial challenges. Organizations seeking to engage members of the *Disengaged* must think creatively about how to make the issue more relevant for

Table 5. Support for Emission Reduction Policies by Audience Segment.

Survey Questions	Audience Segment					
	Alarmed	Con-cerned	Cautious	Dis-engaged	Doubtful	Dis-missive
1. Establish a special fund to help make buildings more energy efficient and teach Americans how to reduce their energy use. This would add a $2.50 surcharge to the average household's monthly electric bill.	3.25	2.91	2.48	2.54	2.09	1.56
2. Provide a government subsidy to replace old water heaters, air conditioners, light bulbs, and insulation. This subsidy would cost the average household $5 a month in higher taxes. Those who took advantage of the program would save money on their utility bills.	3.44	3.07	2.81	2.79	2.23	1.78
3. Regulate carbon dioxide (the primary greenhouse gas) as a pollutant.	3.67	3.22	2.93	2.86	2.43	1.84
4. Require electric utilities to produce at least 20% of their electricity from wind, solar, or other renewable energy sources, even if it cost the average household an extra $100 a year.	3.50	3.14	2.76	2.60	2.36	2.10
5. Sign an international treaty that requires the United States to cut its emissions of carbon dioxide 90% by the year 2050.	3.51	3.07	2.64	2.68	1.98	1.49
6. Require automakers to increase the fuel efficiency of cars, trucks, and SUVS, to 45 mpg, even if it means a new vehicle will cost up to $1,000 more to buy.	3.64	3.32	3.12	2.73	2.68	2.33
7. Fund more research into renewable energy sources, such as solar and wind power.	3.84	3.57	3.31	3.16	3.14	2.96
8. Provide tax rebates for people who purchase energy-efficient vehicles or solar panels.	3.60	3.33	3.12	2.78	2.91	2.60
9. Increase taxes on gasoline by 25 cents per gallon and return the revenues to taxpayers by reducing the federal income tax.	2.50	2.14	2.00	1.97	1.69	1.37
10. Policy support index (mean of 9 measures; α =.86)	3.44	3.09	2.80	2.68	2.39	2.00

(All items measured on 4-point scales, where 1 = strongly oppose & 4 = strongly support; $p < .001$ for all differences).

Table 6. Policy Support Predicted by Socio-Demographics, Political Orientation & Audience Segment.

	Model 1: Socio-demographics	Model 2: Political orientation	Model 3: Audience segment	Model 4: Full model
Age	.01			.01
Education	.06*			.00
Household Income	.00			.01
Gender (2 = F)	.05*			-.02
Marital status (2 = married or w/partner)	-.02			.01
Work status (2 = working)	-.02			-.02
Race: white	-.14			-.07
Race: black	-.06			-.06
Race: Hispanic	-.01			-.04
Race: other	-.04			-.05
Political ideology (5 = very liberal).		.33***		.10***
Audience segment (6 = Alarmed)			.64***	.60***
Adjusted R²	.01	.12	.41	.41
F	2.8**	266.8***	1,411.7***	120.8***
N	2,067	2,052	2,062	2,052

*p<.05;
**p<.01;
***p<.001.
Note: Cell entries are standardized regression weights. For dummy variables, the excluded race category was "mixed race, non-Hispanic."

them. As with the *Cautious* segment, narrative-based communication, and reframing the issue in terms of human health may be productive approaches. Activating new voices to explain the relevance of climate change – such a health professionals [21], members of the faith community [23], and organizations serving low-income families – may be helpful as well.

Members of the *Doubtful* segment are important because – although they currently doubt that global warming is real or harmful, and are disinclined to support actions to address it – they remain open to learning more about this issue. Because the *Doubtful* tend to be politically conservative, organizations that have the ability to work effectively across the political spectrum should consider developing activities to further engage the *Doubtful*.

As a result of their strongly held belief that global warming is not happening or is not human caused, members of the *Dismissive*

segment are highly involved in the issue as adamant opponents to taking any form of action against global warming. Like members of the *Alarmed* segment, however, they are supportive of taking both personal and societal actions to reduce energy use. Thus, while they are likely not a productive audience for a global warming public engagement campaigns per se, they may be an attractive audience for energy-efficiency campaigns because they are receptive to such appeals.

It is important to note that the three classes of variables included in our segmentation – motivations, behaviors, and policy preferences – did not include structural and contextual factors (e.g., the availability of public transportation options, and local or state government incentives to reduce energy use) that previous research has shown to be important in influencing adoption of energy efficiency and conservation actions [24]. The implications

Table 7. Prevalence of Audience Segments in 2008 Based on Three Methods of Identification.

Segment	Latent Class Analysis	Full Discriminant Model		Reduced Discriminant Model	
		Proportion of Sample In Segment	Accuracy of Discriminant Analysis	Proportion of Sample In Segment	Accuracy of Discriminant Analysis
1. Alarmed	18.0%	18.0%	92.6%	17.1%	85.6%
2. Concerned	33.3%	33.4%	91.3%	33.5%	85.8%
3. Cautious	18.7%	17.6%	87.5%	18.0%	80.9%
4. Disengaged	12.2%	13.6%	98.9%	14.9%	96.7%
5. Doubtful	10.6%	9.5%	79.2%	8.0%	60.1%
6. Dismissive	7.2%	8.0%	93.2%	8.5%	89.9%

of this decision are evident in the fact that the between segment differences on energy use and conservation actions are relatively small (albeit significant), whereas the between segment differences on global warming advocacy actions are more pronounced (see Table 3). Thus, this segmentation system is optimized for efforts to educate or engage the public about global warming per se, and less optimized for campaigns intended to promote changes in energy use behavior.

An integral part of strategic planning for a public engagement campaign involves selecting the target audiences that are the best fit for the organization's public engagement goals and resources [25]. Depending on their goals and resources, some organizations might be well served to focus their entire effort on a single target audience. Other organizations might be best served by targeting several audiences, if feasible. Regardless, campaigns that target specific audiences and tailor their materials accordingly are more likely to achieve their public engagement objectives than campaigns that do not [26].

For any given organization, the optimal target audiences are those that are likely to maximize the return on investment in campaign planning and execution. The three most relevant considerations in making that determination are the size of the audience segment, the likelihood that members of the segment will respond in the intended manner, and the organization's ability reach that segment with its current resources [27].

It remains to be seen whether or not organizations involved in global warming public engagement campaigns will be capable of – or interested in – targeting one or more of the identified segments in the ways we describe. A number science-based organizations – including science academies [28], science museums [29], and natural resource and conservation organizations – are currently considering their targeting and tailoring options using the audience segments identified. These developments may be evidence that the method possesses the final necessary attribute of utility: organizations must be capable of targeting one or more of the identified segments [19].

The "one size fits all" approach to global warming communication appears to be the default mode for most organizations, despite the fact that non-targeted approaches are at odds with best practices in campaign management [25]. National global warming education campaigns, for example, tend not to target well-defined audiences, but focus instead on the general public [9].

That non-targeted approaches remain common suggests that many organizations can't – or aren't willing to – bear the added costs of a targeted approach. Non-targeted campaigns are, without question, easier to implement than targeted campaigns. Segmenting and targeting multiple audiences can involve making changes to the organization's structure, information and decision-making systems [19]. At very least, a sustained effort to understand and engage more than one distinct target audience requires a campaign team to divide its planning and program development activities among each audience under consideration. A more intensive approach involves creating a campaign team to focus on each targeted audience [27]. These more intensive methods are common in consumer marketing organizations, yet they remain largely unknown or underutilized outside of the for-profit sector.

To monitor the stability of the audience segments identified in this research over time, we used the 36-item tool on three subsequent national surveys conducted in 2009 and 2010. Pronounced shifts in the size of the segments were evident across the three years of these surveys; for example, the Alarmed segment contracted sharply and the Dismissive segment grew markedly between fall 2008 and late 2009, but both regressed somewhat toward their prior sizes by mid-2010 [30]. We are currently exploring the reasons for these shifts, but our preliminary investigations suggest that meaningful exogenous factors – including a pronounced downturn in the economy, negative media coverage associated with the illegal release of email between climate scientists which became known as "Climategate," and escalation of industry funded global warming denial campaigns [31] – were responsible for the shifts rather than inherent instability in the segmentation method. Indeed, two of these national surveys were conducted within one month of each other [30,32]. Results from these surveys show only small differences between segment sizes when they are measured more-or-less contemporaneously: Alarmed, 13 vs. 14%; Concerned, 28 vs. 31%; Cautious, 24 vs. 23%; Disengaged, 10 vs. 10%; Doubtful, 12 vs. 12%; and Dismissive, 12 vs. 11%.

Our 15- and 36-item survey-based audience segment identification tools – as well as SAS & SPSS syntax to process the data – are available at: http://www.climatechangecommunication.org/SixAmericasManual.cfm. We encourage global warming campaign organizations and social science researchers to examine and evaluate them for their potential utility. To assess the robustness of this method to cultural and other contexts, we particularly encourage social science researchers to adapt these tools and assess their validity in nations other than the U.S.

Materials and Methods

Survey Method

In September through October of 2008, we conducted a nationally representative survey of American adults using KnowledgePanel, an online panel operated by Knowledge Networks. Recruited nationally using random-digit dialing (RDD) telephone methodology, KnowledgePanel is representative of the U.S. population. The panel tracks closely the December 2007 Current Population Survey (published jointly by the U.S. Census Bureau and the Bureau of Labor Statistics) on age, race, Hispanic ethnicity, geographic region, employment status, and other demographic variables.

The length of our questionnaire – a 50-minute completion time for the average respondent – exceeded what most respondents are willing to answer in a single session. As a result, we divided the content of the instrument into two separate questionnaires. An invitation to participate in the first survey was emailed to 3,997 randomly selected panel members. A total of 2,496 completed the questionnaires, a 62% cooperation rate. Two weeks after administration of the first survey was ended, respondents to the first survey received an invitation to participate in the second survey. Completed questionnaires were received from 2,164 respondents, an 87% cooperation rate, leading to an overall 54% within panel completion rate for the study. The period of administration for each survey – from invitation to termination of data collection – was approximately 10 days, during which one reminder email was sent to non-respondents.

To reduce the effects of any non-response and non-coverage bias in the overall panel membership, a post-stratification adjustment was applied using demographic distributions from the most recent data from the Current Population Survey (CPS). Benchmark distributions for Internet Access among the U.S. population of adults are obtained from KnowledgePanel recruitment data since this measurement is not collected as part of the CPS. The post-stratification variables were: Gender (Male/Female); Age (18–29, 30–44, 45–59, and 60+); Race/Hispanic ethnicity (White/Non-Hispanic, Black/Non-Hispanic, Other/Non-Hispanic, 2+ Races/Non-Hispanic, Hispanic); Education (Less than High School, High School, Some College, Bachelor and beyond); Census Region (Northeast, Midwest, South, West); Metropolitan Area (Yes, No); Internet Access (Yes, No).

Measures

We measured a total of 306 variables with the two instruments; 36 of those variables were used in the audience segmentation analysis. Specifically, the 36 items were developed to assess four categories of global warming- and energy-related constructs: global warming beliefs (Table 1), global warming issue involvement (Table 2), global warming and energy efficiency and conservation behaviors (Table 3), and preferred societal response to global warming (Table 4). An index of support for nine specific federal greenhouse gas reduction policies was constructed and used to assess the validity of the segmentation results (Table 5).

Segmentation Analysis

To identify the audience segments, the 36 variables were subjected to Latent Class Analysis using LatentGold 4.5 software [33,34]. LCA is a modeling technique for analyzing case level data with the objective of identifying groups of respondents (segments or latent classes) with similar characteristics. LCA assigns cases into clusters using model-based posterior membership probabilities estimated by maximum likelihood methods. One advantage of LCA is it can handle nominal, ordinal, and continuous variables as well as any combination of these [33]. In addition, unlike cluster analysis, LCA is not highly sensitive to missing data. Respondents with 80% or more complete data on the 36 variables were included in the analysis; this resulted in a sample size of 2,129 for modeling purposes.

The 36 variables in our model were a mixture of ordinal and nominal variables. We submitted five, six, and seven segment solutions to the analyses. One potential problem in estimating latent class models is the possibility of obtaining a local maximum solution rather than a globally-based solution: an estimation algorithm may converge on a local maximum solution, which is the best solution in a neighborhood of the parameter space, but not necessarily the best global maximum. As models become more complex this problem increases. To guard against local maximum solutions, the estimation algorithm should be run several times with different parameter start values. Thus, to address this issue and to ensure the validity and stability of the findings, we conducted the analyses using 5,000 random sets of start values and replicated each solution to ensure model stability. All three models (5-, 6-, and 7-segments) replicated exactly. The three models had similar fit statistics (see Table 8). We examined the profile data for all three models and determined that the six-segment solution offered the highest face validity. Although the seven-segment solution had slightly lower fit statistics (which indicates a better model fit), the difference was small and the six segments described above were more interpretable.

To create an easy-to-use tool for others to categorize survey respondents in new, independent samples, we created a linear discriminant function using the output from the Latent Class Analysis. In contrast to Latent Class Analysis, discriminant analysis does not permit missing data. We therefore used mean substitution for missing data, and then applied this linear function using all 36 variables to our data set. The 36 variable linear function correctly classified 90.6% of the sample (ranging from 79 to 99% in the six segments; see Table 7). Elsewhere in this paper we refer to the 36-variable linear function as the "full discriminant model tool."

Table 8. Model Fit Statistics.

	L^2	$BIC(L^2)$	Npar	$P(L^2)$
5 classes	146560.858	133330.136	402	<.0001
6 classes	145443.384	132695.443	465	<.0001
7 classes	144595.960	132330.799	528	<.0001

Brief Screening Tool Development

To develop a shorter – and therefore more easily used – screening questionnaire capable of classifying members of independent samples into the six audience segments with 80% accuracy or better, we eliminated the 21 least predictive variables from the discriminant function. The resultant 15-item "brief" screening instrument, when applied to our dataset, correctly classifies 83.8% of the sample (ranging from 60 to 97% in the six segments; see Table 7). Elsewhere in this paper we refer to the 15-variable linear function as the "reduced discriminant model tool."

Validation of the Segments

To validate that the segments account for variance in important measures above and beyond variance accounted for by other common explanatory measures, we conducted a series of linear regressions. The dependent measure for these analyses was an index of support for a series of nine federal policies that, if enacted, should reduce national levels of greenhouse gas emissions. Responses to each of these questions were combined into a simple index; the Cronbach's alpha for this policy support scale was 0.86.

In the first analysis, the demographic variables of age, gender, income, education, marital status, work status, and race were regressed against the GHG reduction policy support measure. In the second analysis, a five-point political ideology scale (very liberal, somewhat liberal, moderate, somewhat conservative, very conservation) was added into the regression. In the final analysis, audience segment status was added into the regression.

Human Subjects Approval and Informed Consent

This research was approved by the Human Subjects Review Board at George Mason University and Yale University. Written informed consent was obtained from all participants involved in this research.

Acknowledgments

The authors wish to thank Matthew Nisbet for conceptual input, Guoqing Diao and Katya Seryakova for statistical advice, Russell Shaddox for graphic design support, Karen Akerlof for data and editorial assistance, and Colleen Redding, Norbert Mundorf, Saffron O'Neill and an anonymous reviewer for reviewing and providing useful input on the manuscript.

Author Contributions

Conceived and designed the experiments: EWM AL CRR. Performed the experiments: EWM AL CRR. Analyzed the data: EWM AL CRR CKM. Wrote the paper: EWM AL CRR CKM.

References

1. Rittel HWJ, Webber MM (1973) Dilemmas in a general theory of planning. Policy Sciences 4: 155–69.
2. Australian Public Service Commission (2007) Tackling wicked problems: A public policy perspective. Available: http://www.apsc.gov.au/publications07/wickedproblems.htm. Accessed 28 Jan 2010.
3. Conklin J (2006) Dialogue mapping: Building shared understanding of wicked problems. HobokenNJ: John Wiley & Sons. 242 p.
4. Leiserowitz A (2006) Climate change risk perception and policy preferences: The role of affect, imagery, and values. Climatic Change 77: 45–72.
5. Maibach E, Roser-Renouf C, Leiserowitz A (2008) Communication and marketing as climate change intervention assets: A public health perspective. Am J Prev Med 35: 488–500. DOI: 10.1016/j.amepre.2008.08.016.
6. Maibach E, Priest S (2009) No more "business as usual:" Addressing climate change through constructive engagement. Sci Comm 30: 299–304.
7. Moser S (2010) Communicating climate change: History, challenges, process and future directions. WIREs Clim Change 1: 31–53.
8. Ockwell D, Whitmarsh L, O'Neill S (2009) Reorienting climate change communication for effective mitigation: Forcing people to be green or fostering grass-roots engagement? Sci Comm 30: 305–327.
9. Akerlof K, Maibach E (2008) "Sermons" as a climate change policy tool: Do they work? Evidence from the international community. Global Studies Rev 4: 4–6.
10. Kotler P, Keller K (2008) Marketing management. 13th ed. Englewood Cliffs, NJ: Prentice Hall. 816 p.
11. Kotler P, Lee N (2008) Social marketing: Influencing behaviors for good. Los Angeles: Sage Publications. 444 p.
12. Hornik R (2002) Public health communication: Evidence for behavior change. Mahwah, NJ: Lawrence Erlbaum Associates. 435 p.
13. Sosnik D, Dowd M, Fournier R (2006) Applebee's America: How successful political, business, and religious leaders connect with the new American community. New York: Simon & Schuster. 260 p.
14. Slater M (1995) Choosing audience segmentation strategies and methods for health communication. In: Maibach E, Parrott R, eds. Designing health messages. Los Angeles: Sage Publications. pp 186–198.
15. Dibb S (1999) Criteria guiding segmentation implementation: Reviewing the evidence. J Strategic Marketing 7: 107–129.
16. Maibach E, Weber D, Massett H, Price S, Hancock G (2006) Segmenting health audiences based on their information use and decision-making preferences: Development and initial validation of a brief screening instrument. J Health Comm 11: 717–36.
17. Weir M, Maibach E, Bakris G, Black H, Chawla P, et al. (2000) Implications of a health lifestyle and medication analysis for improving hypertension control. Arch Intern Med 160: 470–480.
18. Maibach E, Maxfield A, Ladin K, Slater M (1996) Translating health psychology into effective health communication: The American Healthstyles Audience Segmentation Project. J Health Psych 1: 261–277.
19. McDonald M, Dunbar I (1995) *Market Segmentation.* Macmillan. 15 p.
20. Dahlstrom M (2010) The role of causality in information acceptance in narratives: An example from science communication. Communication Research First published on June 16, 2010, doi:10.1177/0093650210362683.
21. Maibach EW, Nisbet M, Baldwin P, Akerlof K, Diao G (2010) Reframing climate change as a public health issue: An exploratory study of public reactions. BMC Public Health 10: 299.
22. Protecting health from climate change: Connecting science, policy and people, 2009. World Health Organization. http://whqlibdoc.who.int/publications/2009/9789241598880_eng.pdf (accessed May 24, 2010).
23. Hitzhusen G (2010) Climate change education for faith-based groups. Background paper for the workshop on climate change education for the public and decision makers. Washington, DC: The National Academies, http://www7.nationalacademies.org/bose/Climate_Change_Education_Workshop1_Table_of_Contents.html (accessed October 29, 2010).
24. Dietz T, Stern P (2002) New tools for environmental protection: Education, information, and voluntary measures. Washington, DC: National Academies Press.
25. Smith R (2009) Strategic planning for public relations. 3rd ed. New York: Routledge. 436 p.
26. Noar SM, Benac CN, Harris MS (2007) Does tailoring matter? Meta-analytic review of tailored print health behavior change interventions. Psych Bull 133: 673–693.
27. Andreasen A (1995) Marketing social change. San Francisco: Jossey-Bass. 348 p.
28. The National Academies (2010) Workshop on climate change education for the public and decisionmakers. Board on Science Education and Committee on Human Dimensions of Climate Change, Division of Earth and Life Studies. October 21-22, Washington, DC, http://www7.nationalacademies.org/bose/Climate_Change_Education_Workshop1_Table_of_Contents.html (accessed October 29, 2010).
29. Phipps M (2010) Global Warming's Six America's: A Science Museum of Minnesota Audience Segmentation Analysis. Unpublished manuscript. St. Paul, MN: Science Museum of Minnesota.
30. Leiserowitz A, Maibach E, Roser-Renouf C, Smith N (2010) *Global Warming's Six Americas, June 2010.* Yale University and George Mason University. New Haven, CT: Yale Project on Climate Change, http://environment.yale.edu/uploads/SixAmericasJune2010.pdf.
31. Leiserowitz A, Maibach E, Roser-Renouf C, Smith N, Dawson E (2010) *Climategate, Public Opinion, and the Loss of Trust.* Working paper posted to the Social Science Research Network. http://papers.ssrn.com/sol3/papers.cfm?abstract_id=1633932.
32. Leiserowitz A (2010) Connections between climate literacy and audience's climate change beliefs and attitudes. Background paper for the workshop on climate change education for the public and decision makers. Washington, DC: The National Academies, http://www7.nationalacademies.org/bose/Climate_Change_Education_Workshop1_Table_of_Contents.html (accessed October 29, 2010).
33. Magidson J, Vermunt JK (2002) Latent class models for clustering: A comparison with K-means. Can J Marketing Research 20: 37–44.
34. Magidson J, Vermunt JK (2002) Latent class models. In Kaplan D, ed. The Sage Handbook of Quantitative Methodology for the Social Sciences. Thousand Oaks, CA: Sage. pp 175–198.

7

The Environmental Price Tag on a Ton of Mountaintop Removal Coal

Brian D. Lutz[1]*, Emily S. Bernhardt[2], William H. Schlesinger[3]

1 Department of Biology, Kent State University, Kent, Ohio, United States of America, **2** Department of Biology, Duke University, Durham, North Carolina, United States of America, **3** Cary Institute of Ecosystem Studies, Millbrook, New York, United States of America

Abstract

While several thousand square kilometers of land area have been subject to surface mining in the Central Appalachians, no reliable estimate exists for how much coal is produced per unit landscape disturbance. We provide this estimate using regional satellite-derived mine delineations and historical county-level coal production data for the period 1985–2005, and further relate the aerial extent of mining disturbance to stream impairment and loss of ecosystem carbon sequestration potential. To meet current US coal demands, an area the size of Washington DC would need to be mined every 81 days. A one-year supply of coal would result in ~2,300 km of stream impairment and a loss of ecosystem carbon sequestration capacity comparable to the global warming potential of >33,000 US homes. For the first time, the environmental impacts of surface coal mining can be directly scaled with coal production rates.

Editor: Matteo Convertino, University of Florida, United States of America

Funding: No external funding received for this study.

Competing Interests: The authors have declared that no competing interests exist.

* E-mail: blutz6@kent.edu

Introduction

Mountaintop removal coal mining (MTR) is a particularly invasive mining practice developed in the United States (US) capable of producing low-cost coal. The process of MTR uses explosives and heavy machinery to remove entire mountain ridges in order to access near surface coal deposits, producing vast quantities of mine spoil that fills valleys and buries streams. MTR has expanded dramatically in recent decades and is now the dominant driver of land-use change across the Central Appalachian region [1].

Growing scientific evidence demonstrates that these surface mining activities present severe, negative environmental consequences including widespread destruction of forest habitat, long-term impairment of ecosystem carbon and nutrient cycling, and regional deterioration of stream water quality [2]. Yet, despite this, MTR remains highly controversial. This is not because the scientific evidence is equivocal, but rather because the environmental costs of these surface mining activities must be weighed against the economic benefits of coal production. Industry employment data [3], coal revenues [4], and severance taxes [5,6] have all historically been calculated on a per-unit-coal basis, making the economic effects of policy decisions to increase or decrease production directly and transparently quantifiable. The same has not been true for the environmental costs; while there have been many assessments of MTR impacts we have yet to translate our understanding of these effects into units of coal produced [7]. It is difficult, if not impossible, for policymakers to weigh the costs and benefits of MTR if the environmental impacts are not conveyed in terms of coal production.

During the process of mining, forests are cleared and the soil and bedrock overlying coal seams (overburden) are removed. The land surface is dramatically reengineered, leaving behind a new topography constructed of reshaped mine spoil. Post-mining soil profiles often have higher bulk densities, lower water infiltration rates, and lower nutrient contents [2]. Most reclaimed mines are seeded with grasses and support little woody vegetation regrowth, even many years after site reclamation – representing a long-term loss of forest habitat [2], as well as a loss of ecosystem carbon (C) sequestration potential as forests are converted to grasslands [8,9]. Additionally, surface coal mining in this region produces alkaline mine drainage containing high concentrations of ions and various solutes that can be harmful to aquatic biota in receiving streams [2]. Recent studies have demonstrated that all of these environmental impacts – habitat loss, reduced ecosystem C sequestration, and stream water quality impairment – are directly related to the amount of land area disturbed by mining activities [8–10]. Thus, in order to quantify these environmental impacts in units of coal produced, it is necessary to know how much coal is produced per unit area of mining disturbance.

While adequate data are available describing the amount of land area *permitted* for mining, not all permitted areas are ultimately mined; single mines occupy large tracts of land (up to 88 km^2) and topographic variation makes some coal difficult to access [7]. Regulatory agencies have not recorded reliable data on the total permitted area that is ultimately mined [7], requiring that this information be gained through alternative methods. For 47 counties in southern West Virginia and eastern Kentucky – a study area occupying ~82% of the Central Appalachian coal region (59,569 km^2) [11] – we use estimates of surface mining disturbances derived from historical satellite imagery for the period 1985–2005 [10], and regress the aerial extent of mining disturbances against cumulative county coal production over this period. Using this newly derived estimate of coal produced per unit disturbance, we then convert previously published estimates of the environmental impacts of MTR into units of coal production. Further, we bring new perspective to the ongoing debate surrounding MTR by placing these environmental costs in terms of regional coal production rates, as well as in terms of total US coal demand.

Methods

Mined areas were delineated at decadal intervals (1976, 1985, 1995, and 2005) using previously published methods [10]. Briefly, digital images from the Multispectral Scanner (MSS, 80 m resolution) and Thematic Mapper (TM, 30 m resolution) taken during mid-summer (in order to minimize seasonal variation in illumination and maximize contrast between disturbed areas from surrounding forests) were reviewed to ensure no interference from haze, smog or cloud cover. Historical topographic data (pre-mining) were obtained from the Defense Mapping Agency (U.S. Department of Defense; Digital Terrain Elevation Data, Level I; https://www1.nga.mil/ProductsServices). The hyperspherical direction cosine (HSDC) method [12] was used to reduce albedo-related variations in illumination, and training samples were selected for each imagery date to classify land-cover based on the Anderson Level II system [13]. Erdas IMAGINE and GIS software was used for image analysis based on a two-stage classification process: (1) pixel-based spectral signatures using the supervised maximum likelihood technique [14] were identified for different land cover types, and (2) decision tree analysis was used to classify mined areas. Mined areas were defined as any bare rock or soil land cover that was not within a 400 m buffer surrounding highways, rivers, or agricultural areas. Final products were raster datasets of 30 m resolution for each time step with binary values indicating mining disturbance or not. For more details on mine delineation methodology, see Bernhardt et al. [10].

Surface mines have long lifecycles often exceeding 10 years [7], thus decadal-scale imagery was adequate for capturing most mining activity [10]. Annual county-level coal production data were obtained for the 33 counties within the study area located in eastern Kentucky [15] and for the 14 counties in southern West Virginia [16]. While satellite imagery and mine delineations extended back to 1976, coal production data were only available for all counties beginning in 1985. Therefore, the cumulative extent of mining disturbance over the period 1985–2005 was calculated as the areas that were observed to have been disturbed by surface mining in 1995 and 2005 that were not already disturbed in 1985 or earlier.

Simple linear regression (SLR) was used to estimate the relationship between county-level coal production and aerial mining disturbance extent (lm function, Stat package, CRAN-R statistical software) [17]. One county (Pike, KY) had more than twice the mining disturbance compared to any other county (Table S1 in File S1), and including this county in the analysis substantially decreased the regression slope estimate of how much coal is typically produced per unit disturbance. Although this decrease was not statistically significant, this county was omitted from the regression analysis due to the high leverage exerted by this single data point. Because omitting this data point increases our estimate of how much coal is produced per unit area disturbance, our estimates of the environmental impacts of MTR may be slightly conservative.

In a previous study, we estimated the threshold value of watershed mining impacts at which a receiving stream is likely to be classified as biologically impaired (as defined by significant losses of pollution intolerant stream macroinvertebrates) [10]. We estimated the extent of mining within every watershed throughout 14 counties in southern West Virginia using the same satellite-derived mine delineations used here, and through multiple modeling approaches we assessed macroinvertebrate responses across sites spanning a gradient of mining intensity. Using two separate models we assessed the responses of commonly used biotic indices as a function of catchment mining intensity, the West Virginia Stream Condition Index (WVSCI) [18] and the Genus-Level Index of Most Probable Stream Status (GLIMPSS) [19]. We used a third model, Threshold Indicator Taxa Analysis (TITAN) [20], to assess the response of individual taxa along the mining gradient. These different models were highly consistent with one another, indicating significant impairment of the stream macro-invertebrate community occurs once 2.2–6.3% of the surface area in their watersheds is converted to mines (Table 1) [10]. We were unable to include data for KY in this previous study due to differences between state data availability and sampling protocols, though there are no distinguishing characteristics that would suggest macroinvertebrate communities in KY streams would respond differently to mining intensity than those in WV. In our previous analysis we estimated cumulative stream impairment at decadal intervals over the period 1976–2005; here we truncate the values to only consider the extent of stream impairment having occurred after 1985 to be temporally consistent with the available coal production data (Table 1). Over the period 1985–2005 an estimated 653 km^2 was mined across the 14 WV counties. Based on the different thresholds for stream impairment predicted by the different models, we estimate this resulted in 1,763–1,968 km of stream channel becoming impaired, or 27.0–30.2 m of stream

Table 1. Stream impairment estimates per unit coal produced.

Metric	Mining threshold (%)[a]	Cumulative impairment through 2005 (km)[b]	Impairment from 1985 and prior (km)[c]	Impairment for period 1985–2005 (km)	Stream impairment per unit disturbance (m ha^{-1})[d]	Stream impairment per unit coal (cm ton^{-1})[e]
WVSCI	5.4	2,834	940	1,894	29.0	0.25
GLIMPSS	6.3	3,390	1,422	1,968	30.2	0.26
TITAN	2.2	4,308	2,545	1,763	27.0	0.24

[a]Values correspond to upper 95% CI values reported by Bernhardt et al. (see Table 1 their publication) [10].
[b]Values from Bernhardt et al. [10] for cumulative stream impairment over the period 1976–2005.
[c]Data provided in Fig. 4 of ref.[10], as well as by the authors.
[d]Quotient of stream impairment (1985–2005) divided by mining disturbance (1985–2005 WV only =65,211ha; from Table S1 in File S1).
[e]Converted to stream impairment per unit coal using slope estimate (11,500 tons coal/ha).

Table 2. Foregone ecosystem C sequestration per unit coal produced.

Study	Ecosystem Type	Soil C Sequestration Rate (g C m^{-2} yr^{-1})	Biomass C Sequestration Rate (g C m^{-2} yr^{-1})	Ecosystem C Sequestration Rate (g C m^{-2} yr^{-1})
Simmons et al.	Reclaimed Grassland	159[a]	8[a]	167
Amichev et al.	Reclaimed Forests	274[b]	83[b]	357
Amichev et al.	Unmined Forests	237[b]	152[b]	389

[a]Values are from Simmons et al. Table 5 [8]. Our reference to "Soil" is considered "total belowground" by the authors; our use of "biomass" is considered "total aboveground, as reported by the authors. Reported values for C pools were divided by stand age (15 yr) and converted to g C m^{-2} yr^{-1}.
[b]Values are calculated from Amichev et al. Table 1 [9]. Reported values for C pools for all sites were divided by stand age and converted to gC m^{-2} yr^{-1}. Our reference to "Soil" is considered "SOC" + "Litter Layer C", as reported by the authors; our reference to "Biomass" is considered "Total Tree C", as reported by the authors.

channel length lost for each hectare of surface coal mining disturbance (Table 1). We use the SLR slope estimate relating coal production to disturbance (tons coal ha^{-1}) to translate this estimate of stream impairment (m ha^{-1}) into units of coal produced (cm tons coal^{-1}).

Clearing of forests prior to mining can result in large initial losses of soil and biomass C pools. These ecosystems will re-accumulate C following mine reclamation, but likely at a slower rate because post-mining soils are often compacted with low nutrient content and reduced fertility [2] – all of which affect vegetation re-establishment and, thus, limit C sequestration rates. This has been especially true since the passage of the Surface Mining Control and Reclamation Act (SMCRA) of 1977, which has favored severe soil compaction and reseeding with grasses in order to reduce sediment losses from mined areas [21]. We estimate the amount of foregone C sequestration by comparing previously published data for ecosystems recovering from mining to the C sequestration potential of unmined sites. Simmons et al. [8] inventoried soil and biomass C pools 15 years after a forested site in the Appalachian coal region was cleared, mined and restored to grassland following SMCRA regulations. We divide the reported C pool values by the number of years since mining to estimate annual ecosystem C sequestration rates. Biomass and soil

carbon in this system accumulated at rates of 8 and 159 g C m^{-2} yr^{-1}, respectively (Table 2). Amichev et al. [9] reported ecosystem C soil and biomass pool estimates for 14 pre-SMCRA mining sites across the region exhibiting forest regrowth ("reclaimed forests"), as well as 8 non-mined reference forests ("unmined forests"). We estimate C sequestration rates for these reclaimed and unmined forests by dividing their soil and biomass pool estimates by the site-specific stand ages. The reclaimed forest sites accumulated biomass and soil carbon at 274 and 83 g C m^{-2} yr^{-1}, while the unmined forest sites showed C accumulation rates of 237 and 152 g C m^{-2} yr^{-1} for biomass and soil pools (Table 2). Thus, the net ecosystem production rate (soil + biomass C accumulation) for unmined forests (389 g C m^{-2} yr^{-1}) exceeds that for reclaimed mine sites converted to grasslands (167 g C m^{-2} yr^{-1}) or sites supporting forest regrowth (357 g C m^{-2} yr^{-1}). Our estimate of foregone ecosystem C sequestration resulting from mining, then, ranges from 32–222 g C m^{-2} yr^{-1}. Because most sites mined post-SMCRA (after 1977) have been converted to grasslands [21], the amount of foregone ecosystem C sequestration for the mining disturbance that occurred over our study period is likely at the upper end of the reported range (222 g C m^{-2} yr^{-1}). We divide this estimate of foregone ecosystem C sequestration per unit area disturbance by the regression value of coal produced per unit area

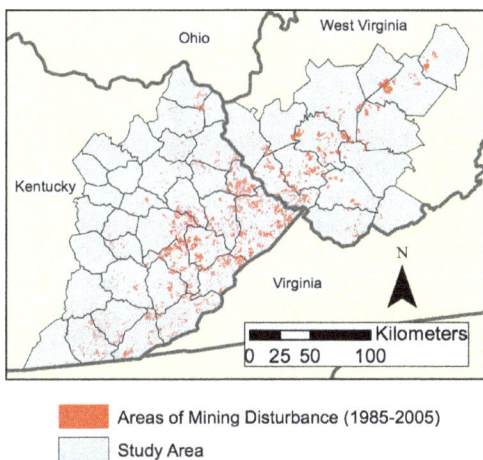

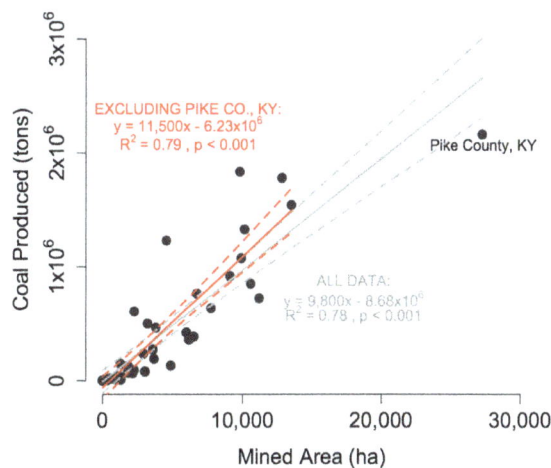

Figure 1. Extent of mining disturbance and relationship to coal production. Red areas show the cumulative extent of surface mining disturbance between 1985–2005 across the study area (left). The amount of disturbance estimated within each county was regressed against cumulative county-level coal production reported for this period (right). Although Pike County, KY had more than twice the mining disturbance compared to any other county, confidence intervals around the slope estimates with the Pike County data included (95% CI = 8,320–11,300) overlapped with the confidence interval estimates for the slope value when this data point was omitted (95% CI = 9,650–13,340; confidence interval estimates calculated using the *confint* function in CRAN-R statistical software). Nonetheless, given this single data point exerts high leverage on the regression results, we use the values for the analysis in which this data point is excluded.

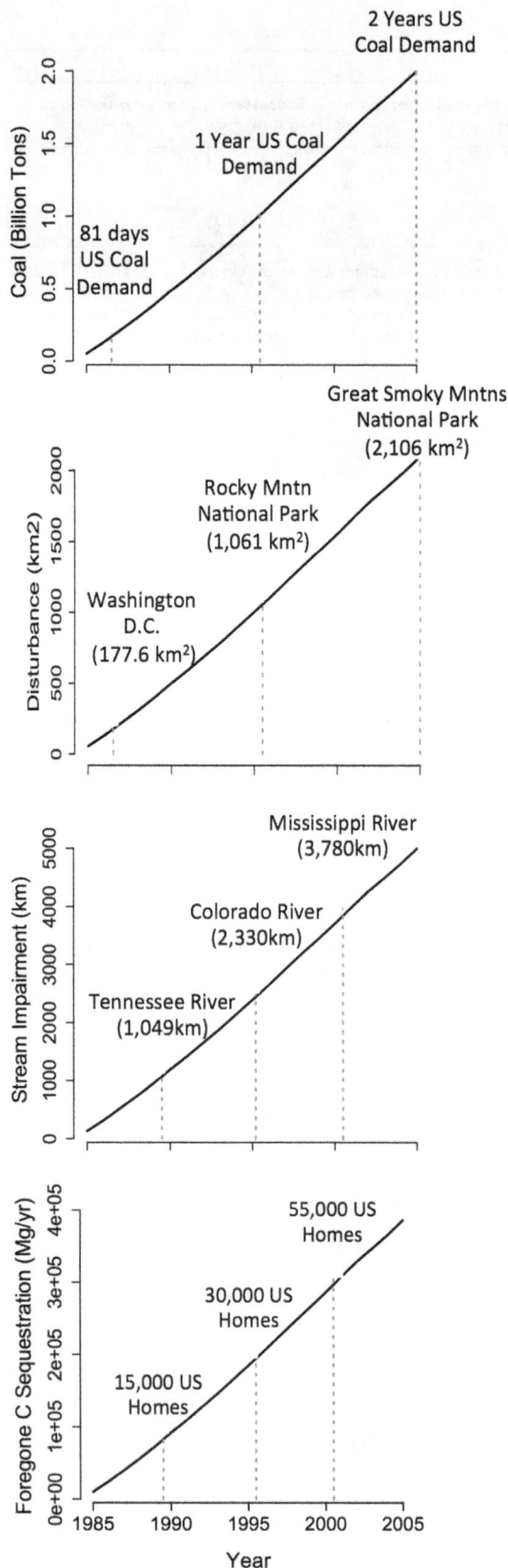

Figure 2. Scaling the environmental costs of coal production. Cumulative surface coal production across the 47 county study area is shown for the period 1985–2005 (A). Cumulative surface mining disturbance (B), stream length impairment (C), and forgone C

sequestration (D) were estimated by multiplying the respective per-unit-coal values by the cumulative coal production data. Vertical dashed lines indicate comparative values for contextualization (see Table S2 in File S1).

disturbance to derive our final estimate of foregone ecosystem C sequestration in units of coal produced.

To bring perspective to these per-unit-coal estimates of the environmental costs of MTR, we calculate the extent of environmental impacts across the region by multiplying the per-unit-coal values by the cumulative annual coal production data across all counties. While we could also estimate the extent of these impacts directly from the amount of land area identified as having been mined in the satellite imagery analysis, the imagery analyses are only of decadal resolution while the coal production data are reported annually. Thus, using the coal production data allow us to represent the expansion of these environmental impacts with higher temporal resolution.

Results and Discussion

We found a surprisingly strong relationship where for every hectare of disturbance 11,500 short tons of coal are produced (1 short ton $= 907.2$ kg; $y = 11,500 \times -6.23 \times 10^6$, $R^2 = 0.79$) or, inversely, for each ton of coal 0.87 m^2 of landscape is disturbed (fig. 1). Bituminous coal from the Central Appalachians has an average density of 1.32 g cm^{-3} [22]. If 11,500 tons of coal are produced per hectare disturbed, this would translate to an average coal seam thickness of only ~ 1 m, compared to reported measurements of coal seam thicknesses for this area of ~ 1.5 m [23]. However, our aerial estimates of disturbance include both the excavation of material overlying coal seams, as well as areas where mining spoil is displaced. Thus, our analysis suggests that the total footprint of MTR disturbances may exceed the spatial extent of underlying coal deposits by a factor of 1.5.

Previous assessments of stream impairment indicated that, on average, 27.0–30.2 m of stream impairment occurs for every hectare of surface mining in the Central Appalachians (Table 1). Given 11,500 tons of coal are produced per hectare disturbed, we estimate 0.25 cm of stream length was impaired, on average, for every ton of coal produced (Table 1).

Previous studies also indicate that mined sites reclaimed to forests and mined sites reclaimed to grasslands had C sequestration rates that were 32 and 222 g C m^{-2} yr^{-1} lower, respectively, compared to unmined forest ecosystems (Table 2). Despite this large range in potential foregone ecosystem C sequestration resulting from mining, the vast majority of reclaimed mines are seeded with grasses and support little woody vegetation regrowth even many years after site reclamation [2]. Because of this, we expect most mined ecosystems to exhibit values towards the upper end of this range. Based on our regression slope of 11,500 ton coal ha^{-1} and using the forgone C sequestration value of 222 g C m^{-2} yr^{-1}, every ton of coal produced from the Central Appalachians is estimated to result in 193 g C ton coal^{-1} yr^{-1} of lost ecosystem C sequestration potential.

The average bituminous coal from this region has a carbon content of 80% [24]. Given 11,500 tons of coal are harvested for every hectare, ~ 0.83 Mg C m^{-2} is released as CO$_2$ through combustion assuming 99% of the C in coal is converted to CO$_2$ [25]. Based on the C sequestration potential of these ecosystems it would take $\sim 5,000$ years for any given hectare of former mines reclaimed to grassland to sequester the carbon released from combustion of the coal removed from that hectare, assuming these ecosystems could persist in an accumulating stage over these time periods. For those rare surface mines where forest regrowth is achieved, it would still take ~2,150 years for these forested hectares to accumulate sufficient C in soils and tree biomass to sequester what was emitted through coal combustion.

While 0.25 cm of stream channel impairment or 193 g yr^{-1} of lost C sequestration per ton of coal may not sound like alarming environmental costs, it is difficult to appreciate these numbers unless they are placed in the context of regional coal production rates. Cumulative coal production within the study area totaled 1.93 billion tons over the 1985–2005 period (fig. 2A), or approximately two years worth of current US coal demand (Table S2 in File S1). To access this coal, nearly 2,000 km^2 of land area was mined, which is comparable in size to the Great Smoky Mountains National Park (2,106 km^2; fig. 2B). If all current US coal demand were to be supplied from surface mines in the Central Appalachians, an area equal in size to Washington DC (177.6 km^2) would need to be mined every 81 days. To supply one year of current US coal demand would require converting 803 km^2 of Appalachian mountains to surface mines, leading to the biological impairment of an estimated ~2,300 km of streams and forgone ecosystem C sequestration of ~185,000 Mg C yr^{-1} (equivalent to the global warming potential of approximately 33,600 average US single family homes) (fig. 2C&D).

Conclusions

Surface coal mining in the Central Appalachians has become an extremely contentious method for producing energy. While the scientific community has adequately demonstrated the *severity* of surface mining impacts [2], considerably less attention has been placed on understanding the *extent* of these environmental impacts and in providing the metrics necessary to compare these environmental costs to the obvious economic benefits of coal. This has been a critical shortcoming, as even the most severe impacts may be tolerable if they are believed to be sufficiently limited in extent. We show, however, that the extent of environmental impacts of these surface mining practices is staggering, particularly in terms of the amount of coal that is produced. Tremendous environmental capital is being spent to achieve what are only modest energy gains.

Acknowledgments

The authors thank J. Amos and D. Campagna for supplying the initial mine delineation data.

Author Contributions

Conceived and designed the experiments: BDL ESB WHS. Analyzed the data: BDL WHS. Wrote the paper: BDL ESB WHS.

References

1. Saylor KL (2008) Land Cover Trends: Central Appalachians. Available: http://landcovertrends.usgs.gov/east/eco69Report.html. Accessed 20 February 2013.
2. Palmer MA, Bernhardt ES, Schlesinger WH, Eshleman KN, Foufoula-Georgiou E, et al. (2010) Mountaintop mining consequences. Science 327: 148–149.
3. EIA (2011) Coal Productivity by State and Mine Type, 2011, 2010. Available: http://www.eia.gov/coal/annual/pdf/table21.pdf. Accessed 20 February 2013.
4. EIA (2011) Coal Prices, Selected Years, 1949–2011. Available: http://www.eia.gov/totalenergy/data/annual/pdf/sec7_21.pdf. Accessed 20 February 2013.
5. WV State Tax Department (2007) West Virginia Coal Severance Tax Estimate. Available: http://www.state.wv.us/taxrev/uploads/sev400c.pdf. Accessed 20 February 2013.
6. KY Office of State Budget (2010) Coal Severance and Processing Tax. Available: http://www.osbd.ky.gov/NR/rdonlyres/52EC9373-1916-401A-AB2F-3F3B52A5F20A/0/0810TEA_CoalSeveranceandProcessingTax.pdf. Accessed 20 February 2013.
7. GAO (2009) Surface Coal Mining: Characteristics of mining in Mountainous Areas of Kentucky and West Virginia. Available: http://www.gao.gov/new.items/d1021.pdf. Accessed: 20 February 2013.
8. Simmons JA, Currie WS, Eshleman KN, Kuers K, Monteleone S, et al. (2008) Forest to reclaimed mine land use change leads to altered ecosystem structure and function. Ecol Apps 18: 104–118.
9. Amichev BY, Burger JA, Rodrigue JA (2008) Carbon sequestration by forests and soils on mined land in the Midwestern and Appalachian coalfields of the US. For Ecol & Manag 256: 1949–1959.
10. Bernhardt ES, Lutz BD, King RS, Fay JP, Carter CE, et al. (2012) How many mountains can we mine? Assessing the regional degradation of Central Appalachian rivers by surface coal mining. ES&T 46: 8115–8122.
11. DOE (1980) Land use and energy. Available: http://www.osti.gov/energycitations/servlets/purl/6300166-L4I340/6300166.pdf. Accessed 20 February 2013.
12. Pouch GW, Campagna DJ (1990) Hyperspherical direction cosine transformation for separation of spectral illumination information in digital scanner data. Photogrammetric Engineering and Remote Sensing 56: 475–479.
13. Anderson JR, Hardy EE, Roach JT, Witmer RE (1976) A land use and land cover classification system for use with remote sensor data. US Geological Survey Professional Paper 964. Available: http://landcover.usgs.gov/pdf/anderson.pdf. Accessed 20 February 2013.
14. Short NM (1982) The remote sensing tutorial. NASA. Available: http://rst.gsfc.nasa.gov. Accessed 20 February 2013.
15. Kentucky Geological Survey (2013) Coal Production Database. Available: http://www.uky.edu/KGS/coal/production/kycoal01.htm. Accessed 20 February 2013.
16. West Virginia Geological and Economic Survey (2010) Coal Summary Statistics. Available: http://www.wvgs.wvnet.edu/www/datastat/coalsummary/coal_summary.asp. Accessed 20 February 2013.
17. R Development Core Team (2010) Available: http://www.R-project.org.Accessed 20 February 2013.
18. Barbour MT, Gerritsen J, Snyder BD, Stribling JB (1999) In Rapid Bioassessment Protocols for Use in Streams and Wadeable Rivers: Periphyton, Benthic Macroinvertebrates and Fish, Second ed.; EPA 841 – B-99-002. Available: http://water.epa.gov/scitech/monitoring/rsl/bioassessment/index.cfm. Accessed 20 February 2013.
19. Pond GJ, Bailey JE, Lowman BM, Whitman MJ (2013) Calibration and validation of a regionally and seasonally stratified macroinvertebrate index for West Virginia wadeable streams. Environmental Monitoring and Assessment. 185:1515–1540.
20. Baker ME, King RS (2010) A new method for detecting and interpreting biodiversity and ecological community thresholds. Methods Ecol. Evol. 1(1): 25–37.
21. Zipper CE, Burger JA, Skousen JG, Angel PN, Barton CD, et al. (2011) Restoring forests and associated ecosystem services on Appalachian coal surface mines. Environmental Management 47: 751–765.
22. Wood GH, Kehn TM, Carter MD, Culbertson WC (1980) Coal Resource Classification System of the US Geological Survey. Available: http://pubs.usgs.gov/circ/c891/table2.htm. Accessed 20 February 2013.
23. EPA (2003) Affected Environment and Consequences of MTM/VF. Available: http://www.epa.gov/region03/mtntop/pdf/III_affected-envt-consequences.pdf. Accessed 20 February 2013.
24. USGS (2009) Coal Resource Availability, Recoverability, and Economic Evaluations in the United States – A Summary. Available: http://pubs.usgs.gov/pp/1625f/downloads/ChapterD.pdf. Accessed 20 February 2013.
25. EPA (1998) Bituminous and subbituminous coal combustion. Available: http://www.epa.gov/ttnchie1/ap42/ch01/final/c01s01.pdf. Accessed 20 February 2013.

8

The Role of Stream Water Carbon Dynamics and Export in the Carbon Balance of a Tropical Seasonal Rainforest, Southwest China

Wen-Jun Zhou[1,2,3,4], **Yi-Ping Zhang**[1,2,3]*, **Douglas A. Schaefer**[1,2], **Li-Qing Sha**[1,2,3], **Yun Deng**[1,2,3], **Xiao-Bao Deng**[1,2,3], **Kai-Jie Dai**[1,2]

1 Key Laboratory of Tropical Forest Ecology, Xishuangbanna Tropical Botanical Garden, Chinese Academy of Sciences, Mengla, Yunnan, China, **2** Xishuangbanna Tropical Botanical Garden, Chinese Academy of Sciences, Kunming, China, **3** Xishuangbanna Station for Tropical Rain Forest Ecosystem Studies, Chinese Ecosystem Research Net, Mengla, Yunnan, China, **4** University of Chinese Academy of Sciences, Beijing, China

Abstract

A two-year study (2009 ~ 2010) was carried out to investigate the dynamics of different carbon (C) forms, and the role of stream export in the C balance of a 23.4-ha headwater catchment in a tropical seasonal rainforest at Xishuangbanna (XSBN), southwest China. The seasonal volumetric weighted mean (VWM) concentrations of total inorganic C (TIC) and dissolved inorganic C (DIC) were higher, and particulate inorganic C (PIC) and organic C (POC) were lower, in the dry season than the rainy season, while the VWM concentrations of total organic C (TOC) and dissolved organic C (DOC) were similar between seasons. With increased monthly stream discharge and stream water temperature (SWT), only TIC and DIC concentrations decreased significantly. The most important C form in stream export was DIC, accounting for 51.8% of the total C (TC) export; DOC, POC, and PIC accounted for 21.8%, 14.9%, and 11.5% of the TC export, respectively. Dynamics of C flux were closely related to stream discharge, with the greatest export during the rainy season. C export in the headwater stream was 47.1 kg C ha^{-1} yr^{-1}, about 2.85% of the annual net ecosystem exchange. This finding indicates that stream export represented a minor contribution to the C balance in this tropical seasonal rainforest.

Editor: David L. Kirchman, University of Delaware, United States of America

Funding: This study was funded by the National Science Foundation of China (40801035, 40571163), the CAS 135 program (XTBG-T03) and Strategic Priority Research Program-Climate Change: Carbon Budget and Related Issues of the Chinese Academy of Sciences(XDA05020302), the Development Program in Basic Science of China (No. 2010CB833501), and the Yunnan Natural Science Foundation of Yunnan Province, China (2008CD167). The funders had no role in study design, data collection and analysis, decision to publish, or preparation of the manuscript.

Competing Interests: The authors have declared that no competing interests exist.

* E-mail: yipingzh@xtbg.ac.cn

Introduction

Streams and small inland rivers are important links between terrestrial and aquatic ecosystems. Cole et al. [1] suggested that inland waters export 1.9 Pg C yr^{-1}, indicating that regional carbon (C) balances can influence transport into large terrestrial rivers and oceans [2]. Recently, several studies have focused on dissolved organic C (DOC), dissolved inorganic C (DIC), particulate inorganic (PIC) and organic C (POC), and even gaseous C (CO$_2$, CH$_4$) in catchment runoff, and on their role in C exports from ecosystems [2–8].

Previous studies showed that the export of dissolved and gaseous C with rivers and streams may vary among forest ecosystems. Shibata et al. [9] found that sum of DIC and DOC export by stream water (7.6 g C m^{-2} yr^{-1}) accounted for only 2% of net ecosystem exchange (NEE) in cool temperate forests of northern Japan, whereas in Canadian boreal forests, C export from surface waters accounted for NEE from 9.5% to 16.4% [10]. In the Amazon, Richey et al. [6] demonstrated that outgassing of CO$_2$ (1.2±0.3 Mg ha^{-1} yr^{-1}) from rivers and wetlands constituted an important C loss. Also Lloret et al. [8] demonstrated the key role of streams in the C balance of forest catchments in the Amazon Basin. Neu et al. [2] showed that C transported by water

comprised about 20% of the total annual C exchange across tropical forest canopies. The roles of surface water in C export vary because of diversity in geographic location, basin-specific soil and vegetation types, catchment topography, climate, and upland-wetland flow paths in forests [4,7,11–16]. As a result, by ignoring the export of CO$_2$, DOC, DIC, PIC and POC via hydrological pathways, terrestrial C budgets are incomplete and net C sequestration could be overestimated [1].

Surface water and wetland play substantial roles in C balance in the Amazon, the largest tropical forest region in the world [2,6,8]. So far, little is known about the importance of C export by headwater streams on the carbon balance of tropical seasonal rain forests (TSRF) at the northern edge of the tropical zone in southwest China (Figure 1). Despite its relatively high latitude, tropical seasonal rain forest has a moist tropical climate due to the influence of the Himalayas. It is unique in terms of forest type, differing from those in the equatorial region of Southeast Asia and has highly diverse and mixed types of floristic compositions due to its unique geographical location between a tropical zone to the south and a subtropical zone to the north [17]. Consequently, the tropical seasonal rain forest in southwest China is an important biogeographic area in Southeast Asia. Tan et al. [18] and Zhang et al. [19] have reported that TSRF in Xishuangbanna

(XSBN) is a small net C source. Accounting for TC export with stream water may make the loss of C from TSRF at XSBN even larger than earlier anticipated. In order to clarify the role of C export by headwater streams, a study was therefore undertaken in TSRF at XSBN. The objectives of this study were (1) to ascertain the seasonal dynamics of different C components (DIC, DOC, PIC, POC, TIC, TOC, and TC), and (2) to assess the contribution of stream export to the C balance in this tropical seasonal rainforest ecosystem.

Materials and Methods

Ethics Statement

All necessary permits were obtained from Xishuangbanna National Nature Reserve for the described field studies which did not involve endangered or protected species.

Study Site

The study area in XSBN (Dai autonomous prefecture), Yunnan province, China (21.16° N, 101.04° E) (Figure 1), is influenced by the Southwest monsoon and dominated by North Tropical Monsoon weather, with annual average temperature 21.5°C, annual average rainfall 1557 mm and average relative humidity 86%. Based on precipitation data, the rainy season (with 84.1% of the total annual precipitation) [20,21], is between May and October. The dry season is between November and April.

The experimental site is located in the centre of the National Forest Reserve in Menglun, Mengla County, Yunnan province, with relatively little human disturbance. The dominant trees are *Terminalia myriocarpa* and *Pometia tomentosa*, which is typical of tropical forest [17]. The total catchment area is 23.4 ha, the slope is 12°~18°, and the soil type is oxisol formed from Cretaceous yellow sandstone with a pH value of 4.5~5.5 and a clay content of 19.5%~29.5% [22].

Experimental Set-up

Hydrological observations. At the watershed outlet, a 90° V-notch weir instrumented with a water-level recorder was installed. The recorder was set to take averaged discharge measurements at 5-min intervals. Daily and monthly discharges were calculated separately from the stream-height data, as follows:

$$Q = 0.014H^{2.5} \tag{1}$$

$$R = QT/1000F \tag{2}$$

Where Q = discharge (m^3/s); H = water head (m); R = runoff (mm); T = time (s); and F = catchment area (km^2). Stream water temperature (SWT) was recorded at the mid-point of stream depth near the stream outlet. Measurements were made every half hour and stored in a data logger.

Water sample sampling and analysis. Stream water was sampled in the middle of the stream outlet. Stream water samples were collected between 8:00 and 9:00 am local time at the sampling site in high-density polyethylene (HDPE) bottles; sampling bottles were completely filled, allowing no headspace. Bottles were rinsed with distilled water after being washed with 3% HCl solution. Bottles were pre-rinsed three times with the stream water before sample collection. The study was done during two full calendar years, from 1.1.2009 to 31.12.2010. During the dry season, stream water samples were collected once per week, in addition to daily samples during three consecutive days following rain events. Stream water was sampled twice per week during the rainy season in 2009, and once per week in 2010. All water samples were immediately transported to the laboratory in insulated bags.

Following the analysis method of Baker et al. [23], all samples were vacuum-filtered through 0.45-μm GFF (Tianjinshi Dongfang Changtai Environmental Protection Technology Co. Ltd., China) pre-rinsed with deionized water and sample water under vacuum. The filtered and unfiltered water samples were analysed for DOC/DIC and TOC/TIC by TOC/TN analyser (LiquiTOC II, Elementar Analyses System GmbH, Germany) respectively, within 24 hours.

The TOC/TN analyser allows particle size up to 200μm. So in this study, the diameter of particulate matter (PIC and POC) was defined from 0.45 μm to 200 μm. TIC and TOC were defined as less than 200 μm for all the water samples. Concentrations of PIC and POC were calculated by subtracting the DIC and DOC concentrations from the TIC and TOC concentrations, respectively.

Calculations and Statistics

The monthly volume-weighted mean (VWM) concentrations were computed as follows:

$$VWM = \frac{\sum_{i=1}^{n} CiVi}{\sum_{i=1}^{n} Vi} \tag{3}$$

Where VWM is the volume-weighted mean concentration, Ci is the concentration (mg L^{-1}), and Vi is the runoff (m^3 s^{-1}) at sampling time.

The monthly C flux was calculated as the monthly VWM C concentration multiplied by monthly discharge. We calculated TIC, DIC, TOC, and DOC flux directly, and calculated PIC, POC, and TC flux as follows:

$$TC\ flux(F_{TC}) = F_{TIC} + F_{TOC} \tag{4}$$

$$PIC\ flux(F_{PIC}) = F_{TIC} - F_{DIC} \tag{5}$$

$$POC\ flux(F_{POC}) = F_{TOC} - F_{DOC} \tag{6}$$

Where F indicates flux.

The correlations among stream discharge, stream water temperature on the one hand, and concentrations of TIC, TOC, DIC, DOC, PIC, and POC on the other, as well as the correlations among different carbon components, were tested using the Pearson correlation (two tailed), employing the software SPSS 15.0.

Results

Seasonal Variations of Rainfall, Stream Runoff, and Stream Water Temperature

The average annual rainfall and runoff for the two years were 1026.1 mm and 326.9 mm, respectively. These values are less than the past 40 years means [24]. Rainfall and runoff were higher during the rainy season (average 848.9 mm and 279.1 mm, respectively) than during the dry season, confirming earlier reports

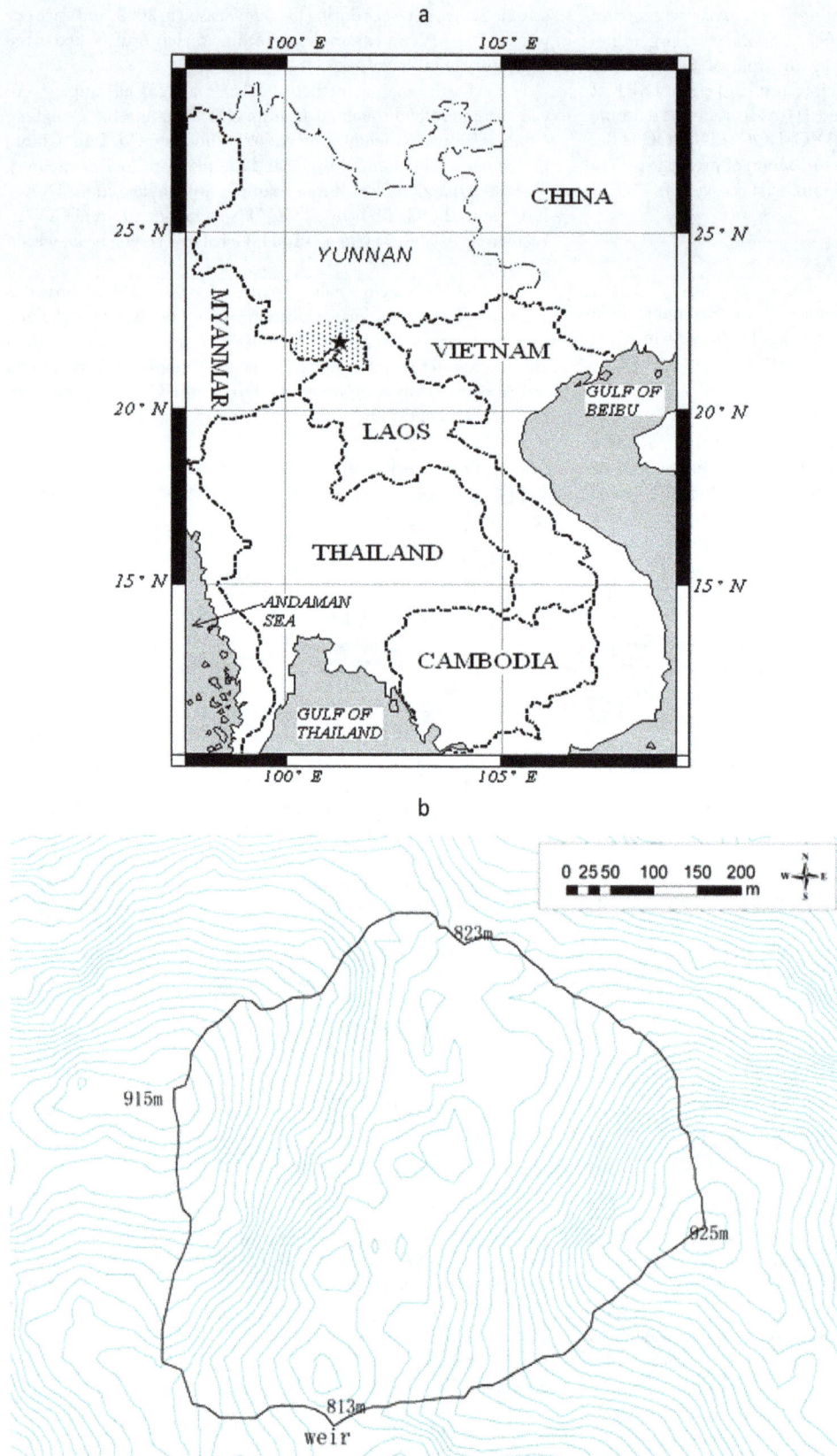

Figure 1. Study site description in Xishuangbanna tropical seasonal rainforest, Southwest China. (a) Location of the study area (indicated by the black star). (b)The catchment site description was from the Advanced Spaceborne Thermal Emission and Reflection Radiometer (ASTER) Global Digital Elevation Model (GDEM) that is a product of METI and NASA.

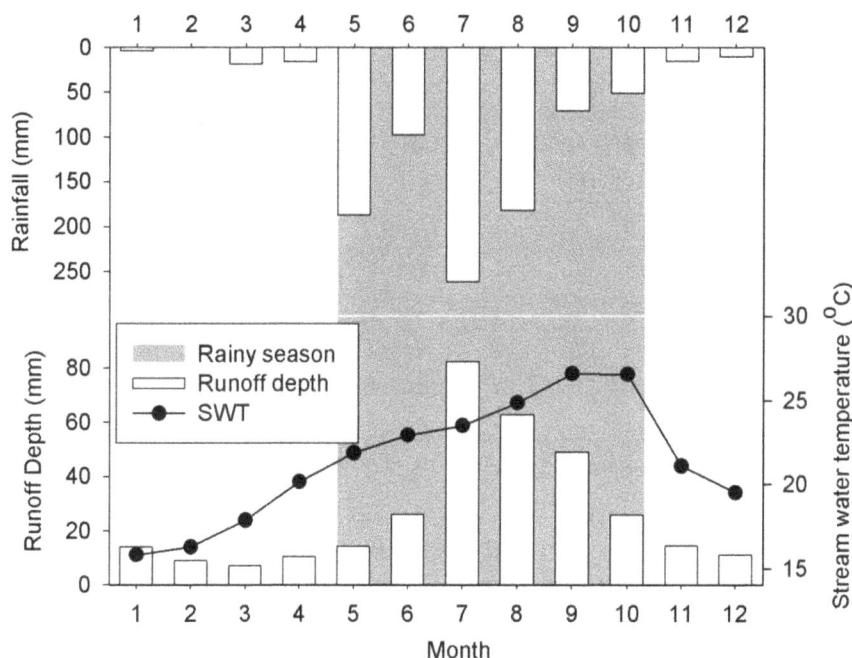

Figure 2. Annual dynamics of rainfall, runoff depth, and stream water temperature in the seasonal tropical rainforest. Values are averages for 2009 and 2010.

on the seasonal dynamics of rainfall and stream discharge [20,25].The seasonal dynamics of rainfall and runoff showed similar patterns and were well correlated ($r = 0.794$, $p<0.001$, $n = 24$; Figures 2, S1).

Average SWT was 21.4°C, with lowest values in January (15.8°C) and peak values in September (26.6°C; Figures 2b, S1).

C concentration Dynamics

DIC was the largest component of TC. The rank order of the overall contribution of different C forms to TC was as follows: DIC (51.8%)>DOC (21.8%) >POC (14.9%) >PIC (11.5%) (Figure 3). However, these contributions differed seasonally: the contribution of DIC to TC was lowest in July (31.0%) and highest in February (64.9%), DOC had the highest contribution to TC in April and May (35.0%) and the lowest in December (10.0%), and the contribution of POC to TC was greatest in July (Figure 3). The DIC: DOC ratio and its monthly variation (2.9 and 67.1%, respectively) were higher than those for PIC: POC (0.8 and 47.3%, respectively).

Seasonal variations in VWM concentration were different among the various C components (Table 1). The VWM concentrations of TIC and DIC were higher during the dry season than during the rainy season, others were similar.

Monthly VWM concentrations of TIC and DIC were significantly correlated ($r = 0.956$, $p<0.001$, $n = 24$). The highest monthly VWM concentration was in March as discharge was the lowest; the lowest concentrations were in June and July while stream discharges were relatively high (Figures 2b, 4a). Both TIC and DIC were negatively correlated to discharge and SWT (Table 2). The highest and lowest monthly VWM concentrations of PIC occurred in August (2.0 mg L^{-1}) and September (1.2 mg L^{-1}) while discharge was high. Although floods increased DOC and POC concentrations, the highest VWM concentrations of TOC, DOC and POC occurred during the beginning of the rainy season (Figure 4b) at intermediate values of discharge (Figure 2b).

The lowest values of TOC and DOC were in September during relatively high discharge, but POC was lowest in February when discharge was low. The seasonal dynamic of DOC was different from DIC ($r = 0.157$, $p = 0.464$, $n = 24$), but those of PIC and POC were similar ($r = 0.515$, $p = 0.010$, $n = 24$). The annual variation of PIC (coefficient of variation (CV) = 19.2%) was less than that of TIC (CV = 23.1%) and DIC (CV = 28.4%). The rank order of the coefficients of variation of organic C forms was as follows: POC (CV = 76.4%), DOC (CV = 58.9%), TOC (CV = 51.2%).

Stream C Flux Dynamics and Distribution

Annual TC export was 53.9 kg ha^{-1} and 40.7 kg ha^{-1} in 2009 and 2010, respectively. The dynamics of the fluxes of C differed between the various compounds. With the exception of POC, the

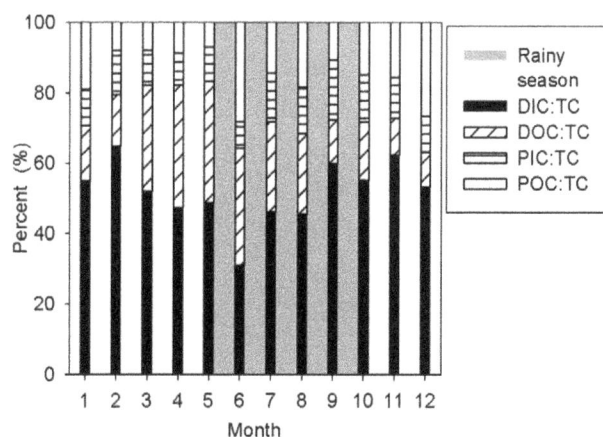

Figure 3. Annual dynamics of percentage difference of various carbon components to TC export, in the headwater stream.

Table 1. Concentration and export of various carbon components in a headwater stream in the tropical seasonal rainforest of Xishuangbanna, Southwest China.

Season	Carbon concentration (mg C L^{-1})			Carbon export (kg C ha^{-1})				
	Dry season	Rainy season	Annual	Dry season	Rainy season	Annual average	2009	2010
TIC	10.1	6.9	7.5	7.6	21.1	28.7	32.5	24.9
TOC	6.6	6.6	6.6	3.9	14.5	18.4	21.2	15.7
DIC	8.9	5.4	6.0	6.4	16.6	22.9	26.0	19.9
DOC	3.7	3.7	3.7	2.2	8.6	10.8	13.4	8.1
PIC	1.6	1.7	1.7	1.2	4.6	5.8	6.5	5.0
POC	3.1	3.3	3.2	1.7	5.9	7.6	7.7	7.5
TC				11.5	35.6	47.1	53.5	40.6

Carbon concentration is the seasonal volumetric weighted mean (VWM) concentration of 2009 and 2010.

greatest flux of all C components occurred in July when discharge was highest (Figures 2, 5). By contrast, POC export was greatest in June when discharge was intermediate. Due to low discharge also C flux was low in February (DOC, TOC) and in March (inorganic C and POC). Measures of seasonal C export (Table 1) showed that most of the C export for all components occurred during the rainy season.

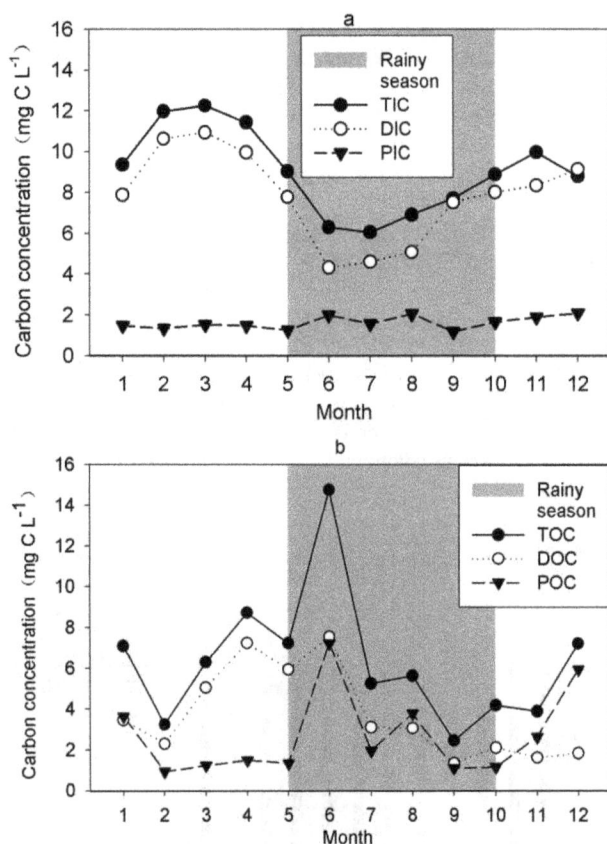

Figure 4. Annual dynamics of monthly volumetric weighted mean concentrations of carbon components in seasonal tropical rainforest. (a) Annual dynamics of monthly volumetric weighted mean concentrations of TIC, DIC, and PIC. (b) Annual dynamics of monthly volumetric weighted mean concentrations of TOC, DOC, and POC.

Contribution Stream Water C Flux in the C Balance of Tropical Seasonal Rainforest

Based on seasonal NEE dynamics (Figure 6a) [19], TC export resulted in an increased net C export ratio (TC: NEE) from 0.54% to 3.30% from April to August, and in a decrease from September to March, ranging from 0.20% to 1.80%. In total, stream C export represented a mean 2.85% (2009, 3.25%; 2010, 2.45%) increase in the annual net carbon export (1660 kg C ha^{-1} yr^{-1}) [19] in the TSRF. The ratios of DIC, DOC, PIC, and POC export to NEE showed the highest fractions of the C components' flux to NEE occurred in August (DIC: NEE = 1.50%, DOC: NEE = 0.75%, PIC: NEE = 0.44%, POC: NEE = 0.60%). The lowest absolute value of ratios of DIC, DOC, and PIC to NEE were observed in November (DIC: NEE = 0.13%, DOC: NEE = 0.024%, PIC: NEE = 0.025%), whereas the lowest value of POC: NEE occurred in February (0.024%; Figure 6b).

Discussion

C Dynamics in a Headwater Stream in the Tropical Seasonal Rainforest of Xishuangbanna

Headwater stream C originates from surface soil, ground water, vegetation (dead and alive), roots and microbial biomass in the forest [8,26,27]. Stream C dynamics depended on rainfall and discharge dynamics in this study and other small catchments [13,16,28]. Accurate calculation of C export with stream water requires representative sampling [12,16]. Our sampling campaigns were throughout 2 years, and included both base flow and storm flow conditions (Figure S1), so that carbon components in stream

Table 2. Pearson correlations for monthly average stream discharge, monthly average SWT, and monthly VWM concentration of carbon components in 2009 and 2010 in a headwater stream in the tropical seasonal rainforest in Xishuangbanna, Southwest China.

Parameters	TIC	TOC	DIC	DOC	PIC	POC
Monthly Q (n = 24)	−0.658**	−0.140	−0.661**	−0.139	0.057	−0.079
SWT (n = 24)	−0.520**	−0.166	−0.482*	−0.103	−0.150	−0.186

*Correlation is significant at the 0.05 level (2-tailed);
**Correlation is significant at the 0.01 level (2-tailed).

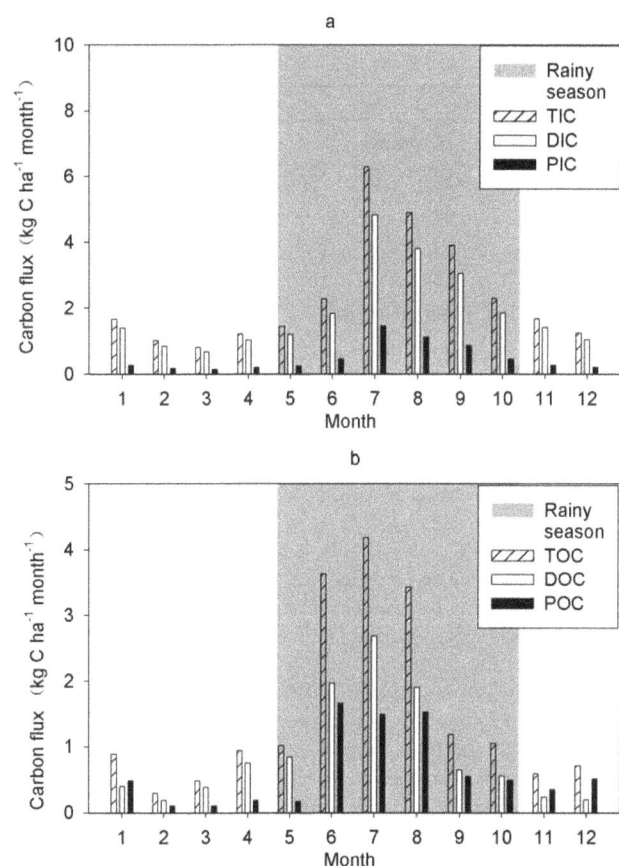

Figure 5. Annual dynamics of flux of carbon components in a headwater stream in tropical seasonal rainforest. (a) Annual dynamics in flux of TIC, DIC, and PIC. (b) Annual dynamics in flux of TOC, DOC, and POC.

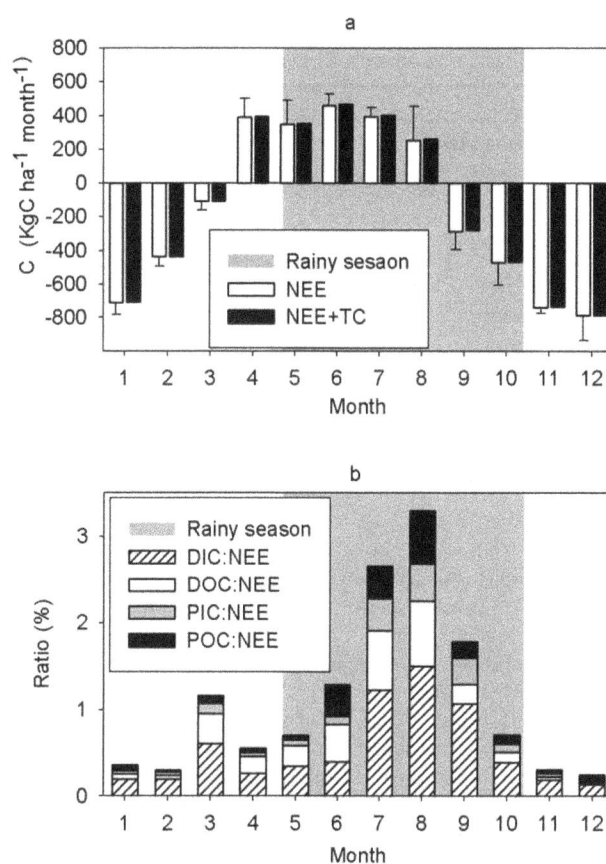

Figure 6. Annual dynamics of ratio of stream carbon export to net ecosystem CO₂ exchange (NEE). (a) Annual dynamics of NEE [19] and the sum of NEE and TC exported by headwater stream. (b) Annual dynamics of ratios of annual flux of DIC (DIC: NEE), DOC (DOC: NEE), PIC (PIC: NEE), and POC (POC: NEE) to NEE.

water were sampled across a wide range of discharge rates (Figure S2). Otherwise, our sampling revealed little particle matter larger than 200 μm only in March when discharge was relatively low (Figure 1). Therefore, this approach ensures the relatively high accuracy of C export calculations, which includes all C matter less than 200 μm in size. Accordingly, our calculations of TC and PC export excluded coarse C particles (>200 μm) in this study.

DIC was the most important C component (Figure 4, Table 1) of C export in the headwater stream of TSRF. This is consistent with the observation that the mineral soil has little organic C (23.88 g kg^{-1}) [22] and that there are few storm events (Figure S1) [2,8], which tend to be higher in DIC than in DOC. Ground water flow, which dominates base flow, is a continuous C conduit from landscape to stream in Amazonian [2,7,8,29,30] and British forests [23]. In streams, groundwater-derived DIC is significantly diluted by surface runoff and interflow [24] and by PIC transferred to the stream by surface runoff [31,32] and lateral movement of soil water during rain events [8,23]. In addition, TIC and DIC concentrations in stream water may decrease with increases in SWT (Table 2) during the rainy season, due to its microbial transformation to DOC or gaseous C [13,31,32]. Similar observations were reported for small streams in the Amazon [32] and in northern California [31].

DOC was the dominated organic carbon form in the headwater stream (Figure 5, Table 1). TOC and DOC dynamics were similar (Figures 4b, S2, Table 1). The annual variance of DOC

(CV = 58.9%) concentration was more than that of DIC (CV = 28.4%), and correlations between DOC and stream water discharge differ from those for DIC (Figures 2, S2, Table 2) suggesting different sources for DOC and DIC. Litter fall peaks in late March [33,34], releasing dissolved organic matter (DOM) through decomposition [7]. Lower stream discharge in March and April further increased the DOM concentration. At the beginning of the rainy season, organic C in surface soil and litter leachate is transported to the headwater stream by surface runoff, interflow water, and canopy throughfall [26,29,35] due to persistent rain events [20]. In addition, a large share of DOC from soil is "flushed" during the last rainstorms of the dry season and the first storms in the rainy season [36]. Also, stream DOC concentration peaked in June (Figures 4b, S2), and was lowest in September (Figure 4b) when litter had been decomposed and the store of DOC in the soil had leached gradually during the rainy season [36] (Figure S2). This explains the weak relationship between discharge and DOC concentration (Table 2), but disagrees with the strong positive or negative correlations of streams having varying agricultural land-use intensities in their catchments [28]. The VWM concentrations were similar in the rainy season and the dry season, which contrasts other studies showing DOC concentrations higher during the rainy season [8,16,32].

Stream C export increased as stream discharge increased (Figures 2b, 4). This result supports the notion that organic and inorganic C export in watersheds is always dominated by runoff

Table 3. Ratios of total carbon export to different components of the carbon cycle in a tropical seasonal rainforest stream in Xishuangbanna, Southwest China.

Links of carbon cycle	kg C ha^{-1} yr^{-1}	TC Ratio%	Reference
NEE	1660	2.85	[19]
Soil surface efflux	14564	0.32	[43]
Soil respiration	9491	0.50	
Litter respiration	3245	1.45	
Stem respiration	14~47	100.31~332.39	[42]
Litterfall mass	7180~12850	0.37~0.66	[44]
Fine root mass of 0–20 cm depth	6124	0.77	[45]
Living fine root mass of 0–20 cm depth	5418	0.87	
Dead fine root mass of 0–20 cm depth	707	6.66	

The TC ratio (%) indicates the ratio of TC export by stream to the amount of carbon in different components of the carbon cycle.

amount despite sometimes even smaller concentrations in stream water [3,27]. TOC flux in this study (18.4 kg C ha^{-1} yr^{-1}) was much less than that in Amazonian rivers and streams (100 kg ha^{-1} yr^{-1}) [3], reflecting higher rainfall and stream discharge in the Amazon tropical region [3,20]. In addition to stream discharge, C export is influenced by vegetation, soil type, and soil inorganic and organic carbon content [3,8,26,28]. DIC flux was higher in this study than in the Southern Amazon (11.3 kg C ha^{-1} yr^{-1}) [2], and was within the range of fluxes reported for Basse-Terre Island catchments, Lesser Antilles, during periods of low flow (1.7±0.9 to 14.8±9.4 kg C ha^{-1} yr^{-1}) and flood (7.3±4.2 to 75.7±36.9 kg C ha^{-1} yr^{-1}) [8] where runoff were higher than XSBN. TOC flux in XSBN was higher than that in subtropical forest in China (7.2 to 9.3 kg C ha^{-1} yr^{-1}) [37], although the subtropical forest catchment had more soil carbon stock (164 t C ha^{-1}) [38] than our site (87.0 t C ha^{-1}) [39]. DOC export (12.0, and 9.5 kg C ha^{-1} yr^{-1} in 2010 and 2009, respectively) at XSBN was less than that measured in Juruena headwater catchments in Brazil (31.5 kg C ha^{-1} yr^{-1}) [7], tropical volcanic islands in Guadeloupe (16.0±9.0 to 57.0±26.0 kg C ha^{-1} yr^{-1}). Also, primary tropical forest (20.7±1.89 kg C ha^{-1} yr^{-1}), secondary forest (18.9±1.4 kg C ha^{-1} yr^{-1}), pine reforestation (17.9±9.0 kg C ha^{-1} yr^{-1}) and cabbage cultivation (14.8±1.0 kg C ha^{-1} yr^{-1}) in tropical highlands in northern Thailand [40], and a Wisconsin stream in a peat land catchment (25.0 kg ha^{-1} yr^{-1}) [41] had higher C export rates than XSBN. The differences in C dynamics among these regions may reflect the soil type, stream chemistry, vegetation, or hydrology. Therefore, future studies should consider the complex mechanisms that underlie regional differences in C dynamics.

Role of Stream Water Export in the C Balance

The contributions of the fluxes of all C components to the net C loss (as determined by the NEE) were determined (Figures 5, 6). Stream export increased C output from April to August but the NEE indicated that TSRF was C source in this period. From September to March carbon accumulation of TSRF was smaller than indicated by the C sink suggested by the NEE, due to organic C export with stream water (Figure 6a). In comparison with NEE, all stream water C components were small (Figure 6b; Table 3), even for TC. TC export in TSRF in XSBN (Table 3) matched or exceeded C emission due to stem respiration [42]. Compared to C emission due to soil respiration (SR) [43], and C sinks represented by litter fall [44], and fine root biomass production [45], the

contribution of stream export was even smaller (Table 3). So, stream C export by headwater stream is negligible in the overall C balance (NEE, SR, litterfall, and root biomass production) of TSRF at XSBN.

Our study contrasts many others, who have suggested that surface waters are an important export pathway for C in tropical regions [2,6,8] and boreal forests [10,46]. Studies in the southern Amazon state of Mato Grosso showed TC (sum of DIC, DOC and fine-particle carbon) export by stream was 7.34% of NEE (1.5 Mg C ha^{-1} yr^{-1}) [2], which is higher than the ratio of TC: NEE 2.85% (Table 3) in this study. In contrast, Shibata et al. [9] found that DIC and DOC output by stream water (7.6 g C m^{-2} yr^{-1}) accounted for only 2% of NEE in cool-temperate forests of northern Japan, which is less than this study. Reason for relatively small C export by TSRF in the headwater stream at XSBN is that stream discharge was less than that of Amazon tropical regions [2,6,8], although NEE [19] and litter input [34] were similar in Amazon regions and XSBN. Furthermore, DOC export to the ecosystem C balance is small if adsorption to the soil matrix is strong [47,48]. Clays particular oxides have greater potential to adsorb DOC compared with the clay-poor sandy podzolic soils found in Amazonian forest [26]. Hence, fluvial export of C from XSBN's clay rich soils is likely to be lower in the present study area.

Based on the discussions above, the sources of different C components varied, leading to differences in the relative influence of stream discharge and SWT on C concentration and seasonal patterns. The relationship between stream discharge and C concentrations, and the distribution of C fractions differed in their influence on the C budget. A comparison of TC: NEE values in tropical seasonal rainforest at Xishuangbanna and in Amazon tropical forest and boreal forest indicates that stream export represents only a small component of the overall forest C balance in tropical seasonal rainforest.

Supporting Information

Figure S1   Sampling date of stream water 2009–2010 in tropical seasonal rainforest at Xishuangbanna, Southwest China. (a) Rainfall and stream water temperature dynamic during 2009 and 2010. (b) Sampling date and the runoff dynamic during 2009 and 2010.

Figure S2   Dynamics of different carbon components of sampling time during 2009–2010. (a) Dynamics in concentration of TIC, DIC, and PIC. (b) Dynamics in concentration of TOC, DOC, and POC.

Acknowledgments

We thank the staffs and technicians of the Xishuangbanna Station for Tropical Rain Forest Ecosystem who assist with field measurements. The authors greatly appreciate Wen-Jun Liu for the catchment figure drawing, Zhi-Hua Zhou for the sampling assistance, and Zhi-Ling Chen, Wu-Fei Liu, San-Mei A and Li-Fang Ou for laboratory works. Jan Mulder provides helpful comments and English writing during the preparation of the manuscript.

Author Contributions

Contributed to writing the manuscript: L-QS YD X-BD K-JD. Conceived and designed the experiments: W-JZ Y-PZ DAS L-QS. Performed the experiments: W-JZ YD X-BD. Analyzed the data: W-JZ DAS Y-PZ YD. Wrote the paper: W-JZ Y-PZ DAS.

References

1. Cole JJ, Prairie YT, Caraco NF, McDowell WH, Tranvik LJ, et al. (2007) Plumbing the global carbon cycle: integrating inland waters into the terrestrial carbon budget. Ecosystems 10: 172–185.
2. Neu V, Neill C, Krusche AV (2011) Gaseous and fluvial carbon export from an Amazon forest watershed. Biogeochemistry 105: 133–147.
3. Meybeck M (1982) Carbon, nitrogen, and phosphorus transport by world rivers. American Journal of Science 282: 401–450.
4. McDowell WH, Asbury CE (1994) Export of carbon, nitrogen, and major ions from 3 tropical tropical montane watersheds. Limnology and Oceanography 39: 111–125.
5. Richey JE, Brock JT, Naiman RJ, Wissmar RC, Stallard RF (1980) Organic carbon oxidation and transport in the Amazon river. Science 207: 1348–1351.
6. Richey JE, Melack JM, Aufdenkampe AK, Ballester VM, Hess LL (2002) Outgassing from Amazonian rivers and wetlands as a large tropical source of atmospheric CO_2. Nature 416: 617–620.
7. Johnson MS, Lehmann J, Selva EC, Abdo M, Riha S, et al. (2006) Organic carbon fluxes within and streamwater exports from headwater catchments in the southern Amazon. Hydrological Processes 20: 2599–2614.
8. Lloret E, Dessert C, Gaillardet J, Albéric P, Crispi O, et al. (2010) Comparison of dissolved inorganic and organic carbon yields and fluxes in the watersheds of tropical volcanic islands, examples from Guadeloupe (French West Indies). Chemical Geology 230: 65–78.
9. Shibata H, Hiura T, Tanaka Y, Takagi K, Koike T (2005) Carbon cycling and budget in a forested basin of southwestern Hokkaido, northern Japan. Ecological Research 20: 325–331.
10. Benoy G, Cash K, McCauley E, Wrona F (2007) Carbon dynamics in lakes of the boreal forest under a changing climate. Environmental Reviews 15: 175–189.
11. Aldrian E, Chen CTA, Adi S, Prihartanto SN, Sudiana N, et al. (2008) Spatial and seasonal dynamics of riverine carbon fluxes of the Brantas catchment in East Java. Journal of Geophysical Research 113: 10.1029/2007jg000626.
12. Buffam I, Galloway JN, Blum LK, McGlathery KJ (2001) A stormflow/baseflow comparison of dissolved organic matter concentrations and bioavailability in an appalachian stream. Biogeochemistry 53: 269–306.
13. Butturini A, Sabater F (2000) Seasonal variability of dissolved organic carbon in a Mediterranean stream. Biogeochemistry 51: 303–321.
14. Doctor DH, Kendall C, Sebestyen SD, Shanley JB, Ohte N, et al. (2008) Carbon isotope fractionation of dissolved inorganic carbon (DIC) due to outgassing of carbon dioxide from a headwater stream. Hydrological Processes 22: 2410–2423.
15. Goller R, Wilcke W, Fleischbein K, Valarezo C, Zech W (2006) Dissolved nitrogen, phosphorus, and sulfur forms in the ecosystem fluxes of a montane forest in Ecuador. Biogeochemistry 77: 57–89.
16. Raymond PA, Saiers JE (2010) Event controlled DOC export from forested watersheds. Biogeochemistry 100: 197–209.
17. Cao M, Zhang J, Feng Z, Deng J, Deng X (1996) Tree species composition of a seasonal rain forest in Xishuangbanna, Southwest China. Tropical Ecology 37: 183–192.
18. Tan Z, Zhang Y, Yu G, Sha L, Tang J, et al. (2010) Carbon balance of a primary tropical seasonal rain forest. Journal of Geophysical Research 115, D00H26, doi: 10.1029/2009JD012913.
19. Zhang Y, Tan Z, Song Q, Yu G, Sun X (2010) Respiration controls the unexpected seasonal pattern of carbon flux in an Asian tropical rain forest. Atmospheric Environment 44: 3886–3893.
20. Wang X, Zhang YP. (2005) An analysis of the characteristics of rainfall and linear tread in the Menglun area of Xishuangbanna, SW China. Journal of Tropical Meteorology 21: 658–664.
21. Zhang YP, Wang X, Wang YJ, Liu WJ, Liu YH (2003) Comparison research on hydrological effect of the canopy of the tropical seasonal rainforest and rubber forest in Xishuangbanna, Yunnan. Acta Ecologica Sinica 23: 2653–2665.
22. Tang YL, Deng XB, Li YW, Zhang SB (2007) Research on the difference of soil fertility in the different forest types in Xishuangbanna. Journal of Anhui Agricultural Sciences 35: 779–781.
23. Baker A, Cumberland S, Hudson N (2008) Dissolved and total organic and inorganic carbon in some British rivers. Area 40: 117–127.
24. Liu WJ, Liu WY, Lu HJ, Duan WP, Li HM (2011) Runoff generation in small catchments under a native rain forest and a rubber plantation in Xishuangbanna, southwestern China. Water and Environment Journal 25: 138–147.
25. Tan ZH, Zhang YP, Song QH, Liu WJ, Deng XB, et al. (2011) Rubber plantations act as water pumps in tropical China. Geophysical Research Letters 38: L24406.
26. McClain ME, Richey JE, Brandes JA, Pimentel TP (1997) Dissolved organic matter and terrestrial-lotic linkages in the central Amazon basin of Brazil. Global Biogeochemical Cycles 11: 295–311.
27. Rantakari M, Mattsson T, Kortelainen P, Piirainen S, Finér L, et al. (2010) Organic and inorganic carbon concentrations and fluxes from managed and unmanaged boreal first-order catchments. Science of the Total Environment 408: 1649–1658.
28. Wilson HF, Xenopoulos MA (2008) Ecosystem and seasonal control of stream dissolved organic carbon along a gradient of land use. Ecosystems 11: 555–568.
29. Palmer SM, Hope D, Billett MF, Dawson JJC, Bryant C (2001) Sources of organic and inorganic carbon in a headwater stream evidence from carbon isotope studies. Biogeochemistry 52: 321–338.
30. Saunders TJ, McClain ME, Llerena CA (2006) The biogeochemistry of dissolved nitrogen, phosphorus, and organic carbon along terrestrial-aquatic flowpaths of a montane headwater catchment in the Peruvian Amazon. Hydrological Processes 20: 2549–2562.
31. Finlay JC (2003) Controls of streamwater dissolved inorganic carbon dynamics in a forested watershed. Biogeochemistry 62: 231–252.
32. dos Santos Sousa E, Salimon CI, de Oliveira Figueiredo R, Krusche AV (2011) Dissolved carbon in an urban area of a river in the Brazilian Amazon. Biogeochemistry 105: 159–170.
33. Ren YH, Cao M (1999) Comparative study on litterfall dynamics in a seasonal rain forest and a rubber plantation in Xishuangbanna, SW China. Acta Phytoecologica Sinica 23: 418–425.
34. Tang JW, Cao M, Zhang JH, Li MH (2010) Litterfall production, decomposition and nutrient use efficiency varies with tropical forest types in Xishuangbanna, SW China: a 10-year study. Plant and Soil 335: 271–288.
35. Meyer JL, Wallace JB, Eggert SL (1998) Leaf litter as a source of dissolved organic carbon in streams. Ecosystems 1: 240–249.
36. Haaland S, Mulder J (2010) Dissolved organic carbon concentrations in runoff from shallow heathland catchments: effects of frequent excessive leaching in summer and autumn. Biogeochemistry 97: 45–53.
37. Qiao Y, Yin G, Luo Y, Liu Y (2009) Dynamics of total organic carbon (TOC) in hydrological processes and its contributions to soil organic carbon pools of three successional forest ecosystems in southern China. Ecology and Environmental Sciences 18: 2300–2307.
38. Zhou GY, Zhou CY, Liu SG, Tang XL, Ouyang XJ, et al. (2006) Belowground carbon balance and carbon accumulation rate in the successional series of monsoon evergreen broad-leaved forest. Science in China Series D-Earth Sciences 49: 311–321.
39. Lue X-T, Yin J-X, Jepsen MR, Tang J-W (2010) Ecosystem carbon storage and partitioning in a tropical seasonal forest in Southwestern China. Forest Ecology and Management 260: 1798–1803.
40. Möller A, Kaiser K, Guggenberger G (2005) Dissolved organic carbon and nitrogen in precipitation, throughfall, soil solution, and stream water of the tropical highlands in northern Thailand. Journal of Plant Nutrition and Soil Science 168: 649–659.
41. Elder JF, Rybick NB, Carter V, Victoria W (2000) Sources and yields of dissolved carbon in northern Wisconsin stream catchments with differing amounts of peatland. Wetlands 20: 113–125.
42. Yan YP, Sha LQ, Cao M (2008) Stem respiration rates of dominant tree species in a tropical seasonal rain forest in Xishuangbanna, Yunnan, southwest China. Journal of Plant Ecology 32: 23–30.
43. Sha LQ, Zheng Z, Tang JW, Wang YH, Zhang YP, et al. (2005) Soil respiration in tropical seasonal rain forest in Xishuangbanna, SW China. Science in China Series D-Earth Sciences 48: 189–197.
44. Zheng Z, Shanmughavel P, Sha L, Cao M, Warren M (2006) Litter decomposition and nutrient release in a tropical seasonal rain forest of Xishuangbanna, southwest China. Biotropica 38: 342–347.

45. Fang QL, Sha LQ (2005) Study of fine roots biomass and turnover in the rubber plantation of Xishuangbanna. Journal of Central South Forestry University 23: 488–494.

46. Butman D, Raymond PA (2011) Significant efflux of carbon dioxide from streams and rivers in the United States. Nature Geoscience 4: 839–842.

47. Fujii K, Uemura M, Hayakawa C, Funakawa S, Sukartiningsih, etal. (2009) Fluxes of dissolved organic carbon in two tropical forest ecosystems of East Kalimantan, Indonesia. Geoderma 152: 127–136.

48. Fujii K, Hartono A, Funakawa S, Uemura M, Kosaki T (2011) Fluxes of dissolved organic carbon in three tropical secondary forests developed on serpentine and mudstone. Geoderma 163: 119–126.

Biomass Partitioning and Its Relationship with the Environmental Factors at the Alpine Steppe in Northern Tibet

Jianbo Wu[1], Jiangtao Hong[1,2], Xiaodan Wang[1]*, Jian Sun[1,2], Xuyang Lu[1], Jihui Fan[1], Yanjiang Cai[1]

1 The Key Laboratory of Mountain Environment Evolution and Its Regulation, Institute of Mountain Hazard and Environment, CAS, Chengdu, China, 2 University of Chinese Academy of Sciences, Beijing, China

Abstract

Alpine steppe is considered to be the largest grassland type on the Tibetan Plateau. This grassland contributes to the global carbon cycle and is sensitive to climate changes. The allocation of biomass in an ecosystem affects plant growth and the overall functioning of the ecosystem. However, the mechanism by which plant biomass is allocated on the alpine steppe remains unclear. In this study, biomass allocation and its relationship to environmental factors on the alpine grassland were studied by a meta-analysis of 32 field sites across the alpine steppe of the northern Tibetan Plateau. We found that there is less above-ground biomass (M_A) and below-ground biomass (M_B) in the alpine steppe than there is in alpine meadows and temperate grasslands. By contrast, the root-to-shoot ratio ($R:S$) in the alpine steppe is higher than it is in alpine meadows and temperate grasslands. Although temperature maintained the biomass in the alpine steppe, precipitation was found to considerably influence M_A, M_B, and $R:S$, as shown by ordination space partitioning. After standardized major axis (SMA) analysis, we found that allocation of biomass on the alpine steppe is supported by the allometric biomass partitioning hypothesis rather than the isometric allocation hypothesis. Based on these results, we believe that M_A and M_B will decrease as a result of the increased aridity expected to occur in the future, which will reduce the landscape's capacity for carbon storage.

Editor: Fei-Hai Yu, Beijing Forestry University, China

Funding: This study was financially supported by the Strategic leading science and technology projects, CAS (XDB03030505), the National science and technology support project (2011BAC09B03), Program of the IMDE, CAS (SDS-135-1203-01), and the Science Foundation for Young Scientists of IMDE, CAS. The funders had no role in study design, data collection and analysis, decision to publish, or preparation of the manuscript.

Competing Interests: The authors have declared that no competing interests exist.

* E-mail: wxd@imde.ac.cn

Introduction

Biomass allocation was an important character for the process of characterization of plant physiological ecology [1], moreover, it also was the result of the plant long-term adapted to different environmental conditions [2].The Biomass allocation also reflect show photosynthates are allocated between above-ground and below-ground biomass [3]. Biomass allocation above-ground and below-ground affects plant growth as well as the overall function of the ecosystem and biogeochemical cycles [4,5]. Therefore, the mechanism by which plants respond to variations in the availability of resources in their environment is a major question in plant ecology [6]. Two important hypotheses regarding biomass allocation of plants have been proposed: (i) optimal partitioning and (ii) isometric allocation [2,7,8]. The optimal partitioning hypothesis suggests that plants respond to variations in the environment by partitioning biomass among various plant organs to maximize the plants' growth rate [9,10]. For example, plants in arid regions are rooted deeper than those in humid environments [11,12]. On the contrary, the isometric allocation hypothesis predicts the net primary productivity of the roots vs the net primary productivity of the shoots (BNPP:ANPP) isometrically without considering the differences in plant species or community types [13–15]. Thus far, biomass allocation has been widely

examined: investigations have focused on individual organisms as well as whole ecosystems. However, no conclusion about biomass allocation has yet been presented.

Optimal partitioning theory might explain the effect of environmental factors on the allocation of plants' photosynthetic products, but this theory does not consider the size of the individual plants [8,16]. The allometric biomass partitioning theory, on the other hand, may resolve biomass allocation patterns in terms of plant size by using standardized major axis (SMA) regression [8,17]. However, this theory does not provide quantitative descriptions about how environmental factors affect biomass allocation. It also cannot explain the mechanism behind how photosynthates are allocated to different organs [18]. Furthermore, it is still hotly debated whether a uniform biomass allocation pattern is applicable to different ecosystems [19].

The alpine steppe is the largest grassland type in the Tibetan Plateau, which contributes significantly to the global carbon cycle [1]. In the alpine grassland ecosystem, few soil nutrients, aridity, and low temperatures limit plant growth [20,21]. According to the optimal partitioning hypothesis, environmental factors likely affect how plant biomass is allocated. At the individual plant level, fewer soil nutrients (particularly nitrogen and water) results in an increase in root biomass. On the contrary, root biomass decreases and shoot biomass increases as soil nutrients increase. This

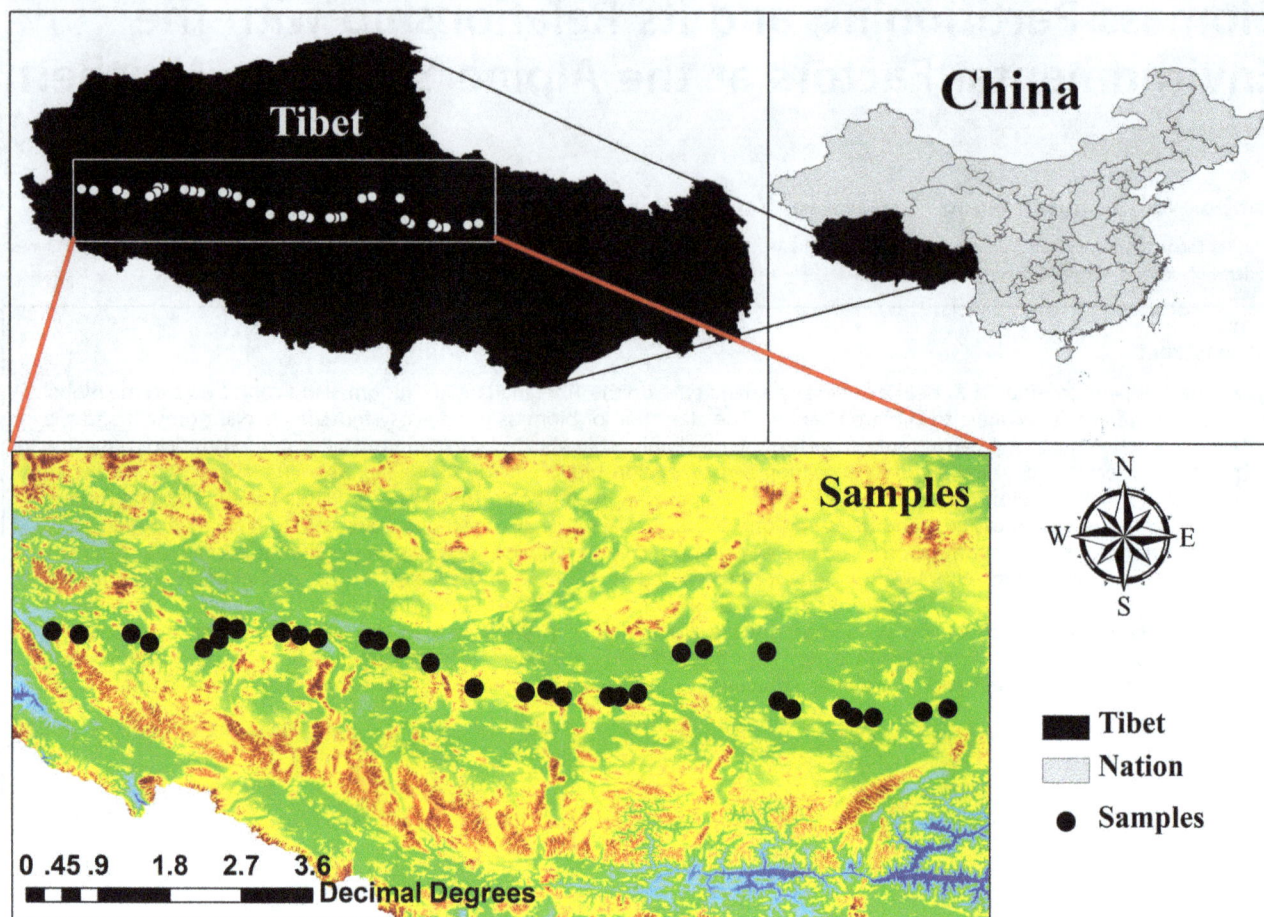

Figure 1. Spatial distribution of the sampling sites in *S. Purpurea* alpine steppe in northern Tibet.

partition model is appropriate for different types of vegetation and life forms of plants [22–26]. Studies have shown that plants allocate more biomass to their roots when water and nutrients in grassland ecosystems are limited [27,28]. Moreover, studies have also suggested that plants allocate photosynthates to root in low-temperature environments, which may increase the rate of nutrient absorption and help the plants adapt to environmental conditions [29–31]. However, Yang et al. (2009a) reported that on the Tibetan alpine grasslands, the relationship between roots and shoots supports the isometric allocation hypothesis [32]. They also

found that this isometric relationship is independent of soil nitrogen and moisture [32]. These results indicate that the mechanism of biomass allocation in the alpine steppe is still misunderstood and unverified in alpine and arid environments. Therefore, this subject requires further investigation. In the present study, we investigated (i) the mechanism behind allocating root and shoot biomass in the Tibetan alpine grassland and (ii) the main factors that affect biomass allocation in the alpine steppe of northern Tibet.

Table 1. Site description of *S. purpurea* alpine steppe.

Site	County	Dominant species	Mean annual precipitation (MAP,mm)	Mean annual temperature (MAT,°C)
S1	Nakchu	*S. purpurea Kobre siahumilis*	428.1	−1.5
S2–S7	Baingoin	*S. purpurea Carex moorcroftii*	321.7	−0.8
S8	Xainza	*S. purpurea C.moorcroftii*	304.5	−0.4
S9–S17	Nyima	*S. purpurea C.m oorcroftii*	200	−0.4
S18–S24	Gêrzê	*S. purpurea*	170.1	0.10
S25–S30	Gêgyai	*S. purpurea*	120	0.45
S31–S32	Gar	*S. purpurea*	72.1	0.7

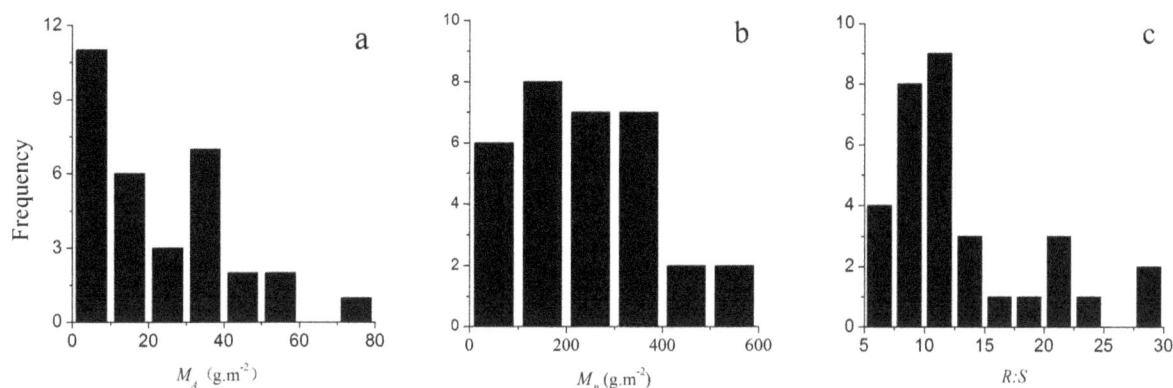

Figure 2. Frequency distributions of (a) above-ground biomass (M_A), (b) below-ground biomass (M_B), and (c) root-to-shoot ratio ($R{:}S$) in *S. purpurea* alpine steppe.

Materials and Methods

Collecting Biomass and Soil Samples

In August 2012, 32 sites were selected on *Stipa purpurea* alpine steppe from Nagqu County to Gar County in northern Tibet. Sampling sites were established at intervals of 30 km (Fig. 1, Table 1). In each site, no specific permits were required for collecting samples and the field studies did not involve endangered or protected species. We selected flat sites with well-protected vegetation. We harvested the aboveground biomass (M_A) and the belowground biomass (M_B) from three blocks of 0.5 m×0.5 m in each site. We collected M_B from soil depths of 0 cm to 15 cm, where most of belowground biomass is located [33,34]. The root samples obtained from the blocks were immediately placed in a cloth bag and then soaked in water to remove the residual soil using a 0.5 mm sieve. Biomass was oven-dried at 65°C until a constant weight was reached, and then it was weighed to the nearest 0.01 g.

Soil samples were collected from two different depths (0–15 cm and 15–30 cm), air-dried, and sieved (2 mm mesh). The fine roots were extracted by hand picking for physical and chemical analyses. The total nitrogen content (TN; TN1:0–15 cm, TN2:15–30 cm) of the soil was determined using the micro-Kjeldahl digestion method. The available nitrogen content (AN; AN1:0–15 cm, AN2:15–30 cm) of the soil was determined using the alkaline hydrolysis diffusion method. All of the element concentrations were expressed as $mg{\cdot}g^{-1}$ on a dry weight basis.

Data Analysis

M_A in grasslands can be considered as annual aboveground net primary productivity (ANPP). Blow-ground net primary productivity (BNPP) was calculated using Gill's method:

$$BNPP = M_B \times \left(\frac{\text{live}M_B}{M_B}\right) \times \text{turnover} \qquad (1)$$

where (live M_B/M_B) = 0.6 and turnover = $0.0009(g{\cdot}m^{-2}) \times M_A$ +0.25 [35,36]. In the present study, the value for (live M_B/M_B) was 0.79, which was measured by Zhou (2001) in the Qinghai region [37]. The relationship between log ANPP and log BNPP was constructed using Model II regression [14,15]. The slope (α) and y-intercept (log b) of the allocation function were determined by standardized major axis (SMA) tests [38]. The heterogeneity between slopes was determined by performing a permutation test and was rejected if P>0.05 [15]. We analyzed the correlations between environmental factors and the measured M_A, M_B, and root-to-shoot ratios ($R{:}S$) using the Pearson correlation. We also examined relationships between M_A, M_B, $R{:}S$, and environmental factors using regression and ordination space partitioning to find the main environmental factors that affected M_A, M_B, and $R{:}S$. Analyses were performed using SPSS software version 16.0 (IBM; Armonk, NY).

Results

Variations in the Chemical Properties of the Soil as well as MA, MB, and R:S

Small variations in the chemical properties of the soil along the sampled transect were found. There also was not significance in available nitrogen and total nitrogen between the two soil layers (Table 2). We found large variations in M_A, M_B, and $R{:}S$ along the sampled transects (Fig. 2). M_A ranged from 2.32 $g{\cdot}m^{-2}$ to 73.6 $g{\cdot}m^{-2}$, while M_B ranged from 22.40 $g{\cdot}m^{-2}$ to 587.32 $g{\cdot}m^{-2}$. $R{:}S$ ranged from 6.19 to 29.15 (Table 3). The median values of M_A, M_B, and $R{:}S$ in *S. purpurea* alpine steppe were 17.16 $g{\cdot}m^{-2}$, 233 $g{\cdot}m^{-2}$, and 11.83, respectively (Table 3).

Biomass Allocation for *S.purpurea* Alpine Steppe

The slope (α) of the plotted relationship between log ANPP and log BNPP of *S. purpurea* alpine grasslands was 0.87 with 95% confidence intervals of 0.75 and 1.01 (Fig. 3). The slope (α) was significantly different from the slope obtained from SMA analysis when the isometric hypothesis was used.

Table 2. Chemical properties of soils in *S. purpurea* alpine steppe.

	Min	Max	Mean	Std. Error	Std. Deviation
AN1 mg·g⁻¹	0.013	0.110	0.057a	0.004	0.025
AN2 mg·g⁻¹	0.008	0.095	0.051a	0.004	0.023
TN1 mg·g⁻¹	0.386	1.630	1.008a	0.061	0.342
TN2 mg·g⁻¹	0.292	1.921	0.980a	0.064	0.362

Table 3. Descriptive statistics of above-ground biomass (M_A), below-ground biomass (M_B), and root-to-shoot (R:S) ratio in *S. purpurea* alpine steppe.

	M_A (g·m^{-2})			M_B (g·m^{-2})			R:S ratio		
	Min	**Max**	**Median**	**Min**	**Max**	**Median**	**Min**	**Max**	**Median**
Present study	2.32	73.6	17.16	22.4	587.32	233	6.19	29.15	11.83
Yang et al. (2009)	9.8	267.4	42.8	44.6	1934.8	206	0.8	13	5.2

Effects of Soil Nitrogen and Environmental Factors on Biomass and R:S

Using the Pearson correlation analysis, we found that M_A and M_B exhibited a significantly positive correlation with available nitrogen in the soil. However, M_A and M_B did not exhibit a significant correlation with total nitrogen (Table 4). The R:S ratio also did not exhibit a significant correlation with soil nitrogen (total or available). M_A, M_B, and R:S did correlate with the MAP of the sampling sites, while these correlations differed from the ones found with MAT (Table 3). In this study, we found that the regression analysis showed the same results as the Pearson correlation analysis (Fig. 4). Using the ordination space partitioning method, we found that MAP was the main factor that affected M_A, M_B, and R:S (Fig. 5).

Discussion

M_A, M_B, and R:S in the Alpine Steppe

In the present study, amounts of M_A and M_B in the alpine steppe (mean = 23.20 g·m^{-2}) were found to be lower than those in the alpine meadows [32] and in the temperate grasslands of China [39]. By contrast, R:S in the alpine steppe was found to be higher than it is in China's alpine meadows [32] and temperate grasslands [39] as well as in temperate grasslands of other regions [1]. These results show that precipitation and temperature affect plant growth and biomass allocation [1,40]. Slower root turnover in colder environments might also results in higher R:S ratios [41–43]. M_A, M_B, and R:S values found in the present study are not consistent with results reported by Yang et al. (2009a), who performed a field investigation from 2001 to 2004 [32]. R:S values have the potential to vary greatly as a result of climate change and anthropogenic activities [44–48].

Mechanism of Biomass Allocation in the Alpine Steppe

Based on the results of our SMA analysis, we found that biomass allocation on the alpine steppe does not fit the isometric hypothesis. By contrast, Yang et al. (2009a) previously reported that biomass allocation on the alpine steppe is supported by the isometric allocation hypothesis [32]. In the harsh alpine ecosystem, scarce precipitation and low temperatures allow plants to allocate more biomass to the roots, which helps plants survive [29–31]. Moreover, roots have also been found to store carbohydrates in alpine grasslands [49,50]. Therefore, biomass allocation in the alpine steppe may reflect the allometric biomass partitioning hypothesis rather than the isometric allocation hypothesis.

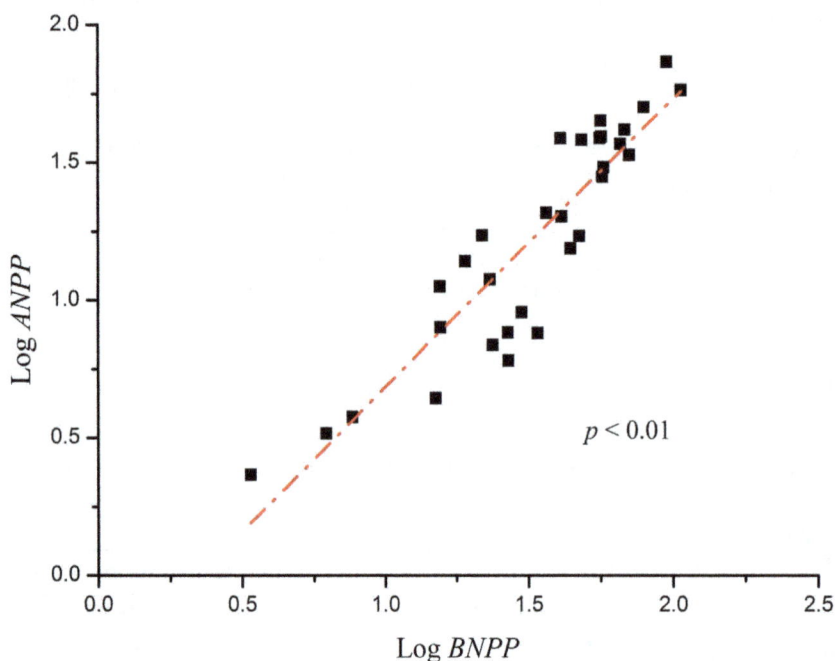

Figure 3. Relationships between above-ground net primary production (ANPP) and below-ground net primary production (BNPP) in alpine steppe by SMA analysis.

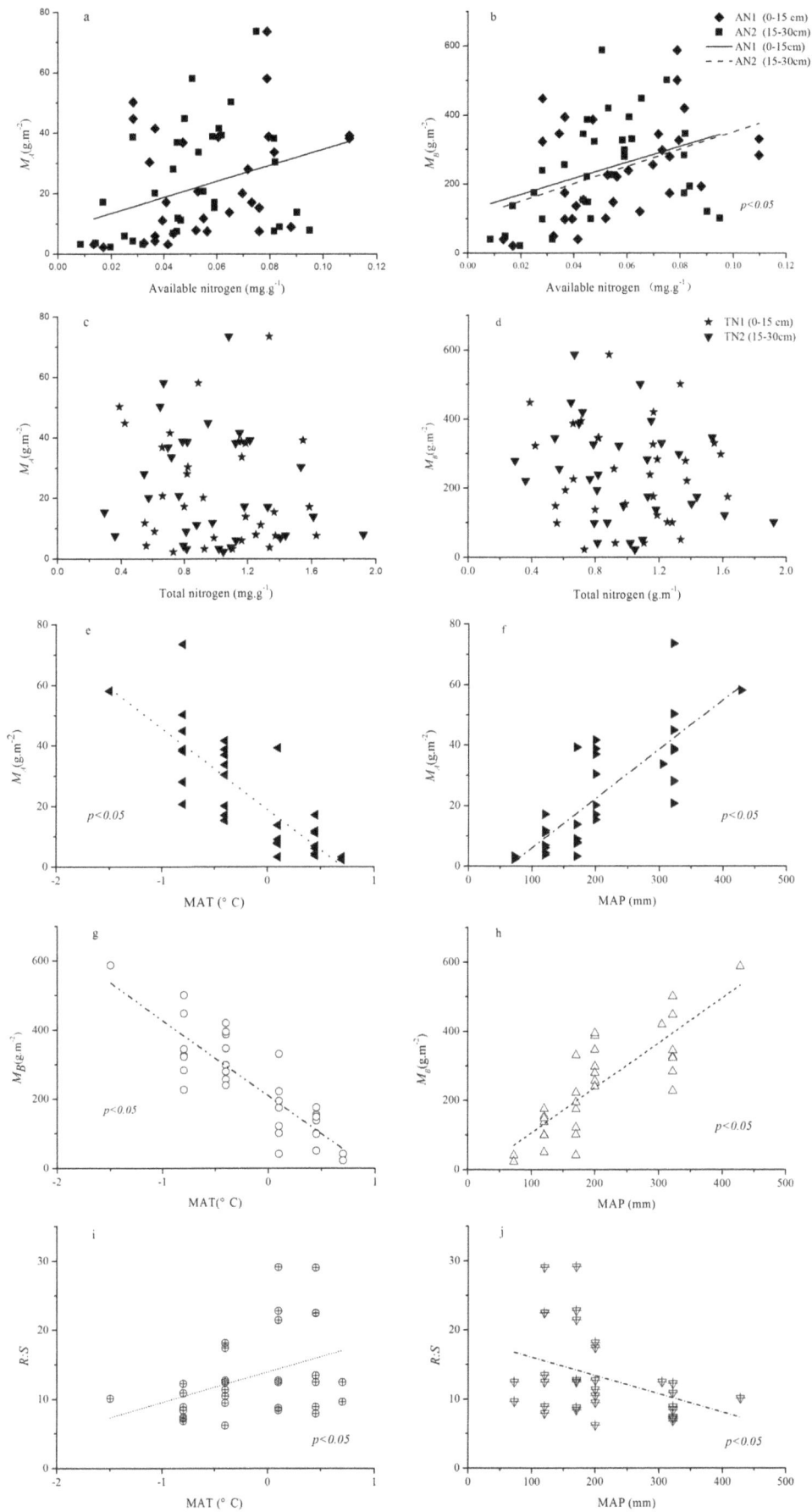

Figure 4. Relationships between biomass allocation (M_A, M_B, and R:S) and environmental factors in alpine steppe. Regressions are shown: (a) M_A versus available nitrogen, (b) M_B versus available nitrogen,(c) M_A versus total nitrogen, (d) M_B versus total nitrogen,(e) M_A versus MAT,(f) M_A versus MAP,(g) M_B versus MAT, (h) M_B versus MAP, (i) R:S ratio versus MAT, and (j) R:S ratio versus MAP.

Relationships between Environmental Factors and MA, MB, and R:S

Precipitation and temperature are considered to be the limiting factors for the growth and distribution of vegetation over the long term [51,52]. In the present study, M_A, M_B, and R:S were mainly affected by the environmental factor of precipitation (MAP), as revealed by ordination space partitioning analysis. These results are consistent with those of other reports about the alpine steppe [46,53–56]. The low temperature in the growing season did not limit the growth of alpine plants because these plants shave evolved to survive in the cold alpine climate [57]. The amounts of aboveground and belowground biomass are higher in sites with higher humidity, but the MAT is also relatively low on the alpine steppe. Precipitation is an essential factor that controls the functions of ecosystems in terrestrial biomes, particularly in arid and semiarid ecosystems [58]. Therefore, precipitation is the main factor that influences amounts of biomass in the alpine steppe.

Moreover, in the present study, we found that amounts of M_A and M_B on the alpine steppe were affected by the available

Table 4. Pearson's correlation between M_A, M_B, and R:S with the environmental factors.

	AN1(0 to 15 cm)	AN2(15 to 30 cm)	TN1(0 to 15 cm)	TN2(15 to 30 cm)	MAT(°C)	MAP(mm)
M_A	0.351*	0.317	−0.126	−0.165	−0.809**	0.791**
M_B	0.429*	0.372*	−0.088	−0.285	−0.853**	0.817**
R:S	−0.047	−0.083	0.207	−0.082	0.392*	−0.378*

**Correlation is significant at the 0.01 level (two-tailed).
*Correlation is significant at the 0.05 level.

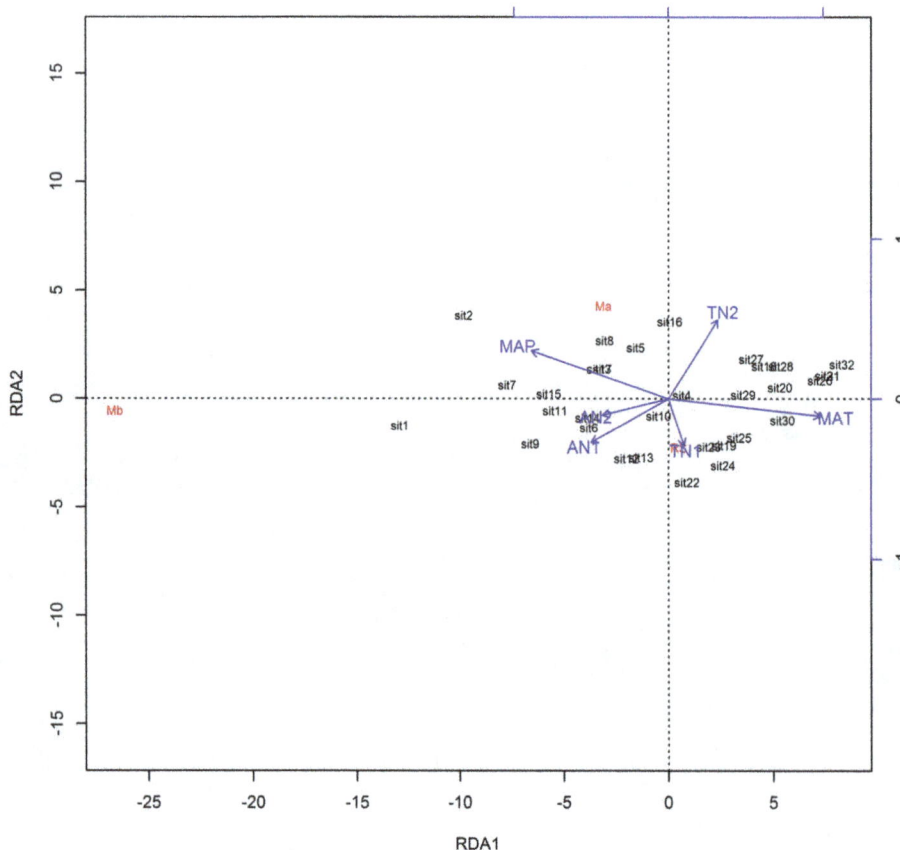

Figure 5. Analysis of the relationship of above-ground biomass (M_A), below-ground biomass (M_B), and root-to-shoot ratio (R:S) with the environmental factors by ordination space partitioning method.

nitrogen content in the soil but not by the total nitrogen content of the soil. These results are inconsistent with those from previous studies, which have showed that M_A and M_B are positively related to total nitrogen content [32,59]. Because available nitrogen can be used to approximate the relative supply of nutrients, nitrogen may be another factor that controls ecosystem processes in regions with abundant water resources [60].

Conclusion

As the climate changes, the degree of aridity has been consistently increasing in northern Tibet [61]. Changes in biomass allocation on the alpine steppe are likely to affect the carbon cycle and the general functioning of the alpine ecosystem. In the present study, we found that the $R{:}S$ ratio in the alpine steppe was higher than that of other grassland systems. The amounts of aboveground and belowground biomass as well as the $R{:}S$ ratio were primarily affected by precipitation. The observed biomass allocation was found to follow the allometric biomass partitioning theory rather than the isometric allocation hypothesis. These results suggest that the landscape's capacity to store carbon will potentially decrease as the degree of aridity in northern Tibet increases.

Acknowledgments

This study was financially supported by the Strategic leading science and technology projects, CAS (XDB03030505), the National science and technology support project (2011BAC09B03), Program of the IMDE, CAS (SDS-135-1203-01), and the Science Foundation for Young Scientists of IMDE, CAS.

Author Contributions

Conceived and designed the experiments: JBW XDW. Performed the experiments: JBW JTH XYL JHF YJC. Analyzed the data: JBW JS. Contributed reagents/materials/analysis tools: JBW JS. Wrote the paper: JBW XDW. Figure: JS.

References

1. Mokany K, Raison RJ, Prokushkin AS (2006) Critical analysis of root: shoot ratios in terrestrial biomes. Global Change Biol 11: 1–13.
2. Bazzaz FA (1997) Allocation of resources in plants: state of the sciences and critical questions. In: Bazzaz FA, Grace J, editors. Plant Resource Allocation. San Diego: Academic Press. 1–37.
3. Shipley B, Meziane D(2002)The balanced-growth hypothesis and the allometry of leaf and root biomass allocation. Funct Ecol 16: 326–331.
4. Kuzyakov Y, Domanski G (2000) Carbon input by plants into the soil. J Plant Nutr Soil Sci 163: 421–431.
5. Litton CM, Raich JW, Ryan MG (2007) Carbon allocation in forest ecosystems. Global Change Biol 13: 2089–2109.
6. Lacointe A (2000) Carbon allocation among tree organs: a review of basic processes and representation in functional-structural tree models. Ann Forest Sci 57: 521–533.
7. Müller I, Schmid B, Weiner J (2000) The effect of nutrient availability on biomass allocation patterns in 27 species of herbaceous plants. Perspect Plant Ecol 3: 115–127.
8. McCarthy MC, Enquist BJ (2007) Consistency between an allometric approach and optimal partitioning theory in global patterns of plant biomass allocation. Funct Ecol 21: 713–720.
9. Bloom AJ, Chapin FS, Mooney HA (1985) Resource limitation in plants an economic analogy. Ann Rev Ecol Syst 16: 363–392.
10. Chapin FS III, Bloom AJ, Field B, Waring RH (1987) Plant responses to multiple environmental factors. BioScience 37: 49–57.
11. Scheck HJ, Jackson RB (2005) Mapping the global distribution of deep roots in relation to climate and soil characteristics. Geoderma 126(1–2): 129–140.
12. Mony C, Koschnick TJ, Haller WT, Muller S (2007) Competition between two invasive Hydrocharitaceae (Hydrilla verticillata (L. f.) (Royle) and Egeria densa (Planch)) as influenced by sediment fertility and season. Aquat Bot 86: 236–242.
13. Enquist BJ, Niklas KJ (2002) Global allocation rules for patterns of biomass partitioning in seed plants. Science 295: 1517–1520.
14. Niklas KJ (2005) Modelling below- and aboveground biomass for non-woody and woody plants. Ann Bot 95: 315–321.
15. Cheng DL, Niklas KJ (2007) Above- and belowground biomass relationships across 1534 forested communities. Ann Bot 99: 95–102.
16. Marcelis LFM, Heuvelink E (2007) Concepts of modelling carbon allocation among plant organs. In: Vos J, Marcelis LFM, Visser PHBd, Struik PC, Evers JB, editors. Functional structural plant modelling in crop production. Dordrecht: Springer. 103–111.
17. Solow AR (2005) Power laws without complexity. Ecol Lett 8: 361–363.
18. Génard M, Dauzat J, Franck N, Lescourret F, Moitrier N, et al.(2008) Carbon allocation in fruit trees: from theory to modelling. Trees 22: 269–282.
19. Han WX, Fang JY(2003) Allometry and its application in ecological scaling. Acta Sci Natur Univ Pekin 39: 583–593. (in Chinese with English abstract).
20. Zhou XM (2001) Kobresia Meadow in China. Beijing: Science Press. 136–139 p.
21. Gugerli F, Bauert MR (2001) Growth and reproduction of Polygonum viviparum show weak responses to experimentally increased temperature at a Swiss alpine site. Bot Helv 111: 169–180.
22. Warembourg FR, Estelrich HD (2001) Plant phenology and soil fertility effects on below-ground carbon allocation for an annual (Bromus madritensis) and a perennial (Bromus erectus) grass species. Soil Biol Biochem 33(10): 1291–1303.
23. Cronin G, Lodge DM. (2003) Effects of light and nutrient availability on the growth, allocation, carbon/nitrogen balance, phenolic chemistry, and resistance to herbivory of two freshwater macrophytes. Oecologia 137: 32–41.

24. Glynn C, Herms DA, Egawa M, Hansen R, Mattson WJ (2003) Effects of nutrient availability on biomass allocation as well as constitutive and rapid induced herbivore resistance in poplar. Oikos 101: 385–397.
25. Vanninen P, Makela A (2005) Carbon budget for Scots pine trees: effects of size, competition and site fertility on growth allocation and production. Tree Physiol 25: 17–30.
26. Grechi I, Vivin P, Hilbert G, Milin S, Robert T, et al. (2007) Effect of light and nitrogen supply on internal C:N balance and control of root-to-shoot biomass allocation in grapevine. Environ Exp Bot 59: 139–149.
27. Dukes JS, Chiariello NR, Cleland EE, Moore LA, Shaw MR, et al. (2005) Responses of grassland production to single and multiple global environmental changes. PLoS Biol 3: 1829–1837.
28. Gao YZ, Chen Q, Lin S, Giese M, Brueck H (2011) Resource manipulation effects on net primary production, biomass allocation and rain-use efficiency of two semiarid grassland sites in Inner Mongolia, China. Oecologia 65: 855–864.
29. Körner CH, Renhardt U (1987) Dry matter partitioning and root length/leaf area ratios in herbaceous perennial plants with diverse altitudinal distribution. Oecologia, 74: 411–418.
30. Wang CT, Cao GM, Wang QL, Jing ZC, Ding LM, et al. (2008) Changes in plant biomass and species composition of alpine Kobresia meadows along altitudinal gradient on the Qinghai-Tibetan Plateau. Sci China Life Sci 51: 86–94.
31. Fan JW, Wang K, Harris W, Zhong HP, Hu ZM, et al. (2009) Allocation of vegetation biomass across a climate-related gradient in the grassland of Inner Mongolia. J Arid Environ 73: 521–528.
32. Yang YH, Fang JY, Ji C, W Han (2009a) Above-and belowground biomass allocation in Tibetan grasslands. J Veg Sci 20: 177–184.
33. Yan Y, Zhang JG, Zhang JH, Fan JR, Li HX (2005) The belowground biomass in alpine grassland in Nakchu Prefecture of Tibet. Acta Ecol Sinica 11: 2818–2823. (in Chinese with English abstract).
34. Li X, Zhang X, Wu J, Shen Z, Zhang Y, et al. (2011) Root biomass distribution in alpine ecosystems of the northern Tibetan Plateau. Environ Earth Sci 64(7): 1911–1919.
35. Gill R A, Kelly RH, Parton WJ, Day KA, Jackson RB, et al. (2002) Using simple environmental variables to estimate below2ground productivity in grasslands. Global Ecol Biogeogr 11: 79–86.
36. Bradford JB, Lauenroth WK, Burke I C (2005) The impact of cropping on primary production in the US Great Plains. Ecol 86: 1863–1872.
37. Zhou XM (2001) Chinese Kobresia Meadow. Beijing: Science Press. 158 p.
38. Falster DS, Warton DI, Wright IJ (2003) (S)MATR: standardized major axis tests and routines. Version 1.0. http://www.bio. mq.edu.au/ecology/SMATR.
39. Ma W, Yang Y, He J, Zeng H, Fang J (2008) Above-and belowground biomass in relation to environmental factors in temperate grasslands, Inner Mongolia. Sci China Life Sci 51(3): 263–270.
40. Hui DF, Jackson RB (2005) Geographical and inter-annual variability in biomass partitioning in grassland ecosystems: a synthesis of field data. New Phytol 169: 85–93.
41. Gill RA, Jackson RB (2001) Global patterns of root turnover for terrestrial ecosystems. New Phytol 147: 13–31.
42. Gao YZ, Giese M, Lin S, Sattelmacher B, Zhao Y, et al. (2008) Belowground net primary productivity and biomass allocation of a grassland in Inner Mongolia is affected by grazing intensity. Plant Soil 307: 41–50.
43. Giese M, Gao YG, Zhao Y, Pan QM, Lin S, et al. (2009) Effects of grazing and rainfall variability on root and shoot decomposition in a semi-arid grassland. Appl Soil Ecol 41: 8–18.

44. Wang GX, Hu HC, Wang YB, Chen L (2007) Response of alpine cold ecosystem biomass to climate changes in permafrost regions of the Tibetan Plateau. J Glaciol Geocryol 29: 671–679. (in Chinese with English abstract).

45. Yu H, Luedeling E, Xu J (2010) Winter and spring warming result in delayed spring phenology on the Tibetan Plateau. Proc Natl Acad Sci U S A 107: 22151–22156.

46. Shi FS, Chen H, Wu Y, Wu N (2010) Effects of livestock exclusion on vegetation and soil properties under two topographic habitats in an alpine meadow on the Eastern Qinghai-Tibetan Plateau. Polish J Ecol 58: 125–133.

47. Li W, Huang HZ, Zhang ZN, Wu GL (2011) Effects of grazing on the soil properties and C and N storage in relation to biomass allocation in an alpine meadow. J Soil Sci Plant Nutr 11: 27–39.

48. Wu JZ, Zhang XZ, Shen ZX, Shi PL, Yu CQ, et al. (2012) Species richness and diversity of alpine grasslands on the Northern Tibetan Plateau: effects of grazing exclusion and growing season precipitation. J Resour Ecol 3: 236–242.

49. Wang JS, Zhang XZ, Zhao YP, Shen ZX, Shi PL, et al. (2008) Spatio- temporal pattern of climate changes in Northern Tibet's Qiangtang Plateau. Resour Sci 12: 1852–1859. (in Chinese with English abstract).

50. Wang L, Niu KC, Yang YH, Zhou P (2010) Patterns of above- and belowground biomass allocation in China's grasslands: evidence from individual-level observations. Sci China Life Sci 53: 851–857.

51. Woodward FI, Williams BG (1987) Climate and plant distribution at global and local scales. Plant Ecol 69: 189–197.

52. Stephenson NL (1990) Climatic control of vegetation distribution: the role of the water balance. Am Nat 5: 649–670.

53. O'Connor TG, Haines LM, Snyman HA (2001) Influence of precipitation and species composition on phytomass of a semiarid African grassland. J Ecol 89: 850–860.

54. Huxman TE, Smith MD, Fay PA, Knapp AK, Shaw MR, et al. (2004) Convergence across biomes to a common rain-use efficiency. Nature 429: 651–654.

55. Yang YH, Fang JY, Pan YD, Ji CJ (2009b) Above-ground biomass in Tibetan grasslands. J Arid Environ 73(1): 91–95.

56. Zhang BC, Cao JJ, Bai YF, Zhou XH, Ning ZG, et al. (2013) Effects of rainfall amount and frequency on vegetation growth in a Tibetan alpine meadow. Clim Chang 118: 197–212.

57. Körner C (2003) Alpine Plant Life: Functional Plant Ecology of High Mountain Ecosystems. Berlin:Springer. 101–111 p.

58. Reynolds JF, Stafford SDM (2002) Do humans cause deserts? In Reynolds JF, Stafford SDM, editors. Berlin: Dahlem University Press. 1–21.

59. Sun J, Cheng GW, Li WP (2013) Meta-analysis of relationships between environmental factors and aboveground biomass in the alpine grassland on the Tibetan Plateau. Biogeosciences 10: 1707–1715.

60. Burke IC, Lauenroth WK, Parton WJ (1997) Regional and temporal variation in net primary production and nitrogen mineralization in grasslands. Ecol 78(5): 1330–1340.

61. Liu X, Chen B (2000) Climatic warming in the Tibetan Plateau during recent decades. Int J Climatol 20: 1729–1742.

10

Comparisons of Three Methods for Organic and Inorganic Carbon in Calcareous Soils of Northwestern China

Xiujun Wang[1,2*], **Jiaping Wang**[1,3], **Juan Zhang**[1,3]

1 State Key Laboratory of Desert and Oasis Ecology, Xinjiang Institute of Ecology and Geography, Chinese Academy of Sciences, Urumqi, Xinjiang, China, **2** Earth System Science Interdisciplinary Center, University of Maryland, College Park, Maryland, United States of America, **3** Graduate University of Chinese Academy of Sciences, Beijing, China

Abstract

With increasing interest in the carbon cycle on arid land, there is an urgent need to quantify both soil organic carbon (SOC) and inorganic carbon (SIC) thus to assess various methods. Here, we present a study employing three methods for determinations of SOC and SIC in the Yanqi Basin of northwest China. We use an elemental analyzer for both SOC and SIC, the Walkley-Black method for SOC, a modified pressure calcimeter method for SIC, and a simple loss-on-ignition (LOI) procedure for determinations of SOC and SIC. Our analyses show that all three approaches produce consistently low values for SOC (1–14 g kg^{-1}) and high values for SIC (8–53 g kg^{-1}). The Walkley-Black method provides an accurate estimate of SOC with 100% recovery for most soil samples. The pressure calcimeter method is as accurate as the elemental analysis for measuring SIC. In addition, SOC and SIC can be accurately estimated using a two-step LOI approach, i.e., (1) combustion at 375°C for 17 hours to estimate SOC, and (2) subsequent combustion at 800°C for 12 hours to estimate SIC. There are strong linear relationships for both SOC and SIC between the elemental analysis and LOI method, which demonstrates the capability of the two-step LOI technique for estimating SOC and SIC in this arid region.

Editor: David L. Kirchman, University of Delaware, United States of America

Funding: This study is financially supported by the Program of 100 Talented Young Scientists with the Chinese Academy of Sciences (O972021001)(http://www.cas.cn/ggzy/rcpy/brjh/). The funders had no role in study design, data collection and analysis, decision to publish, or preparation of the manuscript.

Competing Interests: The authors have declared that no competing interests exist.

* E-mail: wwang@essic.umd.edu

Introduction

Soil carbon storage, as the third largest carbon pool in the Earth System, plays an important role in the global carbon cycle and climate change [1]. The majority of carbon, in most soils, is held as soil organic carbon (SOC) whereas in soils of the arid and semiarid regions, the most common form is inorganic carbon, primarily carbonate [2]. Quantifying both SOC and soil inorganic carbon (SIC) is essential to our understanding of the carbon cycle at regional to global scales.

Soil organic carbon is commonly measured by dry combustion with automated analyzers, or a wet chemical oxidation method, i.e., the Walkley and Black method [3]. The automated technique is simple and accurate, but the cost is high and may not be feasible. In addition, this approach often involves pretreatment with acid to remove carbonate for calcareous soils, which may erode the instrument and destroy organic matter in soil samples [4]. On the other hand, the Walkley and Black method has been used as one of the standard methods to determine SOC [5], particularly in China [6]. This technique is less expensive than dry combustion. However, this procedure may lead to variable recovery of SOC [7,8], and hazard Cr due to the use of dichromate.

Soil inorganic carbon can be accurately measured by several methods [9]. Commonly, SIC is determined by measuring CO_2 production after adding HCl acid into soil [10–13]. Sherrod et al. [12] demonstrated that a modified pressure calcimeter method, with addition of $FeCl_2$ to minimize oxidation of organic matter, provided reliable measurement of SIC. Alternatively, dry combustion with automated analyzers can also be utilized to determine SIC. A direct measurement involves dry combustion of soils that are pre-combusted to remove organic matter in an O_2 stream [14]. An indirect measurement is to determine total carbon (without pretreatment with acid to remove carbonate) and SOC separately, then to calculate the difference. Although the automated approaches would provide accurate estimates for SIC, they might not be used widely because of the high costs associated with purchase and maintenance of automated analyzers.

Apart from these traditional methods, loss-on-ignition (LOI) provides an alternative approach, which involves heating soil samples at high temperature to combust soil organic matter (SOM) or carbonate and measuring weight losses [15,16]. The LOI techniques are simple, less expensive and less labor intensive relative to the automated techniques and chemical methods. Thus, the LOI approaches have been used widely to estimate SOM or SOC in agricultural and forest soils and sediments [16–21], and to determine carbonate content in sediments [13,15,22,23]. However, this technique is rarely applied to low fertility soils for SOM or SOC measurement, or used to measure SIC despite some studies showing that the LOI method can give accurate estimates of carbonate for sediments [13,24]. Moreover, there is little utilization of the LOI methods for estimating SOC and SIC in Chinese soils.

The arid and semiarid lands cover approximately 35% of the Earth's land surface, and may play an important role in the global carbon cycle and climate mitigation [1,25,26]. With recent interest in studying the carbon cycle on arid land [27], there is a need to quantify both SOC and SIC of arid soils thus to assess various methods. Here, we select three techniques to determine SOC and SIC contents in the calcareous soils at the Yanqi Basin of northwest China. We employ an elemental analyzer for both SOC and SIC measurements, a wet oxidation for SOC, a modified pressure calcimeter method for SIC, and a simple LOI procedure for determinations of SOC and SIC. The objective of this study is to examine the relationships between these methods, and to test the hypothesis that the LOI technique has capacity of accurately determining SOC and SIC in arid soils.

Materials and Methods

Site Description

There were no specific permits required for the described field studies. We confirm that the location was not privately-owned or protected in any way, and the field studies did not involve endangered or protected species. Our studying area is along the Kaidu River in the Yanqi Basin (Figure 1), which is located on the southeastern flank of the Tianshan Mountain. Brown desert soil is the main soil type, which was developed from limestone parent material and classified as a Haplic Calcisol [28]. The sampling area (approximately 2,200 km) spans both sides of the Kaidu River, which has various land uses, including desert land, shrub land, cropland and grassland. The area has an annual precipitation of less than100 mm and annual evaporation of approximately 2000 mm. Agricultural lands have irrigation systems that extract water from the Kaidu River, whereas the other lands rely on rainfall and underground waters. Annual mean temperature ranges from 7°C to 10°C. On average, the annual maximum temperature is 38°C, and the minimum temperature −35°C.

Soil Sampling and Analyses

We collected 70 soil samples from 15 profiles in November, 2010. While these profiles vary in texture and color because of land use history, we sampled five layers at most profiles, i.e., 0–5 cm, 5–15 cm, 15–30 cm, 30–50 cm, and 50–100 cm, and determined bulk density (BD). For each layer, soils were air-dried, thoroughly mixed, and sieved to pass a 2-mm screen for pH and electric conductivity (EC) measurements. Basic soil properties are given in Table 1. In brief, soils in this region are characterized by high pH (from 8 to 9.4) and high sand/silt contents. Bulk density is slightly lower in the surface soils (0–15 cm: 1.38–1.41 g cm^{-3}) than in below (1.46–1.54 g cm^{-3}). Electric conductivity shows a decreasing trend over depth, from 1.25–1.32 ms cm^{-1} in the 0–15 cm to 0.7 ms cm^{-1} in the 50–100 cm. Representative subsamples were crushed to pass a 0.25 mm screen for SOC and SIC measurements by the following procedures.

Modified Walkley-Black Method for SOC

This method is modified from the traditional Walkley-Black method [3]. A soil sample (~0.4 g) is treated with 5 ml concentrated H_2SO_4 for 4 hours, then mixed with 5 ml 0.5 M $K_2Cr_2O_7$. The mixture is heated at 150–160°C for 5 minutes, and then cooled at room temperature. The solution is transferred into a triangular flask with 100 ml deionized water. Unreacted $K_2Cr_2O_7$ is determined by titrating with 0.25 M $FeSO_4$. Soil organic carbon content is calculated, without a recovery factor,

from the difference in $FeSO_4$ used between a blank and a soil solution.

Pressure Calcimeter Method for SIC

We use a procedure similar to that of Sherrod [12]. A subsample of 1.0 g soil and a tiny bottle with 2 ml 6 M HCl containing 3% by weight of $FeCl_2 \cdot 4H_2O$ are placed at the bottom of a 100-ml reaction vessel. After sealing the vessel with a butyl rubber stopper that has an aluminum tearoff seal cap, we mix the acid with soil by turning the vessel sideways. Two hours later, we remove the aluminum tearoff seal cap, and then insert an 18-gauge hypodermic needle that is attached to the pressure transducer and voltage meter. We record the voltage reading to two decimal places after 3 to 5 seconds, and then calculate CO_2 concentration using a calibration curve that is developed by mixing reagent grade $CaCO_3$ with oven-dried laboratory sand. Standards are made based on a final weight of 1.0000 g mixture of sand and $CaCO_3$ to obtain inorganic carbon concentrations of 1.2, 3.6, 6.0, 12.0, 18.0, 24.0, 30.0, 36.0, 42.0, 48.0, 54.0 and 60.0 g C kg^{-1}, respectively.

Elemental Analyzer for SOC and SIC

We used a CNHS-O analyzer (EuroEA3000) for SOC and SIC estimates. For SOC measurement, 20 mg soil is pretreated with 10 drips of H_3PO_4 for 12 hours to remove carbonate. The sample is combusted at 1020°C with constant helium flow carrying pure oxygen to ensure complete oxidation of organic materials. CO_2 production is determined by a thermal conductivity detector. Total soil carbon is measured, using the same procedure without pretreatment with H_3PO_4. Soil inorganic carbon is calculated as the difference between total soil carbon and SOC.

Loss-on-ignition (LOI) Procedure for SOM and SIC

Following Wang et al. [16] and Wang et al. [15], we first place 5.0000 g soil in a quartz glass and heated at 105°C for 12 hours to remove soil moisture. We then combust soil in a programmable muffle furnace (S1849, KOYO LINDBEERE LTD) at 375°C for 17 hours and 800°C for 12 hours. Soil organic matter is calculated as the weight loss between 105°C and 375°C, and SIC as the weight loss between 375°C and 800°C:

$$\text{SOM}_{\text{LOI}} (\text{g kg}^{-1}) = \left(\frac{\text{Weight}_{105°C} - \text{Weight}_{375°C}}{\text{Weight}_{105°C}} \right) \times 1000 \quad (1)$$

$$\text{SIC}_{\text{LOI}} (\text{g kg}^{-1}) = \left(\frac{\text{Weight}_{375°C} - \text{Weight}_{800°C}}{\text{Weight}_{105°C}} \right) \times 0.273 \times 1000. \quad (2)$$

The conversion constant of 0.273 in equation (2) is applied to convert mass of CO_2 to mass of carbon.

Results

Soil Organic Carbon

We first compare measured SOC between the automated CNS analysis and the modified Walkley-Black method, and find a strong correlation (r^2 = 0.93, P<0.001) between the two methods (Table 2). As shown in Figure 2, SOC content is below 15 g kg^{-1} in this region. It appears that the modified Walkley-Black method over-estimates SOC when soil contains extremely low SOC (<5 g

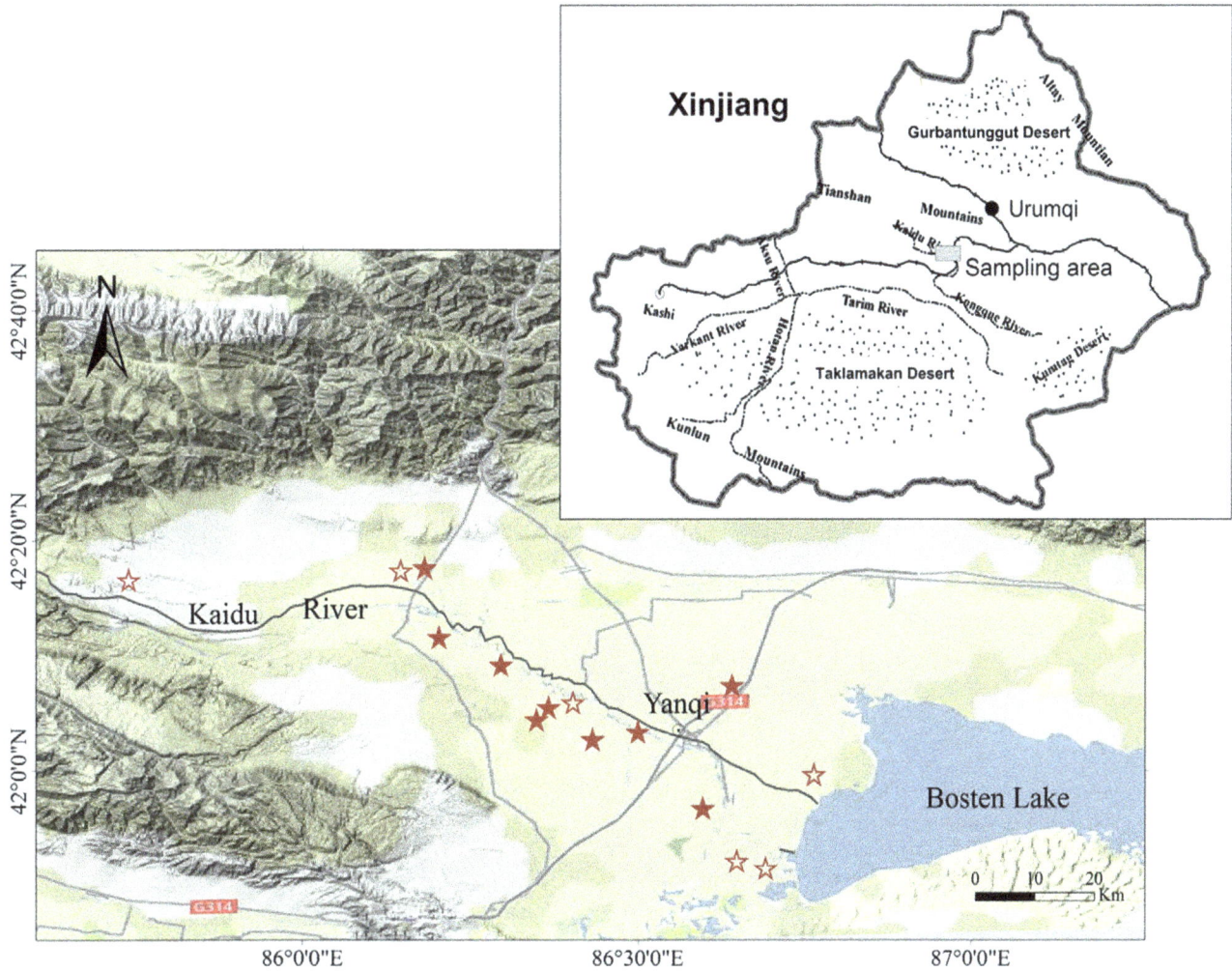

Figure 1. Sampling area with locations of 15 soil profiles. Solid stars denote the sites where topsoil texture was determined.

kg^{-1}). Excluding these data yields an averaged recovery of 0.993. These results indicate that almost 100% SOC in the brown desert soil can be oxidized, which is much higher than previous published recovery rates (less than 82%) for forest soils across Flanders region [7], but in agreement with those ($\sim$100%) in most Tasmanian soils [16]. The high recovery of SOC and large correlation coefficient in our study suggest that the modified Walkley-Black method is as

accurate as the automated CNS technique except for extremely low SOC soils.

We then compare measured SOM by the LOI method with SOC by the automated CNS technique (Figure 3). As expected, SOM values (0–27 g kg^{-1}) are higher than SOC values

Table 1. Means and standard deviations (in brackets) of soil pH, bulk density (BD, g kg^{-1}) and electrical conductivity (EC, ms cm^{-1}) for each layer, and clay, silt and sand contents (%) of topsoil (0–30 cm) from 8 randomly selected soil profiles.

	pH	BD	EC	Clay	Silt	Sand
0–5	8.92 (0.3)	1.41 (0.1)	1.25 (2.4)	5.6 (2.9)	57.3 (23.0)	37.1 (25.1)
5–15	8.77 (0.3)	1.38 (0.1)	1.32 (1.5)			
15–30	8.78 (0.4)	1.50 (0.2)	1.07 (1.2)			
30–50	8.88 (0.4)	1.54 (0.1)	0.98 (1.2)	–	–	–
50–100	9.00 (0.2)	1.46 (0.1)	0.70 (1.2)	–	–	–

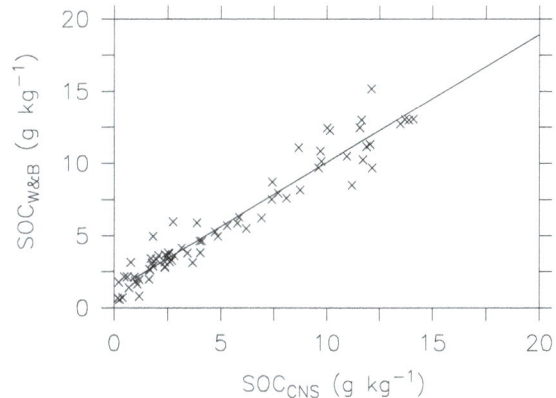

Figure 2. Linear regression of measured SOC by the CNS analyzer and Walkley-Black method.

Table 2. Relationships of soil organic carbon (SOC) and inorganic carbon (SIC) among different methods.

	n	Slope	Intercept	r^2	P *value*
SOC$_{W\&B}$ vs. SOC$_{CNS}$	70	0.886	1.199	0.92	<0.001
SOM$_{LOI}$ vs. SOC$_{CNS}$	70	1.792	4.189	0.90	<0.001
SOC$_{LOI}$ vs. SOC$_{W\&B}$	31	0.949	0.443	0.77	<0.001
SIC$_{PC}$ vs. SIC$_{CNS}$	70	1.025	−1.718	0.99	<0.001
SIC$_{LOI}$ vs. SIC$_{CNS}$	70	1.043	2.582	0.96	<0.001
SIC$_{LOI}$ vs. SIC$_{PC}$	70	1.021	4.241	0.98	<0.001

(<15 g kg^{-1}). The large intercept value of 4.189 indicates that there are some non-SOM weight losses (e.g., from structural water and carbonate) during ignition. The conversion factor of 1.792 is slightly larger than the traditional values of 1.724. We find that the conversion factor is 2.71 for those samples with low SOC content (<5 g kg^{-1}) that are mainly subsoils. Based on all 70 soil samples tested, SOC may be calculated from LOI at 375°C for 17 hours, using the following equation:

$$SOC_{LOI} = (SOM_{LOI} - 4.189)/1.792 \qquad (3)$$

Figure 4 shows comparison of measured SOC by the Walkley-Black method and calculated SOC by the LOI method using the equation (3). To avoid negative values in calculated SOC, we exclude samples that contain less than 4.189 g kg^{-1} SOM$_{LOI}$ in this comparison. The slope value of 0.949 and significantly high correlation coefficient ($r^2 = 0.77$, p<0.001) indicate that the LOI method has potential for estimating SOC in this arid region.

Soil Inorganic Carbon

We assess measured SIC by the pressure calcimeter method and the automated CNS analysis (Figure 5). Clearly, SIC content is much higher than SOC (<15 g kg^{-1}) in this region. The pressure calcimeter method produces slightly lower SIC values (6–51 g kg^{-1}) than the automated CNS technique (9–53 g kg^{-1}). However, there is a high correlation ($r^2 = 0.99$, p<0.001) between the two methods. The slope value of 0.996 and large correlation

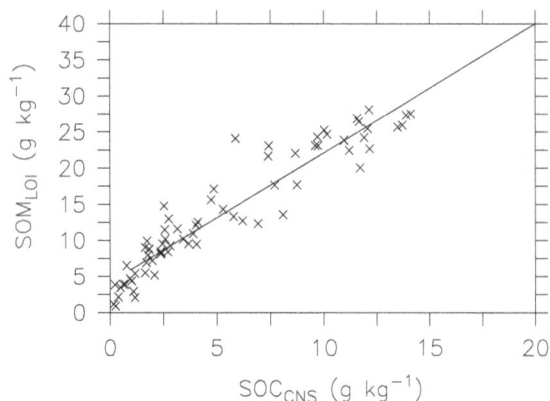

Figure 4. Relationship between SOC calculated by the LOI method and SOC measured by the Walkley-Black method.

coefficient indicate that the pressure calcimeter method is as accurate as the automated CNS technique for the calcareous soils.

Figure 6 presents comparison of measured SIC between the LOI method and the automated CNS analysis. Both methods give a very similar range, and form a linear relationship:

$$SICLOI = 1.043 SICCNS + 2.582 \ (r^2 = 0.96, P < 0.001) \qquad (4)$$

The intercept value of 2.582 is significantly different from 0, suggesting that there may be some non-SIC weight losses (e.g., dehydration from other compounds) during combustion [22,23,29]. Nevertheless, the factors of slope value close to 1 and significant correlation indicate that the LOI technique can produce accurate estimates for SIC.

As shown in Figure 7, there is also a linear relationship for SIC between the pressure calcimeter method and the LOI technique:

$$SIC_{LOI} = 1.021 SIC_{PC} + 4.241 \qquad (5)$$

which shows a similar slope to that in the regression between the LOI method and the CNS analysis (i.e., equation 4). The relatively larger intercept in equation (5) reflects the slightly lower SIC values by the pressure calcimeter method, relative to those by the automated CNS analysis (see Table 2).

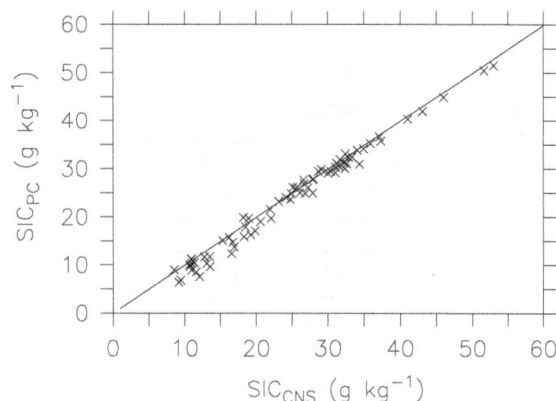

Figure 3. Relationship between SOC measured by the CNS analyzer and SOM determined by the LOI method.

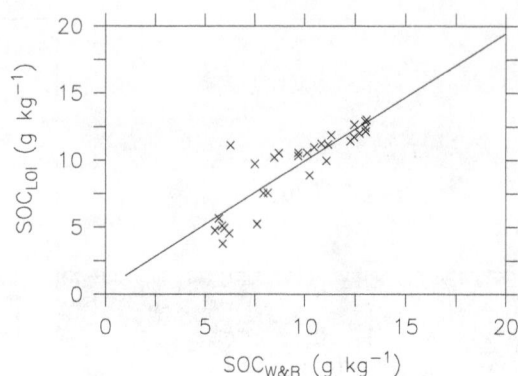

Figure 5. Liner relationship for SIC by the CNS analyzer and pressure calcimeter method.

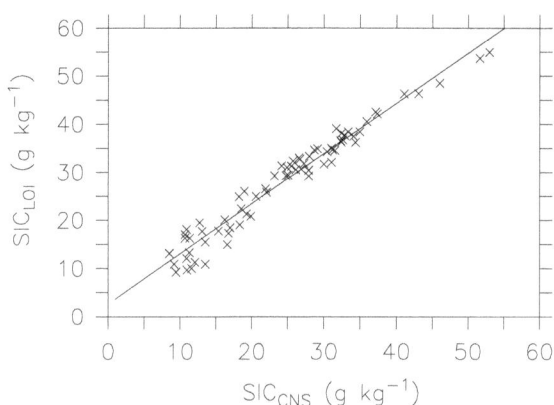

Figure 6. Liner relationship for SIC between the CNS analyzer and LOI method.

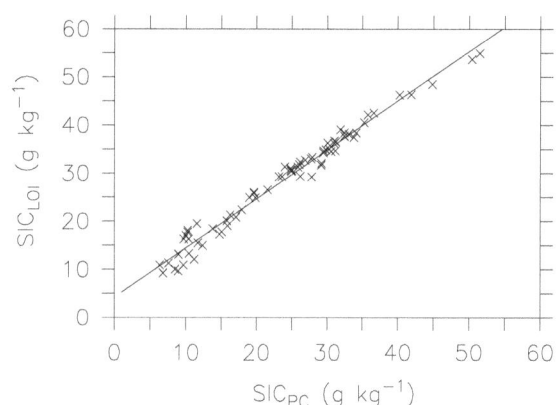

Figure 7. Liner relationship for SIC by the pressure calcimeter and LOI methods.

Discussion

Estimate of SOC

Measurement of SOC has been a common procedure in soil science. Many methods are available, each with advantages and disadvantages in terms of accuracy, expense and convenience [5,30]. The Walkley-Black procedure is considered to be cheap and easy to perform, but may exhibit variable recovery thus a correction factor is often applied to determine the total SOC content for a soil sample. For example, Sleutel et al. [31] applied a correction factor of 1.33 for estimated SOC by the Walkley-Black without external heating. However, an early study [16] showed that without external heating, the Walkley-Black method produced approximately 100% recovery for SOC in most Tasmanian soils. Many modified dichromate oxidation techniques that involve external heating do not require a correction factor [32]. Our study demonstrates that compared with the automated CNS technique, the modified Walkley-Black method with external heating yields almost 100% recovery for SOC in the calcareous soils we tested. However, the Walkley-Black method may over-estimate SOC in soils containing extremely low SOC [8], which is probably attributed to reactions of $K_2Cr_2O_7$ with inorganic soil constituents [33].

While the LOI method has been widely used to estimate SOM and SOC, there is no standard procedure in terms of heating duration and temperature. There have been various combinations with temperature ranging from 150°C to 900°C and duration from 2 hours to 17 hours in the literature, thus a regression between LOI and SOC (measured by an automated analyzer or the Walkley-Black method) varies widely [16,34,35]. In general, with increasing temperature, LOI increases whereas correlation coefficient tends to decrease [35].

One criticism of the LOI technique is that it may result in weight losses from structural water and carbonate during heating [30,34]. These non-SOM weight losses can significantly affect estimates of SOC, particularly in soils containing low SOM and high clay content. As the amount of SOM decreases, dehydroxylation of clays becomes more likely [36]. However, clay content is low in our soils (see Table 1). Thus, weight loss due to dehydroxylation of clays may be small. On the other hand, temperature is the dominant factor affecting LOI, and its relationship with SOC [35]. Therefore, it is critical to choose a temperature that is high enough to completely remove SOM but low enough to prevent dehydroxylation of clay minerals and oxidation of carbonate.

Several studies have shown that a combination of 375°C and 16–17 hours results in a good LOI-SOC relationship, thus it can

provide accurate estimates of SOC in many soil groups [16,37,38]. Our study demonstrates that there is a strong relationship between SOC and LOI at 375°C for 17 hours in the calcareous soils tested. While the non-zero intercept may reflect some non-SOM weight losses, the strong relationship indicates the capacity of using the LOI method for SOC estimation.

Estimate of SIC

Soil carbonate is usually quantified by acid dissolution in the reaction below:

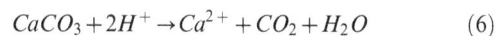

$$CaCO_3 + 2H^+ \rightarrow Ca^{2+} + CO_2 + H_2O \qquad (6)$$

with the determination of either H^+ consumption or CO_2 production. The methods of H^+ consumption that involve reaction with a strong acid, such as HCl addition and back titration of the unreacted acid, may be problematic owing to H^+ consumption by other soil components [9]. Indeed, our own analyses using the H^+ consumption approach lead to significant underestimates of SIC (by 20%) for the soils we tested (data not shown).

There have been many methods involving determination of CO_2 production following acid dissolution [9]. While these methods are much preferred, precautions must be taken to ensure that there is no interference from organic matter oxidation. Thus, an oxidation inhibitor (e.g., $FeCl_2$) is often used to limit the oxidation of organic matter and the subsequent evolution of CO_2 from this source. Our study shows a strong correlation between the pressure calcimeter method and the automated CNS analysis. Given that measured SIC contents by the pressure calcimeter method are slightly lower than, but not significantly different from those by the automated CNS analysis, we conclude that the pressure calcimeter approach can provide an accurate measurement of SIC for the calcareous soils of this arid region.

In contrast to wide utilization of the LOI method for SOC estimates, there is little in the literature regarding application of the LOI method for SIC estimates. However, the LOI method has been used to measure carbonate content in sediments, following combustion of organic carbon at lower temperature [13,15,22,23]. The simple two-step LOI procedure for estimates of organic carbon and carbonate in sediments [15] is similar to that for determination of organic carbon and carbonate in calcareous soils by dry combustion [14]. Thus, it could be a convenient practice for estimating SOC and SIC.

To the best of our knowledge, this study is the first attempt to evaluate the LOI method for SIC measurement. Our study shows a significantly ($r^2>0.96$, $P<0.001$) linear relationship between estimated SIC by the LOI method and determined SIC by direct measurements, i.e., the automated element analyzer and the pressure calcimeter method. Despite the non-zero intercepts and slightly higher values in SIC by the LOI method, the slope values are close to one, which are consistent with those reported for sediment analyses [15]. Our results demonstrate that the LOI method has potential for easily and accurately estimating SIC in calcareous soils of arid regions.

Conclusions

We have compared three methods for determinations of SOC and SIC concentrations in calcareous soils of the Yanqi Basin. In spite of wide ranges in soil properties, SOC is generally low (1–14 g kg^{-1}) whereas SIC is considerably high (9–53 g kg^{-1}). Our study shows that the Walkley-Black method with external heating can provide accurate estimates of SOC (with almost 100% recovery) for most arid soils. The pressure calcimeter method is as accurate as the automated elemental technique for measuring SIC. In addition to these traditional methods, we have also evaluated a two-step LOI approach, i.e., (1) combustion at 375°C for 17 hours to estimate SOC, and (2) subsequent combustion at 800°C for 12 hours to estimate SIC. Our study shows that SOC and SIC can be estimated from LOI for calcareous soils except those with extremely low SOC or SIC contents. This study demonstrates that the two-step LOI technique has potential to be used as a sufficiently accurate technique for estimating SOC and SIC in arid soils. Further testing on a wider range of arid and semi-arid soils is warranted.

Acknowledgments

We are grateful for the reviewer's constructive comments.

Author Contributions

Conceived and designed the experiments: XW JW. Performed the experiments: JW JZ. Analyzed the data: XW. Contributed reagents/materials/analysis tools: XW. Wrote the paper: XW.

References

1. Lal R, Kimble JM (2000) Pedogenic carbonate and the global carbon cycle. In: Lal R, Kimble JM, Eswaran H, Stewart BA, editors. Golbal Climate Change and Pedogenic. Boca Raton, FL, USA.: CRC press. 1–14.
2. Eswaran H, Reich PF, Kimble JM, Beinroth FH, Padmanabhan E, et al. (2000) Global carbon stocks. In: R L, Kimble JM, Eswaran H, Stewart BA, editors. Global climate change and pedogenic carbonates. Boca Raton, FL: Lewis Publishers. 15–27.
3. Walkley A, Black IA (1934) An examination of the Degtjareff method for determining soil organic matter and a proposed modification of the chromic acid titration method. Soil Sci 37: 29–38.
4. Byers SC, Mills EL, Stewart PL (1978) A comparison of methods of determining organic carbon in marine sediments, with suggestions for a standard method. Hydrobiologia 58: 43–47.
5. Nelson DW, Sommers LE, Sparks D, Page A, Helmke P, et al. (1996) Total carbon, organic carbon, and organic matter. Methods of soil analysis Part 3-chemical methods: 961–1010.
6. Zhang WJ, Wang XJ, Xu MG, Huang SM, Liu H, et al. (2010) Soil organic carbon dynamics under long-term fertilizations in arable land of northern China. Biogeosciences 7: 409–425.
7. De Vos B, Lettens S, Muys B, Deckers JA (2007) Walkley–Black analysis of forest soil organic carbon: recovery, limitations and uncertainty. Soil Use and Management 23: 221–229.
8. Conyers MK, Poile GJ, Oates AA, Waters D, Chan KY (2011) Comparison of three carbon determination methods on naturally occurring substrates and the implication for the quantification of 'soil carbon'. Soil Research 49: 27–33.
9. Loeppert R, Suarez D, Sparks D, Page A, Helmke P, et al. (1996) Carbonate and gypsum. Methods of soil analysis Part 3-chemical methods: 437–474.
10. Presley B (1975) A simple method for determining calcium carbonate in sediment samples. J Sediment Petrol 45: 745–C746.
11. Dreimanis A (1962) Quantitative gasometric determination of calcite and dolomite by using Chittick apparatus. Journal of Sedimentary Research 32: 520.
12. Sherrod LA, Dunn G, Peterson GA, Kolberg RL (2002) Inorganic Carbon Analysis by Modified Pressure-Calcimeter Method. Soil Sci Soc Am J 66: 299–305.
13. Heiri O, Lotter AF, Lemcke G (2001) Loss on ignition as a method for estimating organic and carbonate content in sediments: reproducibility and comparability of results. Journal of Paleolimnology 25: 101–110.
14. Rabenhorst MC (1988) Determination of organic and carbonate carbon in calcareous soils using dry combustion. Soil Science Society of America Journal 52: 965–969.
15. Wang QR, Li YC, Wang Y (2011) Optimizing the weight loss-on-ignition methodology to quantify organic and carbonate carbon of sediments from diverse sources. Environmental Monitoring and Assessment 174: 241–257.
16. Wang XJ, Smethurst PJ, Herbert AM (1996) Relationships between three measures of organic matter or carbon in soils of eucalypt plantations in Tasmania. Australian Journal of Soil Research 34: 545–553.
17. David MB (1988) Use of loss-on-ignition to assess soil organic carbon in forest soils Communications in Soil Science and Plant Analysis 19: 1593–1599.
18. De Vos B, Vandecasteele B, Deckers J, Muys B (2005) Capability of loss-on-ignition as a predictor of total organic carbon in non-calcareous forest soils. Communications in soil science and plant analysis 36: 2899–2921.
19. Wright AL, Wang Y, Reddy KR (2008) Loss-on-Ignition Method to Assess Soil Organic Carbon in Calcareous Everglades Wetlands. Communications in soil science and plant analysis 39: 3074–3083.
20. Konen ME, Jacobs PM, Burras CL, Talaga BJ, Mason JA (2002) Equations for predicting soil organic carbon using loss-on-ignition for north central US soils. Soil Science Society of America Journal 66: 1878–1881.
21. Ball DF (1964) Loss-on-ignition as an estimate of organic matter and organic carbon in non-calcareous soils. Journal of Soil Science 15: 84–92.
22. Dean WE (1974) Determination of carbonate and organic matter in calcareous sediments and sedimentary rocks by loss on ignition: comparison with other methods. Journal of Sedimentary Petrology 44: 242–248.
23. Santisteban JI, Mediavilla R, López-Pamo E, Dabrio CJ, Zapata MBR, et al. (2004) Loss on ignition: a qualitative or quantitative method for organic matter and carbonate mineral content in sediments? Journal of Paleolimnology 32: 287–299.
24. Wang Q, Li Y, Wang Y (2011) Optimizing the weight loss-on-ignition methodology to quantify organic and carbonate carbon of sediments from diverse sources. Environmental Monitoring and Assessment 174: 241–257.
25. Scharpenseel HW, Mtimet A, Freytag J (2000) Soil inorganic carbon and global change. In: Lal R, Kimble JM, Eswaran H, Stewart BA, editors. Global Climate Change and Pedogenic Carbonates. Boca Raton, FL: CRC Press. 27–42.
26. Xie JX, Li Y, Zhai CX, Li CH, Lan ZD (2009) CO$_2$ absorption by alkaline soils and its implication to the global carbon cycle. Environmental Geology 56: 953–961.
27. Stone R (2008) Have desert researchers discovered a hidden loop in the carbon cycle? Science 320: 1409–1410.
28. FAO-UNESCO (1988) soil map of the world: revised legend. World Soil Resources Report No 60 FAO: Rome.
29. Allison L, Black C (1965) Methods of soil analysis. Part 2. Chemical and microbiological properties. Agronomy monograph 9: 1367–1378.
30. Chatterjee A, Lal R, Wielopolski L, Martin MZ, Ebinger MH (2009) Evaluation of Different Soil Carbon Determination Methods. Critical Reviews in Plant Sciences 28: 164–178.
31. Sleutel S, De Neve S, Singier B, Hofman G (2007) Quantification of organic carbon in soils: A comparison of methodologies and assessment of the carbon content of organic matter. Communications in Soil Science and Plant Analysis 38: 2647–2657.
32. Lettens S, De Vos B, Quataert P, van Wesemael B, Muys B, et al. (2007) Variable carbon recovery of Walkley-Black analysis and implications for national soil organic carbon accounting. European Journal of Soil Science 58: 1244–1253.
33. Walkley A (1947) A critical examination of a rapid method for determining organic carbon in soils- effect of variations in digestion conditions and inorganic soil constituents. Soil Science 63: 251–264.
34. Szava-Kovats R (2009) Re-analysis of the Relationship between Organic Carbon and Loss-on-Ignition in Soil. Communications in Soil Science and Plant Analysis 40: 2712–2724.
35. Abella S, Zimmer B (2007) Estimating organic carbon from loss-on-ignition in northern Arizona forest soils. Soil Science Society of America Journal 71: 545–550.
36. Pribyl DW (2010) A critical review of the conventional SOC to SOM conversion factor. Geoderma 156: 75–83.
37. McKeague J (1976) Manual of soil sampling and methods of analysis. Land Resource Research Inst., Res. Br. Agriculture Canada, Ottawa, Canada.
38. Beaudoin A (2003) A comparison of two methods for estimating the organic content of sediments. Journal of Paleolimnology 29: 387–390.

11

Life Cycle Assessment of Metals: A Scientific Synthesis

Philip Nuss[1]*, Matthew J. Eckelman[2]

1 Center for Industrial Ecology, School of Forestry and Environmental Studies, Yale University, New Haven, Connecticut, United States of America, **2** Department of Civil and Environmental Engineering, Northeastern University, Boston, Massachusetts, United States of America

Abstract

We have assembled extensive information on the cradle-to-gate environmental burdens of 63 metals in their major use forms, and illustrated the interconnectedness of metal production systems. Related cumulative energy use, global warming potential, human health implications and ecosystem damage are estimated by metal life cycle stage (i.e., mining, purification, and refining). For some elements, these are the first life cycle estimates of environmental impacts reported in the literature. We show that, if compared on a per kilogram basis, the platinum group metals and gold display the highest environmental burdens, while many of the major industrial metals (e.g., iron, manganese, titanium) are found at the lower end of the environmental impacts scale. If compared on the basis of their global annual production in 2008, iron and aluminum display the largest impacts, and thallium and tellurium the lowest. With the exception of a few metals, environmental impacts of the majority of elements are dominated by the purification and refining stages in which metals are transformed from a concentrate into their metallic form. Out of the 63 metals investigated, 42 metals are obtained as co-products in multi output processes. We test the sensitivity of varying allocation rationales, in which the environmental burden are allocated to the various metal and mineral products, on the overall results. Monte-Carlo simulation is applied to further investigate the stability of our results. This analysis is the most comprehensive life cycle comparison of metals to date and allows for the first time a complete bottom-up estimate of life cycle impacts of the metals and mining sector globally. We estimate global direct and indirect greenhouse gas emissions in 2008 at 3.4 Gt CO_2-eq per year and primary energy use at 49 EJ per year (9.5% of global use), and report the shares for all metals to both impact categories.

Editor: Paul Jaak Janssen, Belgian Nuclear Research Centre SCK•CEN, Belgium

Funding: This research was funded by the Yale Criticality Project. The funders had no role in study design, data collection and analysis, decision to publish, or preparation of the manuscript.

Competing Interests: The authors have declared that no competing interests exist.

* Email: philip@nuss.me

Introduction

Metals are ubiquitous in today's society; there are few materials or products where metals are absent or have not played a role in their production. While a century ago, the diversity of metals employed was limited to perhaps a dozen in common uses such as infrastructure and durable goods, today's technologies utilizes virtually the entire periodic table [1,2]. For example, the number of elements employed in integrated circuits used in most electronics products has increased from only twelve elements in 1980 to more than sixty elements today [1], while electronic products themselves are used in an increasing number of applications [2]. Similarly, the elemental complexity of superalloys, which are a class of materials to allow the operation of turbines and jet engines at high temperatures and under corrosive environments, has increased over time as new alloying elements (e.g., rhenium, tantalum, hafnium) are added.

Future global demand for metals is expected to increase further as a result of urbanization and new infrastructure construction in developing countries, widespread use of electronics, and transitions in energy technologies [3]. The use of renewable energy technologies, such as photovoltaic and wind power, is expected to result in an increased demand for both bulk metals (e.g., iron, copper) and specialty metals (e.g., rare earths) when compared to today's largely fossil-based systems [4,5]. While increased future demand for primary metals could be reduced through demateri-alization, substitution with other metallic or non-metallic resources

[6], and increased metals end-of-life recycling rates, a recent study by Graedel et al. [7] indicates that current end-of-life recycling rates for only eighteen metals (out of a total of sixty) are above 50%. These include silver, aluminum, gold, chromium, copper, iron, manganese, niobium, nickel, lead, palladium, platinum, rhenium, rhodium, tin, titanium, and zinc. For many of the specialty metals, such as scandium and yttrium, as well as the rare earth elements, end-of-life recycling rates were found to be less than 1%.

The production of primary (virgin) metal typically includes ore mining and concentrating, smelting or separation, and refining to obtain the element in its metallic form [8,9], with a variety of processing routes available. In each stage, impurities and by-products are separated and the concentration of the metal in the final product increases. Metal refining to sufficient purities frequently requires energy-intensive and precisely-controlled melting stages, often based on the use of fossil-fuel inputs directly as a reductant or indirectly for heat and electricity. In 2007, iron and steel production accounted for 30% and aluminum for 2% of global industrial carbon dioxide (CO_2) emissions (out of a total of 7.6 Gt CO_2) [10]. Pyrometallurgy involves treatment of metal concentrates at high temperatures, in order to strip the metal from its associated mineral constituents, using fossil fuels in heating furnaces or electricity to power an electric arc furnace [8,9]. Hydrometallurgy consists of treating metal ores or concentrates in liquid solution to separate metals from their associated minerals

[8,9]. While high temperatures are not usually required, treatment may take place at high pressures which requires energy to maintain and the provision of liquid agents. A number of studies indicate that the energy intensity of the mining and beneficiation process is likely to increase over time as mines shift from high- to lower-grade metal ores and start mining more complex deposits (downstream metal extraction and refining, however, is likely to be unaffected) [11–13]. This trend may be partially offset by increased process efficiencies. It is expected that future supplies of metals will increasingly consist of non-traditional geological metal resources such as seabed nodules and crustal rock [14].

The metals production system is highly interconnected, consisting of various processes in which the production or recovery of multiple metals occurs simultaneously, as in lead-zinc ores for example [15]. While the environmental implications of the major industrial metals (e.g., iron and copper) have been extensively studied [3], the environmental burdens of many of the minor metals (e.g., niobium, rhenium, hafnium) are essentially unknown, even though they are increasingly employed by industry. Materials scientists and product developers have a growing number of tools available that allow them to consider the environmental implications of their choices in materials, but in general these tools consider a small number of environmental endpoints, and many data gaps remain. However, given the expected increase in global future demand for metals and their importance in today's technologies, it is important that high-fidelity data for life cycle based environmental burdens of metals production are available and that the implications of co-production are clearly understood. Quantifying the environmental burdens per life cycle stage and the interconnectedness of the metals production systems is required in modeling global changes in technology, material substitution and metals criticality in terms of their supply chain vulnerability and supply risk [16–22]. A comprehensive understanding allows us to better manage the impacts and benefits of metals and inform sustainable resource use.

Life cycle assessment (LCA) [23,24] can be used as a tool to quantify the system-wide (cradle-to-gate or cradle-to-grave) environmental burdens of products, services, and technologies. A significant body of research on the life-cycle wide energy use and wider environmental impacts of metals provision is available from various life cycle inventory (LCI) databases [25–38], scientific reports, and publications [3]. However, many LCI data are reported in an aggregated form (either pre-allocated or at system process level), and this makes it difficult to make robust comparisons or to take co-production issues into account. LCI data are also not always representative of global metals production routes and the chemical forms of an element going into use (a metal used in its metallic or mineral form), and may be outdated if representing older technologies. Furthermore, to our knowledge the conventional LCI databases do not report metals data for a number of elements including Be, Sc, Ge, Sr, Zr, Nb, Ru, Ba, Hf, Re, Os, Ir, Bi, and Th, which are increasingly applied in today's technologies.

Against this background, the goal of this study is to present an overview of the cradle-to-gate environmental burdens of 63 metals (plus helium) in their major use forms, including (by increasing atomic number) helium (He), lithium (Li), beryllium (Be), boron (B), magnesium (Mg), aluminum (Al), calcium (Ca), scandium (Sc), titanium (Ti), vanadium (V), chromium (Cr), manganese (Mn), iron (Fe), cobalt (Co), nickel (Ni), copper (Cu), zinc (Zn), gallium (Ga), germanium (Ge), arsenic (As), selenium (Se), strontium (Sr), yttrium (Y), zirconium (Zr), niobium (Nb), molybdenum (Mo), ruthenium (Ru), rhodium (Rh), palladium (Pd), silver (Ag), cadmium (Cd), indium (In), tin (Sn), antimony (Sb), tellurium

(Te), barium (Ba), lanthanum (La), cerium (Ce), praseodymium (Pr), neodymium (Nd), samarium (Sm), europium (Eu), gadolinium (Gd), terbium (Tb), dysprosium (Dy), holmium (Ho), erbium (Er), thulium (Tm), ytterbium (Yb), lutetium (Lu), hafnium (Hf), tantalum (Ta), tungsten (W), rhenium (Re), osmium (Os), iridium (Ir), platinum (Pt), gold (Au), mercury (Hg), thallium (Tl), lead (Pb), bismuth (Bi), thorium (Th), and uranium (U), for reference year 2008.

Materials and Methods

In this study, LCA is used as a tool to quantify and compare the cradle-to-gate environmental burdens of each element on the basis of a functional unit of 1 kg of each element at the factory gate. In addition, a bottom-up estimate of global impacts using year 2008 production data is provided. An LCA model following the ISO 14040 and 14044 standards [23,24] is developed for each element using SimaPro8 software. LCI data used in this study are based on a combination of existing data sets [25,27,30,31,34–37] and an extensive literature research [39–65,22,66–102] as described in detail in Supporting Information S1. The modeled metals production system represents the global mining and metals sector in year 2008, using global production figures for each element, 2006 to 2010 price data for allocation (multi-output processes), and the most recent LCI data available from the literature. More details on the data set are provided in the sections below and Supporting Information S1. The comparative LCA model was constructed using the following steps: First, the 2008 production mix for each element is determined. For example, Ag is produced from Cu ores (17%), Pb ores (28%), Au-Ag ores (34%) and secondary sources (21%) [103], while all Cr is mined on its own (from chromite ore) but is used in different forms, i.e., as ferrochromium (74%), Cr metal (1%), sodium dichromate (17%) and chromite (8%) [25]. Results are then based on a production-weighted average over all production routes with a functional unit of 1 kg of each element at the factory gate. Data collection steps include: (a) Determining whether LCI data are available in existing databases; (b) If no LCI data exists, data are obtained either by direct data collection, or by integrating it with existing LCI data sets (e.g., Ge is a byproduct of Zn smelting and integrated into an existing smelting data set [25] based on the Ge content in the Zn concentrate and 2006–2010 price averages [104]) (see Supporting Information S1); (c) If LCI data exists, the data sets were used if representative of ≥80% of global production and the metal is produced on its own (bachelor). If the metal is co-produced, price data for allocation of environmental burdens were updated to 2006–2010 price averages given in United States Geological Survey (USGS) Mineral Commodity Summaries [104].

The resulting LCIs are linked to Ecoinvent 2.2 [27] data sets in the SimaPro8 LCA software. Impact assessment was performed for the categories of global warming potential (GWP) (IPCC 2007 GWP 100a v1.02 [105]), cumulative energy demand (CED) (Cumulative Energy Demand v1.08 [105]), terrestrial acidification, freshwater eutrophication, potential impacts to human health and ecosystem damage (World ReCiPe H/H Midpoint/Endpoint v1.08 [106]), and human toxicity (USETox v1.02 with recommended and interim characterization factors [107]). Monte Carlo simulation is used to assess the uncertainty associated with the results of the environmental assessment taking into account uncertainties of the LCI data (see Supporting Information S1).

Metals Dataset

The metal life cycles examined are pictured in Figure 1 for the 64 elements, covering virtually all of the metals used in modern

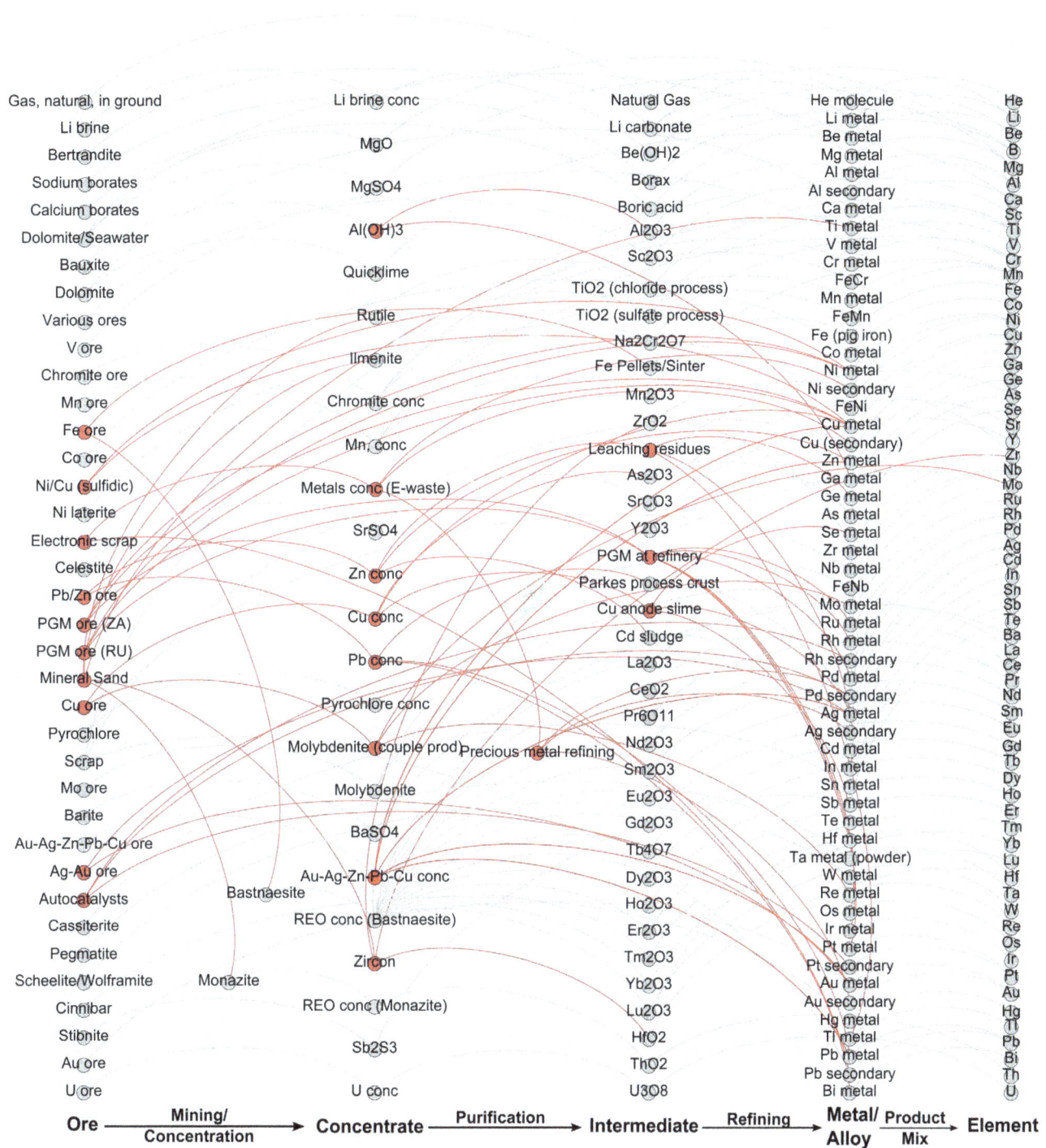

Figure 1. Diagram of the cradle-to-gate production of the minerals and metals (starting with ores from the left) analyzed in this study, in order of atomic number. Nodes representing mineral and metal products and intermediates, and the edges (arrows) indicating the physical transformation of metals via industrial processes from one chemical form (material) into another. Nodes and edges in red color represent joint production and illustrate the interconnectedness of the metals production system. Although not a metal, helium is includes in the assessment as it is sometimes regarded as a critical element required for the cooling of nuclear power plants. The figure was created using Gephi v0.8.2 [113].

products, from infrastructure applications to emerging technologies.

Starting from the left in Figure 1, our assessment includes the physical transformation of metal ores into mineral concentrates (mining and concentration), further transformation into mineral products and intermediates (purification), and subsequent conver-

sion into the final metal or alloy (refining). Graph nodes represent the chemical form in which an element is present during each life cycle stage (i.e., ore, concentrate, intermediate, metal), while edges connecting two nodes indicate the physical transformation taking place as metals move through subsequent production steps. Since each of the 64 elements may be used in various chemical forms (e.g., as metal, oxide, sulfide), the overall environmental burdens for each elements consists of the weighted average of each chemical form supplied globally in 2008 (e.g., globally about 79% of lithium is used as Li_2CO_3 and 21% as Li metal [40]). This is indicated with the edges titled "product mix" and the nodes termed "element" in Figure 1.

In our data set, several metal production routes are interlinked with each other and form an intricate network of industrial processes (highlighted in red color in Figure 1). For metals obtained in multi-output processes (joint production), it is necessary to divide the environmental impacts from the process and all upstream processes among all metal co-product(s). Assigning each metal co-production an appropriate proportion of environmental impacts can be done in multiple, standardized ways, for example by applying mass allocation or economic allocation [24] as specified in the ISO standards for LCA. Economic allocation is used in the metals industry and is carried out by multiplying the price of each product output in dollars per kilogram by its quantity. In this paper, the 2006 to 2010 moving average prices from USGS Mineral Commodity Summaries [104] are used to allocate all environmental burdens across the metals production system shown in Figure 1. A 5-year price average is used to control for price volatility that may occur in a single year. For intermediate products not traded on a market, it is often difficult to obtain robust price information. Following Classen et al (2009) [25], 10% of the average market price is used for intermediate products (representing a 10% profit margin) with further details provided in Supporting Information S1. The choice of allocation is further investigated in a sensitivity analysis (see the discussion section below). In a few cases, disaggregation of LCI data was not possible for each life cycle stage and total impacts for the whole production chain are therefore reported. The complete dataset appears in Table S38 in Supporting Information S1 with additional information for each element provided in sections 2 to 67 of Supporting Information S1.

Results

From our analysis, we show results separately for the three main stages of (1) mining/concentration, (2) purification, and (3) refining. Environmental burdens are examined on the basis of one kilogram of each element at the factory gate as well as on the basis of their annual production in year 2008.

The GWP per kilogram of each element (2008 supply mix) is given in Figure 2, together with an illustration of the interconnected lead-zinc production system and uncertainties associated with the results as an example of how the data set may be used to illustrate joint metals production. Figure 3 summarizes the impacts to CED, acidification, eutrophication, and human toxicity associated with each element.

Several features of the periodic tables are readily apparent. First, precious metals including platinum group elements (Ru, Rh, Pd, Os, Ir, Pt) and Au, display the highest environmental burdens on a per kilogram comparison, while many of the major industrial metals (e.g., Fe, Mn, Ti) are found at the lower end of the environmental impacts scale. For joint production systems (i.e., where one or more precious metals are co-produced with base metals such as copper or nickel) this is partly a result of the economic allocation chosen in this assessment, in which more of the environmental burdens is attributed to the expensive precious metals. Precious metals are those that are generally used in small quantities in products and industrial applications (e.g., catalysts, spark plugs, jewelry). Second, as expected from the use of mostly fossil energy carriers during metals production, there is a strong

correlation between the cumulative energy use (Figure 3a) and GWP (Figure 2). Third, the impacts to acidification, eutrophication, and human toxicity (Figure 3b to d) reveal that for some elements, such as arsenic, copper, molybdenum, mercury and others, environmental burdens may not only be governed by the energy intensity and fuel mix of the metals production process, but may also be due to other factors including the disposal of sulfidic tailings or emissions of toxic or acidifying pollutants to air, soil, and water. Impacts other than GWP and CED could not be further investigated in this study due to the limited availability of LCI data for many of the metals investigated and models that describe, for instance, how mine tailings behave in the environment after disposal [108], and future research is required to do so. While USEtox [107] is recommended as the latest model for modeling human toxicity in LCA [3], metals are included only with interim characterization factors with high uncertainties (several orders of magnitude). The results of Figure 3d should therefore be treated as a first indication of potential impacts to human toxicity. Fourth, the above results imply that a comparison of metals on the basis of their annual global production may yield useful information regarding their absolute impacts, and an investigation into allocation of the environmentally relevant flows of the whole metals production network may elucidate the methodological challenges of reporting a single environmental burden value for each element.

While the periodic tables in Figure 2 and Figure 3 reveal information about the overall impact on a per kilogram basis, they do not break impacts down by life cycle stage, which is useful in analyzing opportunities to reduce system wide environmental pressures of the metals production process. A breakdown by life cycle stage for GWP, CED, human health, and ecosystem damage is illustrated in Figure 4.

As illustrated in Figure 4, for elements in their metallic form (either metals or alloys) the environmental burden are largely due to the purification (i.e., smelting) and refining stages required to obtain the final metal product (see Li, Be, Al, Ferrochromium (FeCr), Cr, Ferromanganese (FeMn), Fe, Cu (combined metal production, SE), Zn, Ge, Se, Zr, Ferroniobium (FeNb), Ru, Ag (from Pb), Cd, In, Te, Pb (from Pb-Zn), and Bi).

For Ge, In, and Ag, the purification stage contributes more to overall impacts than subsequent refining. This is because for metals the smelting process, producing intermediates such as anode slime or leaching residue, is included in the purification stage. For example, the purification stage of Ge and In includes the environmental burdens of Zn smelting, yielding a leaching residue from which both elements are recovered as co-products (together with Bi, and Tl) (see section 20 of Supporting Information S1). Similarly, Ag (from Pb) is recovered from Parkes process crust yielded during the smelting of primary Pb [25] with the environmental burdens of producing the intermediate accounted for in the purification stage.

For Ru (representative in terms of the contribution analysis for the platinum group metals (PGMs)), the aggregated nature of the data set did not allow the separation of the mining and concentration stage from other life cycle stages, which was therefore reported together with purification. The combined step includes the mining and concentration, leaching, hydrometallurgical treatment, and subsequent electrolytic precipitation of the PGM-containing fraction, yielding a dross that is then treated in a refinery [25].

A few metals display approximately similar impacts coming from both mining and concentration, and subsequent purification and refining (see Mn, Cu (from Mo), and Zn (from Au-Ag), and Pb (from Au-Ag). However, the figure also shows that for a number of

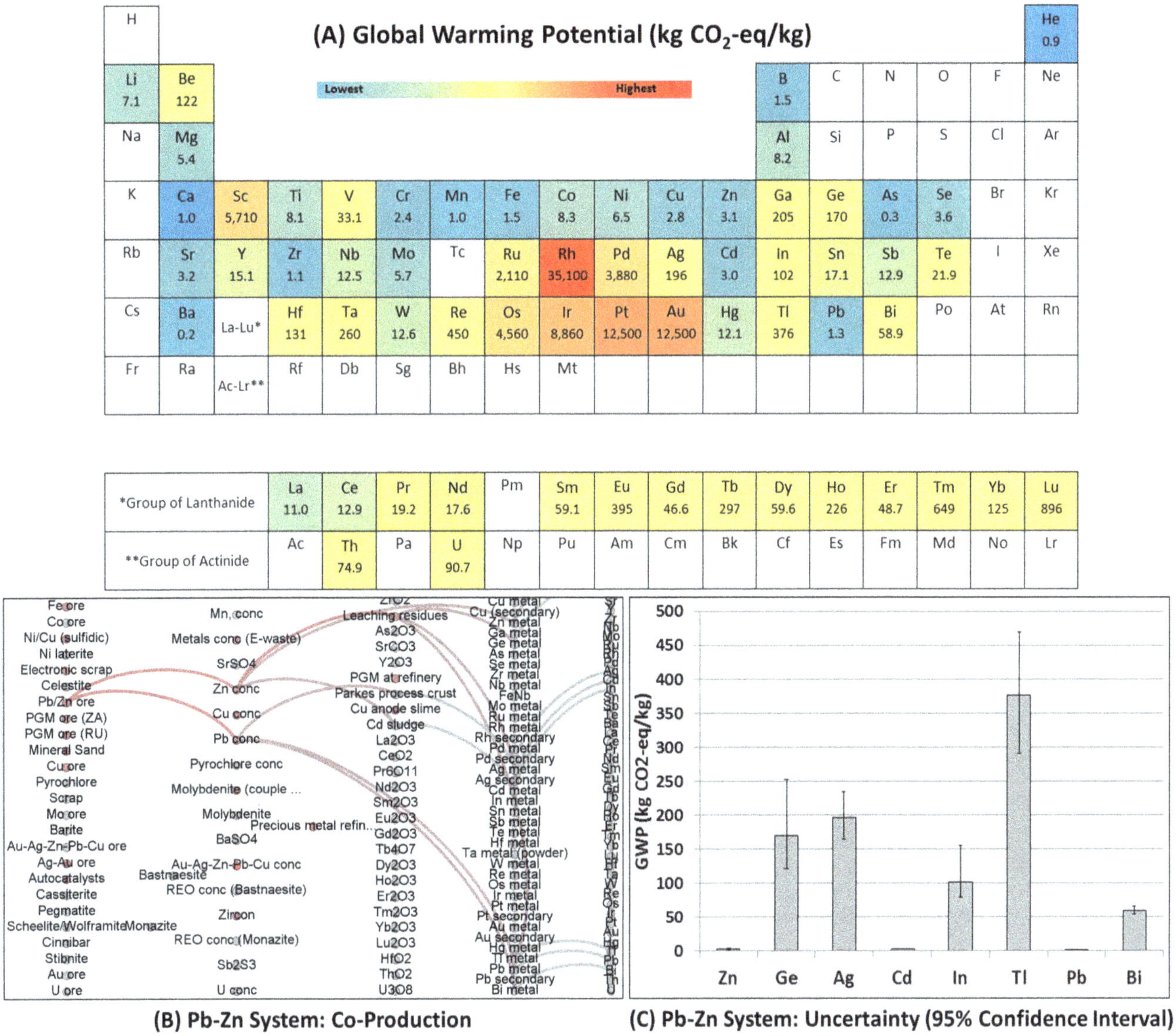

Figure 2. Periodic table of global warming potentials (GWPs). (A) The cradle-to-gate GWP per kilogram of each element (kg CO_2-eq per kg) colored according to the color ramp above. GWPs shown are weighted by the 2008 supply mix for each element (see Table S38 in Supporting Information S1). (B) Illustration of the Pb-Zn system as an example of a joint production scheme (red color) from which Ge, Ag, Cd, In, Tl, and Bi are recovered as co- or by-products. (C) Uncertainty estimates (95% confidence interval) for the elements of the Pb-Zn system were derived using Monte-Carlo analysis.

metals (i.e., Mo, Ag (from Cu), Sb, Re, and Pb (from Au-Ag) the mining and concentration step may be a larger contributor to environmental pressures than subsequent purification and refining.

Secondary metals production (i.e., Ni, Cu, Pd, Ag, and Au from electronic scrap) is found to be dominated by the initial processing (smelting) of electronic scrap. It should be noted, however, that the estimates for secondary metals production shown in Figure 4 are based on inventory data from a single plant in Sweden only, as reported in Classen et al (2009) [25], and may not be representative of the global situation.

For elements in their mineral form ($LiCO_3$, TiO_2, $Na_2Cr_2O_7$, Mn_2O_3, As_2O_3, $SrCO_3$, Y_2O_3 [representative in terms of the contribution analysis for the rare earth oxides], ZrO_2 ThO_2, and UO_2), the environmental burdens are dominated by the purification stage.

A breakdown of human health implications and ecosystem damage on the right hand side of Figure 4 reveals that for some elements, the mining and concentration stage may increase in significance as toxicity related emissions during the mining process and, e.g., the disposal of overburden and tailing wastes are taken into account. As mentioned above, this analysis does not aim to provide a detailed assessment of toxicity-related implications of metals production and we recommend that further research may be carried out to refine the existing data set in this regard.

The per-kilogram impacts discussed above can be multiplied with their respective annual global production quantity in 2008 from USGS Mineral Yearbooks [53] to derive at an estimate of global greenhouse gas emissions and energy use of metals production. The results of this exercise are shown in Figure 5.

Figure 5 illustrates the fact that, if taking into account the global annual production, the major industrial metals of Fe, Al, and Cu

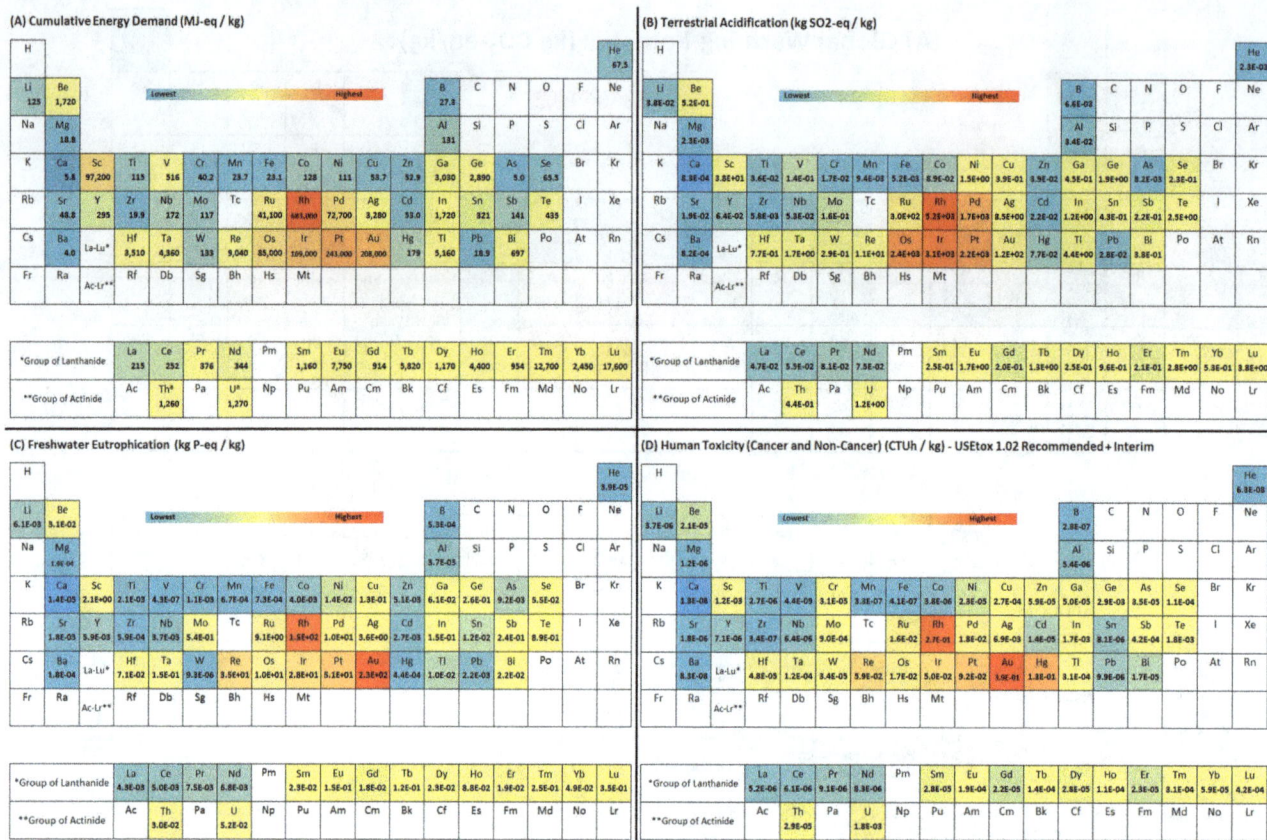

Figure 3. Periodic table of environmental impacts (colored according to the color ramp above). (A) Cradle-to-gate cumulative energy demand (CED) (MJ-eq/kg) per kilogram of each element. [a]Cumulative energy demand for Th and U does not include non-renewable nuclear energy demand of U and Th in ground. (B) Cradle-to-gate terrestrial acidification (kg SO_2-eq/kg). (C) Cradle-to-gate freshwater eutrophication (kg P-eq/kg). (D) Cradle-to-gate human toxicity (cancer and non-cancer) (CTUh/kg). Impacts to acidification and eutrophication were derived using the ReCiPe Midpoint method 1.08 H/H for the globe [106]. Human toxicity was calculated using the USETox v1.02. method with recommended and interim characterization factors [107].

display the highest environmental impacts, while many of precious metals (with the exception of Au) are in the lower third of the impacts magnitude axis. A nice feature of this analysis is that global annual GWP and CED for each of the 63 metals investigated can be estimated with this bottom-up approach. This is done in Figure 6, and Table 1 compares our results with publicly available but highly aggregated, top-down data from the International Energy Agency (IEA).

Table 1 shows that our bottom-up estimates are with 3.4 Gt CO_2 per year approximately in line with the 3.1 Gt CO_2 per year reported in IEA reports [10,109]. This is reasonable because, with the exception of the mining and quarrying category, the IEA estimates include CO_2 emissions from direct and indirect sources, in particular energy generation. In contrast, our estimates for CED of 49 PJ per year are 40% higher than the 35 PJ per year reported by the IEA, because the latter estimate only includes final energy use and does not capture primary energy use of the metals sector. An exception is the non-metallic minerals category, for which CED in this study is slightly lower than the IEA estimate of final energy use. The reason is that our CED estimate only includes limestone production, while the IEA number includes additional product categories such as clay brick and tile, building ceramics, and others [110].

Results from our study for the non-ferrous metals sector, consisting of a total of 61 metals (i.e., 63 metals minus Fe and steel

and Al), are further broken down by metal type in Figure 6 and show that, with regard to global annual greenhouse gas emissions, Cu, Ti, Au, and Zn production account for the largest emissions within this sector. This information is occasionally provided for some of the major non-ferrous metals, e.g., Al, Cu, Cr, Mn, Ni [110], but is generally difficult to obtain for many of the minor metals such as Co, Ga, Re, and others.

Of the 63 metals included in our data set, 42 are obtained as by- or co-products in multi-output processes (Figure 1), and environmental burdens are allocated using 2006 to 2010 moving average prices from USGS statistics [104]. Figure 7 compares impacts to GWP for each of the 42 metals using 2006 to 2010 price data for allocation with (a) mass allocation (Figure 7a), and (b) price-based allocation using 1995 to 2010 price averages (Figure 7b). The diagonal line indicates where results from mass and economic allocation are equal.

The wider spread on the y- than on the x-axis in Figure 7a shows that economic allocation results in more of the environmental burden being allocated, e.g., to the platinum group metals (Rh, Pt, Ir, Os, Pd, Ru), some rare earth elements (Lu, Tm, Eu, Tb, Ho, Sm, Er, Gd), Re, Ge, Hf, and In when compared to mass allocation. These elements are present in small quantities in the ores or intermediate products from which they are recovered and have high market prices (see, e.g., Table S25 in Supporting Information S1 for the rare earth elements and Table S30 in

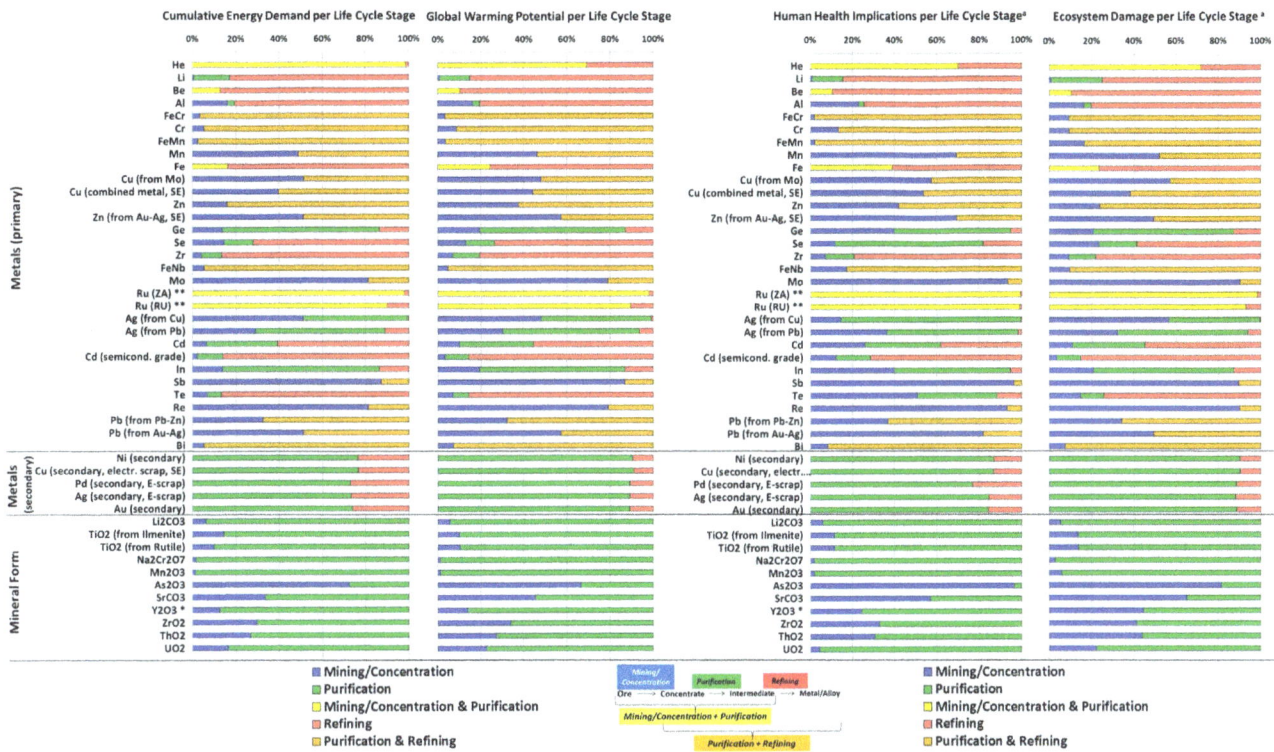

Figure 4. Relative environmental implications by life cycle stage. Only materials for which data for more than one life cycle stage was available are shown. Due to the aggregated nature of many of the data sets, in some instances the figure includes cumulative contributions from two life cycle stages (see legend). [a]The human health implications (DALY/kg) and ecosystem damage (species.yr/kg) were calculated using the World ReCiPe Endpoint (H) impact assessment method v1.08 [106]. They represent potential damages prior to normalization and weighting. *The relative breakdown of the environmental burden of Y_2O_3 production is similar to other rare earths (i.e., La, Ce, Pr, Nd, Sm, Eu, Gd, Tb, Dy, Ho, Er, Tm, Yb, and Lu), which are therefore not shown in this figure. ** The breakdown of the environmental burden of Ru production is similar to other platinum group metals (i.e., Rh, Pd, Os, Ir, and Pt) from primary ore, which are therefore not shown in this figure. FeCr = Ferrochromium. FeMn = Ferromanganese. FeNb = Ferroniobium. Elements in brackets indicate the host metal from which the metal is obtained as a co-product. SE = Sweden.

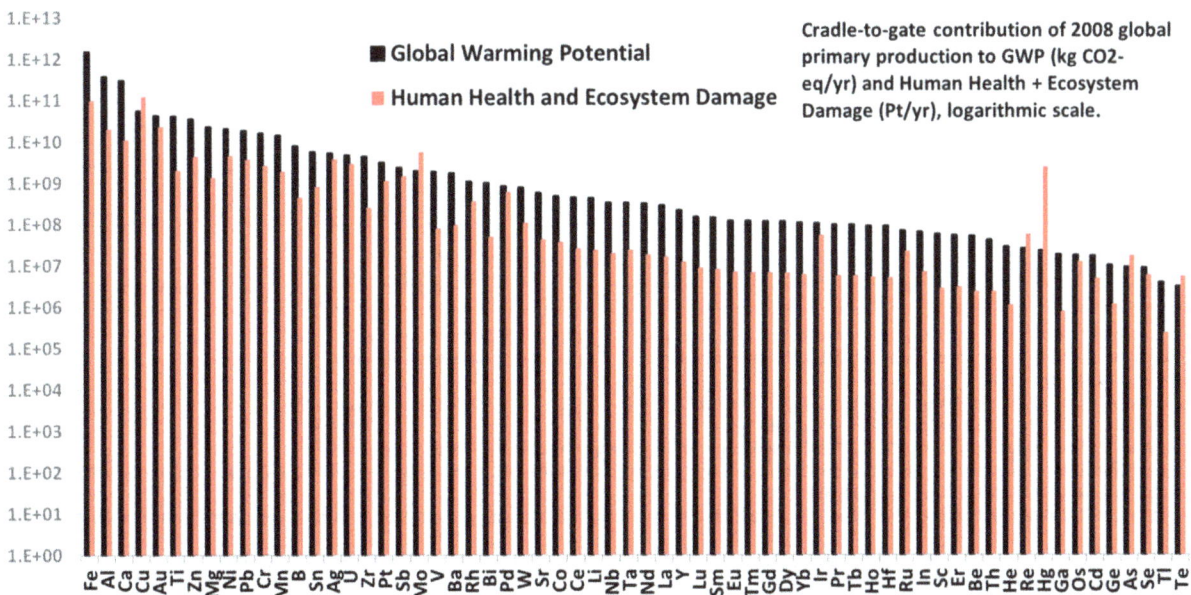

Figure 5. Global annual environmental implications of metals production in 2008. Per-kilogram impacts were multiplied with their respective production figures for year 2008 from USGS Mineral Yearbooks [53]. The cradle-to-gate human health and ecosystem damage (Pt/yr) were derived using the ReCiPe Endpoint method 1.08 H/H for the globe [106].

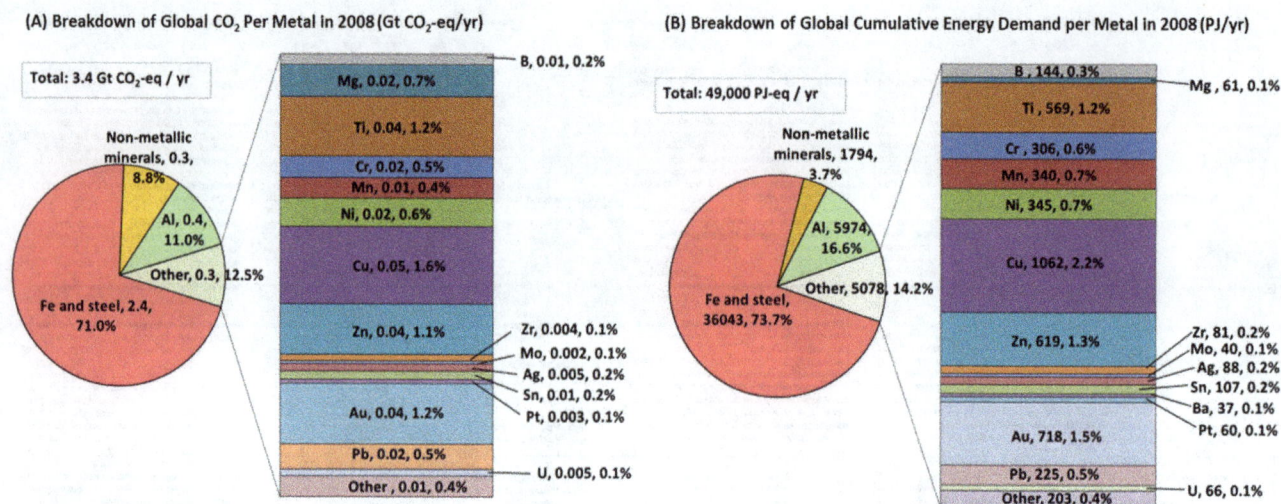

Figure 6. Breakdown of Global CO2 Emissions and Cumulative Energy Demand Per Metal in 2008.

Supporting Information S1 for the platinum group metals). As a result, more of the environmental burdens are attributed to these metals using economic allocation than if mass allocation is applied. This can result in up to two orders of magnitude difference in the impact reported (y-axis in Figure 7a). The reason that mass allocation results in similar GWPs for the rare earth elements and the platinum group metals on a per kilogram basis (Figure 7a) is because in both cases the environmental impacts are derived from single inventory data sets, using mass allocation based on the elemental distribution in the metal concentrate. On the other hand, mass allocation results in higher GWP associated with Ag, Te, Se, Cd, and As. For example, in cases where Ag is obtained

together with Au (about 27% ot total Ag production) more of the environmental burden is attributed to Ag if mass allocation is used because of its larger production volume, while economic allocation places more of the upstream burden on Au. Both Se and Te are obtained from anode slime during copper/silver production with more of the impact allocated to Ag (and less to Se and Te) taking into account price (see Table S12 to S14 in Supporting Information S1). As and Cd are obtained as low-valued by-products from copper and zinc production respectively with less of the environmental burden allocated to the elements taking into account price (see Table S9 in Supporting Information S1 for As, and Table S8b in Supporting Information S1 for Cd). Finally,

Table 1. Comparison of the Breakdown of Global CO$_2$ Emissions and Cumulative Energy Demand Per Metal to Other Studies.

Study	International Energy Agency (IEA)[a]	This study[b]	International Energy Agency (IEA)[a]	This study[b]
Unit	Gt CO$_2$/yr	Gt CO$_2$/yr	PJ/yr	PJ/yr
Fe and steel	2.5[c]	2.4[d]	23,446[j]	36,043[d]
Al	0.2[e]	0.4	3,894[k]	5,974
Other (non-ferrous metals)	0.2[e]	0.3	3,643[j]	5,078
Non-metallic minerals	0.1[f, g]	0.3[i]	1,872[e,j]	1,794[i]
Mining and quarrying	0.1[h]	-	2,219[j]	-
Total	**3.1**	**3.4**	**35,073**	**48,889**

[a]If not stated otherwise, the carbon dioxide (CO$_2$) estimates include CO$_2$ emissions from direct and indirect sources (i.e., upstream electricity production) [109]. Energy use is based on figures for final energy consumption [109], which refer to the energy supplied to the consumer, but do not include the transformation from primary energy carriers and feedstock energy.
[b]Derived by multiplying the per kg global warming potential (GWP) and cumulative energy demand (CED) for each element with their global annual production in year 2008. See Table S38 in Supporting Information S1 for more details.
[c]IEA (2008) [109] as reported in Allwood et al. (2009) [114]. Derived from Figure 16.6, page 483, in IEA (2008) [109].
[d]Based on the average of the ecoinvent 2.2. unit processes "Steel converter, unalloyed, at plant/RER U" and "Steel converter, low alloyed, at plant/RER U", multiplied with USGS global raw steel production figures for 2008 [53].
[e]IEA (2007) [110] and IEA (2008) [109] as reported in Allwood et al. (2009) [114]. Assuming that aluminum accounts for 60% of CO$_2$ emissions in the non-ferrous metals sector.
[f]Emissions from cement production are not counted, which equal 83% of total energy use and 94% of CO2 emissions in the production of non-metallic minerals (Chapter 16, page 490 of IEA (2008) [109]).
[g]Based on Figure 16.9, page 490, in IEA (2008) [109].
[h]Based on Table 16.4, page 481, in IEA (2008) [109]. Only direct industrial energy and process CO$_2$ emissions included.
[i]Only limestone production.
[j]Based on Table 16.2, page 477, in IEA (2008) [109].
[k]Based on page 194 in IEA (2010) [10].

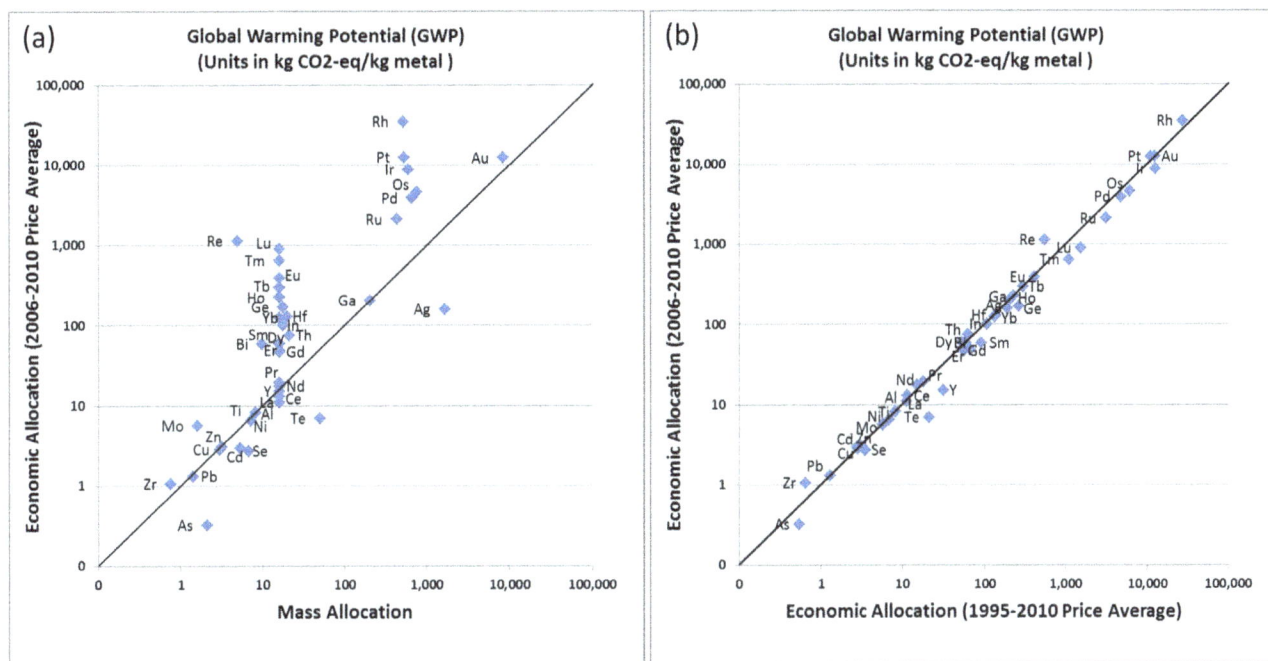

Figure 7. Choice of allocation rationale and implications on metals environmental burdens.

Figure 7b indicates that results of our model are stable to price changes over time which was tested by comparing our data set using a 5-year price average (2006 to 2010) with average prices over a 15-year period (1995 to 2010).

Discussion

Metals are important to maintain the materials base of modern technology and are expected to be of increasing importance in the transition toward sustainable technologies. Because of their widespread use in modern life, it is important to understand their life cycle wide environmental implications to make informed design choices and avoid shifting of environmental burdens. Quantifying the environmental burdens of metals production also gives a more systemic basis for evaluating benefits of material efficiency. This review of existing metals inventory data and collection of new data for several elements (discussed in Supporting Information S1) provides a detailed comparison of the cradle-to-gate environmental burdens of all of the metals used in modern technologies in their major use forms. This dataset can be used to help to inform product designers and decision makers in making environmentally preferred choices when considering alternative elements for use in new products, or when looking into metals substitution options within existing applications.

Perhaps the most obvious but highly significant result from this study is that the modern metals production system is a highly interconnected system with many of the metals being derived from multiple ores and as co-products with other metals along the stages of mining, purification, and refining (Figure 1). Because of this, it is difficult to treat each metal as an independent product and report environmental burdens independently of each other. By unifying the existing metals data, co-production issues can be better illuminated and sensitivity analysis, e.g., regarding different ways of allocating environmental burdens can be made, or assumptions regarding ore grades, recovery efficiencies, and different power mixes in the background system carried out in the future.

Our work showed that for the majority of elements in their metallic form, the cradle-to-gate environmental burdens are largely a result of the purification and refining stages (Figure 4). This trend may be partially altered in the future, given the fact that mines may gradually shift to lower grade ores which would increase the energy intensity of the mining and beneficiation process [11–13]. On the global level in 2008, the primary energy use of the metals and mining sector is estimated at 49 PJ per year (Table 1), which equals about 9.5% of global primary energy demand [111]. While on a per kilogram basis, the major industrial metals (e.g., Fe, Al) are found at the lower end of the environmental impacts scale (Figure 2 and Figure 3), they contribute significantly to global annual energy use and CO_2 emissions simply because of their large annual production (Figure 6 and Table 1). The iron and steel sector alone is responsible for about 74% of primary energy demand within the mining and metals sector. Within the category of non-ferrous metals, consisting of 61 elements, our data set allows a breakdown by each metal; information useful when concerned with optimizing non-ferrous metals systems and life cycles from a point of view of environmental impacts. Considering such issues of scale and co-production information with each metal is important to acquire a better picture of the overall effectiveness and soundness of improvement and substitution options to overall environmental burdens [3].

A sustainable metals management system has to consider many aspects of a metals life cycle and include the environmental, social, and economic spheres of sustainability. One of the limitations of this study is the cradle-to-gate focus, thereby not considering use and end-of-life stages, which are particularly important for metals, being durable and readily recyclable. Several impact categories of immediate interest for metals systems (e.g., toxicity, abiotic resource depletion, and land use) [112], could not (or only partially) be examined in this study, and should be considered in the future as more site-specific inventory data become available and LCA methods for modeling impacts to human health become more sophisticated. Our data set is based on the best available information from the public literature and, given the breadth of the assessment, naturally includes certain data gaps and limitations. However, we hope that by providing detailed information with each element (Supporting Information S1) this dataset can be built upon and further improved in the future. This type of information will be of increasing importance, given the fact that the environmental burdens of metals are likely to become more visible in the future due to the steadily increasing demand [3].

Acknowledgments

We thank Thomas E. Graedel for reviewing and commenting on an early draft of this manuscript, and the whole Yale criticality team for their manifold support in drafting this manuscript.

Author Contributions

Conceived and designed the experiments: PN MJE. Performed the experiments: PN. Analyzed the data: PN MJE. Wrote the paper: PN MJE.

References

1. Greenfield A, Graedel TE (2013) The omnivorous diet of modern technology. Resour Conserv Recycl 74: 1–7. Available: http://www.sciencedirect.com/science/article/pii/S0921344913000396. Accessed 29 October 2013.
2. Graedel TE, Erdmann L (2012) Will metal scarcity impede routine industrial use? MRS Bull 37: 325–331. doi:10.1557/mrs.2012.34.
3. Van der Voet E, Salminen R, Eckelman M, Mudd G, Norgate T, et al. (2013) Environmental Risks and Challenges of Anthropogenic Metals Flows and Cycles. A Report of the Working Group on the Global Metal Flows to the International Resource Panel. United Nations Environment Programme (UNEP).
4. Kleijn R, van der Voet E, Kramer GJ, van Oers L, van der Giesen C (2011) Metal requirements of low-carbon power generation. Energy 36: 5640–5648. Available: http://www.sciencedirect.com/science/article/pii/S0360544211004518. Accessed 24 July 2013.
5. Elshkaki A, Graedel TE (2013) Dynamic analysis of the global metals flows and stocks in electricity generation technologies. J Clean Prod. Available: http://www.sciencedirect.com/science/article/pii/S0959652613004575. Accessed 24 July 2013.
6. Graedel TE, Harper EM, Nassar NT, Reck BK (2013) On the materials basis of modern society. Proc Natl Acad Sci: 201312752. Available: http://www.pnas.org/content/early/2013/11/27/1312752110. Accessed 16 December 2013.
7. Graedel TE, Allwood J, Birat J-P, Buchert M, Hagelüken C, et al. (2011) What Do We Know About Metal Recycling Rates? J Ind Ecol 15: 355–366. Available: http://onlinelibrary.wiley.com/doi/10.1111/j.1530-9290.2011.00342.x/abstract. Accessed 1 October 2012.
8. Chapman PF, Roberts F (1983) Metal resources and energy. Butterworths. 260 p.
9. Gupta CK (2004) Chemical Metallurgy. Wiley-VCH Verlag GmbH & Co. KGaA. Available: http://onlinelibrary.wiley.com/doi/10.1002/3527602003.fmatter/summary. Accessed 24 July 2013.
10. IEA (2010) Energy Technology Perspectives 2010: Scenarios and Strategies to 2050. Paris: International Energy Agency (IEA). Available: http://www.iea.org/publications/freepublications/publication/name,26100,en.html. Accessed 30 October 2013.
11. Norgate T, Jahanshahi S (2011) Reducing the greenhouse gas footprint of primary metal production: Where should the focus be? Miner Eng 24: 1563–1570. Available: http://www.sciencedirect.com/science/article/pii/S089268751100286X. Accessed 19 March 2013.
12. Norgate TE, Jahanshahi S, Rankin WJ (2007) Assessing the environmental impact of metal production processes. J Clean Prod 15: 838–848. Available: http://www.sciencedirect.com/science/article/pii/S0959652606002320. Accessed 24 July 2013.
13. Norgate T, Jahanshahi S (2010) Low grade ores – Smelt, leach or concentrate? Miner Eng 23: 65–73. Available: http://www.sciencedirect.com/science/article/pii/S0892687509002568. Accessed 24 July 2013.
14. Laznicka P (2006) Giant Metallic Deposits: Future Sources of Industrial Metals. Berlin Heidelberg: Springer. Available: http://link.springer.com/book/10.1007/978-3-642-12405-1/page/1. Accessed 24 July 2013.
15. Verhoef EV, Dijkema GPJ, Reuter MA (2004) Process Knowledge, System Dynamics, and Metal Ecology. J Ind Ecol 8: 23–43. Available: http://onlinelibrary.wiley.com/doi/10.1162/1088198041269382/abstract. Accessed 24 July 2013.
16. Erdmann L, Graedel TE (2011) Criticality of Non-Fuel Minerals: A Review of Major Approaches and Analyses. Environ Sci Technol 45: 7620–7630. Available: http://dx.doi.org/10.1021/es200563g.
17. Graedel TE, Barr R, Chandler C, Chase T, Choi J, et al. (2012) Methodology of Metal Criticality Determination. Environ Sci Technol 46: 1063–1070. Available: http://dx.doi.org/10.1021/es203534z.
18. Nuss P, Harper EM, Nassar NT, Reck BK, Graedel TE (2014) Criticality of Iron and Its Principal Alloying Elements. Environ Sci Technol 48: 4171–4177. Available: http://dx.doi.org/10.1021/es405044w. Accessed 10 April 2014.
19. EC (2010) Critical raw materials for the EU. Brussels, Belgium: European Commission (EC). Available: http://ec.europa.eu/enterprise/policies/rawmaterials/documents/index_en.htm. Accessed 27 December 2012.
20. NRC (2008) Minerals, Critical Minerals, and the U.S. Economy..Washington, D.C: The National Academies Press. 264 p.
21. EC (2014) Report on Critical Raw Materials for the EU. Brussels, Belgium: European Commission (EC). Available: http://ec.europa.eu/enterprise/policies/raw-materials/files/docs/crm-report-on-critical-raw-materials_en.pdf. Accessed 10 June 2014.
22. Nassar NT, Barr R, Browning M, Diao Z, Friedlander E, et al. (2012) Criticality of the Geological Copper Family. Environ Sci Technol 46: 1071–1078. Available: http://dx.doi.org/10.1021/es203535w.
23. ISO (2006) Environmental Management - Life Cycle Assessment - Principles and Framework, ISO14040. Geneva, Switzerland: ISO.
24. ISO (2006) Environmental Management - Life Cycle Assessment - Requirements and Guidelines, ISO 14044. Geneva, Switzerland: ISO.
25. Classen M, Althaus H-J, Blaser S, Scharnhorst W, Tuchschmidt M, et al. (2009) Life Cycle Inventories of Metals, Data v2.0. Dübendorf, CH: Ecoinvent Centre, ETh Zurich.
26. Althaus H-J, Classen M (2005) Life Cycle Inventories of Metals and Methodological Aspects of Inventorying Material Resources in ecoinvent (7 pp). Int J Life Cycle Assess 10: 43–49. Available: http://www.springerlink.com/content/x2118371742730hp/abstract/. Accessed 8 August 2012.
27. Ecoinvent (2010) Ecoinvent Life Cycle Inventory database v2.2. Swiss Centre for Life Cycle Inventories. Available: http://www.ecoinvent.ch/. Accessed 7 November 2010.
28. European Commission (2011) ELCD core database version II. Joint research Centre - LCA Tools, Services and Data. Available: http://lca.jrc.ec.europa.eu/lcainfohub/datasetArea.vm. Accessed 31 May 2011.
29. NREL (National Renewable Energy Laboratory) (2008) U.S. Life Cycle Inventory Database (U.S. LCI), v1.6.0. National Renewable Energy Laboratory (NREL). Available: http://www.nrel.gov/lci/database/. Accessed 26 January 2010.
30. TU . Delft (2001) IdEMAT Life Cycle Inventory Database. Available: http://www.idemat.nl/. Accessed 27 August 2012.
31. UBA (2010) ProBas - Lebenszyklusdatenbank [Life Cycle Inventory Database]. Dessau (Germany); Freiburg (Germany): Umweltbundesamt (German Federal Environmental Agency) and Öko-Institut. Available: http://www.probas.umweltbundesamt.de/php/index.php. Accessed 22 November 2010.
32. CPM (Centrum för Produktrelaterad Miljöanalys) (2010) CPM LCA Database. Goteborg, Sweden: Center for Environmental Assessment of Product and Material Systems (CPM), Chalmers University of Technology. Available: http://www.cpm.chalmers.se/CPMDatabase/. Accessed 22 November 2010.
33. PE International (2012) GaBi 6 life cycle inventory database. Leinfelden-Echterdingen, Germany: PE International. Available: http://database-documentation.gabi-software.com/america/support/gabi/gabi-6-lci-documentation/. Accessed 29 October 2013.
34. Granta Design (2012) Granta: CES Selector materials selection software. Available: http://www.grantadesign.com/products/ces/. Accessed 13 August 2012.
35. Sutter J (2007) Life Cycle Inventories of Highly Pure Chemicals. Dübendorf, CH: Ecoinvent Centre, ETh Zurich.
36. Althaus H-J, Hischier R, Osses M, Primas A, Hellweg S, et al. (2007) Life Cycle Inventories of Chemicals Data v2.0. Dübendorf, CH: Ecoinvent Centre, ETh Zurich.
37. Kellenberger D, Althaus HJ, Kuenniger T, Lehmann M (2007) Life Cycle Inventory of Building Products. Duebendorf, Switzerland.
38. Giegrich J, Liebich A, Lauwigi C, Reinhardt J (2012) Indikatoren/Kennzahlen für den Rohstoffverbrauch im Rahmen der Nachhaltigkeits-diskussion. Dessau-Roßlau, Germany: UMWELTBUNDESAMT. Available: http://www.uba.de/uba-info-medien/4237.html. Accessed 5 November 2012.
39. Häussinger P, Glatthaar R, Rhode W, Kick H, Benkmann C, et al. (2000) Noble Gases. Ullmann's Encyclopedia of Industrial Chemistry. Wiley-VCH

Verlag GmbH & Co. KGaA. Available: http://onlinelibrary.wiley.com/doi/10.1002/14356007.a17_485/abstract. Accessed 11 October 2013.

40. Gruber PW, Medina PA, Keoleian GA, Kesler SE, Everson MP, et al. (2011) Global Lithium Availability. J Ind Ecol 15: 760–775. Available: http://onlinelibrary.wiley.com/doi/10.1111/j.1530-9290.2011.00359.x/abstract. Accessed 9 May 2013.

41. USGS (2011) Mineral Commodity Summaries 2011. Reston, VA: U.S. Geological Survey. Available: http://minerals.er.usgs.gov/minerals/pubs/commodity/. Accessed 31 August 2012.

42. Cunningham LD (2004) Beryllium Recycling in the United States in 2000. In: Sibley SF, editor. Flow Studies for Recycling Metal Commodities in the United States. Reston, VA: U.S. Department of the Interior, U.S. Geological Survey (USGS). Available: http://pubs.usgs.gov/circ/c1196p/c1196p.pdf. Accessed 28 August 2013.

43. Petzow G, Aldinger F, Jönsson S, Welge P, van Kampen V, et al. (2000) Beryllium and Beryllium Compounds. Ullmann's Encyclopedia of Industrial Chemistry. Wiley-VCH Verlag GmbH & Co. KGaA. Available: http://onlinelibrary.wiley.com/doi/10.1002/14356007.a04_011.pub2/abstract. Accessed 15 December 2012.

44. BeST (2013) Beryllium Facts and Figures. Beryllium Sci Technol Assoc BeST. Available: http://beryllium.eu/about-beryllium-and-beryllium-alloys/facts-and-figures/beryllium-extraction/. Accessed 28 August 2013.

45. Mc Neil D (2013) Beryllium. Roskill Mineral Services. Available: http://beryllium.eu/?wpfb_dl = 30. Accessed 28 August 2013.

46. Materion (2013) Personal communication. Materion Corporation.

47. USGS (2008) 2008 Minerals Yearbook. Reston, VA: U.S. Geological Survey.

48. Smith RA (2000) Boric Oxide, Boric Acid, and Borates. Ullmann's Encyclopedia of Industrial Chemistry. Wiley-VCH Verlag GmbH & Co. KGaA. Available: http://onlinelibrary.wiley.com/doi/10.1002/14356007.a04_263/abstract. Accessed 10 October 2013.

49. Kenny M, Oates T (2007) Lime and Limestone. Ullmann's Encyclopedia of Industrial Chemistry. Wiley-VCH Verlag GmbH & Co. KGaA. Available: http://onlinelibrary.wiley.com/doi/10.1002/14356007.a15_317.pub2/abstract. Accessed 9 September 2013.

50. Hluchan SE, Pomerantz K (2006) Calcium and Calcium Alloys. Ullmann's Encyclopedia of Industrial Chemistry. Wiley-VCH Verlag GmbH & Co. KGaA. Available: http://onlinelibrary.wiley.com/doi/10.1002/14356007.a04_515.pub2/abstract. Accessed 9 September 2013.

51. Vrana LM (2000) Calcium and Calcium Alloys. Kirk-Othmer Encyclopedia of Chemical Technology. John Wiley & Sons, Inc. Available: http://onlinelibrary.wiley.com/doi/10.1002/0471238961.0301120308090202.a01.pub3/abstract. Accessed 4 September 2013.

52. World Bank (2012) World Data Bank. World Dev Indic WDI Glob Dev Finance GDF. Available: http://databank.worldbank.org/ddp/home.do?Step = 12&id = 4&CNO = 2. Accessed 31 October 2012.

53. USGS (2010) 2010 Minerals Yearbook. Reston, Virginia: U.S. Department of the Interior, U.S. Geological Survey.

54. InfoMine (2011) Calcium Metal: Production, Market and Forecast in Russia and in the World. Moscow: InfoMine Research Group. Available: http://eng.infomine.ru/files/catalog/380/file_380_eng.pdf. Accessed 21 October 2013.

55. USGS (2009) Mineral Commodity Summaries 2009. Reston, VA: U.S. Geological Survey. Available: http://minerals.er.usgs.gov/minerals/pubs/commodity/. Accessed 31 August 2012.

56. McGill I (2000) Rare Earth Elements. Ullmann's Encyclopedia of Industrial Chemistry. Wiley-VCH Verlag GmbH & Co. KGaA. Available: http://onlinelibrary.wiley.com/doi/10.1002/14356007.a22_607/abstract. Accessed 8 November 2012.

57. USGS (2012) 2012 Minerals Yearbook. Reston, VA: U.S. Geological Survey.

58. Wang W, Pranolo Y, Cheng CY (2011) Metallurgical processes for scandium recovery from various resources: A review. Hydrometallurgy 108: 100–108. Available: http://www.sciencedirect.com/science/article/pii/S0304386X11000648. Accessed 6 September 2013.

59. USGS (2013) Mineral Commodity Summaries 2013. Reston, VA: U.S. Geological Survey. Available: http://minerals.er.usgs.gov/minerals/pubs/commodity/. Accessed 31 August 2012.

60. Ashby MF, Miller A, Rutter F, Seymour C, Wegst UGK (2012) CES EduPack for Eco Design — A White Paper. Cambridge, UK: Granta Design Ltd.

61. Woolery M (2000) Vanadium Compounds. Kirk-Othmer Encyclopedia of Chemical Technology. John Wiley & Sons, Inc. Available: http://onlinelibrary.wiley.com/doi/10.1002/0471238961.2201140123151512.a01.pub2/abstract. Accessed 18 September 2012.

62. Liddell N, Spring S, Welch C, Lucas A (2011) Vanadium Sector Review. Ocean Equities Ltd. Available: http://www.niplats.com.au/media/articles/Investor---Research/20110719-Ocean-Equities-Vanadium-Sector-Review---July-2011-261/Vanadium-Thematic-Ocean-Equities-Research-July-2011.pdf. Accessed 18 September 2012.

63. RMG (2006) Raw Materials Data Software - The Mining Database. Stockholm, Sweden: Raw Materials Group.

64. Greber JF (2000) Gallium and Gallium Compounds. Ullmann's Encyclopedia of Industrial Chemistry. Wiley-VCH Verlag GmbH & Co. KGaA. Available: http://onlinelibrary.wiley.com/doi/10.1002/14356007.a12_163/abstract. Accessed 10 November 2012.

65. Grund SC, Hanusch K, Breunig HJ, Wolf HU (2000) Antimony and Antimony Compounds. Ullmann's Encyclopedia of Industrial Chemistry. Wiley-VCH

Verlag GmbH & Co. KGaA. Available: http://onlinelibrary.wiley.com/doi/10.1002/14356007.a03_055.pub2/abstract. Accessed 8 September 2012.

66. MacMillan JP, Park JW, Gerstenberg R, Wagner H, Köhler K, et al. (2000) Strontium and Strontium Compounds. Ullmann's Encyclopedia of Industrial Chemistry. Wiley-VCH Verlag GmbH & Co. KGaA. Available: http://onlinelibrary.wiley.com/doi/10.1002/14356007.a25_321/abstract. Accessed 9 September 2012.

67. Hibbins SG (2000) Strontium and Strontium Compounds. Kirk-Othmer Encyclopedia of Chemical Technology. John Wiley & Sons, Inc. Available: http://onlinelibrary.wiley.com/doi/10.1002/0471238961.1920181508090202.a01.pub2/abstract. Accessed 9 September 2012.

68. Coope B (2006) Spain's celestite celebration. Industrial Minerals (IM). Available: http://www.indmin.com/pdfs/697/67081/200607042.pdf. Accessed 3 September 2013.

69. Nielsen RH, Wilfing G (2000) Zirconium and Zirconium Compounds. Ullmann's Encyclopedia of Industrial Chemistry. Wiley-VCH Verlag GmbH & Co. KGaA. Available: http://onlinelibrary.wiley.com/doi/10.1002/14356007.a28_543.pub2/abstract. Accessed 27 August 2012.

70. Gambogi J (2011) Zirconium and Hafnium. 2010 Minerals Yearbook. U.S. Geological Service (USGS).

71. Nielsen RH, Wilfing G (2000) Hafnium and Hafnium Compounds. Ullmann's Encyclopedia of Industrial Chemistry. Wiley-VCH Verlag GmbH & Co. KGaA. Available: http://onlinelibrary.wiley.com/doi/10.1002/14356007.a12_559.pub2/abstract. Accessed 27 August 2012.

72. Lundberg M (2012) Environmental analysis of zirconium alloy production. Available: http://uu.diva-portal.org/smash/record.jsf?pid = diva2:475527. Accessed 27 August 2012.

73. Cardarelli F (2008) Materials Handbook: A Concise Desktop Reference. Springer. 1365 p.

74. Miller GL (1954) Arc melting Kroll zirconium sponge. Vacuum 4: 168–175. Available: http://www.sciencedirect.com/science/article/pii/S0042207X54800066. Accessed 22 August 2012.

75. Munnoch S (2012) Hafnium. Minor Metals Trade Association (MMTA). Available: http://www.mmta.co.uk/uploaded_files/Hafnium%20MJ.pdf. Accessed 27 August 2012.

76. USGS (2011) Zirconium and Hafnium. 2011 Mineral Commodity Summaries. U.S. Geological Survey. Available: http://minerals.usgs.gov/minerals/pubs/commodity/zirconium/mcs-2011-zirco.pdf. Accessed 27 August 2012.

77. IAMGOLD (2009) Niobec Tour Presentation. IAMGOLD Corporation. Available: http://www.iamgold.com/Theme/IAmGold/files/presentations/Niobec%20Tour%202009%20-%20%20Niobec%20Presentation%20FINAL.pdf. Accessed 17 September 2012.

78. Albrecht S, Cymorek C, Eckert J (2000) Niobium and Niobium Compounds. Ullmann's Encyclopedia of Industrial Chemistry. Wiley-VCH Verlag GmbH & Co. KGaA. Available: http://onlinelibrary.wiley.com/doi/10.1002/14356007.a17_251.pub2/abstract. Accessed 16 September 2012.

79. IPPC (2001) Reference Document on Best Available Techniques in the Non Ferrous Metals Industries. Integrated Pollution Prevention and Control (IPPC), European Commission. Available: http://circa.europa.eu/Public/irc/env/ippc_brefs/library?l = /bref_ferrous_metals_1/ferrous_metals_enpdf/_EN_1.0_&a = d. Accessed 15 September 2012.

80. BIO Intelligence Service (2010) Environmental Impacts of some Raw Materials through LCA Methods. Paris.

81. Graf GG (2000) Tin, Tin Alloys, and Tin Compounds. Ullmann's Encyclopedia of Industrial Chemistry. Wiley-VCH Verlag GmbH & Co. KGaA. Available: http://onlinelibrary.wiley.com/doi/10.1002/14356007.a27_049/abstract. Accessed 10 November 2012.

82. Kresse R, Baudis U, Jäger P, Riechers HH, Wagner H, et al. (2000) Barium and Barium Compounds. Ullmann's Encyclopedia of Industrial Chemistry. Wiley-VCH Verlag GmbH & Co. KGaA. Available: http://onlinelibrary.wiley.com/doi/10.1002/14356007.a03_325.pub2/abstract. Accessed 9 September 2012.

83. Chinese Society of Rare Earths (2010) Chinese Rare Earth Yearbook 2010 (in Chinese). Beijing, China.

84. Du X, Graedel TE (2011) Global In-Use Stocks of the Rare Earth Elements: A First Estimate. Environ Sci Technol 45: 4096–4101. Available: http://dx.doi.org/10.1021/es102836s. Accessed 22 September 2013.

85. Gupta CK, Krishnamurthy N (2004) Extractive Metallurgy of Rare Earths. CRC Press. 537 p.

86. Dahmus J, Gutowski T (2004) An environmental analysis of machining Anaheim, California, USA. Available: http://stuff.mit.edu/afs/athena.mit.edu/course/2/2.810/www/ASME2004-62600.pdf. Accessed 8 April 2013.

87. Millensifer TA (2000) Rhenium and Rhenium Compounds. Kirk-Othmer Encyclopedia of Chemical Technology. John Wiley & Sons, Inc. Available: http://onlinelibrary.wiley.com/doi/10.1002/0471238961.1808051420180509.a01.pub3/abstract. Accessed 23 July 2013.

88. Vermaak CF (1995) Platinum-group metals: a global perspective. Mintek. 312 p.

89. CPM (2012) The CPM Platinum Group Metals Yearbook 2012. 26th ed. New York: Euromoney Books. Available: http://www.cpmgroup.com/shop/product/precious-metals/cpm-platinum-group-metals-yearbook-2012. Accessed 24 September 2013.

90. Johnson Matthey (2013) Platinum 2013. Royston, United Kingdom: Johnson Matthey.

91. Naldrett AJ (2011) Fundamentals of Magmatic Sulfide Deposits. In: Li C, Ripley EM, editors. Magmatic Ni-Cu and PGE deposits: geology, geochemistry, and genesis. Reviews in economic geology. Littleton, CO: Society of Economic Geologists. pp. 1–50.

92. USBM (1985) Mineral Facts and Problems: 1985 Edition. Washington D.C.: U.S. Department of the Interior, U.S. Bureau of Mines.

93. DeVito SC, Brooks WE (2000) Mercury. Kirk-Othmer Encyclopedia of Chemical Technology. John Wiley & Sons, Inc. Available: http://onlinelibrary.wiley.com/doi/10.1002/0471238961.1305180304052209.a01.pub2/abstract. Accessed 8 September 2012.

94. EC (2004) Mercury Flows in Europe and the World: The Impacts of Decommissioned Chlor-Alkali Plants. Brussels: European Commission. Available: http://ec.europa.eu/environment/chemicals/mercury/pdf/report.pdf. Accessed 8 September 2012.

95. Micke H, Wolf HU (2000) Thallium and Thallium Compounds. Ullmann's Encyclopedia of Industrial Chemistry. Wiley-VCH Verlag GmbH & Co. KGaA. Available: http://onlinelibrary.wiley.com/doi/10.1002/14356007.a26_607/abstract. Accessed 8 September 2012.

96. USGS (1999) Metal Prices in the United States Through 1998. Reston, VA: U.S. Geological Survey. Available: http://minerals.usgs.gov/minerals/pubs/metal_prices/metal_prices1998.pdf. Accessed 5 November 2012.

97. Sutherland CA, Milner EF, Kerby RC, Teindl H, Melin A, et al. (2000) Lead. Ullmann's Encyclopedia of Industrial Chemistry. Wiley-VCH Verlag GmbH & Co. KGaA. Available: http://onlinelibrary.wiley.com/doi/10.1002/14356007.a15_193.pub2/abstract. Accessed 31 May 2013.

98. Naumov A (2007) World market of bismuth: A review. Russ J Non-Ferr Met 48: 10–16. Available: http://www.springerlink.com/content/n300673l6p7n271n/abstract/. Accessed 7 September 2012.

99. Andrae ASG, Itsubo N, Yamaguchi H, Inaba A (2008) Life Cycle Assessment of Japanese High-Temperature Conductive Adhesives. Environ Sci Technol 42: 3084–3089. Available: http://dx.doi.org/10.1021/es0709829.

100. Stoll W (2000) Thorium and Thorium Compounds. Ullmann's Encyclopedia of Industrial Chemistry. Wiley-VCH Verlag GmbH & Co. KGaA. Available: http://onlinelibrary.wiley.com/doi/10.1002/14356007.a27_001/abstract. Accessed 27 August 2012.

101. Cuney M (2012) Uranium and Thorium: The Extreme Diversity of the Resources of the World's Energy Minerals. In: Sinding-Larsen R, Wellmer F-W, editors. Non-Renewable Resource Issues. International Year of Planet Earth. Springer Netherlands. pp. 91–129. Available: http://link.springer.com/chapter/10.1007/978-90-481-8679-2_6. Accessed 21 March 2013.

102. Schmidt G (2013) Description and critical environmental evaluation of the REE refining plant LAMP near Kuantan/Malaysia. Darmstadt, Germany: Oeko Institute e.V. Available: http://www.oeko.de/oekodoc/1628/2013-001-en.pdf. Accessed 18 March 2013.

103. Thomson Reuters (2012) World Silver Survey 2012 - A Summary. Washington D.C.: The Silver Institute. Available: https://www.silverinstitute.org/site/wp-content/uploads/2012/07/wss12sum.pdf. Accessed 22 October 2013.

104. USGS (2012) Mineral Commodity Summaries 2012. Reston, VA: U.S. Geological Survey. Available: http://minerals.er.usgs.gov/minerals/pubs/commodity/. Accessed 31 August 2012.

105. Goedkoop M, Oele M, de Sch A, Vieira M (2008) SimaPro Database Manual - Methods library. Netherlands: PRé Consultants. Available: http://www.pre.nl/download/manuals/DatabaseManualMethods.pdf. Accessed 30 November 2010.

106. Goedkoop M, Heijungs R, Huijbregts M, De Schryver A, Struijs J, et al. (2009) ReCiPe 2008, A life cycle impact assessment method which comprises harmonised category indicators at the midpoint and the endpoint level; First edition Report I: Characterisation. Available: http://www.lcia-recipe.net. Accessed 27 March 2010.

107. Rosenbaum RK, Bachmann TM, Gold LS, Huijbregts MAJ, Jolliet O, et al. (2008) USEtox—the UNEP-SETAC toxicity model: recommended characterisation factors for human toxicity and freshwater ecotoxicity in life cycle impact assessment. Int J Life Cycle Assess 13: 532–546. Available: http://link.springer.com/article/10.1007/s11367-008-0038-4. Accessed 6 April 2014.

108. Doka G (2008) Life Cycle Inventory data ofmining waste:Emissions from sulfidic tailings disposal. Zürich, Switzerland: Doka Life Cycle Assessments. Available: http://www.doka.ch/SulfidicTailingsDisposalDoka.pdf. Accessed 1 December 2013.

109. IEA (2008) Energy Technology Perspectives 2008: Scenarios and Strategies to 2050. Paris: International Energy Agency (IEA). Available: http://www.iea.org/publications/freepublications/publication/name,3771,en.html. Accessed 29 November 2013.

110. IEA (2007) Tracking Industrial Energy Efficiency and CO2 Emissions. ISBN: 978-92-64-03016-9. Paris: International Energy Agency. Available: http://www.iea.org/textbase/nppdf/free/2007/tracking_emissions.pdf. Accessed 3 January 2010.

111. IEA (2010) World Energy Outlook 2010. Paris, France: International Energy Agency (IEA). Available: http://www.worldenergyoutlook.org/media/weo2010.pdf. Accessed 24 January 2014.

112. Yellishetty M, Haque N, Dubreuil A (2012) Issues and Challenges in Life Cycle Assessment in the Minerals and Metals Sector: A Chance to Improve Raw Materials Efficiency. In: Sinding-Larsen R, Wellmer F-W, editors. Non-Renewable Resource Issues. Dordrecht: Springer Netherlands. pp. 229–246. Available: http://link.springer.com/chapter/10.1007/978-90-481-8679-2_12?null. Accessed 8 October 2012.

113. Bastian M, Heymann S, Jacomy M (2009) Gephi: an open source software for exploring and manipulating networks. Int AAAI Conf Weblogs Soc Media.

114. Allwood JM, Cullen JM, Milford RL (2010) Options for Achieving a 50% Cut in Industrial Carbon Emissions by 2050. Environ Sci Technol 44: 1888–1894. Available: http://dx.doi.org/10.1021/es902909k. Accessed 13 May 2013.

12

Environmental Management in Small and Medium-Sized Companies: An Analysis from the Perspective of the Theory of Planned Behavior

Agustín J. Sánchez-Medina[1,2]*, **Leonardo Romero-Quintero**[1,2], **Silvia Sosa-Cabrera**[1]

1 Department of Economics and Management, University of Las Palmas de Gran Canaria, The Canary Islands, Spain, **2** University Institute of Cybernetic Science and Technology, University of Las Palmas de Gran Canaria, The Canary Islands, Spain

Abstract

In the business context, concern for the environment began to develop when pressure from the public administration and environmental awareness groups raised the specific requirements for companies. The Theory of Planned Behavior considers that people's conduct is determined by the intention of carrying out a certain behavior. Thus, the individual's intent is determined by three factors related to the desired outcome of the behavior: the Personal Attitude toward the Results, the Perceived Social Norms, and the Perceived Behavioral Control over the action. Therefore, the objectives of this paper are to clarify the attitudes of the managers of Canarian small and medium-sized companies about taking environmental measures, and try to demonstrate whether there is a relationship between the proposed factors and the intention to take these measures.

Editor: Enrico Scalas, Universita' del Piemonte Orientale, Italy

Funding: These authors have no support or funding to report.

Competing Interests: The authors have declared that no competing interests exist.

* E-mail: asanchez@dede.ulpgc.es

Introduction

During the past decade, the globalization of economic activity has led to an intensification of competition on a worldwide scale. This situation is causing companies that do not adapt to this new scenario to see reduced profit margins. They are then forced to devise strategies to achieve competitive advantages that would allow them to recover or improve their profitability. Therefore, the study of the determinants of business competitiveness becomes a matter of great importance. The key is to know what the origins of the competitive advantages in a market are, and what to do to maintain and/or improve these advantages [1]. Until well into the eighties, knowledge of the environment meant that the main areas of interest in the study of business competiveness were the sectorial analysis and competition, focusing on these non controllable factors of the company [2–6]. However, since the end of the same decade, many studies have highlighted the need to examine not only the markets, but also the behavior of the organization [7–11].

On the other hand, growing concern about environmental deterioration has resulted in pressure for companies to incorporate more respectful behavior toward the environment, and this can be used as a strategic issue for making businesses competitive. In this context, small and medium-sized enterprises (hereinafter SMEs), which provide the economic and business structure of many territories, play an important role. In fact, in the European Union, SMEs account for 99% of the existing companies [12]. Therefore, their contribution to the creation of value and employment, as well as their environmental impact, can be considerable. Although it is clear that the contribution of only one SME to the sustainable development of a region is relatively small, considered as a group, they have a significant influence on the quality of the development of a particular geographical area, even exceeding that of the larger companies on occasion. In this respect, the greater the presence of SMEs in the economy or in a territory, the greater their influence is on the level of sustainable economic development [13–15].

The academic field has found it difficult to establish a precise and consensual definition of the concept of an SME. Thus, the specialized literature provides numerous criteria to characterize SMEs, such as the number of employees, turnover, total assets, etc. [16–20]. To the disadvantages of this lack of consensus, one must add the existing differences among the characteristics of SMEs, depending on the sector in which they operate. Due to these discrepancies, researchers have applied different definitions in their studies.

All of this has led governments and organizations from different countries to establish different criteria for defining what types of companies can be considered SMEs [21]. Thus, since 2005, the Commission of the European Communities states that for a company to be considered an SME, the maximum limits are as follows: 250 employees, 50 million Euros turnover, and 43 million Euros in assets [22]. Spain's General Accounting Plan is the legal text that regulates the accounting within companies. It specifically indicates that "[...] 1. All companies will be able to apply this General SME Accounting Plan, whatever their juridical form (individual or corporate), which at the end of two consecutive fiscal years must meet at least two of the following criteria: a) The total assets must not exceed 2,850,000 Euros; b) The net amount of their annual turnover must not exceed 5,700,000 Euros; c) The average number of employees during the fiscal year must not exceed 50." [23].

The purpose of this paper is to examine the perceptions of managers of SMEs, as defined in the criteria established in the Spanish regulations, about the need to implement environmental measures within their companies. Thus, based on the Theory of Planned Behavior [24], in which intent is one of the best predictors of people's planned behavior, it is interesting to determine whether an SME manager intends to carry out measures to improve the company's environmental behavior. Therefore, the overall objective of this study is to test the ability of the factors proposed in Ajzen's model to predict the intention of undertaking environmental measures in SMEs. An empirical analysis was carried out in companies located in Gran Canaria (Canary Islands – Spain) whose high energy consumption or high waste production could cause major environmental problems in their daily management.

Company and Environment

Although it is true that for a long time the economic and productive activities of mankind have had a strong relationship with the environment, only in recent years has society become concerned about the negative effects this relationship might have on the environment.

Until a few decades ago, from an analytical perspective, the business activity was considered a closed system in itself, where economic agents, consumers or producers behaved in rational ways, seeking to maximize their welfare, without taking into account the impact of their actions on the social and physical environment. While in the past there was no need to worry about possible environmental damages because nature itself solved the majority of the problems arising from production processes, distribution and consumption through recycling and biological processes [25], the situation has changed significantly. Recent social and institutional concerns about environmental deterioration have produced pressure on companies to introduce more respectful behavior toward their natural environment.

Nevertheless, before introducing these new environmental perspectives into the business, one should be aware that the environment has three significant functions [26]. First, it is the fundamental source of resources, as the environment is a supplier of the natural inputs for the production process. Second, nature provides recreational services related to enjoying the environment, such as scenic beauty, clean air, etc. And finally, the natural environment assimilates the waste and residues generated in this production and consumption. However, it must be acknowledged that only when the amount of waste discharged into the environment is within the limits set by its assimilative capacity can nature maintain its role as a repository of this waste (see Figure 1).

Traditionally, businesses and the environment have been, and in certain aspects continue to be, conflicting elements: business is a threat to the environment, and environmental concern presents obstacles to business development and job creation. In recent years, however, this opposition has been overcome, while the concept of sustainable development has been imposed. As indicated by Ruesga and Duran [27], businesses and the environment are destined to understand each other: the company plays a leading role in investigating and contributing technological solutions for environmental problems; and for the company, the environment constitutes a rapidly expanding market, a business opportunity, and an opportunity to create employment.

All business organizations, regardless of their size, activity or scope, generate environmental problems to a greater or lesser degree, and they must meet the challenge of complying with the requirements of the natural environment in which they operate.

These requirements are demonstrated in the form of pressure from its stakeholders - any group or individual who can affect or be affected by the achievement of the organization's objectives or its actions [29].

The literature includes empirical research confirming the stakeholder demands that determine the companies' environmental strategies [30–35]. Figure 2 summarizes the major stakeholders who have the ability to influence companies' environmental policies.

It seems clear that the different environmental obligations will increase in the future, with more demanding legislation and increasingly strong pressure from society and the market. Therefore, companies will need to adapt to this situation by adopting certain ecological principles. This adaptation process will generally require a transformation of the companies, their products, their production systems, and their management practices [37], which will translate into a reduction in companies' impact on the natural environment. In short, the environment has become a strategic issue for businesses [38,39], and it requires constant attention and a suitable integration of all its aspects in the company's strategy. As a result, company managers must take steps to bring about environmental improvements.

Intention Model to Undertake Environmental Measures

Research in social psychology has widely used Fishbein and Ajzen's theory of rational action [40] to investigate the relationship between motivation and a wide variety of behaviors [41,42], proving successful in many cases [43].

The fact that not all individuals behave in the same way in similar situations implies that their behavior is influenced by internal variables. Managers are no exception; therefore, they do not all have the same willingness to undertake certain actions to pursue a specific purpose. Proof of this is that the individual's psychological attitudes form a key part of the research on the entrepreneurial phenomenon [44], both within and outside the company.

This study will focus specifically on environmental initiatives within a company. It is assumed that such actions can be considered planned behavior. As can be deduced from the academic literature on psychology, intentions have been shown to be the best predictor of planned behavior. Thus, according to Ajzen [24], intentions help to understand the act itself, leading to the question of whether the factors that make up the intentions model encourage the act of carrying out environmental measures. Managers' intentions are of great interest in this sense, as they correspond to their state of mind and focus their attention on this objective.

The intention begins the process of starting an action. Therefore, models that explain the cognitive process that leads managers to act, based on these intentions, are presented as an alternative to stimulus-response models in attempting to understand their behavior. Moreover, social psychology offers intention models that can be used to explain or predict social and managerial behaviors. These models provide a theoretical framework that specifically shows the nature of the process underlying the intentional behavior. According to Krueger et al. [45], intent-based models have been applied to explain the manager's behavior in several studies [46,47].

Some authors have based their papers on the search for the existence of certain personality traits associated with entrepreneurial activity [48]. Others have focused on demonstrating the importance of demographic characteristics, such as age, sex, place

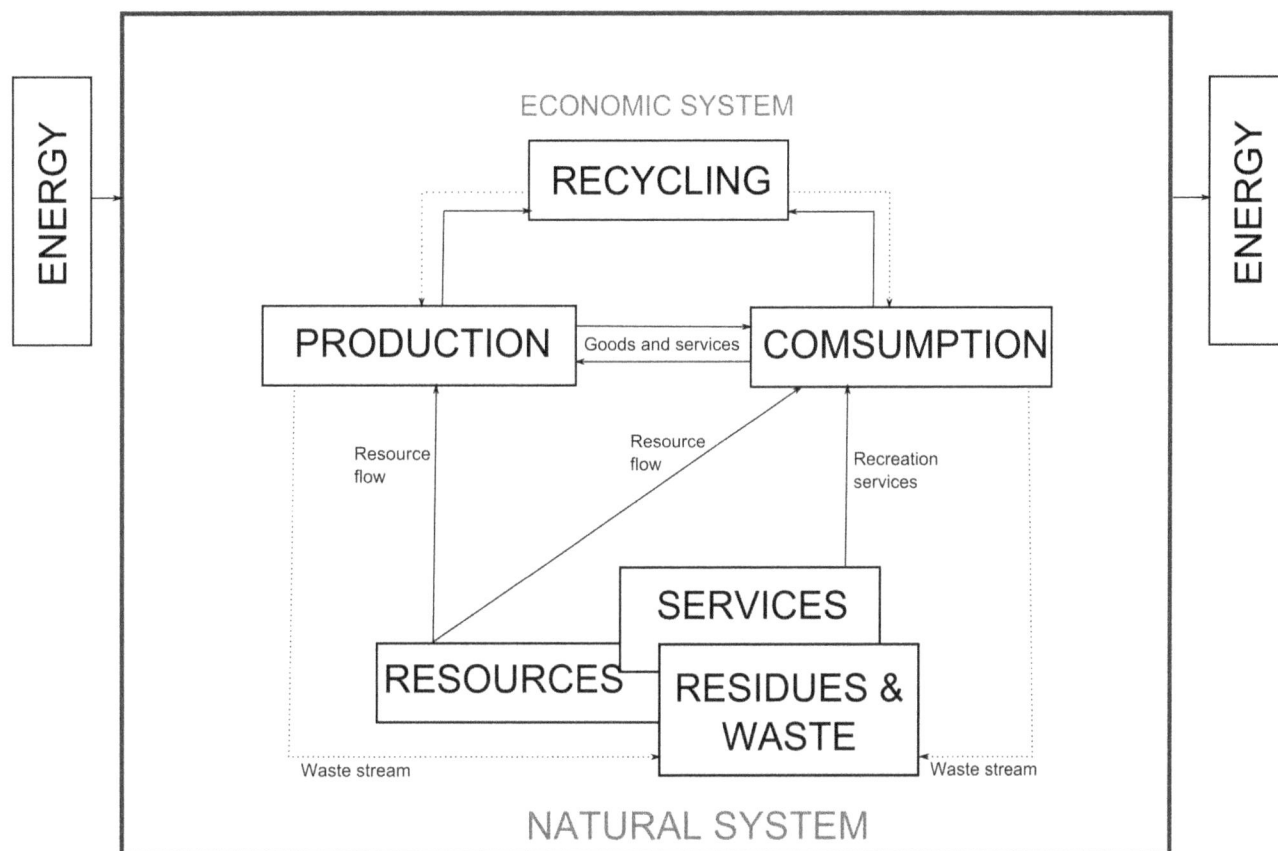

Figure 1. Environment and Economy. Source: Adapted from Ruesga and Durán [27] and Del Brio and Junquera [28].

of origin, religion, education, work experience etc. [49]. These two types of analysis have led to the identification of significant relationships between certain traits or demographic characteristics and the performance of managerial behaviors. Nevertheless, their predictive ability has been very limited [50]. With regard to the theoretical aspect, many authors have criticized both the methodological and conceptual problems presented [24,51,52]. Along the lines of the studies by Shapero and Sokol [51] on the Entrepreneurial Event model and Fishbein and Ajzen [40] on the Theory of Reasoned Action, Ajzen [24] contributes his Theory of Planned Behavior. The intention of performing a behavior

depends on the subject's attitude toward that behavior [24]. Therefore, it seems reasonable that if a manager has a favorable attitude toward performing a certain behavior, he/she is much more likely to do so. Thus, this approach based on individuals' attitudes is superior to others based on their traits or demographic aspects [45,53].

The Theory of Planned Behavior has been used in many fields, particularly in the managerial area. Many studies have linked this theory to the decision to create a firm [45,46,54–60]. Alternatively, this theory has also been applied to the study of pro-environmental behavior, focusing mainly on the attitudes of individuals and households toward recycling, their environmentally aware attitudes, and reducing pollution [61–69]. However, we found no research in the literature that applied the factors used in the Ajzen model to implementing environmental measures in SMEs. Therefore, important work remains to be done on this topic. The present paper attempts to contrast whether the factors proposed by Ajzen [24] in his theory influence the decisions made by SME managers about their company's environmental performance. Intentionality is said to be influenced by three aspects (see figure 3) whose relative importance is expected to vary for different situations and behaviors. Therefore, in this paper, we attempt to verify whether the social norms about taking environmental measures, the ability of managers to do so, and their attitudes toward these norms influence their intention of carrying them out. These decisions, as discussed throughout this paper, are becoming increasingly relevant for both companies and society in general.

In the context studied in this paper (see figure 4), the perceived social norm or the subjective norm can also help us to explain the

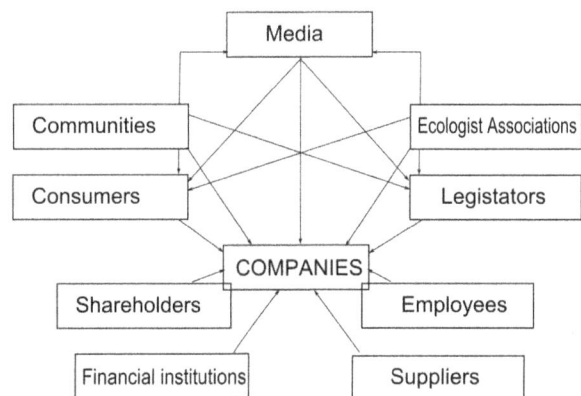

Figure 2. Environmental Stakeholders. Source: Garcés [36].

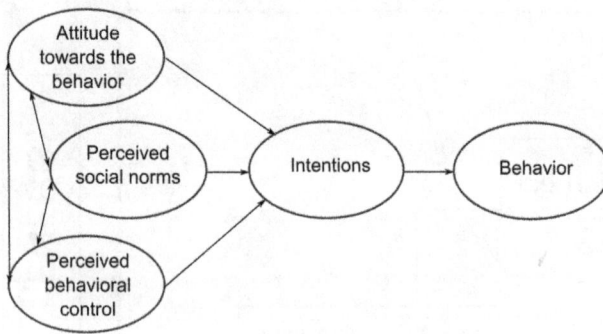

Figure 3. Ajzen's model of planned behavior. Source: Ajzen [24].

intention of carrying out environmental measures. In particular, the idea of the subjective norm means that the personal perceptions of the company's important reference groups, who believe that the behavior in question must be carried out, could influence both the Attitude toward the Behavior and the Perceived Behavioral Control, all of which lead to the following hypothesis:

H1: The Perceived Social Norms positively influence the Attitude toward the Behavior of undertaking environmental measures.
H2. The Perceived Social Norms positively influence the Perceived Behavioral Control for undertaking environmental measures.

The attitude-intention relationship used in the Ajzen model [24] should also be present. Therefore, the intention of taking environmental actions should increase if the manager has a positive attitude toward taking these measures. This relationship is reflected in the following hypothesis:

H3: The Attitude toward the Behavior positively influences the intention to undertake environmental measures.

Finally, greater perceived control, defined as the perception of ease or difficulty of performing the behavior that the manager wants to perform, favors an increase in the intention of performing environmental actions. This hypothesis has also been extensively validated in studies that use the theory of Planned Behavior, and in the environmental context it translates into the following relationship:

H4: The Perceived Behavioral Control positively influences the intention to undertake environmental measures.

According to Liñán and Chen [47], the external variables only have a direct influence on the antecedents of intention. Therefore, in the present study the control variables are included to explain the Attitude toward the Behavior and the Perceived Behavioral Control. Two types of control variables have been used, those related to the demographic information of the business owner (gender, educational level and work experience) and those related to the organizational characteristics of the company (size and activity sector).

The demographic variables included in this study have commonly been used in the literature on intentionality [46,47,70–72]. Work experience has been included, for example, in studies like those by Liñán and Chen [47], Robinson et al. [53], Kibler [70], Nabi and Liñán [71]. Another variable related to the demographic characteristics of the business owner that has frequently been employed in this type of studies is the educational level [47,73,74].

On the other hand, given that social responsibility actions, among them environmental ones, are conditioned by characteristics of the company [75–79], two control variables were also included that have to do with these characteristics. The first was company size, measured by the number of employees. This variable is relevant because it has traditionally been positively associated with social performance [76,77]; as companies grow, they focus more attention on the stakeholders and need to more effectively respond to their demands [78,80]. The other variable considered was the activity sector to which the company belongs. It was included because, as stated by Waddock and Graves [76] and Graves and Waddock [81], it is relevant to control differences stemming from pertaining to a certain activity sector.

Design and Research Methodology

It is necessary to emphasize that the managers who were chosen as the object of this study were in charge of small businesses located on the Island of Gran Canaria whose activity causes significant environmental problems. The decision to study SMEs is mainly due to the importance that these types of businesses have in the economy. Before the crisis of the 1970's, production and job creation were concentrated in large firms [82]. However, from that decade on, a change in tendency was detected that produced an increase in the importance of SMEs to national economies, as Loveman and Segenberger [83] and Schwalbach [84] confirmed in empirical studies. On the other hand, the competitiveness of these types of companies depends, fundamentally, on the capabilities of the manager or owner, on investments in intangible and technological equipment, and on their flexible innovation capacity [85]. Therefore, this paper focuses on the manager or owner and, more specifically, on the intention and the constraints of environmental measures.

In this research, the method used to obtain the necessary information to fulfill the stated objectives was a survey whose basic observation tool was a questionnaire [86]. The questionnaire was carried out by interviewers who contacted the managers directly. This procedure, although more costly than self-administered questionnaires, ensured that the answers were given by the desired people, and that the task was not given to someone else within the company. The same interviewers made contact with the managers and set up an appointment to complete the questionnaire. However, before proceeding with each questionnaire, an exploration study was carried out in order to obtain an estimate of the

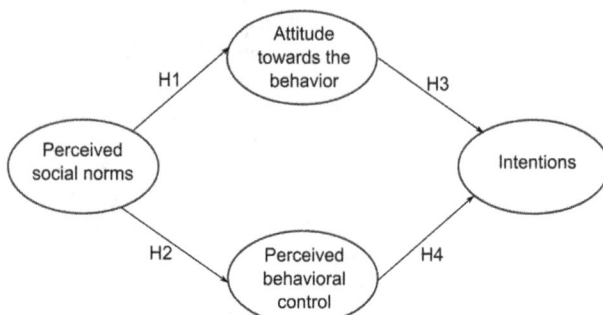

Figure 4. Sample of the conditioning intentionality Model.

actual state of each SME and the managers' perceptions of the importance of intangibles in running their businesses.

Finally, it must be noted that a pretest was conducted on ten managers, which helped us to highlight the questions that would not be clearly understood or that could lead to confusion when responding. After the pretest was carried out, we made alterations in several questions in order to ensure that the respondents were able to fully understand them.

Questionnaire responses were obtained from 201 SMEs. Table 1 summarizes the data from the quantitative research undertaken. The sample error was 6.6%, with a confidence level of 95%. It should be pointed out that, as mentioned above, the chosen spectrum did not include all SMEs in Gran Canaria, but only those whose high energy consumption or high waste production could cause major environmental problems in their daily management. Thus, the companies studied were vehicle repair shops, carpentry businesses, bakeries and restaurants. This research paper has been examined by members of the Human Research Ethics Committee of the University of Las Palmas de Gran Canaria, who have met for this purpose, and a special authorization is not considered to be required.

From the total census of the above-mentioned companies, which was obtained through the SABI database (Bureau Van Dijk – Financial company information and business intelligence for companies in Spain and Portugal), a random sampling was conducted within each group to choose the organizations that participated in the study.

To measure the different constructs, a Likert-type scale was used. The point system followed this format: 1 (strongly disagree) to 7 (strongly agree), adapted from the Entrepreneurial Intention Questionnaire (EIQ) proposed by Liñán and Chen [47].

The control variables were measured in the following way. For gender, a dichotomous variable was used, where a value of 0 was assigned to men and 1 to women. For work experience, a four-level scale was used, with a value of 1 for less than five years of experience, 2 for between 5 and 10 years, 3 for between 11 and 15 years, and 4 for more than 15 years of experience. Educational level was measured on a five-level scale, with a value of 1 for primary studies or no studies, 2 for secondary education, 3 for vocational or technical education, 4 for a university degree, and 5 for a post-graduate degree.

For company size, a five-level scale was used where the following values were assigned according to the number of employees: 1 for less than 10; 2 for 10 to 24; 3 for 25 to 49; 4 for 50 to 99; and 5 for more than 100. Finally, in the activity sector, as six

categories were established (food, automobile repair and maintenance, wood and metal industry, hospitality and restaurant, other small industries, and sales), 5 dichotomous indicators were used, that is, n-1 categories. Each of these dichotomous variables was associated with one of the first 5 categories mentioned previously, with these variables taking a value of 1 when the company was included in the sector linked to this variable and 0 in the opposite case. To represent the sixth category, all the indicators were given a value of 0.

The fieldwork was completed, and the data obtained were codified and set in table form. For this, the statistical program SPSS (Statistical Package for Social Sciences) for Windows version 19 was used. The Analysis of structural equations using the Partial Least Squares (PLS) technique was also applied to study the data. This methodology, which uses the algorithm of the Ordinary Least Square (OLS), was designed to reflect the theoretical and empirical aspects of the social qualities of the behavioral sciences, where there is generally sufficient theoretical support but little available information [87]. The PLS method is considered to be the most suitable when samples are relatively small, as in the case of this research [88]. This study specifically used SmartPls software version 02.00 [89].

The extent of common method bias was assessed with Harman's one-factor test, and it was performed by including all the items in a principal components factor analysis [90]. Evidence of common method bias exists when one factor accounts for most of the covariance. Thus, as suggested by Podsakoff and Organ [91], both the independent variables and dependent variables were included in the factor analysis. The factor analysis produced 5 factors, with none of them explaining the majority of the total variance.

The use of PLS requires two steps [92]. The first consists of the evaluation of the measurement model, in order to determine whether the relationship between the observed variables and the theoretical concepts or constructs being measured is correct. In order to carry out this analysis, the individual reliability of each item, the reliability of the construct, the Average Extracted Variance (AVE), and the discriminate validity of the indicators as measurements of the latent variables or constructs were evaluated. In order to analyze the reliability of the measurement scales, the Cronbach's alpha, among other statistics, was used. Its value ranges from zero to one, with the possibility of adopting negative values, which would imply that some of the items were measuring contrary elements. Thus, the closer the value of this statistic is to

Table 1. Research Factsheet.

Aspects of the Research	Description
Methodological procedure	Survey
Question Types	Categorised
Spectrum	SMEs with problematic environmental issues
Geographical scope	Gran Canaria
Data collection method	Questionnaire carried out by interviewers
Sample size	201 Companies
Confidence level	95% p=q 50%
Sample error	6.6%
Date of completion of the pretest	February 2012
Date of completion of the fieldwork	February 2012 to June 2012

the item, the greater the internal consistency of the indicators on the valued scale [93].

The second step consists of the evaluation of the proposed model. The aim is to confirm the extent to which causal relationships specified by the proposed model are consistent with the data available. In this way, one can endeavor to test how much of the variance in the endogenous variables is explained by the predicted constructs. The measurement of the predictive power of a model is given the value R^2 for the latent dependent variables.

The stability and validity of the estimates were examined using the *t-statistic* obtained by the bootstrap test with 500 subsamples. Finally, to verify the validity of the model, the Stone-Geisser test (Q^2) was carried out. This test was used as a criterion to measure the predictive relevancy of the dependent constructs. In the case of $Q^2 > 0$, the model has predictive relevancy, and in the opposite case, it does not.

Results

The sample in this study consisted of 201 companies on the Island of Gran Canaria. These companies are SMEs that are susceptible to having major environmental problems. Those companies that were not included due to their characteristics had no significant environmental problems and, consequently, no need to implement environmental measures. As a result, footwear and textile companies and small bazaars were not included, but vehicle repair shops, bakeries, carpentry workshops etc., were included. As already noted above, the company selection was random, and the interviews were held with individuals in positions of responsibility within the companies.

In the sample,154 of the interviewees were males (76.6%), leaving a percentage of 23.4% of women. Regarding the educational level, 22.4% had finished primary studies or less, 25.6% had a university degree, and the rest had finished secondary or vocational education.

Regarding work experience, the majority of the business owners had a lot of experience. Thus, only 7.5% had less than 5 years of experience. As far as the legal set-up of the analyzed companies is concerned, approximately 60% of the companies were limited companies, followed by SA Corporations with 22.5%, and, finally, self-employed with only 17,3%.

Regarding the activity sector, 6 categories were established: food, automobile repair and maintenance, wood and metal industry, hospitality and restaurant, other small industries (printing, paint factories, industrial hygiene, etc.) and sales. The highest percentage of business owners surveyed belonged to the food sector (22.4%), and the lowest to the wood and metal industry (10.9%) (See Table 2 for more details).

Analysis of the measurement model

To evaluate the measurement model, we must first note the individual reliability of each item. This procedure was carried out by examining the charges or simple correlations of the measurements or indicators with their respective constructs. According to Carmines and Zeller [94], in order to accept an indicator as part of a construct, it must have a value $>= 0.707$, implying that the variance shared between the construct and its indicators is greater than the error variance. However, other authors [92,95] argue that it should not be so restrictive, and that indicators exceeding 0.65 should not be eliminated. As seen in Table 3, all indicators meet the conditions that exceed the value of 0.707.

A second condition to bear in mind is the internal consistency, in order to assess the rigor with which the manifest variables are measuring the same latent variable. For this purpose, the

composite reliability should be >0.7. As shown in Table 4, all cases exceed the value of 0.90. In the same table, it can also be seen that in each case Cronbach's alpha values superior to 0.87 are presented, which leads us to affirm that our constructs are reliable.

In the third step to evaluate the validity of the scale used, the Average Variance Extracted (AVE) was studied. Fornell and Larcker [96] recommend that the AVE be greater than 0.5, which means that over 50% of the construct variance is due to its indicators. As reflected in Table 3, this requirement is met in all the constructs used. The AVE exceeds 61% in the Perceived Social Norms, and values exceeding 74% can be seen in Perceived Behavioral Control, Intentions, and Attitude toward the Behavior.

Finally, we analyzed the discriminant validity, which tells us to what extent a model construct is different from other constructs that make up the model. One way to verify these circumstances is to demonstrate that the correlations between the constructs are lower than the square root of the AVE. Table 3 shows the correlation matrix of constructs as having replaced the value of the correlation in the diagonal by the square root of the AVE. As the diagonal values are the highest value in each row and column, we can affirm the existence of discriminant validity.

As all the previous tests were positive, it is now possible to state that the measurement model is valid and reliable. In the next paragraph, the evaluation of the proposed model under study will be examined.

Proposed model Evaluation

Once the validity of the measurement model had been studied, the causal relationship proposed in the model was evaluated. Thus, the quantity of the endogenous variances explained by the predicted constructs can be observed. The measurement of a model's predictive power is the value of R^2 for the latent dependent variables. Figure 5 shows that the value of R^2 for Intention is 0.386, indicating that 38% of the variance in this construct is explained by the model. In other words, pending the verification of the validity of this relationship, which will be carried out next, the findings show that almost 38% of the variance of the variable "intention to undertake environmental measures" is determined by the Attitude toward the Behavior and the Perceived Behavioral Control. Moreover, and although they are secondary results, it is also possible to state that the construct variance in the Perceived Behavioral Control is explained by 19.99% of the Perceived Social Norms. Finally, 15.77% of the Attitude toward the Behavior variance is derived from the Perceived Social Norms (see Table 5).

Table 6 shows the path values of the various relationships proposed in the study. To evaluate the validity of these relationships, we used the Bootstrap technique, which provided the standard deviation and the *t-statistic*. This way, the stability of the estimates was examined using a t-Student distribution with a line obtained from the Bootstrap Test with 5000 subsamples [97]. If we observe the values obtained, in every given *path* the value exceeds 3.106, which was established in the *t-statistic* as having a significance level of 0.001. On the other hand, some authors have proposed that, in addition to the analyses mentioned, the confidence interval of the paths should be studied. Thus, "If a confidence interval for an estimated path coefficient w does not include zero, the hypothesis that w equals zero is rejected" [98]. All of the proposed hypotheses meet this criterion, which means they are supported.

In the case of the control variables, which, as mentioned above, are linked to both the Attitude toward the Behavior and the Perceived Behavioral Control, the results only support the existence of relationships between the gender of the business

Table 2. Description of the sample.

GENDER	Frequency	Percent	Valid Percent	STUDIES	Frequency	Percent	Valid Percent
Male	154	76.6%	76.6%	Primary or no studies	45	22.4%	23.4%
Female	47	23.4%	23.4%	Secondary	45	22.4%	22.6%
Total	201		100.0%	Vocational or technical education	53	26.4%	27.0%
				Degree	51	25.4%	26.0%
EXPERIENCE	**Frequency**	**Percent**	**Valid Percent**	Postgraduate degree	2	1.0%	1.0%
Less than 5 years	15	7.5%	7.8%	Total	196		100.0%
Between 5 and 10 years	40	19.9%	20.8%	Missing	5	2.5%	
Between 11 and 15 years	47	23.4%	24.5%				
More than 15 years	90	44.8%	46.9%	**ACTIVITY SECTOR**	**Frequency**	**Percent**	**Valid Percent**
Total	192		100.0%	Food	45	22.4%	23.4%
Missing	9	4.5%		Automobile	29	14.4%	15.1%
				Wood and metal	22	10.9%	11.5%
SIZE	**Frequency**	**Percent**	**Valid Percent**	Hospitality and restaurant	26	12.9%	13.5%
Less than 100 employees	111	55.2%	57.2%	Other small industries	32	15.9%	16.7%
Between 10 and 24 employees	53	26.4%	27.3%	Sales	38	18.9%	19.8%
Between 25 and 49 employees	14	7.0%	7.2%	Total	192		100.0%
Between 50 and 99 employees	6	3.0%	3.1%	Missing	9	4.5%	
More than 100 employees	10	5.0%	5.2%				
Total	194		100.0%	**LEGAL set-up**	**Frequency**	**Percent**	**Valid Percent**
Missing	7	3.5%		Self-employed	33	16.4%	17.3%
				Limited companies	115	57.2%	60.2%
				SA Corporations	43	21.4%	22.5%
				Total	191		100.0%
				Missing	10		

owner and the size of the business with the Attitude toward the Behavior (significance level of 0.05). Thus, it can be said that women and business owners who manage larger organizations have more favorable attitudes toward this type of behaviors.

In addition, to verify the validity of the model, the Stone-Geisser – Cross-validated Redundancy (Q^2) test was performed. This test was used as a criterion to measure the predictive relevance of the dependent constructs. In the case that $Q^2 > 0$, it indicates that the model has predictive relevance. As can be seen in Table 5, in all cases the values of Q^2 were positive, which verifies the predictive relevancy of the model.

Finally, all of the hypotheses were found to be positive.

Hypothesis 1, which stated that *Perceived Social Norms have a positive influence on the Attitude toward the Behavior to undertake environmental measures*, is confirmed ($\beta = 0.396$; $p < 0.001$).

Hypothesis 2, which proposed that *The Perceived Social Norms have a positive influence on the Perceived Behavioral Control to undertake environmental measures*, is accepted ($\beta = 0.433$; $p < 0.001$).

Hypothesis 3, which suggested that *The Attitude toward the Behavior positively influences the Intention to undertake environmental measures*, was validated ($\beta = 0.368$; $p < 0.001$).

Hypothesis 4, which stated that the Perceived Behavioral Control positively influences the Intention to undertake environmental measures, was accepted ($\beta = 0.406$; $p < 0.001$).

Discussion

This paper emphasizes the close relationship between businesses and the environment. In fact, businesses undoubtedly have an impact on the environment. To increase understanding about how small and medium-sized companies develop the intention to engage in actions that minimize their environmental impacts, it is necessary to analyze the determinants of this intention. Thus, it should be pointed out that in this type of businesses, innovative behavior is usually the reflection of the individual entrepreneurial intentions of the manager. In fact, the entrepreneur decisively conditions the development of small businesses, with some authors even considering them an extension of their founders [99–103]. The figure of the business owner and his/her perceptions are vital for carrying out actions within the company.

This research constitutes one of the first steps toward better understanding the importance of the factors in Ajzen's model [24] in explaining SME managers' intentions to carry out environmental actions. Therefore, we must emphasize that the results obtained from the analysis statistically confirm the validity of the relationships proposed in all the previously mentioned hypotheses. Perceived Social Norms have a positive influence on the Attitude

Table 3. Outer model loadings and cross loadings.

	Attitude toward the Behavior	Intentions	Perceived Behavioral Control	Perceived Social Norms
atb1	**0.823**	0.385	0.293	0.383
atb2	**0.875**	0.414	0.225	0.298
atb3	**0.904**	0.397	0.252	0.376
atb4	**0.847**	0.472	0.213	0.309
int1	0.405	**0.851**	0.424	0.301
int2	0.523	**0.782**	0.423	0.382
int3	0.356	**0.826**	0.451	0.373
int4	0.376	**0.913**	0.450	0.410
int5	0.409	**0.934**	0.451	0.421
pbc1	0.210	0.339	**0.801**	0.354
pbc2	0.181	0.418	**0.885**	0.345
pbc3	0.298	0.444	**0.890**	0.479
pbc4	0.291	0.548	**0.899**	0.414
psn1	0.448	0.440	0.388	**0.777**
psn2	0.274	0.273	0.333	**0.841**
psn3	0.358	0.445	0.460	**0.820**
psn4	0.153	0.280	0.337	**0.758**
psn5	0.299	0.257	0.271	**0.769**
psn6	0.257	0.293	0.335	**0.741**

toward the Behavior of undertaking environmental measures and the Perceived Behavioral Control in carrying them out. In turn, both the Perceived Behavioral Control to undertake environmental measures and the Attitude toward this Behavior have a positive influence on the Intention to implement environmental measures. Hence, the theory of planned action is validated in the context of environmental intentions, resulting in an instrument with the potential to improve the environmental management of small companies. On the other hand, it is also interesting that the attitude toward this type of behaviors is influenced by the gender of the business owner and the size of the organization he/she manages. Thus, the attitude is more favorable when the business owner is a woman or when the company has a larger size.

From a practical point of view, in a socially aware society that considers the environment important, it is likely that both the manager and the people around him/her will positively assess the performance of this type of action, thus increasing the manager's intention to carry out these actions. Therefore, beyond the legal requirements that companies may have to follow, it is necessary to encourage environmental awareness in the society, which can help managers of SMEs to engage in more proactive behavior that can improve the company's environmental performance, make it more socially responsible, and create value.

Economic, technological and knowledge resources are not sufficient if the person who manages the company does not have a positive attitude and the capacity to carry out the necessary actions. Furthermore, in small businesses concern for the environment is usually not a high priority, so that any action designed to improve this type of management is beneficial. In this sense, encouraging the environmental training of managers of SMEs can ensure that these individuals feel able to successfully develop activities that improve the environmental performance of

Table 4. Construct reliability, convergent validity and discriminant validity.

	AVE	Composite Reliability	Cronbach's Alpha	Attitude toward the Behavior	Intentions	Perceived Behavioral Control	Perceived Social Norms
Attitude toward the Behavior	0.744	0.921	0.885	**0.862**			
Intentions	0.745	0.936	0.913	0.485	**0.863**		
Perceived Behavioral Control	0.756	0.925	0.893	0.287	0.511	**0.870**	
Perceived Social Norms	0.616	0.906	0.877	0.399	0.440	0.461	**0.785**

Diagonal elements (bold) are the square root of variance shared between the constructs and their measures (AVE).
Off-diagonal elements are the correlations among constructs.
For discriminant validity, the diagonal elements should be larger than the off-diagonal elements.

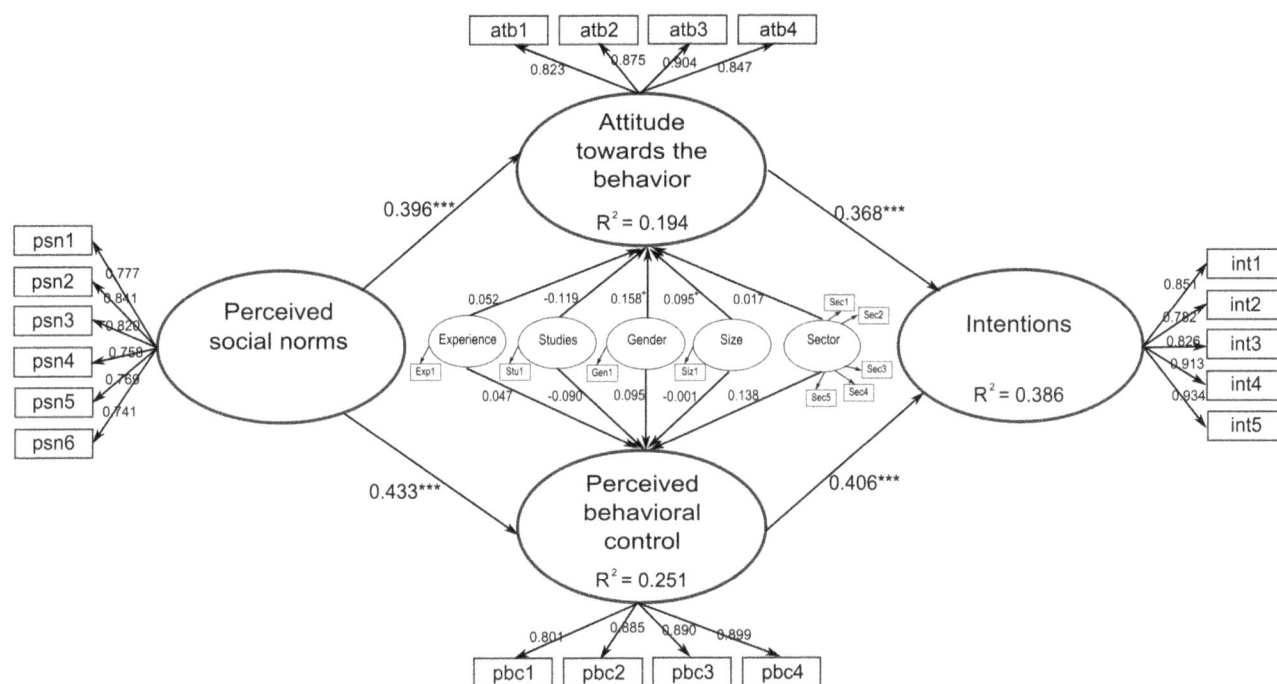

Figure 5. Indicator Charges of different constructs.

their businesses, thus increasing their intentions to carry out such actions.

Therefore, and because the Ajzen model assumes that perceptions are learned [45], fomenting environmental awareness both in society and in managers of SMEs should be a key aspect of development and training programs. Given that one of the characteristics of the modelizations performed with the PLS technique is its predictive capacity [97], one relevant issue

stemming from the present study has to do with fostering improvements in companies' environmental management. Thus, even though the study was carried out in a situation of important economic crisis, the "Attitude toward the Behavior" and the "Perceived Behavioral Control" were found to have the same weight in the entrepreneur's intention to undertake environmental initiatives in his/her company. This circumstance is important because the lack of economic resources is one of the clear

Table 5. Effects on endogenous variables.

	R^2	Q^2	Direct effect	Correlation	Variance explained
Attitude toward the Behavior	0.194	0.122			
H1: Perceived Social Norms			0.396	0.399	15.77%
Sector			0.017	0.120	0.21%
Size			0.095	0.068	0.64%
Studies			−0.119	−0.099	1.19%
Experience			0.052	0.114	0.59%
Gender			0.158	0.064	1.00%
Perceived Behavioral Control	0.251	0.188			
H2: Perceived Social Norms			0.433	0.461	19.99%
Sector			0.138	0.230	3.19%
Size			−0.001	−0.016	0.00%
Studies			−0.090	−0.127	1.15%
Experience			0.047	0.139	0.66%
Gender			0.095	0.015	0.14%
Intentions	0.386	0.273			
H3: Attitude toward the Behavior			0.368	0.485	17.85%
H4: Perceived Behavioral Control			0.406	0.511	20.74%

Table 6. Structural model results.

Hypothesis	Suggested effect	Path coefficients	t-value (bootstrap)	p-Value	Percentile bootstrap 95% confidence level		Support
					Lower	Upper	
H1: Perceived Social Norms→Attitude toward the Behavior	+	0.396***	4.658	0.000	0.233	0.503	Yes
H2: Perceived Social Norms→Perceived Behavioral Control	+	0.433***	7.355	0.000	0.259	0.552	Yes
H3: Attitude toward the Behavior→Intentions	+	0.368***	5.355	0.000	0.229	0.562	Yes
H4: Perceived Behavioral Control→Intentions	+	0.406***	5.429	0.000	0.318	0.549	Yes

Control variables		Path coefficients	t-value (bootstrap)	p-Value	Percentile bootstrap 95% confidence level		Support
					Lower	Upper	
Gender→Attitude toward the Behavior		0.158^{+}	2.166	0.000	0.015	0.300	Yes
Gender→Perceived Behavioral Control		$0.095^{ns'}$	1.340	0.033	−0.044	0.235	No
Sector→Attitude toward the Behavior		$0.017^{ns'}$	0.173	0.863	−0.178	0.212	No
Sector→Perceived Behavioral Control		$0.138^{ns'}$	1.175	0.243	−0.092	0.369	No
Size→Attitude toward the Behavior		0.095^{+}	2.015	0.046	0.003	0.187	Yes
Size→Perceived Behavioral Control		$-0.001^{ns'}$	0.011	0.992	−0.103	0.102	No
Studies→Attitude toward the Behavior		$-0.119^{ns'}$	1.883	0.062	−0.244	0.005	No
Studies→Perceived Behavioral Control		$-0.090^{ns'}$	1.281	0.203	−0.229	0.048	No
experience→Attitude toward the Behavior		$0.052^{ns'}$	0.887	0.377	−0.063	0.167	No
experience→Perceived Behavioral Control		$0.047^{ns'}$	0.741	0.460	−0.077	0.171	No

Relationships among the constructs, T-Bootstrap.
*$p < 0.05$;
**$p < 0.01$;
***$p < 0.001$; ns: not significant (based on t(4999). one-tailed test.
t(0.05; 4999) = 1.64791345; t(0.01; 4999) = 2.333843952; t(0.001; 4999) = 3.106644601.
Relationship between the dependent variable and the control variables, T-Bootstrap.
$^{+}p < 0.05$;
$^{++}p < 0.01$;
$^{+++}p < 0.001$;
$^{ns'}$: not significant (based on t(4999). two-tailed test).
t(0.05; 4999) = 1.96043859; t(0.01; 4999) = 2.57681312; t(0.001; 4999) = 3.29247411.

determinants in business owners' consideration that they cannot carry out certain activities. Therefore, if environmental authorities or organisms want small business owners to invest money in making their companies more environmentally responsible, they can choose between two different routes. The first would be to motivate this type of measures by providing the owner with economic or technical aid; the second would be to create more awareness about this issue. Although the former would have faster results that would be easier to measure, the latter option, in spite of taking longer to produce results, would be more efficient and save the administration money in the long run. Thus, an interesting strategy would be to deploy a combination of both measures. In an initial phase, and in order to start a tendency, the first route would be used. In the following phases, the money invested in the first route would be gradually reduced in order to increase the allotment for the second route.

As a limitation of this paper, we can highlight that the firms analyzed were located on an island that depends mostly on tourism and, consequently, assesses environmental issues to a greater extent. We must also add that the data collection period coincides with a very unfavorable economic situation, which may have had an influence on the responses obtained.

Finally, future lines of investigation could extend the study to other samples (different regions, sectors, economic situations, large companies, etc.) and examine whether the intention is put into action. On the other hand, economic aspects related to investments, cost savings, etc., could be introduced into the study, along with managers' perceptions about the impact of environmental actions on customers.

Acknowledgments

We would like to thank all interviewed managers for their time and patience in answering our inquiries. We greatly appreciate the assistance of editor Enrico Scalas.

Author Contributions

Conceived and designed the experiments: AS. Performed the experiments: AS LR. Analyzed the data: AS. Contributed reagents/materials/analysis tools: AS LR SS. Wrote the manuscript: AS LR SS.

References

1. Calvo M, González ZE, Pérez P, Arbelo A, Chinea A (2002) Origen de las ventajas competitivas en la empresa: un análisis empírico. In: Calvo M, González ZE, Pérez P, Arbelo A, Chinea A, editors. Selección de investigaciones empresariales. Santa Cruz de Tenerife: Fundación FYDE-CajaCanarias. pp. 113–140.

2. Hirshleifer J (1980) Price theory and applications. New Jersey: Prentice Hall.

3. Scherer FM (1980) Industrial market structure and economic performance. Chicago: Rand McNally.

4. Porter ME (1981) The contributions of industrial organization to strategic management. Academic of Management Review 6: 609–620.

5. Rumelt RP (1984) Towards a strategic theory of the firms. In: Lamb RB, editor. Competitive strategic management. Englewood Cliffs, NJ: Prentice Hall. pp. 556–570.

6. Barney JB (1986) Strategic factor markets: Expectations, luck and business strategy. Management Science 42: 1231–1241.

7. Cool K, Schendel D (1988) Performance differences among strategic groups. Strategic Management Journal 9: 207–233.

8. Hansen G, Wertnerfelt B (1989) Determinants of firm performance: The relative importance of economic and organizational factors. Strategic Management Journal 10: 399–411.

9. Rumelt RP (1991) How much does industry matter? Strategic Management Journal 12: 167–185.

10. Powell TC (1996) How much industry matter? An alternative empirical test. Strategic Management Journal 17: 323–334.

11. McGahan AM, Porter ME (1997) How much industry matter? Strategic Management Journal 18: 15–30.

12. European Commission (2011) Small and medium-sized enterprises (SMEs). Enterprise and Industry. Available: http://ec.europa.eu/enterprise/policies/sme/index_en.htm. Accessed 2012 Nov 16.

13. Welford R, Gouldson A (1993) Environmental management and business strategy. London: Pitman Publishing.

14. Vargas CM (2000) Community development and micro-enterprises: Fostering sustainable development. Sustainable Development 8(1): 11–26.

15. Medina D, Medina R, Romero L (2002) Las pequeñas y medianas empresas (PYMES) y su contribución al desarrollo sostenible. In: León C, García A, editors. Empresa y medio ambiente en Canarias. Colección investigación empresarial. Santa Cruz de Tenerife: Fundación FYDE-CajaCanarias. pp. 197–222.

16. De Koning A, Snijders J (1994) Policy on small and medium-sized enterprises in countries of the European Community. International Small Business Journal 13: 25–39.

17. Meredith GG (1994) Small business management in Australia. Sydney: McGraw Hill.

18. Holt D, Anthony S, Viney H (2000) Supporting environmental improvements in small and medium-sized enterprises in the UK. Greener Management International 30: 29–49.

19. Walczuch R, Braven GV, Lundgren H. (2000) Internet adoption barriers for small firms in the Netherlands. European Management Journal 16(5): 561–572.

20. Utomo H, Dodgson M (2001) Contributing factors to the diffusion of it within small and medium-sized firms in Indonesia. Journal of Global Information Technology Management 4(2): 22–37.

21. MacGregor RC, Vrazalic L (2007) E-commerce in regional small to medium enterprises. Hershey: IGI Global.

22. Commission of European Communities (2003) Commission Recommendation 2003/361/EC of 6 May 2003 concerning the definition of micro, small and medium-sized enterprises. Official Journal L 124 of 20.05.2003: 36–41.

23. BOE (2007) Real Decreto 1515/2007, de 16 de noviembre, por el que se aprueba el Plan General de Contabilidad de Pequeñas y Medianas Empresas y los criterios contables específicos para microempresas. BOE 279 (21 de noviembre de 2007): 47560–47566.

24. Ajzen I (1991) The theory of planned behavior. Organizational Behavior and Human Decision Processes 50: 179–211.

25. Gallego JA (1974) Economía del medio ambiente. Madrid: Instituto de Estudios Fiscales, Ministerio de Hacienda.

26. Common M (1988) Environmental and resources economics: An introduction. London: Logman.

27. Ruesga SM, Durán G (1995) Empresa y medio ambiente. Madrid: Pirámide.

28. Del Brío JA, Junquera B (2001) Medio ambiente y empresa: de la confrontación a la oportunidad. Madrid: Civitas.

29. Freeman RE (1984) Strategic Management: A stakeholder approach. Boston: Pitman.

30. Henriques I, Sadorsky P (1999) The relationship between environmental commitment and managerial perceptions of stakeholder importance. Academy of Management Journal 42: 87–99.

31. Buysse K, Verbeke A (2003) Proactive environmental strategies: a stakeholder management perspective. Strategic Management Journal 24: 453–70.

32. Sharma S, Henriques I (2005) Stakeholder influences on sustainability practices in the Canadian forest products industry. Strategic Management Journal 26: 159–80.

33. Murillo-Luna JL, Garcés-Ayerbe C, Rivera-Torres P. (2008) Why do patterns of environmental response differ? A stakeholders' pressure approach. Strategic Management Journal 29: 1225–40.

34. Rueda-Manzanares A, Aragón-Correa A, Sharma S (2008) The influence of stakeholders on the environmental strategy of service firms: The moderating effects of complexity, uncertainty and munificence. British Journal of Management 19: 185–203.

35. Darnall N, Henriques I, Sadorsky P (2010) Adopting proactive environmental strategy: The influence of stakeholders and firm size. Journal of Management Studies 47: 1072–1094.

36. Garcés C (1998) De la economía medioambiental a la teoría de la administración medioambiental. Análisis de la actitud medioambiental de la empresa española, 1990–1995. Zaragoza: Universidad de Zaragoza (thesis).

37. Shrivastava P (1995) Environmental technologies and competitive advantage. Strategic Management Journal 16: 183:200.

38. Beaumont J (1992) Managing the environment: Business opportunity and responsibility. Futures April: 187–205.

39. Taylor S (1992) Green Management: The next competitive weapon. Futures September: 669–680.

40. Fishbein M, Ajzen I (1975) Belief, attitude, intention and behavior: An introduction to theory and research. Reading MA: Addison Wesley.

41. Netemeyer RG, Burton S, Johnston M (1991) A comparison of two models for the prediction of volitional and goal-directed behaviors: A confirmatory analysis approach. Social Psychology Quarterly 54: 87–100.

42. Madden TJ, Ellen PS, Ajzen I (1992) A comparison of the theory of planned behavior and the theory of reasoned action. Personality and Social Psychology Bulletin 18: 3–9.

43. Sheppard BH, Hartwick J, Warshaw PR (1988) The theory of reasoned action: A meta-analysis of past research with recommendations for modifications and future research. Journal of Consumer Research 15: 325–343.

44. Johnson BR (1990) Toward a multidimensional model of entrepreneurship: The case of achievement motivation and the entrepreneur. Entrepreneurship Theory and Practice 14(3): 39–54.

45. Krueger NF, Reilly MD, Carsrud AL (2000) Competing models of entrepreneurial intentions. Journal of Business Venturing 15(5/6): 411–432.

46. Zhao H, Hills GE, Siebert SE (2005) The mediating role of self-efficacy in the development of entrepreneurial intentions. Journal of Applied Psychology 90(6): 1265–1272.

47. Liñán F, Chen Y (2009) Development and cross-cultural application of a specific instrument to measure entrepreneurial intentions. Entrepreneurship Theory and Practice 33(3): 593–617.

48. McClelland DC (1961) The Achieving Society. Priceton, NJ: Van Nostrand.

49. Storey DJ (1994) Understanding the Small Business Sector. London-Boston: International Thomson Business Press,

50. Reynolds P (1987) New firms: Societal contribution versus survival potential. Journal of Business Venturing 2: 231–246.

51. Shapero A, Sokol L (1982) Social dimensions of entrepreneurship. In: Kent CA, Sexton DL, Vesper KH, editors. Encyclopedia of entrepreneurship. Englewood Cliffs, NJ: Prentice Hall. pp. 72–90.

52. Garnet WB (1989) Who is an entrepreneur? Is the wrong question. Entrepreneurship Theory and Practice 13(4): 47–68.

53. Robinson PB, Stimpson DV, Huefner JC, Hunt HK (1991) An attitude approach to the prediction of entrepreneurship. Entrepreneurship Theory and Practice 15(4): 13–30.

54. Autio E, Keeley RH, Klofsten M, Parker GGC, Hay M (2001) Entrepreneurial intent among students in Scandinavia and in the USA. Enterprise and Innovation Management Studies 2(2): 145–160.

55. Peterman NE, Kennedy J (2003) Enterprise education: Influencing students' perceptions of entrepreneurship. Entrepreneurship Theory and Practice 28(2): 129–144.

56. Lee SH, Wong PK (2004) An exploratory study of technopreneurial intentions: A career anchor perspective. Journal of Business Venturing 19(1): 7–28.

57. Veciana JM, Aponte M, Urbano D (2005) University students' attitudes towards entrepreneurship: A two countries comparison. International Entrepreneurship and Management Journal 1(2): 165–182.

58. Fayolle A, Gailly B, Lassas-Clerc N (2006) Assessing the impact of entrepreneurship education programmes: A new methodology. Journal of European Industrial Training 30(9): 701–720.

59. Kolvereid L, Isaksen E (2006) New business start-up and subsequent entry into self-employment. Journal of Business Venturing 21(6): 866–885.

60. Engle RL, Dimitriadi N, Gavidia JV, Schlaegel C, Delanoe S, et al. (2010) Entrepreneurial intent: A twelve-country evaluation of Ajzen's model of planned behavior. International Journal of Entrepreneurial Behaviour & Research 16: 35–57.

61. Boldero J (1995) The prediction of household recycling of newspapers: The role of attitudes, intentions, and situational factors. Journal of Applied Social Psychology 25: 440–462.

62. Taylor S, Todd P (1995) An integrated model of waste management behavior: A test of household recycling and composting intentions. Environment and Behavior 27: 603–630.

63. Taylor S, Todd P (1997) Understanding the determinants of consumer composting behavior. Journal of Applied Social Psychology 27: 602–628.

64. Oom Do Valle P, Rebelo E, Reis E, Menezes J (2005) Combining behavioral theories to predict recycling involvement. Environment and Behavior 37(3): 364–396.

65. Oreg S, Katz-Gerro T (2006) Predicting proenvironmental behavior cross-nationally. Values, the theory of planned behavior, and value-belief-norm theory. Environment and Behavior 38(4): 462–483.

66. Durán M, Alzate M, López W, Sabucedo JM (2007) Emociones y comportamiento pro-ambiental. Revista Latinoamericana de Psicología 39: 287–296.

67. Durán M, Alzate M, Sabucedo JM (2009) La influencia de la norma personal y la teoría de la conducta planificada en la separación de residuos. Medio Ambiente y Comportamiento Humano 10: 27–39.

68. Herranz-Pascual MK, Proy-Rodríguez R, Eguiguren-García JL (2009) Comportamientos de reciclaje: propuesta de modelo predictivo para la CAPV. Medio Ambiente y Comportamiento Humano 10: 7–26.

69. Cordano M, Welcomer S, Scherer R, Pradenas l, Parada V (2010) A cross-cultural assessment of three theories of pro-environmental behavior: A comparison between business students of Chile and the United States. Environment and Behavior 43: 634–657. doi: 10.1177/0013916510378528.

70. Kibler E (2013) Formation of entrepreneurial intentions in a regional context. Entrepreneurship & Regional Development: An International Journal, 25(3–4): 293–323.

71. Nabi G, Liñan F (2013) Considering business start-up in recession time: The role of risk perception and economic context in shaping the entrepreneurial intent. International Journal of Entrepreneurial Behaviour & Research 19(6): 633–655.

72. Schlaegel C, Koenig M (2013) Determinants of Entrepreneurial Intent: A Meta-Analytic Test and Integration of Competing Models. Entrepreneurship Theory and Practice. doi:10.1111/etap.12087.

73. Cooper AC (1993) Challenges in predicting new firm performance. Journal of Business Venturing 8(3): 241–253.

74. Liñán F (2004) Intention-based models of entrepreneurship education. Piccola Impresa/Small Business 2004(3): 11–35.

75. Ullman AA (1985) Data in search of a theory: A critical examination of the relationships among social performance, social disclosure, and economic performance of U.S. firms. Academy of Management Review 10: 540–557.

76. Waddock S, Graves SB (1997) The corporate social performance-financial performance link. Strategic Management Journal 18: 303–319.

77. McWilliams A, Siegel D (2000) Corporate social responsibility and financial performance: correlation or misspecification? Strategic Management Journal 21: 603–609.

78. Hillman AJ, Keim GD (2001) Source shareholder value, stakeholder management, and social issues: What's the bottom line? Strategic Management Journal 22: 125–139.

79. Fini R, Grimaldi R, Marzocchi GL, Sobrero M (2012) The Determinants of Corporate Entrepreneurial Intention Within Small and Newly Established Firms. Entrepreneurship Theory and Practice 36: 387–414.

80. Burke L, Logsdon JM, Mitchell W, Reiner M, Vogel D (1986) Corporate community involvement in the San Francisco Bay Area. California Management Review 28: 122–141.

81. Graves SB, Waddock SA (1994) Institutional owners and corporate social performance. Academy of Management Journal 37: 1034–1046.

82. Piore M, Sabel C (1984) La segunda ruptura industrial. Madrid: Alianza Universidad.

83. Loveman G, Segerberger W (1991) The re-emergence of small scale production: An international compararison. Small Business Economics 3(1): 1–37.

84. Schwalbach J (1994) Small business dynamics in Europe. Small Business Economics 6(1): 21–25.

85. OCDE (1993) Les petites et moyennes entreprises: Technologie et compétitivité. Paris: OCDE.

86. Sierra Bravo R (1991) Técnicas de investigación social. Teoría y ejercicios. Madrid: Paraninfo.

87. Wold H (1979) Model construction and evaluation when theoretical knowledge is scarce: An example of the use of partial least quares. Geneva: Faculté des Sciences Économiques et Sociales-Université de Genève.

88. Roldán J, Real J, Leal A (2005) Organizacional learning enablers and business performance: The role of the oganizational size as an moderador variable. Pls'05 International Symposium. Barcelona: Spad Groupe Test&Go.

89. Ringle CM, Wende S, Wil A (2005) Smartpls for Windows 2005. Version 2.0 (beta). Hamburg: University of Hamburg. Available: http://www.smartpls.de. Accessed 20 April 2012.

90. Podsakoff PM, MacKenzie SB, Lee JY, Podsakoff NP (2003) Common method biases in behavioral research: A critical review of the literature and recommended remedies. Journal of Applied Psychology 88: 879–903.

91. Podsakoff PM, Organ DW (1986) Self-reports in organizacional research: Problems and prospects. Journal of Management 12: 531–544.

92. Barclay D, Higgins C, Thompson R (1995) The Partial Least Squares (PLS): Approach to causal modeling: personal computer adoption and use as an illustration. Technology studies 2: 285–309.

93. George D, Mallery P (1995) SPSS/PC step by step: A simple guide and reference. Belmont, CA: Wadsworth Publishing Company.

94. Carmines EG, Zeller RA (1979) Reliability and validity assessment. London: SAGE University Papers.

95. Chin WW (1998) Issues and opinion on structural equation modeling. MIS Quarterly 22: 7–15.

96. Fornell C, Larcker DF (1981) Evaluating structural equation models with unobservable variables and measurement error: Algebra and stadistics. Journal of Marketing Research XVIII(1): 39–50.

97. Roldán JL, Sánchez-Franco MJ (2012) Variance based structural equation modeling: guidelines for using partial least squares in information systems research. In: Mora M, Gelman O, Steenkamp A, Raisinghani M, editors. Research Methodologies, Innovations and Philosophies in Software Systems Engineering and Information Systems. Hershey, PA: Information Science Reference. pp 193–221.

98. Henseler J, Ringle CM, Sinkovics RR (2009) The use of partial least squares path modeling in international marketing. Advances in International Marketing 20: 277–320.

99. Van de Ven AH, Hudson R, Schoroeder DM (1984) Designing new business start-ups: Entrepreneurial, organizational and ecological considerations. Journal of Management 10: 87–107.

100. Brüderl J, Preisendörfer P, Ziegler R (1992) Survival chances of newly funded business organizations. American Sociological Review 57(2): 227–242.

101. Chandler GN, Jansen E (1992) The founder's self-assessed competence and venture performance. Journal of Business Venturing 7(3): 223–236.

102. Brüderl J, Preisendörfer P (1998) Network support and the success of newly founded business. Small Business Economics 10: 213–225.

103. Zahra SA, Filatotchev I (2004) Governance of the entrepreneurial threshold firm: A knowledge-based perspective. Journal of Management Studies 41: 885–897.

13

Multi-Scale Measures of Rugosity, Slope and Aspect from Benthic Stereo Image Reconstructions

Ariell Friedman*, Oscar Pizarro, Stefan B. Williams, Matthew Johnson-Roberson

Australian Centre for Field Robotics, University of Sydney, Australia

Abstract

This paper demonstrates how multi-scale measures of rugosity, slope and aspect can be derived from fine-scale bathymetric reconstructions created from geo-referenced stereo imagery. We generate three-dimensional reconstructions over large spatial scales using data collected by Autonomous Underwater Vehicles (AUVs), Remotely Operated Vehicles (ROVs), manned submersibles and diver-held imaging systems. We propose a new method for calculating rugosity in a Delaunay triangulated surface mesh by projecting areas onto the plane of best fit using Principal Component Analysis (PCA). Slope and aspect can be calculated with very little extra effort, and fitting a plane serves to decouple rugosity from slope. We compare the results of the virtual terrain complexity calculations with experimental results using conventional *in-situ* measurement methods. We show that performing calculations over a digital terrain reconstruction is more flexible, robust and easily repeatable. In addition, the method is non-contact and provides much less environmental impact compared to traditional survey techniques. For diver-based surveys, the time underwater needed to collect rugosity data is significantly reduced and, being a technique based on images, it is possible to use robotic platforms that can operate beyond diver depths. Measurements can be calculated exhaustively at multiple scales for surveys with tens of thousands of images covering thousands of square metres. The technique is demonstrated on data gathered by a diver-rig and an AUV, on small single-transect surveys and on a larger, dense survey that covers over $3,750\,m^2$. Stereo images provide 3D structure as well as visual appearance, which could potentially feed into automated classification techniques. Our multi-scale rugosity, slope and aspect measures have already been adopted in a number of marine science studies. This paper presents a detailed description of the method and thoroughly validates it against traditional *in-situ* measurements.

Editor: Christopher Fulton, The Australian National University, Australia

Funding: This work is supported by the Australian Research Council (ARC) Centre of Excellence programme, funded by the ARC and the New South Wales State Government and the Integrated Marine Observing System through the DIISR National Collaborative Research Infrastructure Scheme. The funders had no role in study design, data collection and analysis, decision to publish, or preparation of the manuscript.

Competing Interests: The authors have declared that no competing interests exist.

* E-mail: a.friedman@acfr.usyd.edu.au

Introduction

Terrain complexity is strongly correlated to biodiversity in marine environments [1–4]. Even when terrain is represented as digital bathymetry, it is necessary to abstract these digital terrain models into simpler representations in order to perform analytical work. Ecologists typically use indices, such as rugosity, slope and aspect to describe habitat structure [5]. Rugosity is a measurement that provides a notion of terrain complexity. It is a ratio between the actual length (or area) along the undulating terrain and the straight-line distance (or planar projected area). Values of 1 typically indicate flat terrain and the higher the complexity of the terrain, the higher the rugosity value. Fine-scale rugosity is traditionally measured *in-situ* by divers along a single, linear profile using chain-tape methods [1,6,7] or profile gauges [1]. In these methods, rugosity is calculated to be the ratio between the length of the contoured surface profile and the linear distance between the end points. These traditional methods are labour intensive, depth limited and put humans at risk. As a result, surveys tend to be spatially and temporally sparse and not easily repeatable. These measurements are performed using scuba, usually at depths of less than 30 m, which means that the majority of marine habitats cannot be described by this measure. Furthermore, the outputs of

transects using the traditional approach are calculated at a single, predefined resolution and scale imposed by the link-size (or gauge spacing) and the transect length. This is an important limitation since some spatial patterns and processes operate at scales not well resolved by the particular choice of chain or gauge [7]. In addition, using a length measure to capture 3D structure is not well suited to characterise the holistic features of natural landscapes [4], and measurements are prone to dramatic variation with minor changes in chain placement. When handling a physical chain *in-situ*, it may be difficult to lay out in a perfectly straight line from start to end, and this may lead to an over estimate of the rugosity due to side-to-side variation in the chain's path. Draping a chain also has an environmental impact that may lead to modifying or damaging the survey site.

Performing virtual calculations over georeferenced, high-resolution 3D bathymetry deals with these issues. It is also possible to perform calculations that better account for the 3D nature of the terrain in ways that would be impossible to measure in the field. The methods have little to no environmental impact, can be easily repeated for monitoring purposes and can be computed at multiple scales over large spatial extents.

There has been previous work that derives terrain complexity measures from bathymetric maps collected from ship borne surveys [8,9]. However, these methods cannot resolve fine scale structure due to the resolution of the survey data. Other studies have used airborne LIDAR to measure topography [10], but unfortunately these measurements are depth limited due to the poor penetration of the laser in water. In addition, neither of these techniques capture a representation that is easy to interpret visually.

Underwater vehicles, capable of high precision navigation, and equipped with downward-looking stereo cameras can recover bathymetry at fine resolutions over relatively large, contiguous extents of seafloor [11]. Measures derived from these surveys make it possible to obtain dense coverage over larger spatial extents and beyond the depths safely attainable by human divers [12]. Given that the surveys and calculations can be performed without humans, a potential source of measurement bias is eliminated. Furthermore, Autonomous Underwater Vehicles (AUVs) with proper navigation systems provide the ability for easy repeat transects, making it possible to revisit an area of interest for monitoring purposes [11].

Rugosity for a 3D surface is defined as the ratio between the area of the contoured or draped surface and the area of its orthogonal projection onto a plane. A method for calculating rugosity on raster-formatted digital elevation grids has been proposed by Jenness in [5], however forcing an irregular mesh into a raster grid causes reconstructions to be less accurate. Furthermore, Jenness's proposed rugosity calculation is subject to edge-effect problems and by using the horizontal planimetric area, rugosity is affected by slope.

The method proposed in this paper uses the geo-referenced stereo imagery obtained using AUVs or a diver-held stereo-camera rig to generate fine-scale bathymetric reconstructions with centimetre resolution in the form of irregular 3D triangular meshes [12]. Unlike a real chain, conducting measurements on a virtual surface allows for the measurement of complex features such as overhangs and underhangs. It may, however, be useful to note that the downward-looking stereo cameras that were used for this paper, collected imagery from a bird's eye view, with an altitude on the order of $2-4m$. As a result, the terrain reconstructions that we are working with did not generally capture the structure of these occluded features, but with a multi-view camera setup, these measurements would be possible. The use of image-derived bathymetry also provides the potential to combine interpretations based on 3D structure and visual appearance, which has proven useful for deriving descriptors for automated classification of benthic imagery [13–16]. We propose a new method for calculating rugosity, derived from the sum of the area of the triangles that make up the surface and dividing that by the sum of their projections onto the plane of best fit. Fitting a plane to the data ensures that rugosity and slope are decoupled at the scale of the chosen window size. As a consequence of fitting a plane, obtaining slope and aspect is trivial.

There are already a number of ecological and biological studies that have made use of our fine-scale measures of terrain complexity [13,17,18]. This paper builds upon our previous publication [15] and provides a detailed explanation of the calculations, presents multi-scale results on real data and validates the results using an experiment designed to compare our method to the traditional *in-situ* chain-tape survey technique. The code used in this paper can be downloaded from:

http://marine.acfr.usyd.edu.au/permlinks/afri7947/code-trisurfterrainfeats.php.

The remainder of this paper is organised as follows. sec:data outlines the stereo imaging platforms and the data processing pipeline. sec:calcs presents a detailed explanation of how measurements of rugosity, slope and aspect can be calculated from the stereo-derived 3D meshes. sec:validation provides validation results comparing traditional *in-situ* measured chain-tape measurements to equivalent virtual chain-tape and area-based calculations conducted on the 3D reconstructions. We also present results for surveys performed by a diver-rig and an AUV, and then finally sec:conclusion shows conclusions and presents directions for future work.

Materials

The University of Sydney's Australian Centre for Field Robotics (ACFR) develops and operates underwater stereo imaging systems that have been used on a selection of AUVs, Remotely Operated Vehicles (ROVs) [19], manned submersibles and diver-held systems [20]. Photos of example platforms are shown in Figure 1. While AUVs (Figure 1(A) & (B)) are capable of comparatively large spatial coverage, the diver-rig (Figure 1(C)) is useful for performing rapid surveys in shallow water without the need for any additional infrastructure or ship time.

The platforms are all designed for high-resolution, georeferenced survey work and each includes a downward-looking camera pair with a baseline of approximately $7cm$, pixel resolution of 1360×1024 and a field of view of 42×34 degrees. The platforms carry their own light and power sources and typically aim to maintain an altitude of $2-3m$, capturing overlapping stereo image pairs at a frequency of $1-3Hz$, depending on platform speed and altitude. This results in $3-6$ views of each scene point. All of the platforms have a suite of navigation sensors including GPS (for the surface), a pressure/depth sensor, a compass and inclinometers. The AUVs and ROVs are usually also fitted with Doppler Velocity Logs (DVL) and Ultra Short Baseline (USBL) transponders as well a selection of oceanographic and acoustic sensors.

Using the visual-aided navigation pipeline from [21] and the meshing system described in [20], the stereo imagery is combined with pose estimates to deliver fine-scale 3D, texture mapped terrain reconstructions. The processing pipeline for generating the stereo meshes is broken down into the following steps:

1. **Data Acquisition and Preprocessing:** The stereo imagery is acquired by a stereo-imaging platform and preprocessed to partially compensate for vignetting, lighting and wavelength-dependent colour absorption [20].

2. **Visual SLAM:** The platform poses are estimated through a technique called visual Simultaneous Localisation and Mapping (SLAM) [21]. Images are searched for visual loop closures and all the data from various navigational sensors are fused together to make a consistent estimate of the platforms's pose and location at every instant a stereo photo pair is captured. A visual loop closure can be thought of as a recognised landmark identified from the images. When a landmark is observed for a second time, it is possible to correct the estimated platform position to improve its navigation solution.

3. **Stereo Depth Estimation:** 2D features are matched between stereo image pairs and the 3D position is determined by triangulation. The 3D point clouds are converted into Delaunay triangulated meshes.

4. **Mesh Aggregation:** The individual stereo meshes are put into a common reference frame using SLAM-based poses and fused into a single mesh using volumetric range image processing (VRIP) [22]. Discontinuities between integrated meshes are minimised and simplified versions of the mesh are produced to allow for fast visualization at broad scales. The average resolution of the simplified 3D mesh is

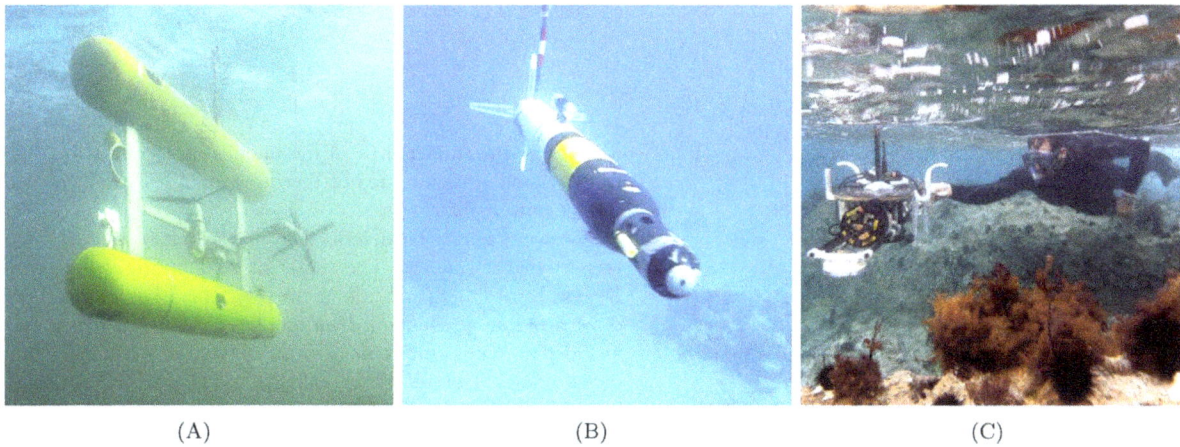

Figure 1. ACFR stereo imaging platforms in action. (A) shows *Sirius* AUV, (B) shows *Iver2* AUV and (C) shows the diver-rig.

$4,214\,vertices/m^2$, with an average triangle edge length of $4.2\,cm$.

5. **Texturing:** The polygons of the complete mesh are assigned textures based on the projection of overlapping imagery, and the result is a large-scale photo-realistic 3D reconstruction of the benthos [12].

Methods

The digital terrain reconstruction is defined by a Delaunay Triangular Irregular Network (TIN) which is made up by a set of triangular faces that connect vertices to make a 3D surface [23]. The vertices of the surface are contained in the set $\mathbf{V} = \{\mathbf{v}_m\}$, such that $\mathbf{v}_m \in \mathbb{R}^3$ and $m = 1,...,m$, where m is the total number of vertices in the surface. $\mathbf{v}_m = (x_m, y_m, z_m)$ represents the vertex m described by its x,y,z coordinates. The triangles of the surface are contained in the set $\mathbf{T} = \{\mathbf{t}_n\}$, where $n = 1,...,n$, such that $\mathbf{t}_n \subset \mathbf{V}$, and n is the total number of triangles contained in the surface. $\mathbf{t}_n = (\mathbf{v}_{1_n}, \mathbf{v}_{2_n}, \mathbf{v}_{3_n})$ represents a triangle defined by three vertices in $\mathbf{V}$.

Virtual chain-tape rugosity

For traditional *in-situ* rugosity assessments, a chain of known length, L_{chain}, is draped over the undulating substrate in a straight line and the linear distance, D_{chain}, between the end points of the chain is measured using a tape measure, as illustrated by Figure 2.

Rugosity, r_{chain}, for that transect is then computed to be the ratio between L_{chain} and D_{chain}, i.e.:

$$r_{chain} = \frac{L_{chain}}{D_{chain}} \qquad (1)$$

The rugosity value can vary depending on the resolution and type of chain that is used, however, it will always be a function of terrain complexity. For a flat area, we would expect $L_{chain} = D_{chain}$ with $r_{chain} = 1$. For more complex terrain, $L_{chain} > D_{chain}$ and therefore $r_{chain} > 1$.

Using the reconstructed fine-scale terrain model it is possible to perform virtual chain-tape measures over the TIN. This can be done by specifying three points to define a vertical plane and linking all the vertices in the mesh that lie on (or very close to) the plane to make a virtual chain. Let the plane be defined by a starting vertex, $\mathbf{v}_S = (x_S, y_S, z_S)$, an ending vertex, $\mathbf{v}_E = (x_E, y_E, z_E)$ and a third vertex directly above one of the others to define a vertical plane $\mathbf{v}_S^* = (x_S, y_S, z_S + \Delta)$, where Δ is some arbitrary non-zero value and $\mathbf{v}_S, \mathbf{v}_E \in \mathbf{V}$. We then define the subset of vertices that make up the virtual chain as, $\mathbf{C} \subseteq \mathbf{V}$. The subset $\mathbf{C}$ is determined by examining the point to plane distance d_m for every vertex in $\mathbf{V}$ and selecting the ones that fall within a threshold, δ, to the plane. The value of δ needs to be selected based on the resolution of the mesh and the point-plane distance is given by the equation,

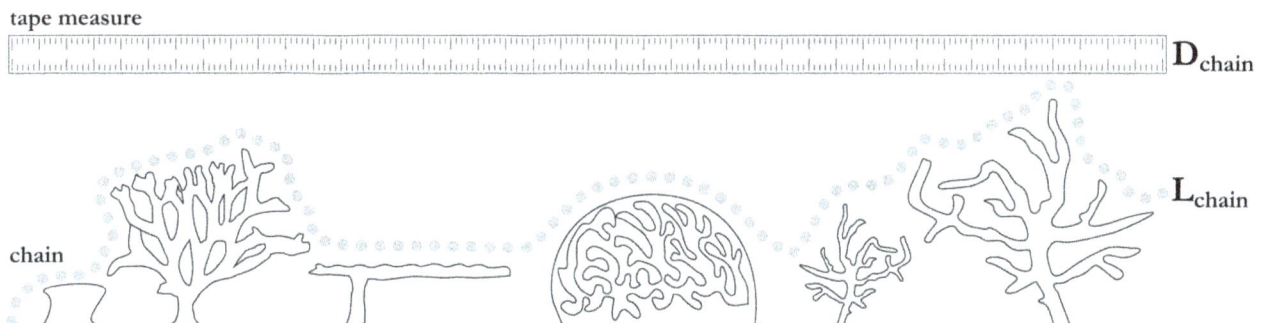

Figure 2. Chain-tape rugosity illustration. Image adapted from [26].

$$d_m = \hat{\mathbf{q}} \cdot \mathbf{v}_m + d_0 \qquad (2)$$

where $\hat{\mathbf{q}}$ is the unit vector normal to the plane and d_0 is the distance of the plane from the origin. The normal vector can be found by taking the normalised cross product of two vectors that lie on the plane:

$$\hat{\mathbf{q}} = \frac{\overrightarrow{\mathbf{v}_S \mathbf{v}_S^*} \times \overrightarrow{\mathbf{v}_S \mathbf{v}_E}}{\|\overrightarrow{\mathbf{v}_S \mathbf{v}_S^*} \times \overrightarrow{\mathbf{v}_S \mathbf{v}_E}\|}$$

and d_0 is a constant that can be calculated from $\hat{\mathbf{q}}$ and a point on the plane, e.g.:

$$d_0 = -\hat{\mathbf{q}} \cdot \mathbf{v}_S$$

We can then compute the Euclidean distance matrix for all the vertices in $\mathbf{C}$ and starting at $\mathbf{v}_S$, we trace out a virtual chain by linking all the adjacent vertices in one direction until we reach $\mathbf{v}_E$. An example of this is shown in Figure 3.

The virtual chain-tape rugosity in equ:chainrgsty can then be computed by dividing the sum of all distances between the adjacent vertices in $\mathbf{C}$, to give L_{chain}, and dividing it by D_{chain} which is simply the straight-line Euclidean distance between $\mathbf{v}_S$ and $\mathbf{v}_E$.

Virtual area-based rugosity

Given that we have a 3D reconstruction of the terrain, we can compute a ratio of areas, as opposed to a ratio of lengths. The rugosity index for a particular location in the terrain mesh can be calculated by dividing the surface area of the undulating terrain by the area of the orthogonal projection of the surface onto a plane. Instead of selecting the length of the chain, we select the size and shape of the bounding box or window with which to do the

calculation. The area-based rugosity index, r, is therefore:

$$r = \frac{A}{A'} \qquad (3)$$

where A is the surface area of the undulating terrain within the window, and A' is the area of the orthogonal projection of that surface onto a plane.

The window can be described by the subset of triangles and vertices that it encloses. The subset of vertices are contained in $\mathbf{X} = \{\mathbf{x}_k\}$, such that $k = 1,...,k$ and $\mathbf{X} \subseteq \mathbf{V}$, where k is total number of vertices that are contained within the window. A vertex is only included in $\mathbf{X}$ if it forms part of a triangle that falls entirely within the window. The subset of triangles within the window are contained in $\mathbf{W} = \{\mathbf{w}_j\}$, where $j = 1,...,j$ and j is the total number of triangles that are contained in the window. $\mathbf{w}_j = (\mathbf{x}_{1_j}, \mathbf{x}_{2_j}, \mathbf{x}_{3_j})$ represents a triangle comprised of three vertices in $\mathbf{X}$, such that $\mathbf{w}_j \subset \mathbf{X}$.

The area of the contoured surface bounded by the window A, is equal to the summation of the areas of all the individual triangles that are contained within the window

$$A = \sum_{j=1}^{j} j a_j. \qquad (4)$$

The area of an individual triangle, a_j, in the contoured surface can be calculated to be half the magnitude of the cross product of the vectors representing two adjacent sides of the triangle. The intuition for this calculation is as follows: let a triangle in the surface, $\mathbf{w}_j$, be defined by the vertices $\mathbf{x}_{1_j} = (x_1, y_1, z_1)$, $\mathbf{x}_{2_j} = (x_2, y_2, z_2)$, $\mathbf{x}_{3_j} = (x_3, y_3, z_3)$, and the adjacent vectors $\overrightarrow{\mathbf{x}_{2_j}\mathbf{x}_{1_j}}$ and $\overrightarrow{\mathbf{x}_{2_j}\mathbf{x}_{3_j}}$ to be:

$$\overrightarrow{\mathbf{x}_{2_j}\mathbf{x}_{1_j}} = [x_1 - x_2]i + [y_1 - y_2]j + [z_1 - z_2]k$$

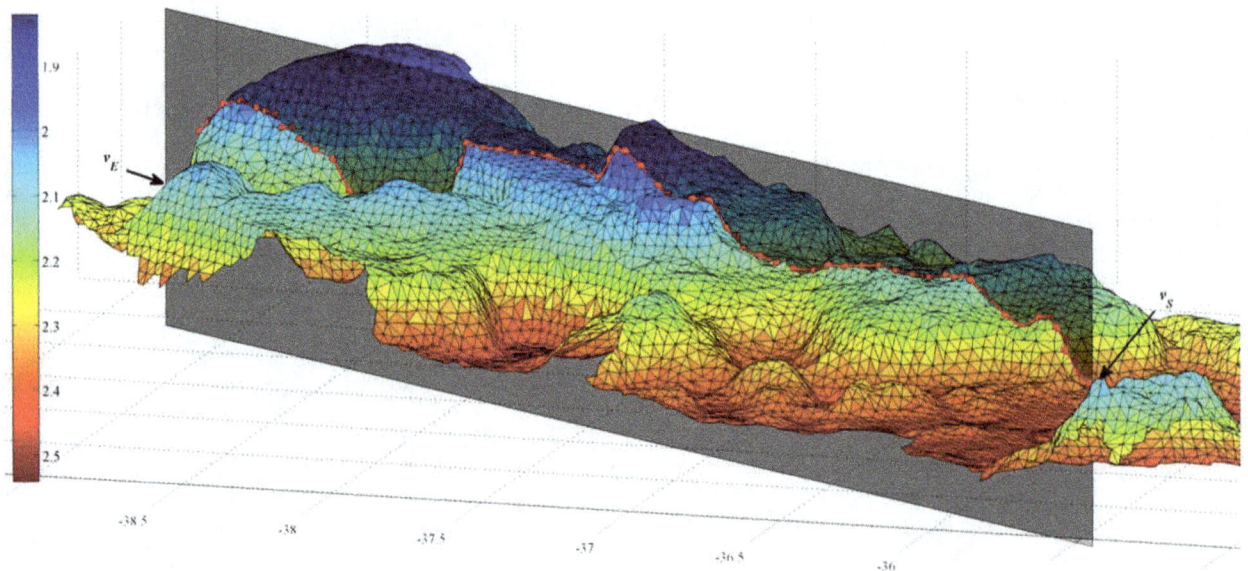

Figure 3. Example of a virtual chain 'draped' over a 3D terrain reconstruction. The coloured surface represents the terrain to be examined. The horizontal axis shows Easting (metres) and the colour bar shows depth (metres). The shaded grey plane represents the plane on which the linear rugosity will be measured while the red line and dots represent the 'chain', which is made up of those points that fall within a distance of $\delta = 5mm$ from the plane. The points $\mathbf{v}_S$ and $\mathbf{v}_E$ show the start and end verticies of the virtual chain.

$$\overrightarrow{x_{2_j}x_{3_j}} = [x_3 - x_2]i + [y_3 - y_2]j + [z_3 - z_2]k$$

The area of a parallelogram with sides $\overrightarrow{x_{2_j}x_{1_j}}$ and $\overrightarrow{x_{2_j}x_{3_j}}$ is equal to the magnitude of the cross product of vectors representing two adjacent sides. The area of an individual triangle a_j is then half of this, and can be expressed as

$$a_j = \frac{1}{2}\|\overrightarrow{x_{2_j}x_{1_j}} \times \overrightarrow{x_{2_j}x_{3_j}}\|. \qquad (5)$$

Next we need to consider the projected area A', which is the area of the orthogonal projection of the surface contained within the window, onto a plane. The correct choice of plane is an important consideration. Simply projecting the points onto the horizontal x,y plane by setting the z components to zero, for example, confounds the rugosity measurement by coupling it with slope. This would mean that flat, steep terrain would exhibit an overstated rugosity index. Ideally, we would like to have rugosity decoupled from slope at the scale of the chosen window size. Therefore, we require the area of the orthogonal projection of the surface onto the plane that best fits its vertices (contained in $\mathbf{X}$). The plane that best represents the data can be obtained using Principal Component Analysis (PCA).

PCA is used to determine the orthogonal projection of the data onto the *principal subspace* (a lower dimensional linear space) such that the variance of the projected data is maximized [24]. It involves evaluating the mean and the covariance matrix of the data $\mathbf{X}$ and then finding the eigenvectors and corresponding eigenvalues of the covariance matrix. By ordering the eigenvectors in the order of descending eigenvalues, an ordered orthogonal basis $\mathbf{u}$ is created containing the eigenvectors

$$\mathbf{u} = (\hat{a}, \hat{b}, \hat{c}) \qquad (6)$$

where $\hat{a}$ is the principal component and has the direction of largest variance of the data, $\hat{b}$ is the secondary component, and $\hat{c}$ is the third component and has the direction of the least variance of the data, and is orthogonal to the principal and secondary components. Consequently, $\hat{c}$ is a direction vector normal to the principal plane of the data, however, it is ambiguous as to whether it is inward or outward facing. Given that the data is obtained from overhead imagery, it is assumed that the outward facing normal will always have an upward facing component. This is enforced by checking the sign of the dot product between $\hat{c}$ and the upward facing unit vector, $\hat{k}$

$$\text{if } \left(\widehat{c} \cdot \widehat{k} > = 0\right)$$

$$\text{then } \left(\widehat{p} = \widehat{c}\right)$$

$$\text{else } \left(\widehat{p} = -\widehat{c}\right)$$

endif

where, $\hat{p}$ is the outward-facing normal to the principal plane of the data.

The projected area A' can now be expressed as a summation of the areas of the individually projected triangles bound by the window

$$A' = \sum_{j=1}^{J} a_j(|\hat{p} \cdot \hat{n}_j|) \qquad (7)$$

where

$$\hat{n}_j = \frac{\overrightarrow{x_{2_j}x_{1_j}} \times \overrightarrow{x_{2_j}x_{3_j}}}{\|\overrightarrow{x_{2_j}x_{1_j}} \times \overrightarrow{x_{2_j}x_{3_j}}\|}$$

is the unit vector normal to the face of triangle j and $|\hat{p} \cdot \hat{n}_j|$ gives a ratio for the projected area of the triangle on the plane to its actual contoured area in 3D space. From this, it is possible to compute the rugosity index shown in Equation 3.

Other virtual terrain measurements

Given that we now have the vector, $\hat{p}$, normal to the plane of best fit, it is relatively straightforward to obtain measurements for the slope and aspect of the same windowed region of the terrain.

Slope. Slope, denoted by θ, refers to the angle between the plane of best fit and the horizontal plane. This angle is equivalent to the angle between the normal vectors of the two planes and can be obtained from their dot product, which is $\hat{p} \cdot \hat{k} = \cos\theta$ (noting that $\hat{p}$ and $\hat{k}$ are both unit vectors). Thus, slope can be calculated as

$$\theta = \cos^{-1}(\hat{p} \cdot \hat{k}). \qquad (8)$$

The slope is a positive angle in the range $(0, \frac{\pi}{2})$.

Aspect. Aspect, denoted by ψ, refers to the direction that the surface slope faces. It is defined as the angle between the positive x axis and the projection of the normal onto the x,y plane. It can be calculated as

$$\psi = \tan^{-1}\left(\frac{p_x}{p_y}\right) \qquad (9)$$

where p_x and p_y are the components of $\hat{p}$ in the x and y directions, respectively, and $\tan^{-1}$, in this case, is the 4-quadrant inverse tangent that outputs an angle in the range $(-\pi, \pi)$. For analytical purposes, it may be useful to split aspect into vector components to eliminate the discontinuity associated with angular wrap-around:

$$\psi_N = \cos\psi$$

$$\psi_E = \sin\psi$$

where ψ_N denotes 'Northness' and ψ_E denotes 'Eastness'.

Results

In this section, we compare the virtual measurements obtained from the reconstructed terrain models to traditional *in-situ* measurement techniques, and we also present results for real data collected by a diver-rig and an AUV. Except for the *in-situ* experiments that used a physical chain, the methods proposed in this study are completely non-contact and do not require physical samples to be collected. The *in-situ* experiments did not involve endangered species or protected areas, and accordingly, no specific permits were required for the described field studies.

(A)

(B)

(C)

Figure 4. Example survey transects showing different bottom types. The figures show the photo-realistic 3D mosaic and also the depth mapped bathymetry for each transect. The small red circles show the start and end points of the chain ($L_{chain} = 5m$) that was laid out over the terrain. (A) shows a highly rugged patch ($D_{chain} = 4m$, $r_{chain} = 1.25$). It also shows the same patch from an oblique perspective. (B) shows a relatively flat patch ($D_{chain} = 5m$, $r_{chain} = 1.00$) and (C) shows a patch with medium relief ($D_{chain} = 4.3m$, $r_{chain} = 1.16$). There is also a zoomed in view of the start and end of the chain shown in (C).

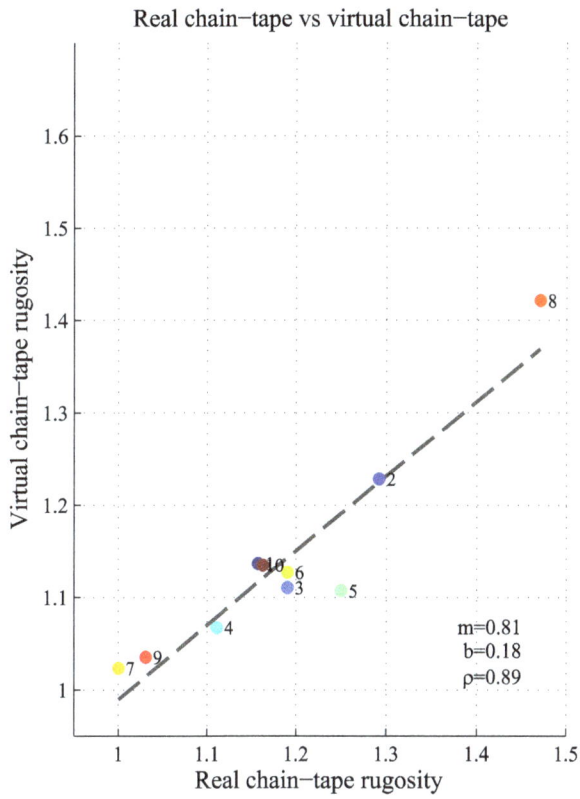

(A)

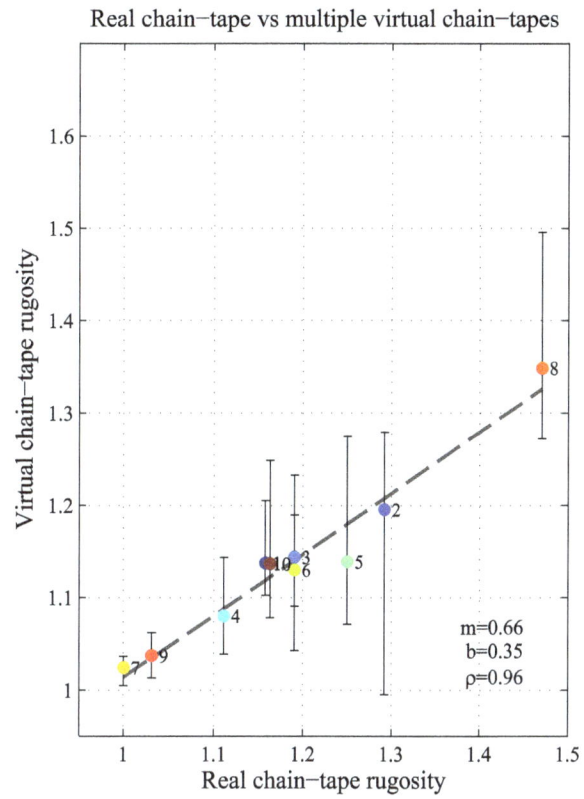

(B)

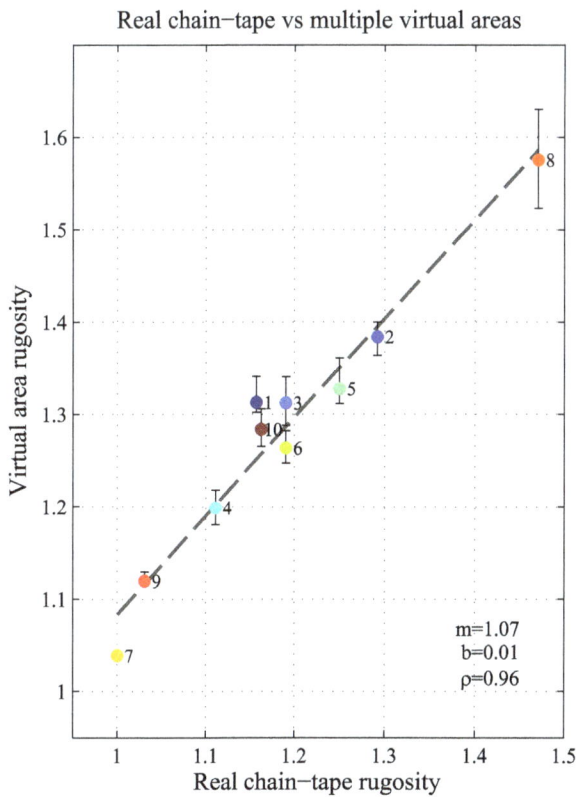

(C)

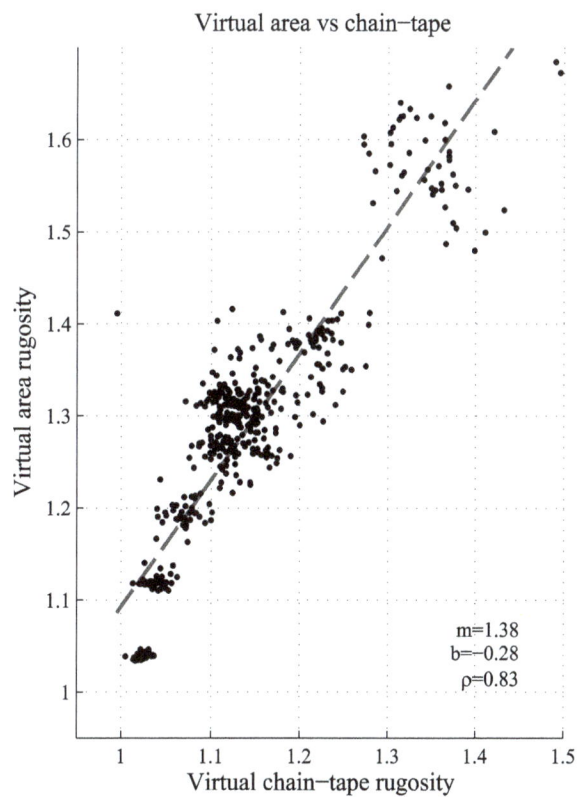

(D)

Figure 5. Comparison of virtual and *in-situ* measured rugosity measurements. (A) shows virtual chain rugosity values vs physical chain rugosity measurements for 10 different transects with varied bottom types. (B) shows the mean, minimum and maximum virtual chain-tape rugosity values for 49 virtual chains translated by less than $40cm$ from the measured location for each of the 10 transects vs the physical, real chain-tape rugosity measurements. (C) shows the mean, minimum and maximum virtual area-based rugosity with $1m \times D_{chain}$ sized windows centred and oriented over the 49 virtual chains for each of the 10 transects vs the physical, real chain-tape rugosity measurements. (D) compares each virtual chain-tape rugosity to the corresponding virtual area-based rugosity for all 490 virtual measurements (49 for each of the 10 transects.) The figures also show the least-squares linear regression fit of the means, ρ: correlation, m: slope and b: intercept per transect.

Field validation experiment

We carried out an experiment that involved laying down and measuring a physical chain ($L_{chain} = 5m$) over a selection of different transects with varying bottom types. We then surveyed each transect with the diver-held stereo imaging platform, shown in Figure 1(C). After processing the data and generating the georeferenced photo-realistic 3D meshes, we were able to pick out the locations of the start and end points of the chain for each transect and then calculate the virtual chain-tape measure explained in sec:virtchain. Figure 4 shows example transects, and Figure 4(C) shows a zoomed in view of the start and end points of the chain. The location of these points was used as the start and end points for draping the virtual chain.

Figure 5(A) shows the virtual chain rugosity measures vs the physical *in-situ* chain rugosity measurements for 10 different transects with varied bottom types. It shows a correlation of about 0.89 between the two measurements. The slope of the line of best fit to the data is 0.81 suggesting the real chain-tape rugosity values are generally higher. Explanations for this may be attributable to the fact that it is quite difficult to lay the chain out in a perfectly straight line when out in the field. Side-to-side variations in the real chain's placement may cause the rugosity to be overestimated. In addition slop in the real chain's links may lead to the chain bunching up in places, which would also cause the *in-situ* chain rugosity measurement to be overestimated.

The results in Figure 5(A) show that it is possible to obtain similar measurements from the reconstructions to what divers would recover out in the field, but without any chains and tapes. This method also allows greater flexibility with regards to the size and positioning of the 'chain' and it is possible to acquire this data using machines without putting humans at risk. In addition, the reconstructions constitute a visual record of the surveyed transect.

In an attempt to determine how much the results vary with minor changes to chain placement, we translated the virtual chain position by varying its start and end locations by a small amount, keeping the chain orientation and measured length, D_{chain}, constant. The start and end points of the virtual chains were

translated about the original measured locations by 5 cm, 10 cm, 20 cm and 40 cm, at 12 different points spanning a full circle with 30° increments (i.e.: it is moved around in a manner similar to the coupling rod connecting the wheels of a train). This results in 48 additional chains per transect, all 'laid out' in parallel with the same orientation, but with minor translations in positioning. Figure 6 illustrates how the virtual chain was translated about the terrain reconstruction.

Figure 5(B) shows the mean, minimum and maximum rugosity values for the 49 virtual chains translated about the same transect. The mean rugosity values of the 49 virtual chains translated about the measured start and end points exhibit an even stronger correlation with the physical chain measurements, of 0.96 (for the means). However, there is a large spread between the minimum and maximum virtual chain-tape rugosity values over each transect. The virtual chain-tape rugosity index varied as much as 0.28 on a single transect which equates to a difference of $1.4m$ in the straight line measurements, D_{chain}. This large variation due to minor changes in virtual chain placement (of less than $40cm$), suggests that a 1D length measure may not be well suited to capture 3D terrain structure and it motivates the need for a measure that is more robust to minor variations in positioning. A 2D area-based measurement of rugosity is less sensitive to this because with small changes in positioning, most of the area within the window is still over the same terrain, compared to the chain that may be draped over completely different terrain features. Consequently, the area based rugosity measurement is a more representative measure of the terrain complexity. Figure 5(C) shows the results of the real chain-tape rugosity vs virtual area based rugosity for $1m$-wide windows centred over the 49 virtual chains, with the lengths and orientations of the windows the same as that of the virtual chains. Even though these measurements are quite different, it is apparent that a strong correlation still exists between the rugosity values for the area-based measurement and the real chain-tape measures (0.96 for the means). However, the area based measurement is taking the structural complexity of a $1m \times D_{chain}$ window into account, and it is apparent that it is far more robust to changes in placement and therefore more

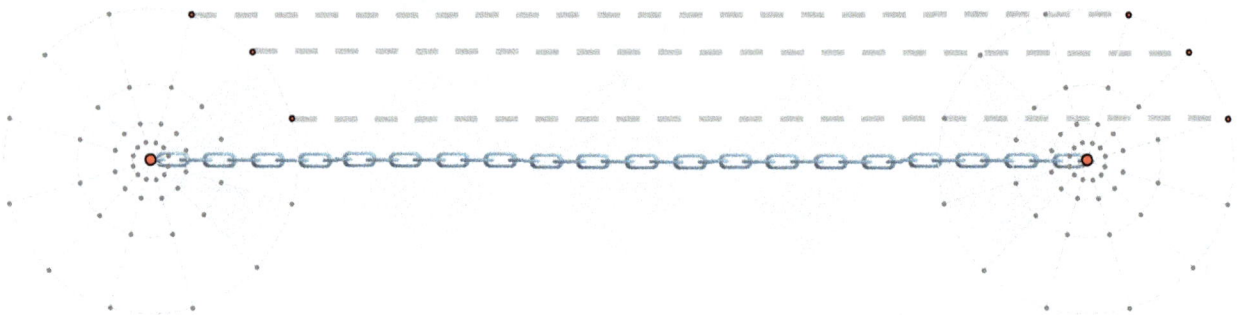

Figure 6. Illustration showing systematic translation of virtual chain placement. The start and end points of the chain were moved from the original measured locations by 5 cm, 10 cm, 20 cm and 40 cm, at 12 different points spanning a full circle with 30° increments. This results in a total of 49 virtual chains per transect, all with similar length and orientation. The figure shows the original measured chain positions (big red points in centre of circles), and three examples of the 48 additional translated virtual chains connecting the corresponding start and end points.

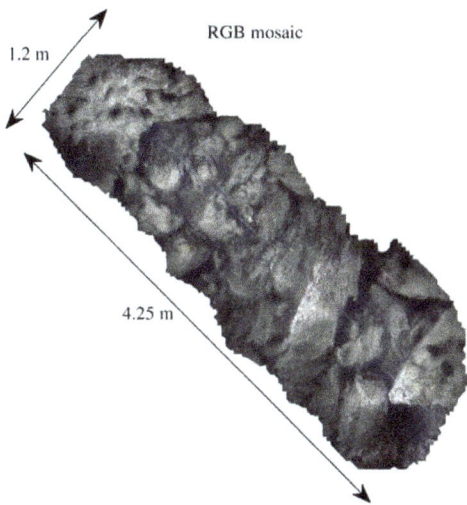

(A)

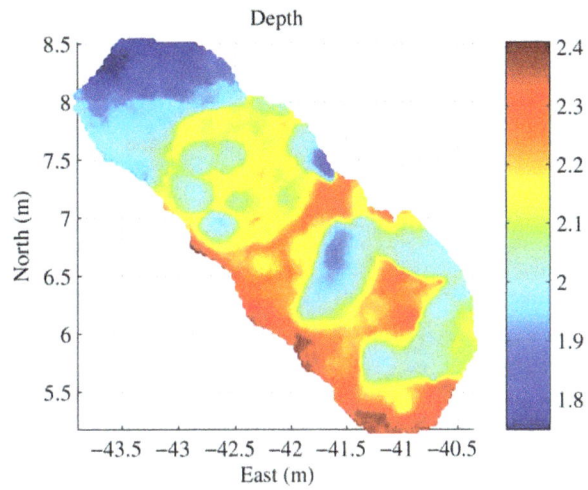

(B)

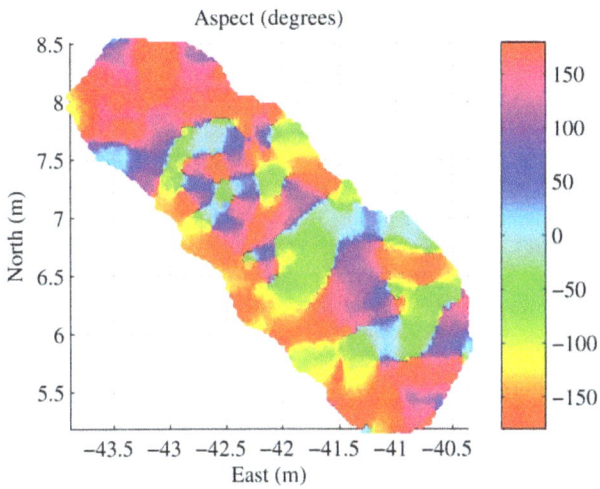

(C)

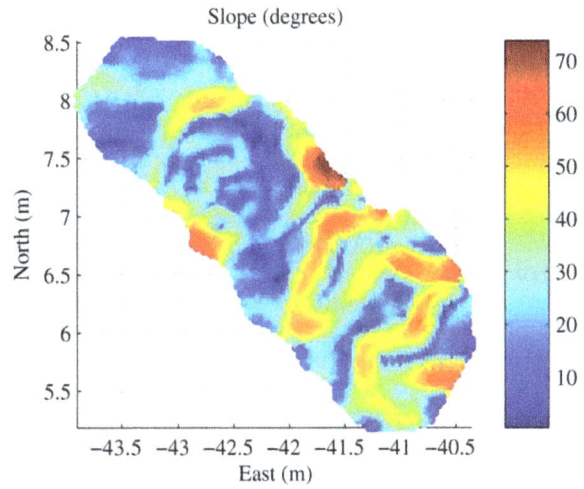

(D)

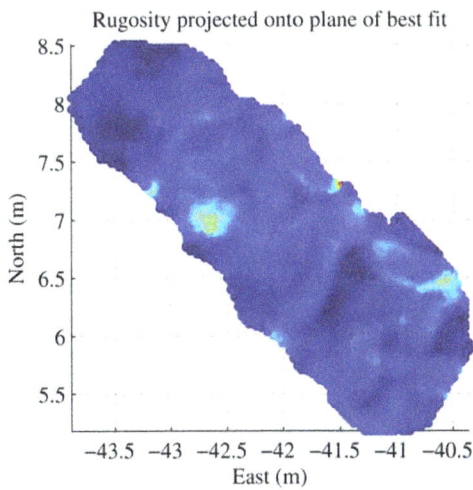

(E)

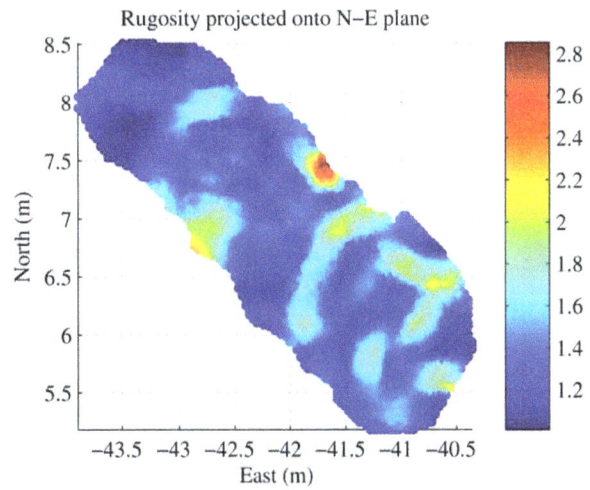

(F)

Figure 7. Fine-scale surface complexity measurements for a small, single transect diver-rig survey. Results were computed with a window size of $30cm \times 30cm$ positioned over every vertex in the mesh. (A) shows the photo-realistic 3D mosaic, (B) shows the depth/bathymetry map, (C) shows aspect, (D) shows slope, (F) shows area-based rugosity projected onto the N-E plane and (E) shows rugosity projected onto the plane of best fit.

repeatable, with a much lower spread between the minimum and maximum values resulting from translating the window over the transect, when compared to translating the virtual chain. Figure 5(D) shows a plot comparing virtual chain rugosity to virtual area rugosity. It shows an increase in variability with increasing rugosity.

Small-scale, single transect diver-rig survey results

The diver-rig can be used to obtain dense reconstructions of a patch of interest, or reconstructions along a single transect, as shown in Figure 4. It is a useful tool for rapid diver-based assessments and does not need the supporting infrastructure required by an AUV or ROV. Figure 7 shows results for a diver-rig survey conducted in Fairlight, New South Wales, Australia. It consists of a single transect spanning approximately $4.25m \times 1.2m$. Figure 7(A) shows an overhead view of the 3D photo-realistic mosaic and Figure 7(B) shows the bathymetry/depth map. The results in Figure 7(C)–(F) show results for aspect, slope and rugosity calculated at a resolution of $5cm$ with a relatively small window size of $30cm \times 30cm$.

The effects of projecting to the plane of best fit. From Figures 7(D) and (F), it is apparent that the rugosity projected onto the N-E horizontal plane appears to be higher at regions of higher slope. Comparison of Figures 7(E) and (F) highlights the effect of projecting the area onto the plane of best fit.

In order to provide an understanding of the results, we ran the calculations on a simple simulated terrain example made up of a peak and a trough with a point of inflection between them that has a high slope. Figure 8 shows results for a simulated surface. From Figures 8(B) and (C) it is apparent that the rugosity projected onto the N-E horizontal plane is highest at the point of maximum slope. Figure 8(D) shows the rugosity projected onto the plane of best fit (PCA plane), and shows the highest values at the stationary points, which are points of zero slope.

This decoupling with slope is supported by examining the correlation matrices for the different calculations. tab:corrmatfairlight shows the correlation matrix for the diver-rig survey and tab:simsurfcorrmat shows the correlation results for the simulated terrain. In both cases we can see that slope angle and the values for rugosity projected onto the N-E horizontal plane are very strongly correlated, and although there is still a mild correlation between

slope and PCA plane rugosity, there is a stronger correlation between PCA plane rugosity and N-E horizontal plane rugosity. It is apparent that fitting a plane serves to decouple rugosity from slope.

Broad-scale, dense AUV survey results

The AUV *Sirius* is part of the Integrated Marine Observing System (IMOS) and is used to collect repeatable, time-series data at various sites around Australia [11]. Figure 9 outlines the current repeat monitoring sites and provides a sense of the scale of the AUV observing program. Figure 10 shows the results for an AUV survey performed at Scott Reef that densely covered an area of $50m \times 75m$ with 9,831 stereo image pairs. This survey featured a partially populated substrate boundary between dense coral and barren sand, as illustrated by Figure 10(B).

Figure 11 shows the effect of different window sizes on the calculation of rugosity, slope and aspect. A larger window provides more spatial smoothing, however too much smoothing causes information loss. It can be seen from Figure 11 that rugosity appears to be a good indicator for the different substrate types and it outlines the boundary between the different substrates shown in Figures 10(A) and (B) quite closely. Consequently, these measures have been found to be useful descriptors for automatically discriminating different habitat types [13–16].

A note on aspect angle. The aspect angle must be considered with reference to the slope, i.e. at regions where the slope is close to zero, the aspect is relatively erratic since the normal vector points almost directly up and the direction of the component of the normal projected onto the N-E plane, changes dramatically with a small change in any of the variables in the calculation. It should also be noticed that aspect is subject to angular wraparound where a value of $-180°$ should be interpreted to be the same as a value of $+180°$. This needs to be taken into consideration when interpreting the results. Consequently in Figures 7 and 11, the aspect plots were displayed using a circular colour map that shows a continuos blend about $+/-180°$. Measurements of aspect are likely to be more useful for classification purposes when framed in context with water currents and environmental conditions to calculate a notion of exposure. It is also possible to weight aspect with slope angle to provide a notion of magnitude.

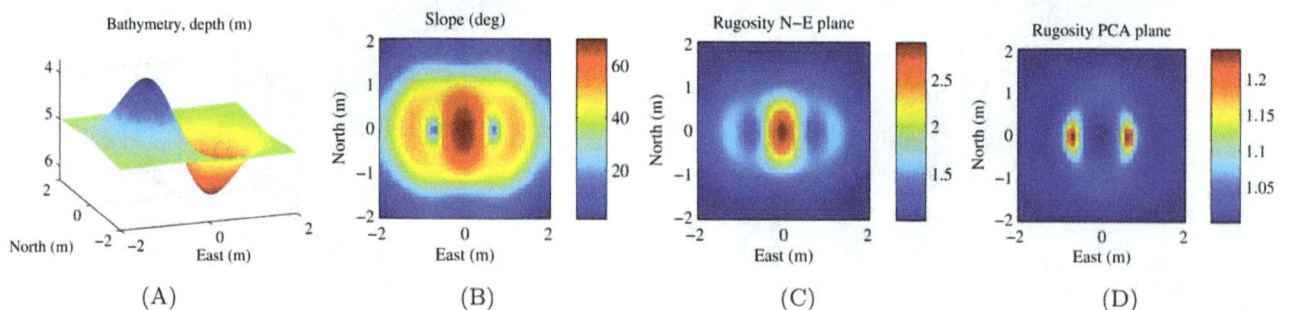

Figure 8. Results for simulated terrain model for exponential function. $D = 3 \times N \times e^{(-N^2 - E^2)} + 5$, where D, N and E are Depth, Northing and Easting in metres. The results are computed with a mesh resolution of $5mm$ and a window size of $1m \times 1m$. (A) shows an oblique view of the 3D bathymetry, (B) shows the slope angle, (C) shows the rugosity projected onto the N-E horizontal plane and (D) shows the rugosity projected onto the plane of best fit.

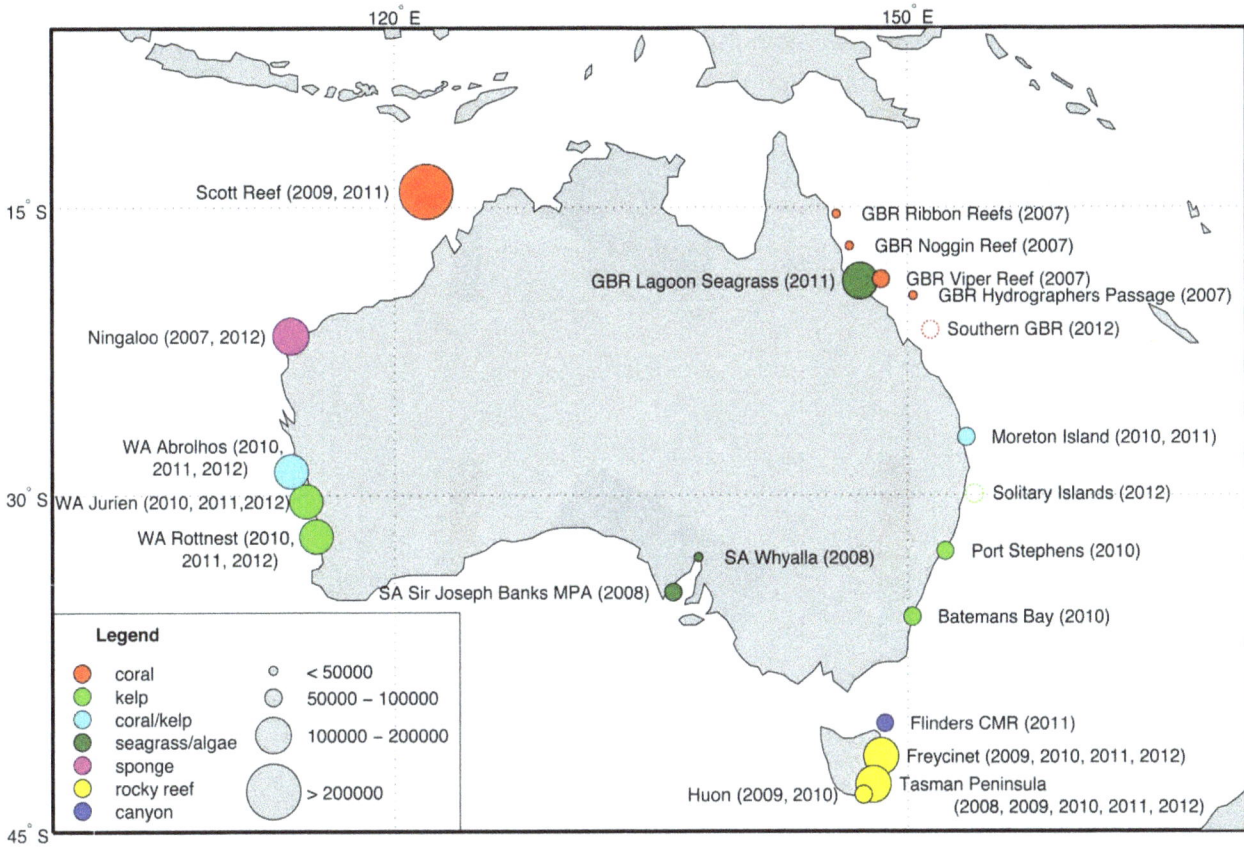

Figure 9. AUV survey locations around Australia [11]. The circles are coloured by dominant habitat type and scaled based on the number of images currently available in the IMOS AUV Facility image archive.

Effects of window size. The window size needs to be chosen with reference to the spatial scales of the environmental features to be considered. It can be likened to the chain/transect length in the conventional chain-tape method, of which the importance of scale has been outlined in [2,3,25]. The window size has an impact on the discriminatory power of the measure as a descriptor. Smaller window sizes do not capture as much variation in the ruggedness of the surface and larger window sizes provide spatial smoothing of the results. This is demonstrated by the results in Figure 11. The

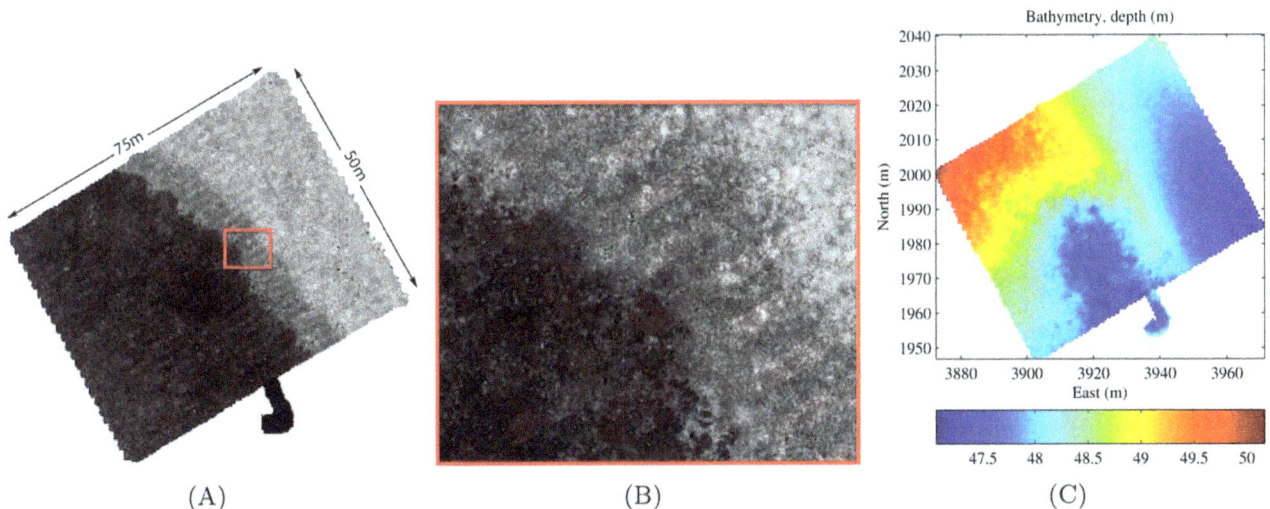

Figure 10. Dense AUV grid at Scott Reef off western Australia covering $50m \times 75m$ **with 9,831 stereo image pairs.** (A) Textured 3D mesh overview of survey site reconstructed using the method outlined in sec:data. (B) Close up of transition zone showing dense coral cover, barren sand and an intermediate, partially populated substrate class. (C) Colour map of mesh depth/bathymetry.

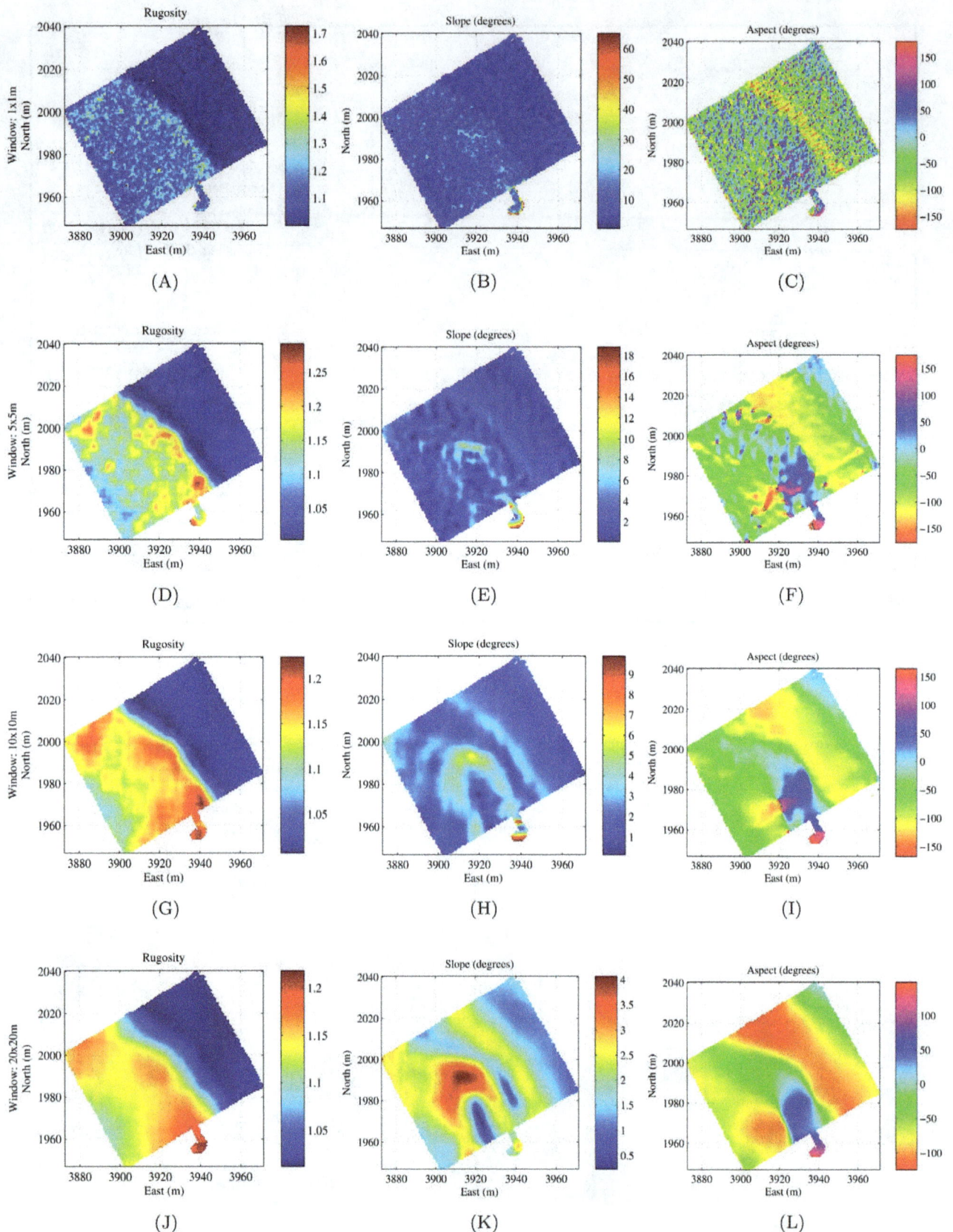

Figure 11. Dense AUV grid completed in Scott Reef showing the effect of different window sizes on the results. (A), (B) and (C) show rugosity, slope and aspect with a window of $1m \times 1m$. (D), (E) and (F) show rugosity, slope and aspect with a window of $5m \times 5m$. (G), (H) and (I) show rugosity, slope and aspect with a window of $10m \times 10m$. (J), (K) and (L) show rugosity, slope and aspect with a window of $20m \times 20m$.

Table 1. Correlation matrix for slope, PCA plane-fit rugosity and horizontal N-E plane rugosity for diver-rig survey.

	SLOPE	RGSTY-PCA	RGSTY-NE
SLOPE	1	0.21	0.85
RGSTY-PCA	0.21	1	0.56
RGSTY-NE	0.85	0.56	1

Results were computed with a window size of $30cm \times 30cm$.

Table 2. Correlation matrix for slope, PCA plane-fit rugosity and horizontal N-E plane rugosity for simulated terrain.

	SLOPE	RGSTY-PCA	RGSTY-NE
SLOPE	1	0.43	0.91
RGSTY-PCA	0.43	1	0.52
RGSTY-NE	0.91	0.52	1

Results were computed with a resolution of $5mm$ with a window size of $1m \times 1m$.

window size needs to be selected in accordance with the scale of processes to be observed.

Effects of mesh resolution. The mesh resolution is analogous to the link-size for the chain-tape method. The importance of link size is explored in [7]. Over the experiments that we performed, coarse mesh resolutions impacted the accuracy of the results, particularly with small window sizes. Resolutions that are too fine may be susceptible to noise in real-world terrain reconstructions that arises from uncertainty in the 2D feature locations and in the estimate of the stereo camera calibration parameters. We found that the *cm*-scale mesh resolutions that we typically work with, coupled with window sizes on the order of metres provide repeatable, robust results. It may also be important to note that just as it would be difficult to compare rugosity values computed with different chain link sizes, it may be difficult to compare virtual terrain complexity measurements computed with different mesh resolutions. The resolution should be chosen such that it is robust to noise, while still maintaining an adequate representation of the variability in the terrain.

Conclusion and Future Work

This paper has demonstrated how multi-scale measures of rugosity, slope and aspect can be derived from fine-scale bathymetric reconstructions created using georeferenced stereo imagery collected by AUVs, ROVs, manned submersibles or diver-held stereo camera systems. We presented a new method for calculating rugosity by considering the area of triangles within a window and their projection onto the plane of best fit, which was found using PCA. Through obtaining the plane of best fit, rugosity is decoupled from slope, and as a consequence of fitting a plane, slope and aspect are calculated with very little extra effort. The results of the virtual terrain complexity calculations were compared to experimental results using conventional *in-situ* measurement methods. It was shown that performing calculations over a digital terrain reconstruction is more robust, flexible and easily repeatable. We showed that using the digital 3D terrain reconstructions, it is possible to perform measurements that are difficult (if not impossible) to obtain manually in the field. In addition, the techniques are completely non-contact, which reduces the environmental impact of the surveying technique, making it more useful for repeat monitoring. Using an autonomous platform, the measurements can be collected without putting a human in the water, and beyond traditional scuba depth limits.

References

1. McCormick MI (1994) Comparison of field methods for measuring surface topography and their associations with a tropical reef fish assemblage. Marine Ecology Progress Series 112: 87–96.
2. Commito JA, Rusignuolo BR (2000) Structural complexity in mussel beds: the fractal geometry of surface topography. Journal of Experimental Marine Biology and Ecology 255: 133–152.

The technique was demonstrated on small single transect surveys gathered by a diver-rig and on a larger AUV survey consisting of tens of thousands of images covering thousands of square metres.

Future work may involve combining slope and aspect with current flow fields inferred using an acoustic doppler current profiler (ADCP), which may provide a good indicator of environmental exposure and a proxy for benthic habitat types. Given the method's computational tractability, it may also prove useful as a virtual 'sensor' to inform adaptive surveying strategies such as delineating zones of significant change in rugosity (i.e. interface between reef and sand or healthy and damaged reef). As mentioned in the paper, the visual information co-registered with the structural complexity can be used for improved descriptors for automated classification. Although we have done some work feeding these measures into automated interpretation tools, more work is needed to properly showcase the potential of the presented measures for this purpose.

Tables

Table 1 shows correlation matrix for slope, PCA plane-fit rugosity and horizontal N-E plane rugosity for diver-rig survey. Results were computed with a window size of 30 $cm \times 30$ cm. Table 2 shows correlation matrix for slope, PCA plane-fit rugosity and horizontal N-E plane rugosity for simulated terrain. Results were computed with a resolution of 5 mm with a window size of 1 $m \times 1$ m.

Acknowledgments

We would like to thank the Australian Institute for Marine Science for making ship time available to support this study. The captains and crews of the R/V Solander were instrumental in facilitating successful deployment and recovery of the AUV. We would also like to acknowledge the help of Navid Nourani-Vatani, Donald Dansereau and Daniel Bongiorno for collecting field rugosity validation and diver-rig data and we would like to thank all those who have contributed to the development and operation of the AUV. This includes Ian Mahon, Stephen Barkby, Ritesh Lal, Paul Rigby, Jeremy Randle, Bruce Crundwell and the late Alan Trinder, Duncan Mercer and George Powell.

Author Contributions

Conceived and designed the experiments: AF OP SBW. Performed the experiments: AF OP. Analyzed the data: AF OP MJR. Contributed reagents/materials/analysis tools: AF OP SBW MJR. Wrote the paper: AF. Research supervisors: SBW OP.

3. Alexander T, Barrett N, Haddon M, Edgar G (2009) Relationships between mobile macroinver-tebrates and reef structure in a temperate marine reserve. Marine Ecology Progress Series 389: 31–44.
4. Sleeman J, Boggs G, Radford B, Kendrick Ga (2005) Using Agent Based Models to Aid Reef Restoration: Enhancing Coral Cover and Topographic Complexity

through the Spatial Arrangement of Coral Transplants. Restoration Ecology 13: 685–694.

5. Jenness JS (2004) Calculating landscape surface area from digital elevation models. Wildlife, Society Bulletin 32: 829–839.

6. Frost NJ, Burrows MT, Johnson MP, Hanley ME, Hawkins SJ (2005) Measuring surface complexity in ecological studies. Limnology and Oceanography: Methods : 203–210.

7. Knudby A, LeDrew E (2007) Measuring structural complexity on coral reefs. In: American Academy of Underwater Sciences, 26th Symposium. Dauphin Island.

8. Dartnell P, Gardner JV (2004) Predicting seaoor facies from multibeam bathymetry and backscatter data. Photogrammetric engineering and remote sensing 70: 1081–1091.

9. Moustier CD, Matsumoto H (1993) Seaoor acoustic remote sensing with multibeam echo-sounders and bathymetric sidescan sonar systems. Marine Geophysical Researches 15: 27–42.

10. Brock JC, Wright CW, Clayton TD, Nayegandhi A (2004) Lidar optical rugosity of coral reefs in biscayne national park, orida. Coral Reefs 23: 48–59.

11. Williams SB, Pizarro O, Jakuba MV, Steinberg D, Friedman A (2012) Monitoring of Benthic Reference Sites: Using an Autonomous Underwater Vehicle. Robotics & Automation Magazine, IEEE 19: 73–84.

12. Johnson-Roberson M, Pizarro O, Williams SB, Mahon I (2010) Generation and visualization of large-scale three-dimensional reconstructions from underwater robotic surveys. Journal of Field Robotics 27: 21–51.

13. Seiler J, Friedman A, Steinberg D, Barrett N (In press) Image-based continental shelf habitat mapping using novel automated data extraction techniques. Continental Shelf Research.

14. Steinberg DM, Williams SB, Pizarro O, Jakuba MV (2010) Towards autonomous habitat classification using Gaussian Mixture Models. In: Intelligent Robots and Systems (IROS), 2010 IEEE/RSJ International Conference on. pp. 4424–4431. doi:10.1109/IROS.2010.5652480.

15. Friedman AL, Pizarro O, Williams SB (2010) Rugosity, Slope and Aspect derived from Bathymetric Stereo Image 3D Reconstructions. In: OCEANS 2010 Sydney. IEEE. URL http://www.google.com.au/search?client=safari&rls=en-us&q=Rugosity+Slope+and+Aspect+derived+from+Bathymetric+Stereo+Image+3D+Reconstructions&ie=UTF-8&oe=UTF-8&redir_esc=&ei=uN3jT7L7FOXumAXc973vCg.

16. Friedman A, Steinberg D, Pizarro O, Williams SB (2011) Active learning using a Variational Dirichlet Process model for pre-clustering and classification of underwater stereo imagery. In: International Conference on Intelligent Robots and Systems (IROS). IEEE, pp. 1533–1539.

17. Bridge TCL, Done T, Beaman R, Friedman A, Williams S, et al. (2011) Topography, substratum and benthic macrofaunal relationships on a tropical mesophotic shelf margin, central Great Barrier Reef, Australia. Coral Reefs 30: 143–153.

18. Bridge TCL, Done TJ, Friedman A, Beaman RJ, Williams SB, et al. (2011) Variability in mesophotic coral reef communities along the Great Barrier Reef, Australia. Marine Ecology Progress Series 428: 63–75.

19. Bell KLC, Fuller SA (2011) New frontiers in ocean exploration. Oceanography Society Virginia Beach, Va., Research ship, Nautilus. URL http://books.google.com.au/books?id=igDeZwEACAAJ&dq=intitle:New+Frontiers+in+Ocean+Exploration&hl=&cd=1&source=gbs_api.

20. Mahon I, Pizarro O, Johnson-Roberson M, Friedman A, Williams SB, et al. (2011) Reconstructing pavlopetri: Mapping the world's oldest submerged town using stereo-vision. In: Robotics and Automation (ICRA), 2011 IEEE International Conference on. IEEE, pp. 2315–2321.

21. Mahon I, Williams SB, Pizarro O, Johnson-Roberson M (2008) Efficient view-based SLAM using visual loop closures. Robotics, IEEE Transactions on 24: 1002–1014.

22. Curless B, Levoy M (1996) A volumetric method for building complex models from range images. ACM SIG-GRAPH: Computer graphics and interactive techniques : 303–312.

23. Lee DT, Schachter B (1980) Two algorithms for constructing a Delaunay triangulation. International Journal of Parallel Programming 9: 219–242.

24. Bishop CM (2006) Pattern Recognition and Machine Learning. Cambridge, UK: Springer Science+Business Media. URL http://trs-new.jpl.nasa.gov/dspace/handle/2014/40298.

25. Kuffner IB, Brock JC, Grober-Dunsmore R, Bonito VE, Hickey TD, et al. (2007) Relationships between reef fish communities and remotely sensed rugosity measurements in biscayne national park, orida, usa. Environmental biology of fishes 78: 71–82.

26. Hill J, Wilkinson C (2004) Methods for ecological monitoring of coral reefs. Technical report.

14

Analysis of Changes in Traumatic Symptoms and Daily Life Activity of Children Affected by the 2011 Japan Earthquake and Tsunami over Time

Masahide Usami[1]*, **Yoshitaka Iwadare**[1], **Kyota Watanabe**[1], **Masaki Kodaira**[2], **Hirokage Ushijima**[1], **Tetsuya Tanaka**[1], **Maiko Harada**[1], **Hiromi Tanaka**[1], **Yoshinori Sasaki**[1], **Kazuhiko Saito**[2]

1 Department of Child and Adolescent Psychiatry, National Center for Global Health and Medicine, Kohnodai Hospital, Chiba, Japan, 2 Department of Child Mental Health, Imperial Gift Foundation, Aiiku Hospital, Tokyo, Japan

Abstract

Background: On March 11, 2011, Japan was struck by a massive earthquake and tsunami. The tsunami caused tremendous damage and traumatized a number of people, including children. This study aimed to compare traumatic symptoms and daily life activity among children 20 months after the 2011 Great East Japan Earthquake and Tsunami with those observed after 8 months.

Methods: The study comprised two groups. The first comprised 12,524 kindergarten, elementary school, and junior high school children in Ishinomaki City, Miyagi Prefecture, Japan, who were evaluated 8 months after the disaster. The second comprised 10,597 children from the same place who were evaluated 20 months after the disaster. The Post Traumatic Stress Symptoms for Children 15 items (PTSSC-15), a self-completion questionnaire on traumatic symptoms, and a questionnaire on children's daily life were distributed to the children. An effective response was obtained from 11,639 (92.9%, 8 months after) and 10,597 (86.9%, 20 months after) children.

Results: The PTSSC-15 score was significantly higher in junior high school girls than in boys. The PTSSC-15 score was significantly higher in 4th–6th grade girls than in boys after 8 months. Elementary and junior high school children evaluated after 20 months had a significantly lower PTSSC-15 score than those evaluated after 8 months. The number of children having breakfast was significantly higher after 8 months than that after 20 months. In both the groups, children of all grades who had breakfast had a significantly lower PTSSC-15 score than those who did not have breakfast.

Conclusions: We conclude that traumatic symptoms and daily life activity of children who survived the earthquake and tsunami improved over time.

Editor: Reury F. P. Bacurau, University of Sao Paulo, Brazil

Funding: This work was supported by the grant of National Center for Global Health and Medicine (24–108), Soroptimist International of the Americas Japan Shimomatsu Region, Ishinomaki Rotary Club, Yaohigashi Rotary Club,and Tokuyama Rotary Club. The funders had no role in study design, data collection and analysis, decision to publish, or preparation of the manuscript.

Competing Interests: The authors have declared that no competing interests exist.

* E-mail: usami.masahide@hospk.ncgm.go.jp

Background

On March 11, 2011, Japan was struck by a massive earthquake and tsunami. The tsunami caused tremendous damage and traumatized a number of people, including children[1–8]. Till date, a number of studies have been conducted on children who have survived disasters [7–20].

After a disaster, posttraumatic stress disorder (PTSD) is the psychiatric diagnosis that should be considered most carefully[11,21,22]. However, traumatic symptoms tend to spontaneously heal over time; therefore, the morbidity of PTSD is dependent on the time, subjects, and methods used in the survey[12,14,23–29].

We collected information on daily life activity, environmental damage conditions, and traumatic symptoms of children who survived the 2011 Japanese Earthquake and Tsunami 8 months after the disaster. That study demonstrated that relationships between environmental damage conditions and traumatic symptoms were dependent on gender, age, house damage, evacuation experience, and bereavement experience[7]. Furthermore, children with house damage or evacuation experiences slept for a significantly shorter duration than those without these experiences[8].

Twenty months following the disaster, all shelters were closed, and the reconstruction, donations, and installation of temporary housing are still progressing. It is necessary to discuss whether traumatic symptoms and daily life activity of the children who survived the disaster has improved 20 months after the disaster in comparison with those after 8 months. Therefore, we collected information on the traumatic symptoms and daily life activity of these children 20 months after the disaster.

This study aimed to evaluate and compare the changes in traumatic symptoms of children 8 and 20 months after the earthquake and tsunami. The main hypothesis of this study was that traumatic symptoms and daily life activities of children who survived the tsunami improved between 8 and 20 months after the disaster. This hypothesis indicates that these children live under stable life environments such as going to school, having breakfast daily, and retaining a comfortable sleep environment in their new homes.

Methods

Our Survey was conducted according to the principles expressed in the Declaration of Helsinki.

Each survey was conducted as a part of the school education program under the initiative of the Board of Education in Ishinomaki City.

Ishinomaki City is the second largest city (population, 162,822) in Miyagi Prefecture, Japan. As of February 15, 2012, the death toll in Ishinomaki City was 3,182, and 557 people were missing. The total number of collapsed houses and buildings, including half-collapsed houses, was 33,378, and 7,298 temporary houses had been constructed.

Surveys were distributed to all children who attended five kindergartens, 43 elementary schools, and 21 junior high schools in Ishinomaki City, Miyagi Prefecture. The survey was conducted in November 2011 and 2012 (8 and 20 months after the earthquake disaster, respectively). Some children who survived the huge tsunami had migrated to another city from Ishinomaki city. The number of children who went to municipal kindergartens, elementary schools, and junior high schools were 12534 in November 2011, and the number was 12193 in November 2012.

The method of administering surveys in 2011 and 2012 was the same. First, the Education Committee of Ishinomaki City explained the survey method to the principals of all of the schools. Following this, the teachers distributed a letter explaining the survey, which had been constructed by the Education Committee, to all children and their parents. The letter clearly stated that by filling the questionnaire, both student and parents were considered to be providing consent to participate in the survey. The letter also specified that the survey results would be used to provide children with psychological care to facilitate their education at the school and that the results would be published as a medical paper. Informed consent was thus obtained when the students filled the questionnaire. The Ethical Committee of the National Center for Global Health and Medicine approved this study, including the consent procedure.

Participants

On November 2011 (8 months after the disaster), PTSSC-15, a self-completion questionnaire on traumatic symptoms, was distributed to 12,524 children registered at municipal schools in Ishinomaki City. A self-completion questionnaire about the daily life, waking time, time of sleep onset, and breakfast consumption was also distributed to 12, 534 children. A questionnaire on the environmental damage experienced by the children was distributed to their teachers. On November 2012 (20 months after the disaster), PTSSC-15 and the questionnaire on daily life were distributed to 12,193 children registered at municipal schools and to their teachers.

Parents of kindergartners and 1st–3rd grade elementary school students were asked to fill the questionnaire while talking to their children. Informed consent for participation in the survey was obtained when the completed questionnaires were received from the children.

Answers were obtained from 12,346 (98.6%) of the 12,524 children (8 months after) and 11,124 (91.2%) of the 12,193 children (20 months after) to whom questionnaires were distributed. An effective response was obtained from 11,639 (92.9%, 8 months after) and 10,597 (86.9%, 20 months after) children. Effective responses at 20 months had no connection with effective responses at 8 months by anonymity.

Answers to the questionnaire on environmental damage 8 months after the disaster with regard to all 12,524 children were obtained from teachers. Table 1 shows data regarding gender, age, and environmental damage conditions (house damage, evacuation conditions, and bereavement experience) in 11,639 children 8 months after the disaster (Table 1)[6,7]. When teachers had no information regarding house damage, evacuation conditions, and bereavement experience, the answer was defined as "unknown."

Measures

This was a paper-based survey with questions regarding traumatic symptoms using a self-report form. The self-report form consisted of PTSSC-15 and a questionnaire on daily life developed by the authors.

PTSSC-15

PTSSC-15 is a self-completion questionnaire on the stress reactions in children after disasters. A similar survey, the Posttraumatic Stress Symptom 10 (PTSS-10)[30], had fewer questions and was used as a screening test after the Great Hanshin Earthquake(Kato et al., 1996). PTSS-10 was administered to 105 Norwegian children (6–17 years old) 10 and 30 months after the 2004 South East Asia Tsunami[31]. Five questions that were considered to reveal important psychosomatic characteristics following disasters (flashback, appetite loss, somatic reactions such as headache and abdominal pain, attention deficit, and anxiety) were added to PTSS-10, and PTSSC-15 consisting of 15 questions was constructed in Japan[32].

Each question in the questionnaire is scored in six levels: 0 = completely disagree, 1 = mostly disagree, 2 = partially disagree, 3 = partially agree, 4 = mostly agree, and 5 = completely agree. Higher scores indicate more severe traumatic symptoms and depressive symptoms. Tominaga and colleagues demonstrated the reliability and validity of PTSSC-15 for Japanese children and adolescents[32].

Questionnaire on daily life

The daily life questionnaire included items related to waking and sleep-onset times. For the total sleep duration on weekdays and holidays, the subject was asked to write the total usual daily sleep hours, usual wake-up time, and the time at which the subject usually goes to sleep.

The daily life questionnaire included items related to having breakfast every morning. Each question was scored in two levels: 0 = did not usually have breakfast daily and 1 = usually have breakfast daily.

Statistical analysis

PTSSC-15, school grades, and gender

Children were divided into four grade groups: kindergartners, 1st–3rd grade (elementary school students), 4th–6th grade (elementary school students), and 7th–9th grade (junior high school students). For grade group and gender, the median PTSSC-15 score and interquartile range were separately determined 8 and

Table 1. Damage condition of children affected by the 2011 Japan Earthquake and Tsunami.

Items			N = 11639	
House damage		No	6986	(60.0%)
	Yes	Total collapse	2243	(19.3%)
		Half collapse	2354	(20.2%)
		Total	4597	(39.5%)
	Unknown		56	(0.5%)
Evacuation experience		No	8228	(70.7%)
	Yes	Currently living in an evacuation center	90	(0.8%)
		Used to live in an evacuation center	2845	(24.4%)
		Living in temporary housing	976	(8.4%)
		Used to live in temporary housing	51	(0.4%)
		Evacuation experience at least once	3248	(27.9%)
	Unknown		163	(1.4%)
Bereavement experience		No	9241	(79.4%)
	Yes	Father	71	(0.6%)
		Mother	66	(0.6%)
		Brothers and sisters	44	(0.4%)
		Grandfather and grandmother	355	(3.1%)
		Classmates	1498	(12.9%)
		Teacher in-charge	32	(0.3%)
		Others	270	(2.3%)
		At least one bereavement experience	2103	(18.1%)
	Unknown		295	(2.5%)

N, number of cases.

20 months after the disaster. The PTSSC-15 score was statistically compared between males and females by the Mann–Whitney U test for each grade group. Effect sizes were then calculated on the basis of the Mann–Whitney statistics.

Sleep duration and breakfast consumption

The average sleep duration on weekdays and holidays was also calculated for each grade and gender separately after 8 and 20 months. For children suffering from traumatic experiences, the average sleep duration on weekdays and holidays after 8 months was compared with that after 20 months.

The number of children who had breakfast after 8 and 20 months was calculated for each grade and gender. The number of children who had breakfast daily after 8 months was compared with that after 20 months for each grade and gender. The PTSSC-15 scores of children who usually had breakfast daily and those who did not usually have breakfast daily were separately compared after 8 and 20 months.

Change in PTSSC-15 scores and daily life activities

The average PTSSC-15 scores in each grade group and gender was separately calculated after 8 and 20 months.

The difference in the average PTSSC-15 score after 8 and 20 months was statistically analyzed by two-factor analysis of variance for each grade group and gender.

The difference in the sleep duration after 8 and 20 months was statistically analyzed by two-factor analysis of variance for each gender and grade group.

The difference in the average PTSSC-15 score of those who had breakfast and those who did not was separately analyzed by two-factor analysis of variance for each gender and grade group after 8 and 20 months.

In all tests, a significance level of 0.05 was used in two-sided tests. Analyses were performed using PASW 18.0.

Results

Descriptive information

The participants evaluated 8 months after the disaster included 11,692 children (5959 males, 5733 females), while those evaluated after 20 months included 10749 children (5402 males, 5347 females) who were exposed to the 2011 Japanese Earthquake and Tsunami. Table 2 shows the gender, The Post Traumatic Stress Symptoms for Children 15 items (PTSSC-15) score, age, sleep duration, and breakfast consumption (Table 2).

PTSSC-15 score after 8 and 20 months

Table 3 shows the PTSSC-15 score after 8 and 20 months in each grade group and gender (Table 3). The PTSSC-15 score was significantly higher (P<0.001) in girls than in boys in the 7th–9th grade (junior high schools). The PTSSC-15 score was significantly higher (P<0.001) in girls than in boys in the 4th–6th grade after 8 months. These effect sizes were under 0.30.

The average PTSSC-15 score was compared in each grade group, gender, and month (Table 4). Children in all grade groups, except kindergartners evaluated after 20 months, had a significantly lower PTSSC-15 score than children evaluated at 8 months

Table 2. Characteristics of children affected by the 2011 Japan Earthquake and Tsunami.

Items		After 8 months		After 20 months	
		N = 11,639		N = 10,597	
Gender	Male	5939	(51.0%)	5302	(50.0%)
	Female	5700	(49.0%)	5295	(50.0%)
Age at the time of the disaster (y) (Mean)		10.9	(SD = 2.7)	10.9	(SD = 2.7)
PTSSC-15 score (Mean)		20.5	(SD = 14.5)	18.8	(SD = 14.0)
Sleep duration on weekdays (Mean)		8.4(h)	(SD = 1.3)	8.5(h)	(SD = 1.2)
Sleep duration on holidays (Mean)		9.0(h)	(SD = 1.4)	9.2 (h)	(SD = 1.3)
Breakfast consumption	yes	10673	(91.7%)	9928	(93.7%)
	no	827	(7.1%)	645	(6.1%)
	unknown	139	(1.2%)	24	(0.2%)

$[F(16811) = 14.97$, $P<0.0001$; $F(17534) = 53.20$, $P<0.0001$; $F(17395) = 35.57$, $P<0.0001]$.

Sleep duration after 8 and 20 months

The average sleep duration on weekdays and holidays was compared in each grade group, gender, and month (Table 5). On weekdays, children in the 4th–6th grade group evaluated after 20 months had a significantly longer sleep duration than those evaluated after 8 months $[F(17534) = 6.484$, $P<0.05]$. On holidays, children in the 4th–9th grade group evaluated after 20 months had a significantly longer sleep duration than those evaluated after 8 months $[(F(16811) = 5.533$, $P<0.05$; $F(17534) = 39.57$, $P<0.0001$; $F(17395) = 8.155$, $P<0.001)$.

Breakfast consumption after 8 and 20 months

The number of children who had breakfast (10673 children, 91.7%) was greater after 8 months than that (9928 children, 93.7%) after 20 months (chi-square test, $P<0.001$).

The average PTSSC-15 scores of children who had and those who did not have breakfast were compared in each grade group, gender, and separately after 8 and 20 months (Table 6). After 8 months, children in all grade groups who had breakfast had a significantly lower PTSSC-15 score than those who did not have breakfast $[F(1235) = 10.19$, $P<0.0001$; $F(13501) = 27.85$,

$P<0.0001$; $F(13929) = 73.31$, $P<0.0001$, $F(13819) = 60.94$, $P<0.0001)$. After 20 months, children in all grade groups who had breakfast had a significantly lower PTSSC-15 score than those who did not have breakfast $[F(1231) = 4.976$, $P<0.05$; $F(13206) = 25.75$, $P<0.0001$; $F(13576) = 36.07$, $P<0.0001$, $F(13544) = 27.47$, $P<0.0001]$.

Discussion

This study showed that the traumatic symptoms and daily life activity of children who suffered damage and loss improved 20 months after this severe natural disaster compared with the traumatic symptoms and daily life activity of children evaluated 8 months after the disaster.

A previous study elucidated that PTSSC-15 scores were related to the environmental damage caused by the 2011 Japanese Tsunami[1]. This study showed that PTSSC-15 decreased over time. It suggested that children suffering traumatic symptoms from disasters such as the massive tsunami may recover over time. However, using only a self-completion questionnaire as a screening tool for PTSD after a huge disaster may result in an inflated number of children who appear to be at a high risk for PTSD. If psychiatrists or clinical psychologists use a self-completion questionnaire as a screening tool for PTSD after a

Table 3. Average PTSSC-15 score (grade and gender).

		Gender										
		Male				Female					Effect size	P value
		M	IR			N	M	IR			N		
After 8 months	Kindergartners	14.0	3.0	–	25.5	119	15.0	3.0	–	25.5	127	0.03	ns
	1st–3rd grade	16.0	6.0	–	28.0	1866	17.0	7.0	–	28.0	1736	0.03	ns
	4th–6th grade	17.0	7.0	–	29.0	1975	20.0	9.0	–	32.0	1973	0.08	<0.0001
	7th–9th grade	20.0	9.0	–	32.0	1979	26.0	14.0	–	37.0	1864	0.15	<0.0001
After 20 months	Kindergartners	12.0	2.0	–	25.0	111	12.0	3.0	–	23.0	127	0.09	ns
	1st–3rd grade	15.0	5.0	–	26.0	1632	16.0	6.0	–	27.0	1581	0.03	ns
	4th–6th grade	15.0	6.0	–	28.0	1792	17.0	7.0	–	28.0	1798	0.05	ns
	7th–9th grade	18.0	8.0	–	29.0	1767	23.0	12.0	–	34.0	1789	0.29	<0.0001

M, median; IR, interquartile range; N, number of cases.

Table 4. Average PTSSC-15 score (grade, gender, and month)

Grade group	Gender	Months after disaster								F	P value
		After 8 months			After 20 months						
		M	SD	N	M	SD	N				
Kindergartners									Gender × months	0.031	ns
	Male	15.1	12.5	119	13.2	11.6	111		Months	2.274	ns
	Female	15.7	12.8	127	14.2	12.5	127		Gender	0.504	ns
1st–3rd grade									Gender × months	0.416	ns
	Male	17.4	12.8	1866	16.4	12.9	1632		Months	14.97	<0.0001
	Female	18.2	12.8	1736	16.8	12.6	1581		Gender	3.741	ns
4th–6th grade									Gender × months	4.526	<0.05
	Male	19.6	14.4	1975	17.9	13.9	1792		Months	53.20	<0.0001
	Female	21.7	14.7	1973	18.6	14.0	1798		Gender	18.10	<0.0001
7th–9th grade									Gender × months	0.529	<0.0001
	Male	21.3	14.7	1979	19.5	14.0	1767		Months	35.57	<0.0001
	Female	26.0	15.5	1864	23.7	14.8	1789		Gender	167.6	<0.0001

M, mean; SD, standard deviation; N, number of cases.

Table5. Average sleep duration on weekdays and holidays (grade, gender, and month).

	Grade group	Gender	Months after disaster								F	P value
			After 8 months			After 20 months						
			M	SD	N	M	SD	N				
Weekday	Kindergartners									Gender × months	2.122	ns
		Male	9.8	0.7	119	9.8	0.9	111		Months	2.122	ns
		Female	10.0	0.6	127	9.8	0.8	127		Gender	2.122	ns
	1st–3rd grade									Gender × months	0.0	ns
		Male	9.3	0.6	1866	9.3	0.7	1632		Months	0.0	ns
		Female	9.3	0.6	1736	9.3	0.6	1581		Gender	0.0	ns
	4th–6th grade									Gender × months	6.484	<0.05
		Male	8.7	0.9	1975	8.8	0.9	1792		Months	6.484	<0.05
		Female	8.7	0.8	1973	8.7	0.8	1798		Gender	6.484	<0.05
	7th–9th grade									Gender × months	0.0	ns
		Male	7.6	1.2	1979	7.6	1.2	1767		Months	0.0	ns
		Female	7.3	1.1	1864	7.3	1.0	1789		Gender	130.2	<0.0001
Holidays	Kindergartners									Gender × months	1.885	ns
		Male	9.9	0.8	119	9.9	0.8	111		Months	1.885	ns
		Female	10.2	0.8	127	10.0	0.8	127		Gender	7.539	<0.01
	1st–3rd grade									Gender × months	5.533	<0.05
		Male	9.3	0.9	1866	9.4	0.9	1632		Months	5.533	<0.05
		Female	9.7	0.8	1736	9.7	0.9	1581		Gender	271.1	<0.0001
	4th–6th grade									Gender × months	9.891	<0.001
		Male	8.6	1.5	1975	8.9	1.4	1792		Months	39.57	<0.0001
		Female	9.4	1.3	1973	9.5	1.3	1798		Gender	484.7	<0.0001
	7th–9th grade									Gender × months	1.85E-10	ns
		Male	8.5	1.6	1979	8.6	1.6	1767		Months	8.155	<0.001
		Female	8.8	1.4	1864	8.9	1.4	1789		Gender	73.40	<0.0001

M, mean; SD, standard deviation; N, number of cases.

Table 6. Average PTSSC-15 score (grade, gender, and breakfast consumption) in 2011 and 2012.

	Grade group	Gender	Eating breakfast								F	P value
			Yes			No						
			M	SD	N	M	SD	N				
2011	Kindergartners									Gender × breakfast consumption	0.4711	ns
		Male	14.1	11.5	110	35.4	13.5	5		Breakfast consumption	4.976	<0.05
		Female	15.3	12.7	120	20.5	10.1	4		Gender	0.073	ns
	1st–3rd grade									Gender × breakfast consumption	1.217	ns
		Male	17.1	12.6	1741	22.5	13.8	79		Breakfast consumption	25.75	<0.0001
		Female	18.0	12.7	1616	23.9	14.7	69		Gender	0.438	ns
	4th–6th grade									Gender × breakfast consumption	0.520	ns
		Male	18.9	13.9	1826	26.3	17.2	142		Breakfast consumption	36.07	<0.0001
		Female	21.1	14.5	1870	30.5	15.9	95		Gender	1.942	ns
	7th–9th grade									Gender × breakfast consumption	0.560	ns
		Male	20.9	14.7	1725	24.6	15.7	244		Breakfast consumption	27.47	<0.0001
		Female	24.9	15.2	1665	33.3	16.5	189		Gender	18.66	<0.0001
2012	Kindergartners									Gender × breakfast consumption	0.4711	ns
		Male	12.4	11.3	105	24.3	9.8	4		Breakfast consumption	4.976	<0.05
		Female	14.1	12.6	121	20.4	7.9	5		Gender	0.073	ns
	1st–3rd grade									Gender × breakfast consumption	1.217	ns
		Male	16.1	12.6	1554	23.1	14.9	77		Breakfast consumption	25.75	<0.0001
		Female	16.6	12.5	1523	21.1	14.2	56		Gender	0.438	ns
	4th–6th grade									Gender × breakfast consumption	0.520	ns
		Male	17.5	13.6	1678	23.0	16.2	108		Breakfast consumption	36.07	<0.0001
		Female	18.2	13.8	1712	25.2	15.8	82		Gender	1.942	ns
	7th–9th grade									Gender × breakfast consumption	0.560	ns
		Male	19.0	13.6	1568	24.2	15.9	192		Breakfast consumption	27.47	<0.0001
		Female	23.4	14.6	1667	27.3	16.6	121		Gender	18.66	<0.0001

M, mean; SD, standard deviation; N, number of cases.

disaster, they are obligated to treat a number of children detected with a high risk for PTSD[7,34]. Therefore, a self-completion questionnaire is insufficient as a screening tool for PTSD after a disaster.

Differences in traumatic symptoms due to gender and age were recognized in this study. According to the effect size, the difference due to gender was negligible; however, the difference due to age was large. The responders in the case of kindergarten and lower-grade elementary school students were the parents, and it is possible that not only the status of the children but also the psychological anxiety and status of their parents affected the responses. Therefore, it is necessary to comprehensively judge the status of children by considering not only their psychological status but also the psychological conditions of their parents.

We found that the sleep duration on holidays in 1st–9th grade group after 20 months was longer than that after 8 months. The sleep duration on weekdays in the 4th–6th grade group after 20 months was longer than that after 8 months. A previous study

elucidated that the relationship between sleep duration and traumatic symptoms displayed low correlations[8]. Thus, sleep duration may not be a good predictor of improvedtraumatic symptoms.

Furthermore, we found that having breakfast was closely related to traumatic symptoms of the children. These results suggest that parents may not be able to afford a good living environment following the tsunami and that the children were severely depressed and/or neglected. After the huge disaster, child abuse, orphans, and poverty are a serious issue in Japan[33,34]. Thus, for children with traumatic symptoms, greater support of the family may be important. This segment of the questionnaire regarding having breakfast daily may be useful for quick screening of children with traumatic symptoms or parents with a reduced family support function.

Limitations

This study was an observation study that compared the same age group at 8 months and at 20 months despite the fact that those examined at 8 months were a year older at 20 months. It was better to compare 1st–3rd graders at 8 months with 2nd–4th graders at 20 months in order to discussed factors affected traumatic symptoms of children who survived huge disaster over time. According previous study using PTSSC-15, the score of PTSSC-15 increased with age of children [7]. Therefore, if we compared 1st–3rd graders at 8 months with 2nd–4th graders at 20 months, we could not determine whether the improvement because of age, or because of the environment change. We had to try another designed study to compare the same age group at 8 months and at 20 months according to the fact that those examined at 8 months were a year older at 20 months. This study was merely to observe the change over time in the trauma symptoms of children living in a tremendous disaster area.

This study was a survey using a self-completion questionnaire conducted in only one district in Japan. It is impossible to calculate the morbidity of PTSD in children after the 2011 Japanese Earthquake and Tsunami based on the results of the survey. Furthermore, this survey did not follow the cause of each individual's traumatic symptom. This study only shows improvement in trauma symptoms of children in Ishinomaki City. Therefore, this study is insufficient as an epidemiological survey and cohort study for psychiatric diagnosis. Examinations by child psychiatrists using operational diagnostic criteria and structured interviews are necessary for accurate psychiatric diagnosis. Moreover, the results of this study on children in Ishinomaki City do not reflect all characteristics of children affected by the 2011 Japanese Earthquake and Tsunami.

Conclusions

This study elucidated that traumatic symptoms and daily life activity of children who survived the earthquake and tsunami improved with time. It is important to not only evaluate the traumatic symptoms using a self-completion questionnaire but also confirm specific information regarding the function of family support for children, such as ensuring that the child has breakfast.

Acknowledgments

My deepest appreciation goes to the educational committee in Ishinomaki City.

Author Contributions

Conceived and designed the experiments: MU YI MK KW. Performed the experiments: MU YI MK KW HT YS TT MH KS. Analyzed the data: MU HU. Contributed reagents/materials/analysis tools: MU YI MK KW. Wrote the paper: MU. Approved the proposal with some revisions: MU. Supervised the final paper: MU. Read and approved the final manuscript: MU YI MK KW HT YS TT HU MH KS.

References

1. Butler D (2011) Fukushima health risks scrutinized. *Nature* 472: 13–14. doi:10.1038/472013a.
2. Mike T (2011) [Caring for children's 'mind' after the earthquake]. No To Hattatsu 43: 342.
3. Procter NG, Crowley T (2011) A Mental Health Trauma Response to the Japanese Earthquake and Tsunami. *Holistic Nursing Practice* 25: 162–164. doi:10.1097/HNP.0b013e31821a6955.
4. Experiences of the great East Japan earthquake March 2011. (2011) Experiences of the great East Japan earthquake March 2011. 58: 332–334. doi:10.1111/j.1466-7657.2011.00924.x.
5. Hayashi K, Tomita N (2012) Lessons Learned From the Great East Japan Earthquake: Impact on Child and Adolescent Health. *Asia-Pacific Journal of Public Health* 24: 681–688. doi:10.1177/1010539512453255.
6. Nakahara S, Ichikawa M (2013) Mortality in the 2011 tsunami in Japan. *J Epidemiol* 23: 70–73.
7. Usami M, Iwadare Y, Kodiara M, Watanabe K, Aoki M, et al. (2012) Relationships between Traumatic Symptoms and Environmental Damage Conditions among Children 8 Months after the 2011 Japan Earthquake and Tsunami. *PLoS ONE* 7: e50721. doi:10.1371/journal.pone.0050721.t007.
8. Usami M, Iwadare Y, Kodiara M, Watanabe K, Aoki M, et al. (2013) Sleep Duration among Children 8 Months after the 2011 Japan Earthquake and Tsunami. *PLoS ONE* 8: e65398. doi:10.1371/journal.pone.0065398.t008.
9. Balsari S, Lemery J, Williams TP, Nelson BD (2010) Protecting the Children of Haiti. *N Engl J Med* 362: e25. doi:10.1056/NEJMp1001820.
10. Becker SM (2007) Psychosocial Care for Adult and Child Survivors of the Tsunami Disaster in India. *J Child Adolesc Psych Nursing*. 20: 148–155. doi:10.1111/j.1744-6171.2007.00105.x.
11. Chemtob CM (2008) Impact of Conjoined Exposure to the World Trade Center Attacks and to Other Traumatic Events on the Behavioral Problems of Preschool Children. *Arch Pediatr Adolesc Med*. 162: 126. doi:10.1001/archpediatrics.2007.36.
12. Hafstad GS, Kilmer RP, Gil-Rivas V (2011) Posttraumatic growth among Norwegian children and adolescents exposed to the 2004 tsunami. *Psychological Trauma: Theory, Research, Practice, and Policy* 3: 130–138. doi:10.1037/a0023236.
13. Manuel Carballo BH, Hernandez M (2005) Psychosocial aspects of the Tsunami. *J R Soc Med*. 98: 396.
14. Mullett-Hume E, Anshel D, Guevara V, Cloitre M (2008) Cumulative trauma and posttraumatic stress disorder among children exposed to the 9/11 World Trade Center attack. *Am J Orthopsychiatry*. 78: 103–108. doi:10.1037/0002-9432.78.1.103.
15. Oncu EC (2010) The effects of the 1999 Turkish earthquake on young children: Analyzing traumatized children's completion of short stories. *Child Dev*. 81: 1161–1175.
16. Piyasil V, Ketumarn P (2008) Psychiatric disorders in children at two year after the tsunami disaster in Thailand. ASEAN J psychiatry 8(9): 99–103.
17. Piyasil V, Ketuman P, Plubrukarn R, Jotipanut V, Tanprasert S, et al.(2007) Post traumatic stress disorder in children after tsunami disaster in Thailand: 2 years follow-up. *J Med Assoc Thai* 90: 2370–2376.
18. Piyasil V, Ketumarn P, Prubrukarn R, Ularntinon S, Sitdhiraksa N, et al. (2011) Post-traumatic stress disorder in children after the tsunami disaster in Thailand: a 5-year follow-up. J Med Assoc Thai 94 Suppl 3: S138–S144.
19. Piyasil V, Ketumarn P, Ularntinon S (2008) Post-traumatic stress disorder in Thai children living in area affected by the tsunami disaster: a 3 years followup study. *ASEAN Journal of Psychiatry* 8: 99–103.
20. Ulartinon S, Piyasil V, Ketumarn P, Sitdhiraksa N, Pityarastian N, et al. (2008) Assessment of psychopathological consequences in children at 3 years after tsunami disaster. J Med Assoc Thai 91 Suppl 3: S69–S75.
21. Kato H, Asukai N, Miyake Y, Minakawa K, Nishiyama A (1996) Post-traumatic symptoms among younger and elderly evacuees in the early stages following the 1995 Hanshin-Awaji earthquake in Japan. *Acta Psychiatr Scand*. 93: 477–481. doi:10.1111/j.1600-0447.1996.tb10680.x.
22. Ma X, Liu X, Hu X, Qiu C, Wang Y, et al. (2011) Risk indicators for post-traumatic stress disorder in adolescents exposed to the 5.12 Wenchuan earthquake in China. *Psychiatry Res*. 189: 385–391. doi:10.1016/j.psychres.2010.12.016.
23. Dyb G, Jensen TK, Nygaard E (2011) Children's and parents' posttraumatic stress reactions after the 2004 tsunami. *Clin Child Psychol Psychiatry*. 16: 621–634. doi:10.1177/1359104510391048.
24. Jia Z, Tian W, He X, Liu W, Jin C, et al. (2010) Mental health and quality of life survey among child survivors of the 2008 Sichuan earthquake. *Qual Life Res*. 19: 1381–1391. doi:10.1007/s11136-010-9703-8.
25. Kim BN, Kim JW, Kim HW, Shin MS, Cho SC, et al. (2009) A 6-Month Follow-Up Study of Posttraumatic Stress and Anxiety/Depressive Symptoms in Korean Children After Direct or Indirect Exposure to a Single Incident of Trauma. *J Clin Psychiatry* 70: 1148–1154. doi:10.4088/JCP.08m04896.
26. Weisz JR, Jensen AL (2001) Child and adolescent psychotherapy in research and practice contexts: Review of the evidence and suggestions for improving the field. *Eur Child Adolesc Psychiatry* 10: S12–S18. doi:10.1007/s007870170003.
27. Wiguna T, Guerrero APS, Kaligis F, Khamelia M (2010) Psychiatric morbidity among children in North Aceh district (Indonesia) exposed to the 26 December 2004 tsunami. Asia-Pacific Psychiatry 2: 151–155. doi:10.1111/j.1758-5872.2010.00079.x.
28. WHO (n.d.) Mental Health Assistance to the Populations Affected by the Tsunami in Asia. Available: wwwwhoint/mental_health/resources/tsunami/en/.
29. WHO (2005) Tsunami Wreaks Mental Health Havoc. 1 June 2005.
30. Weisaeth L (1989) A study of behavioural responses to an industrial disaster. *Acta Psychiatr Scand Suppl*. 355: 13–24.
31. Jensen TK, Dyb G, Nygaard E (2009) A Longitudinal Study of Posttraumatic Stress Reactions in Norwegian Children and Adolescents Exposed to the 2004

Tsunami. *Arch Pediatr Adolesc Med.* 163: 856. doi:10.1001/archpediatrics.2009.151.

32. Tominaga Y, Takahasi T, Yoshida R, Sumimoto K, Kajikawa N (2002) The reliability and validity of Post Traumatic Stress symptoms for Children-15 items(PTSSC-15). *The journal of human development and clinical psychology* 8: 29–36. In Japanese

33. Chance for Children (2011) Annual Report 2011. Available: http://www.cfc.or.jp/CFC_AR_2011.pdf. Accessed 2013 Dec 29.

34. Fukuchi N (2012) The impact that the earthquake disaster has on child nurturing. Japanese. *Journal of Child Abuse and Neglect* 14: 14–19. In Japanese

15

Potential of Best Practice to Reduce Impacts from Oil and Gas Projects in the Amazon

Matt Finer[1]*, Clinton N. Jenkins[2], Bill Powers[3]

1 Biodiversity Program, Center for International Environmental Law, Washington D.C., United States of America, **2** Department of Biology, North Carolina State University, Raleigh, North Carolina, United States of America, **3** E-Tech International, Santa Fe, New Mexico, United States of America

Abstract

The western Amazon continues to be an active and controversial zone of hydrocarbon exploration and production. We argue for the urgent need to implement best practices to reduce the negative environmental and social impacts associated with the sector. Here, we present a three-part study aimed at resolving the major obstacles impeding the advancement of best practice in the region. Our focus is on Loreto, Peru, one of the largest and most dynamic hydrocarbon zones in the Amazon. First, we develop a set of specific best practice guidelines to address the lack of clarity surrounding the issue. These guidelines incorporate both engineering-based criteria and key ecological and social factors. Second, we provide a detailed analysis of existing and planned hydrocarbon activities and infrastructure, overcoming the lack of information that typically hampers large-scale impact analysis. Third, we evaluate the planned activities and infrastructure with respect to the best practice guidelines. We show that Loreto is an extremely active hydrocarbon front, highlighted by a number of recent oil and gas discoveries and a sustained government push for increased exploration. Our analyses reveal that the use of technical best practice could minimize future impacts by greatly reducing the amount of required infrastructure such as drilling platforms and access roads. We also document a critical need to consider more fully the ecological and social factors, as the vast majority of planned infrastructure overlaps sensitive areas such as protected areas, indigenous territories, and key ecosystems and watersheds. Lastly, our cost analysis indicates that following best practice does not impose substantially greater costs than conventional practice, and may in fact reduce overall costs. Barriers to the widespread implementation of best practice in the Amazon clearly exist, but our findings show that there can be great benefits to its implementation.

Editor: Matteo Convertino, University of Florida, United States of America

Funding: This work was primarily supported by the Gordon and Betty Moore Foundation. Additional funding came from Blue Moon Fund, National Geographic Society, and NASA Biodiversity Grant (ROSES-NNX09AK22G). The funders had no role in study design, data collection and analysis, decision to publish, or preparation of the manuscript.

Competing Interests: The authors have declared that no competing interests exist. E-Tech International is a nonprofit organization, E-Tech International analyzes the potential environmental and social impacts of large development projects in less industrialized countries. E-Tech has no commercial interest in any of the best practices technologies discussed in this paper. Thus, the authors declared no competing interests because they did not feel that his affiliation might be perceived as interfering with the full and objective presentation of the research.

* E-mail: mfiner@ciel.org

Introduction

The western Amazon, one of the most biologically and culturally rich regions on Earth [1–3], continues to be an active and controversial zone of hydrocarbon exploration and production [4]. Hydrocarbon blocks – geographic areas delimited by national governments for the exploration and production of oil and gas – cover vast swaths of the region, including protected areas and titled indigenous territories [5]. Moreover, international bidding rounds on new oil and gas blocks in Colombia, Ecuador, and Peru confirm that exploration activities continue expanding deeper into the most remote tracts of the western Amazon. The lone exception is Ecuador's Yasuní-ITT Initiative, a novel government proposal that seeks international compensation in exchange for not drilling sizable oil deposits in the core of the megadiverse Yasuní National Park [1,6].

With governments promoting ever more oil development in the western Amazon, there needs to be greater attention given to minimizing the associated ecological and social risks [7]. Direct impacts include deforestation for access roads, drilling platforms, helipads, and pipeline routes, as well as contamination from spills, leaks and discharges [5]. Indirect effects, which include selective logging, hunting, and deforestation, primarily arise from the human colonization along new access routes [5]. Considerable social conflict, particularly with native communities, may also arise from these direct and indirect impacts [5].

While we strongly support efforts like the Yasuní-ITT Initiative as a potential mechanism to avoid completely the problems of hydrocarbon activities in the Amazon, we also argue for rigorous best practices where projects do move forward. We define a best practice as one that minimizes the environmental impact associated with typical practice, and that has been successfully employed in a commercial oilfield exploration or production project in Latin America.

At least three major obstacles currently impede the advancement of best practice in the western Amazon. First, best practice lacks a precise set of guidelines in applicable regulations. This regulatory gray area allows project proponents to define almost any practice as "best practice," and often results in typical high-impact practice being approved as best practice in environmental impact studies. Second, the lack of easily accessible and precise data on planned activities and infrastructure makes it difficult for policy makers and civil society to evaluate upcoming projects and push for best practice. Much of the currently available information relates to just the geographic extent of the hydrocarbon blocks, and not the more important planned activities within. Third, questions regarding cost, or assumptions that best practice will impose substantially greater costs, are common and likely deter companies from deviating from conventional practices.

We present here a three-part study aimed at overcoming these obstacles and demonstrating the potential of hydrocarbon sector best practice to minimize ecological and social impacts in western Amazonia. Our focus is on the Department of Loreto in northern Peru (Figure 1). Loreto, along with the neighboring Ecuadorian Amazon, is one of the largest and most dynamic hydrocarbon zones in the Amazon [5,8].

Loreto, a vast territory covering nearly 369,000 km^2, makes an ideal case study for a number of reasons. The region possesses extraordinary biological and cultural diversity [1,9], along with vast tracts of largely intact tropical forest, driving an urgency to minimize extractive industry impacts. It is home to a large number of active hydrocarbon blocks spanning the full range of project

stages, from pre-exploration to long-time production. In regards to the latter, a pair of 1970s-era oil operations caused significant contamination by dumping toxic production waters into local waterways for nearly four decades [10]. Therefore, local policy makers and residents are acutely aware of the potential risks from oil development. In addition, a number of recent exploration projects have yielded new oil and gas discoveries in Loreto, greatly increasing the probability that hydrocarbon development will continue as a major issue for the region well into the future.

We first present a set of best practice guidelines designed to minimize the impact of hydrocarbon activity in the Amazon. These guidelines incorporate both engineering-based criteria and key ecological and social factors. E-Tech International originally formulated the engineering guidelines, which are based on both Peruvian law and the latest in global technology [11]. We subsequently added the ecological and social factors to ensure that engineering best practice projects also do not threaten sensitive areas.

Second, we provide a detailed analysis of existing and planned hydrocarbon activities and infrastructure. In doing so, we move

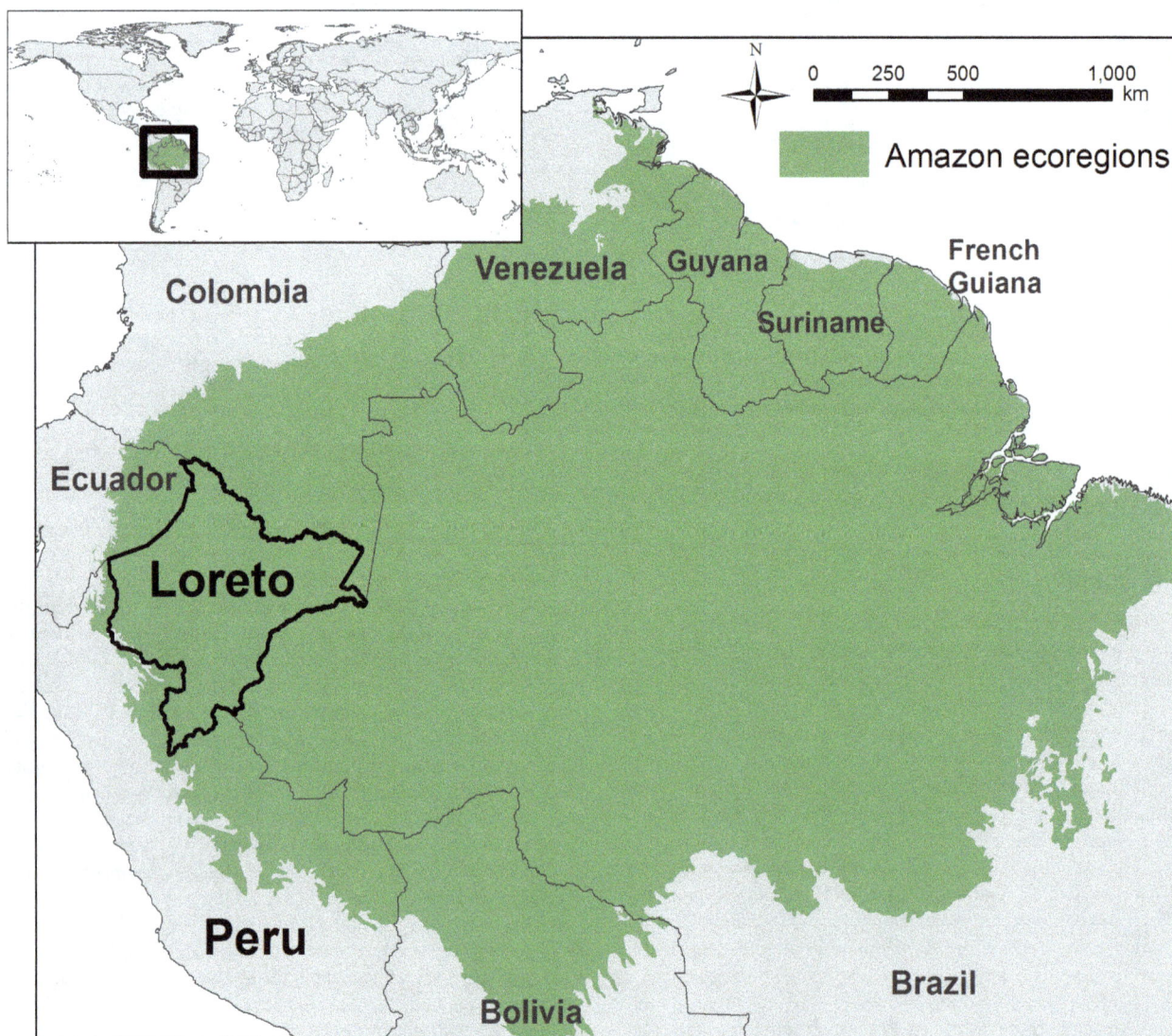

Figure 1. Study focal area. We focus on the Department of Loreto in the northern Peruvian Amazon. Amazon ecoregions are as defined by [61].

beyond evaluation based solely on the extent of hydrocarbon blocks and provide a more comprehensive examination of actual activities. This includes detailed data on existing and planned activities for all field-based phases of a hydrocarbon project, namely seismic exploration, exploratory wells, production wells, access roads, and pipelines.

Third, we evaluate the planned activities and infrastructure with respect to the best practice guidelines from part one. We analyze all planned projects in relation to both the engineering guidelines and the following four ecological and social factors: protected areas, indigenous territories, critical ecosystems, and priority watersheds. This evaluation represents a more strategic, larger-scale analysis than the current system of project-level, local-scale studies, and it would ideally take place within the context of a Strategic Environmental Assessment (SEA) [5]. Since 2008, Peruvian law has required national, regional, and local authorities to undertake SEAs for plans, polices, and programs that may have significant environmental impacts [12,13], but only a handful have been completed to date [14].

We also conduct an initial analysis on the estimated difference in cost between use of best practice and conventional development.

Finally, we discuss our findings in terms of how the use of best practice can minimize negative impacts, particularly deforestation and contamination.

Results

Best practice

The basis of the best practice guidelines was an analysis of both cutting-edge technology and Peruvian regulation (Table 1). To understand the implementation of best practice, it is important to understand first the typical life cycle of a hydrocarbon project in the Peruvian Amazon, which follows several basic steps. The government agency Perupetro creates the blocks ("lotes" in Spanish) and then promotes and auctions them internationally [15]. Recently there have been annual or biannual bidding rounds with one to two dozen blocks promoted and auctioned together. Perupetro ultimately signs the final contract with the selected company for each respective block, but the contract must first be approved by presidential decree [15]. The contract term, which runs 30 years for oil and 40 years for natural gas, includes two phases: exploration and production. The exploration phase is for seven years (with possible extensions) and includes a Minimum Work Program for the required amount of seismic lines and exploratory wells to be carried out by the operating company [15].

Two types of seismic testing are common in the Amazon, 2-dimensional (2D) and 3-dimensional (3D) [5,11]. The former generates an initial 2D cross-section of the subsurface, while the latter generates a 3D model to define in detail the deposit(s). On the ground, 2D is characterized by relatively spread-out linear transects (at least 1 km separation) cut through the forest, whereas 3D lines form tight grids (100s of meters separation) and are typically measured in square kilometers [11]. Seismic lines are typically less than two meters wide and do not require the cutting of large trees. Explosive charges are placed at regular intervals along these lines in holes of six to nine meters, and parallel lines of geophones register the echo patterns of the explosions on subsurface structures. These echo patterns reveal geologic structures that may contain oil or gas and that may warrant further assessment with exploratory wells [11].

If commercially viable quantities of oil or gas are discovered, the concession may proceed to production phase. However, contracts may be, and often are, terminated by the operating company during the exploration phase. Historically, the design of production phase has been characterized by many closely spaced drilling platforms, extensive networks of access roads, and pipeline routes with wide right-of-ways [11]. Moreover, in a number of projects designed during the 1970s, traditional practice included the dumping of toxic production waters directly into local waterways.

Engineering criteria. The first step of best practice, from an engineering perspective, is that the operating company must present an overall conceptual plan based on best practice for all phases of the project before beginning any work on the ground. We recommend that such a best practice conceptual plan be required during the company submission of its Minimum Work Program to the government during the bidding phase. This system would have the dual benefit of incorporating best practice into the bidding competition and subsequently the final contract signed by the company and the government. As a result, the use of best practice would be a formal and binding obligation. This recommendation of incorporating best practice into the Minimum Work Program would require a modification to current regulation.

Following this step, exploration activities should combine remote aerial electromagnetic surveys of subsurface structures with existing field information to create a precise state-of-the-art subsurface computer model of the hydrocarbon structures. The construction of this model involves an integrated approach that uses existing field data from seismic testing and exploratory wells as calibration points for new remote sensing data. A recent project in Brazil demonstrated the utility of this integrated approach to produce a precise subsurface computer model with minimal new intervention on the ground [11,16]. The aim of this innovation is to conduct new seismic testing only in areas where there is a demonstrated potential for commercial deposits. Typically oil companies do not combine the remote sensing data with existing data from earlier exploration programs to refine the study area for the purpose of minimizing the amount of subsequent seismic testing.

At the core of best practice is Extended Reach Drilling (ERD), a technique to reach a larger subsurface area from one surface drilling location. First developed in the late 1980s, ERD is a type of advanced directional drilling where the horizontal reach is at least two times greater than the vertical depth [11]. In practical terms, it means a single drilling platform can reach multiple distant targets in an oil or gas deposit, thereby reducing the total number of required platforms. The U.S. National Petroleum Council [17] recently recognized ERD as a key technology for reducing footprints of drilling operations. The current world record for ERD is 12.4 km, and any horizontal distance up to 8 km is now considered routine for an ERD well [11]. Therefore, there should be a large separation, at least 16 km, between drill sites.

ERD has been used in numerous Latin American exploratory and production drilling projects, but not yet in the Peruvian Amazon. In Argentina, two recent exploration projects employed ERD wells with horizontal displacements of approximately 4 and 5 km, in 2007 and 2008 respectively [11]. Also in Argentina, a production project beginning in 1997 drilled a series of ERD wells of more than 10 km. Most recently, in 2011, an exploration project in Colombia employed an ERD well. Although ERD has not yet seen application in Peru, it is important to note that national hydrocarbon regulation does require that drilling sites disturb the least amount of land possible [18] (see Article 67). Use of ERD would minimize the amount of land disturbed for drilling sites compared to any typical project limited to vertical or directional drilling techniques only.

The use of ERD relates to two additional key best practices: 1) no new access roads, processing facilities, or permanent camps

Table 1. Best practice guidelines.

1. Presentation of an overall project development plan based on best practice prior to initiating the exploration phase.

2. Use of state-of-the-art subsurface computer model that integrates airborne electromagnetic data and existing seismic data to minimize the need for new seismic projects.

3. All exploration and production platforms must be capable of drilling Extended Reach Drilling (ERD) wells with a horizontal displacement of at least 8 km (i.e., minimum distance between platforms of 16 km).

4. New access road construction is prohibited (e.g., no new roads between platforms and processing facilities or in pipeline/flowline rights-of-way).

5. Permanent camps may only be constructed along the banks of navigable rivers, not in the jungle interior.

6. Only permissible means of transport are by air and river, with defined limits on the size of transport vessels and on frequency of movements.

7. The maximum pipeline/flowline right-of-way construction width must be less than 13 m with intervals of canopy bridges at least every 1,000 m.

8. Pipelines should be designed/operated with: increased wall thickness to withstand soil movements and effects of internal erosion; regular internal traverses with intelligent inspection tools to detect internal abnormalities and lateral movement of the pipeline; automatic shut-off valves at each tie-in point of welded pipeline sections; and oil spill rapid response teams.

9. Adequate funds must be reserved for site abandonment that includes removal and/or remediation of contaminated materials, soil, and water sources, and revegetation of cleared areas with native species.

10. Consideration of key ecological and social factors such as protected areas, indigenous territories, key ecosystems, and key watersheds in determining whether oil & gas development should be pursued at all.

beyond the banks of navigable rivers, and 2) transport of people, materials, and equipment must be by air or river (with controls on size and frequency of movements). In other words, companies must operate as if at sea, a roadless development concept known as the offshore model [19]. In addition, production platforms deeper in the jungle and away from navigable rivers must be unmanned, with raw production fluids transported via roadless flowlines to the respective processing facility located along a navigable river. Processing facilities are where the production fluids – oil, gas, and production water – are separated, and the oil is prepared for export via pipeline, the gas burned for onsite use, and the production water re-injected into a subsurface formation. These points related to roadless development are consistent with Peruvian hydrocarbon regulation, which requires preferential use of river and air transport, and which states that road construction can only proceed if it is demonstrated that river and air transport are not possible [18] (see Article 40). For example, the Camisea natural gas project in southern Peru has been in operation since 2004 with no permanent camps away from navigable rivers and no access roads [11].

Regarding pipelines and flowlines, best practice calls for a greatly minimized right-of-way (ROW), with a reduction from the traditional 25 m down to 13 m or less. This "green pipeline" ROW technique, or "ducto verde" in Spanish, also emphasizes conforming the ROW to natural contours and emphasis on manual clearing (instead of heavy machinery) to further reduce impacts, particularly on steep slopes. This type of reduced-impact pipeline corridor was employed on one ROW section of the Camisea Project, in contrast to the higher-impact traditional pipeline ROWs used in other pipeline/flowline sections of the same project. Another major advantage of this type of narrowed ROW corridor is the ability to maintain canopy bridges. Canopy bridges are tree canopy sections along the ROW that remain intact to facilitate the passage of wildlife, at intervals of approximately one kilometer or more [20]. In order to minimize contamination threats related from pipelines, best practice also calls for increased wall thickness (to withstand soil movements and internal erosion), regular internal traverses with intelligent inspection gauges to detect internal abnormalities and lateral movement of the pipeline, automatic shut-off valves at each

welded tie-in point, and establishment of rapid response teams [21].

In terms of site abandonment, companies must set aside adequate funds to assure removal and/or remediation of contaminated materials, soil, and water sources, and revegetation of cleared areas with native species [11].

Ecological and social factors. In addition to the engineering-based best practices, it is critical to consider a range of key ecological and social factors. In other words, using technical best practice is not necessarily a license to operate in sensitive areas. Based on previous evaluations of ecological and social factors to consider in assessing projects in areas of high biodiversity and intact forest [5,8,22,23], we chose five: protected areas, priority watersheds, key ecosystems, indigenous territories, and proposed reserves for indigenous peoples in voluntary isolation.

Loreto has 14 official protected areas as established by the national protected areas agency SERNANP. Of these, 11 are managed nationally (two national parks, four national reserves, two communal reserves, and three reserved zones) and 3 are managed regionally (regional conservation areas). In the IUCN system of protected area categories, Peruvian national parks are considered as category II, national reserves as category VI, and the remaining areas either have no category or it is currently undeclared. Of these five types of protected area designations, just national parks are off-limits to extractive industries according to Peruvian Law. However, the new Güeppi – Sekime National Park (established in October 2012) allows the continuation of previously existing concessions. Therefore, 13 of the 14 protected areas in Loreto do not legally prohibit hydrocarbon activities. However, the national protected areas agency (SERNANP) must provide a technical favorable opinion before the energy ministry will approve activities within protected areas.

For priority watersheds, we focus on the Nanay River, a critical resource that provides drinking water to the departmental capital city of Iquitos. The classification of additional priority watersheds in Loreto is still under review by authorities. For key ecosystems, we focus on white-sand forests. Although low in overall species diversity, this rare and fragile ecosystem contains a high number of endemics and is considered a high conservation priority in Loreto [24].

Loreto is also home to a great abundance of indigenous peoples' territories. According to the latest publicly available data from the Instituto del Bien Común (IBC), there are around 500 titled indigenous territories in Loreto. Data for solicited new territories or solicited extensions of existing territories are more preliminary. The IBC data indicate that there are 24 solicited new territories and 29 solicited extensions of existing territories, although the true figures are likely to be much higher for both. In addition, within Loreto there are five proposed reserves for indigenous peoples in voluntary isolation. The right of indigenous peoples to be consulted in order to obtain their free, prior and informed consent about development decisions that will affect them is established under the International Labor Organization's Convention 169 [25] and the United Nations Declaration on the Rights of Indigenous Peoples [26]. Peru is a signatory of the former and voted in support of the latter. Moreover, Peru promulgated a landmark indigenous consultation law based on ILO 169 in 2011 [27].

Finally, two additional factors to consider, but beyond the scope of this study, are the greenhouse gas emissions and use of royalties from hydrocarbon activities. Regarding the former, carbon emissions arise from project-related forest loss, transportation, and energy generation, and of course the ultimate burning of the extracted hydrocarbons [28]. Indeed, one of the selling points of Ecuador's Yasuní-ITT Initiative is not only the avoided on-site deforestation, but also the maintenance of 410 million metric tons of CO_2 permanently underground [6]. For the latter, it is important to note that over 90% of royalties from hydrocarbon activities go to regional and local governments, and a portion of this money is used for transportation and other development projects that may also have environmental and social impacts [29,30].

Existing and planned activities and infrastructure

Hydrocarbon Blocks. As of October 2012, there were 48 hydrocarbon blocks in Loreto (Figure 2), covering 215,169 km^2 or 57.4% of the department. Of these, 29 are active concessions under contract with multinational energy companies. Four of these active concessions are in production phase (Blocks 1AB, 8, 31B, and 67) and the remaining 25 in exploration phase. The remaining 19 blocks are part of Perupetro's new bidding round.

Of the 25 concessions in the exploration phase, five have approved or pending environmental impact studies for seismic testing, three for exploratory wells, and six for both seismic testing and exploratory wells (Figure S1). The remaining concessions have not yet prepared environmental impact studies or begun exploration work.

Twenty-nine companies were operating or participating in the Loreto concessions during 2012. All but one are multi-nationals based outside of Peru. The 28 multi-nationals originate from 14 countries, including Argentina, Brazil, Canada, Colombia, France, Spain, Vietnam, the United Kingdom, and the United States of America. However, company turnover is relatively high. For example, during the course of this study, the primary concession holder changed in Blocks 64, 67, 123, and 129.

There are two important additional items to emphasize regarding this current state of hydrocarbon blocks in Loreto. First, although at the time of this publication Block 67 was not yet producing oil, the operating company declared this block commercially viable in late 2006, and it is currently officially classified as production phase. Second, many hydrocarbon blocks have previously existed but subsequently been retired and do not appear in Figure 2. Thus, many exploration wells and seismic lines displayed in subsequent figures appear outside the current blocks.

Seismic testing. Oil companies have conducted extensive 2D seismic testing in Loreto over the past 40 years, with a smaller but increasing amount of 3D seismic testing in recent years (Figure 3). This includes 61,403 km of 2D seismic lines (9% conducted since 2007) and 2,565 km^2 of 3D seismic (71% conducted since 2007). As illustrated in Figure 3, testing has been concentrated in southern and central Loreto, while much of northern and eastern Loreto has yet to experience major exploration. In regards to planned testing, five blocks (95, 109, 121, 130, and 135) have pending 2D projects totaling 3,900 km (Figure 3). Two additional blocks (1AB and 39) have pending 3D projects totaling 1,738 km^2.

Exploratory and production wells. Official data indicate that oil companies have drilled 223 exploratory wells in Loreto (Figure 4A), with 12% of them drilled since 1998 (the earliest date for which we have detailed data). Of these wells, nearly half (105) are outside of current production blocks and therefore may provide key field information to create subsurface computer models, potentially minimizing the need for extensive new exploratory campaigns.

Companies operating in Loreto have extracted 1.016 billion barrels of oil [31]. Of this production, Blocks 1AB, 8, and 31 have contributed 68%, 31%, and 1%, respectively. Annual oil production in Loreto peaked at 47 million barrels in 1979 [32] and has steadily fallen to 10.2 million barrels in 2011 [31,33,34], a decrease of 78%. The Peruvian Energy Ministry estimates over 393 million barrels of oil remain in these blocks (72% in Block 1AB, 25% in Block 8, and 3% in Block 31) [35].

There are currently 219 active production wells in Loreto (Figure 4A). Most are in Block 1AB (62.5%) (Figure 4B), with the remainder in Block 8 (20.5%) and Block 31 (17%). According to the Energy Ministry, there are also ~50 active reinjection wells and ~240 inactive and abandoned production wells [31].

Seventeen of the 28 exploratory wells drilled since 1998 have encountered hydrocarbon deposits in Blocks 31E, 39, 64, 67, 95, and 100 (Figure 4A). The type of newly discovered hydrocarbon varies considerably, with light oil in Block 64, medium oil in Block 95, heavy oil in Blocks 39 and 67, and shale gas in Block 31E. The Peruvian Energy Ministry estimates reserves (proven, probable, and possible) of around 928 million barrels in the three blocks with oil (40.5% in Block 39, 35% in Block 67, and 24.5% in Block 64) [35]. An additional 31.6 million barrels is reported from the latest oil discovery in Block 95. In terms of upcoming production, Block 67 is by far the most advanced, with approved environmental impact studies for the pipeline and development wells (Figure 4B).

Environmental impact studies have been submitted for 66 additional exploratory well platforms in Blocks 39, 64, 95, 102, 121, 123, 129, 130, and 135 (Figure 4A).

Roads and pipelines. Over time, the companies operating Blocks 1AB and 8 have constructed an extensive access road and flowline network to service the production wells and processing facilities. In addition, the North Peruvian Pipeline transports oil from these blocks to Peru's Pacific coast. Within Loreto, this flowline/pipeline network extends ~1,156 km (Figure 4A). Transport of crude oil from Block 31 to Pucallpa is via the Ucayali River.

There are plans to extend the existing pipeline network to connect with the new oil discoveries in the region (Figure 4A). The environmental impact study for a new 207 km pipeline to transport heavy crude from Block 67 to the starting point of the existing North Peruvian Pipeline was approved in 2011 (Figure 4B). Completion of this pipeline is scheduled for 2017. Preliminary plans also exist to transport light crude from Block 64 to the existing North Peruvian Pipeline. In August 2012, Peru and

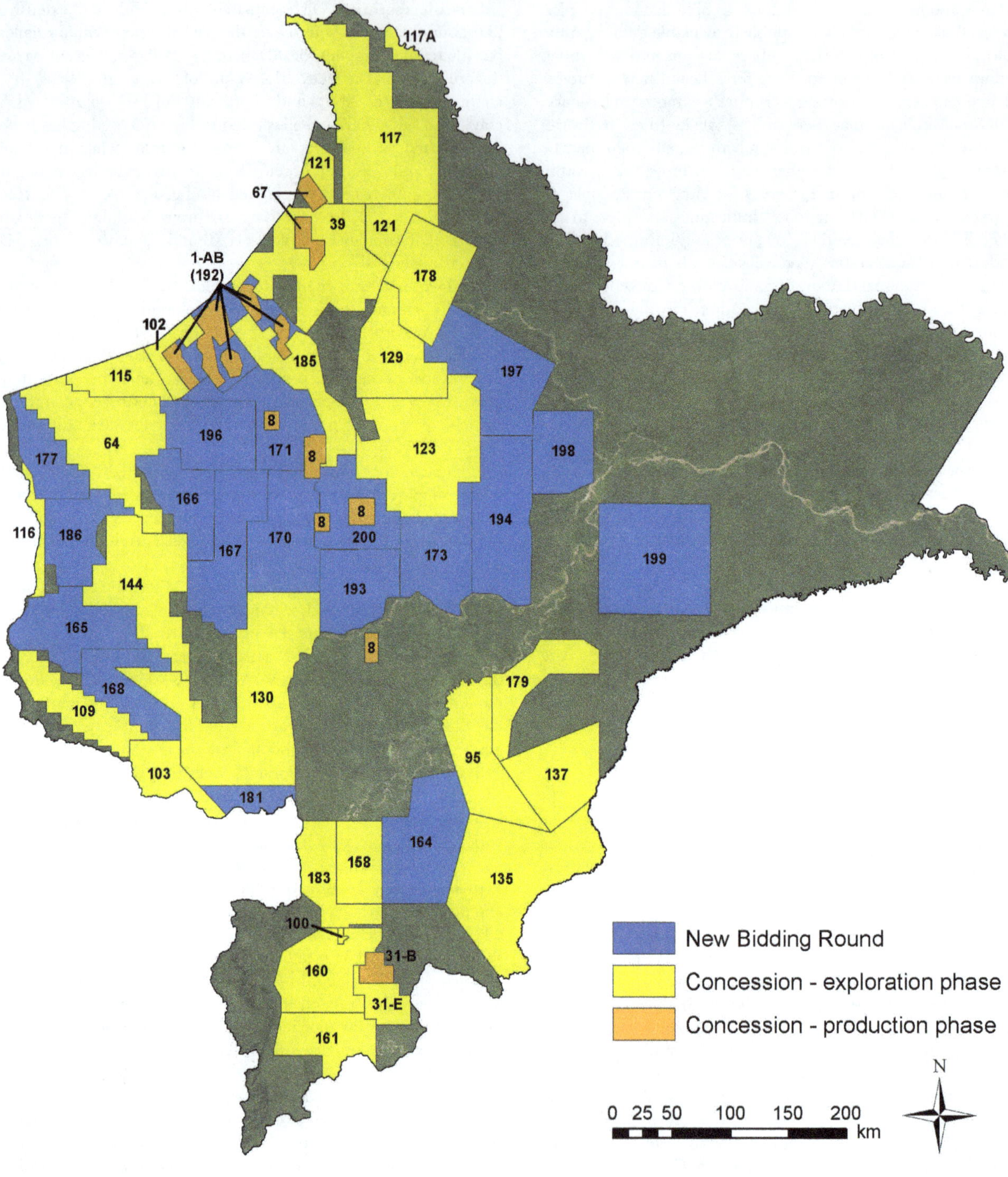

Figure 2. Hydrocarbon blocks in Loreto. There are three general types of blocks based on the contractual agreement between government and a company: concession in exploration phase, concession in production phase, and proposed concession under promotion or negotiation.

Ecuador signed an agreement that would allow the transport of Ecuadorean crude across the border to the North Peruvian Pipeline.

We calculate a cumulative network of 803 km of access roads in Blocks 1AB, 8, and 31 (Figure 4A). The largest access road network by far is in Block 1AB, a sprawling network of 504 km (Figure 4B).

The recently approved environmental impact study for the Block 67 production wells includes plans for a new 85 km access road network adjacent to the internal pipelines during the construction phase (Figure 4B). According to the approved plan, half of this access road network will be eliminated after the construction phase, including the connections between the three oil fields. There are also preliminary plans for construction of a 36 km access road in Block 64 (Figure 4A).

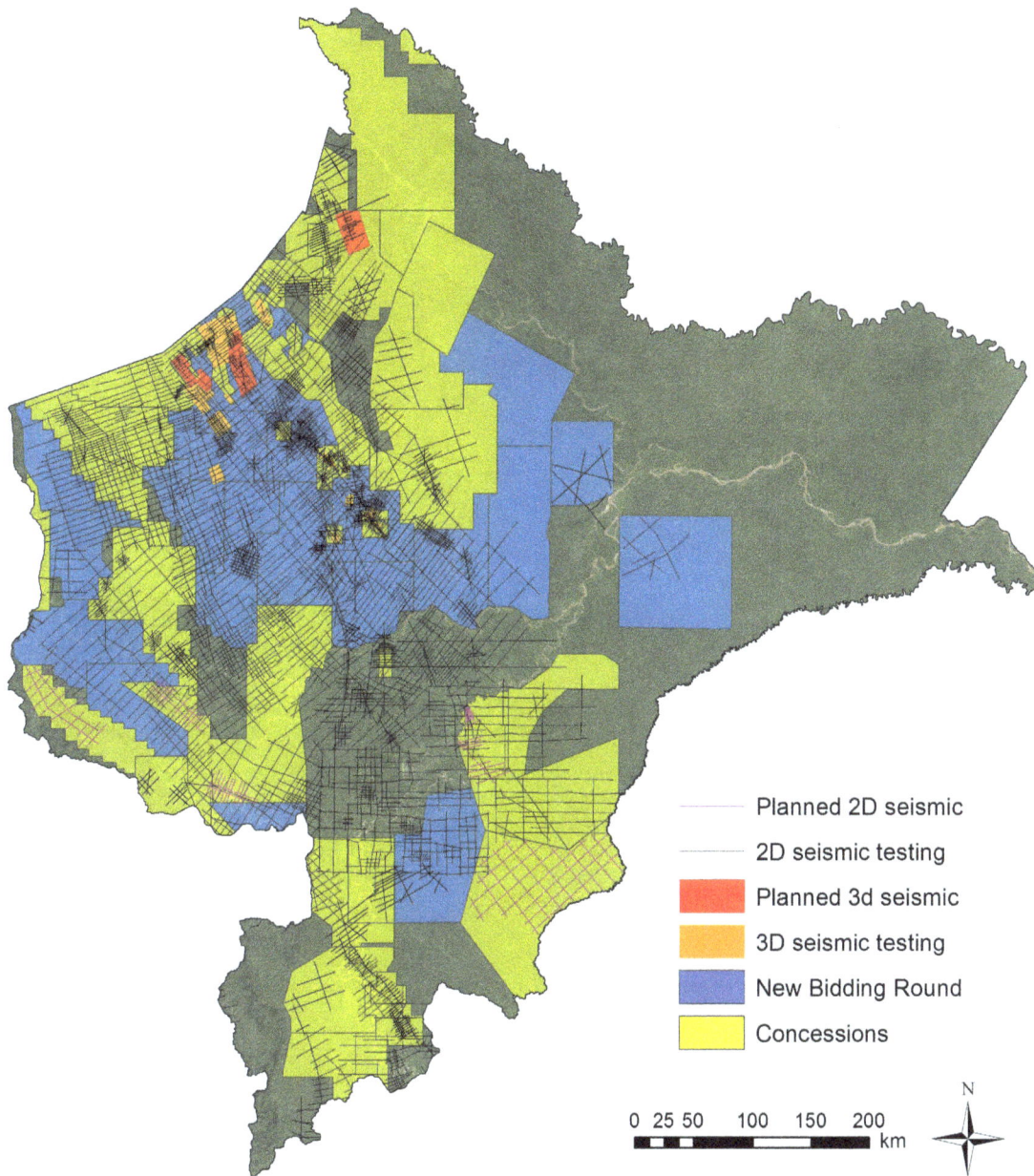

Figure 3. Existing and planned 2D and 3D seismic testing in Loreto. 2D testing is represented by straight lines and is measured in kilometers while 3D testing is represented by polygons and measured in square kilometers. It is important to note that numerous hydrocarbon blocks have previously existed but subsequently been retired. Thus, many seismic lines appear outside the current blocks.

Best practice

Engineering criteria. We analyzed all planned projects in relation to the best practice guidelines presented earlier. Starting with Block 67 as an example, the production plan consists of 21 production well platforms and three processing facilities (Figure 5a). The platforms are distributed among the three major oil deposits (eight in Paiche, six in Dorado, and seven in Piraña) and each major deposit has its own processing facility. Within each oil deposit, the multiple production platforms are located relatively close together, often separated by less than two km. In each case, the proposed drilling platforms are all within eight km of a single hypothetical ERD-capable drilling platform (Figure 5a). Figure 5b illustrates an alternative ERD-based Block 67 production field design using just three ERD-capable drilling platforms, one for

each oil deposit. In addition, note that in Figure 5b, the Dorado processing facility is gone, leaving only the two processing facilities, Paiche and Piraña, located near navigable rivers. This important modification would also mean the elimination of nearly all access roads.

In regards to the planned Block 67 pipeline, the main technical issue is the width of the right-of-way. As noted above, best practice calls for a maximum ROW width of 13 m or less. The Block 67 operating company at the time (Perenco) originally proposed a 25 m ROW width for much of the pipeline length (177 km), with a reduction to 20 m for the 30 km length crossing the Pucacuro National Reserve. Under pressure from the Peruvian government, the company increased the section having a 20 m width to 141 km, but resisted committing to the 13 m width achievable

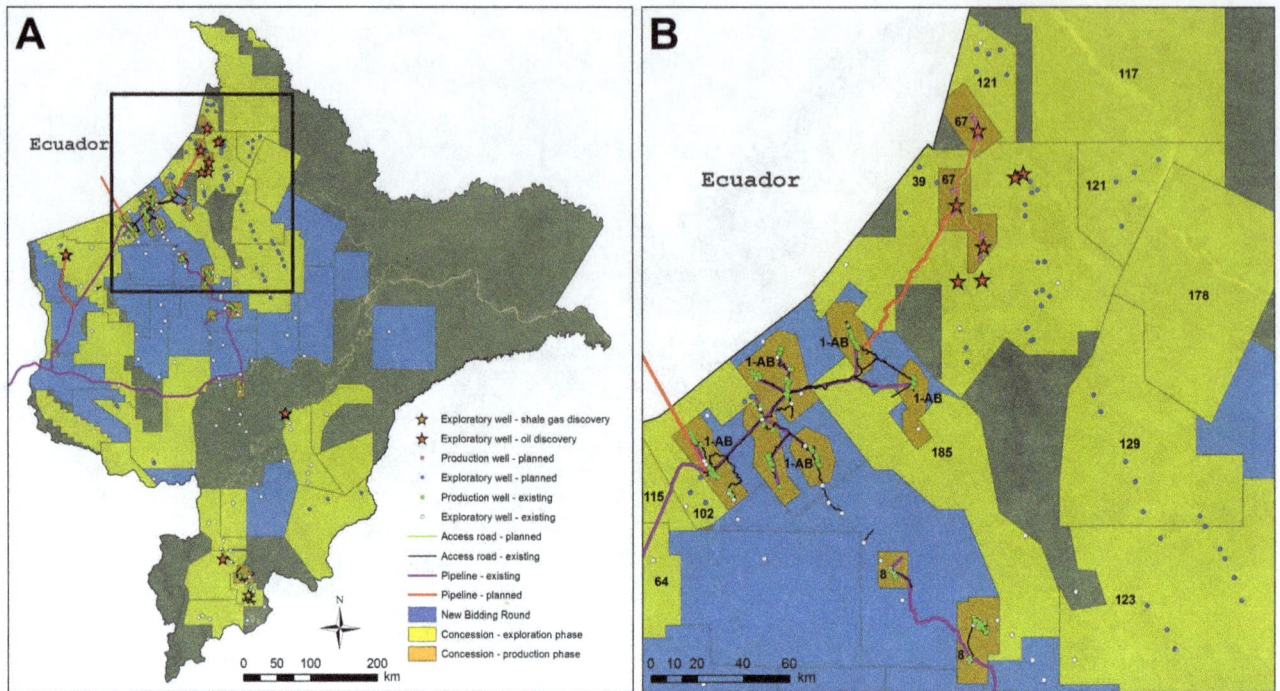

Figure 4. Existing and planned exploratory wells, production wells, access roads, and flowlines/pipelines in Loreto. (A) Map for all of Loreto. Note that stars indicate the Block 64 light crude oil discovery, the Block 95 medium oil discovery, and the Block 31 shale gas discovery. (B) Zoom of high activity zone in and around Blocks 1AB, 39, and 67. Note that stars indicate the Blocks 39 and 67 heavy crude oil discoveries. It is important to note that numerous hydrocarbon blocks have previously existed but subsequently been retired. Thus, many exploration wells appear outside the current blocks.

with the green pipeline technique. The company also agreed to a reduced ROW width of 10 m for 0.68 km of the pipeline corridor within Pucacuro that will include canopy bridges.

We also analyzed all other current plans for new exploratory well drilling platforms to determine how many proposed platforms could be eliminated by employing ERD. Figure 5b illustrates an alternative scenario that eliminates all platforms within 16 km of each other, with exploratory wells being drilled up to 8 km from each platform. This scenario assumes each drilling platform is ERD-capable. Of the 66 planned platforms in Blocks 39, 64, 95, 102, 121, 123, 129, 130, and 135, we estimate that nearly half (31) could be eliminated using ERD (Figure S2).

This reduction in infrastructure would translate directly to a reduction in deforestation. According to a sampling of environmental impact studies, we found that each new drilling platform requires the clearing of 2 to 4.5 hectares of forest and production phase processing stations require around 6 hectares each. For example, the Block 67 development project without best practice – consisting of 3 processing stations and 21 drilling platforms – would require a footprint exceeding 1 km^2 for these facilities. Using best practice to eliminate 18 drilling platforms and one processing facility would reduce forest loss by over 75%. In addition, the new Block 67 access road network and pipeline corridor, without best practice, would result in an additional 7 km^2 of direct forest loss [36,37]. With best practice, total direct forest loss would be significantly less, as the vast majority of the roads would be eliminated and the pipeline corridor would be seven to twelve meters narrower along nearly the entire length.

In addition, a review of environmental impact studies and post-project reports reveals that best practice would result in reduced forest loss during the exploration seismic phase. Most seismic projects require at least 50 heliports (larger projects may call for at

least 200) and literally hundreds of camps and drop zones [8]. Typical area requirements are around 2,400 m^2 for helipads, 300 m^2 for temporary camps, and 20 m^2 for drop zones. For example, a recently completed 1,480 km 2D seismic operation in Blocks 123 and 129 (that constructed 272 heliports, 208 camps, and 4,050 drop zones) had a cumulative 0.85 km^2 footprint [38,39]. A planned 3,700 km 3D seismic operation in Block 39 (calling for 75 heliports, 42 camps, and 3,800 drop zones) projects a 5.99 km^2 footprint [40].

Ecological and social factors. We found that oil blocks overlap 34% (29,000 km^2) of the protected area system in Loreto, with 19 blocks overlapping 10 protected areas (eight national and two regional) (Figure S3). The protected areas that are the most compromised by oil blocks include the Alto Nanay – Pintuyacu – Chambira Regional Conservation Area, Sierra del Divisor Reserved Zone, and Pucacuro National Reserve. A number of blocks cover an additional 17,150 km^2 of officially designated protected area buffer zones, primarily around Cordillera Azul National Park and Pacaya-Samiria National Reserve. A number of currently producing wells in Block 8 are within Pacaya-Samiria. Two of the recent Block 39 oil discoveries are within Pucacuro, as is a 30 km stretch of the planned pipeline from Block 67 to the Northern Peru Pipeline. There are 21 planned exploration wells within three protected areas. Thirteen of these planned wells are within Alto Nanay – Pintuyacu – Chambira (Blocks 123 and 129), seven are within Pucacuro (Block 39), and one is within Sierra del Divisor (Block 135).

The vast majority of blocks (90%) overlap titled or petitioned indigenous territories (Figure S4). Put another way, the oil blocks overlap 68% of these indigenous lands (42,548 km^2). Production wells in Blocks 1AB, 8, and 31 are located around or upstream of indigenous communities. This is also true of the oil discovery in

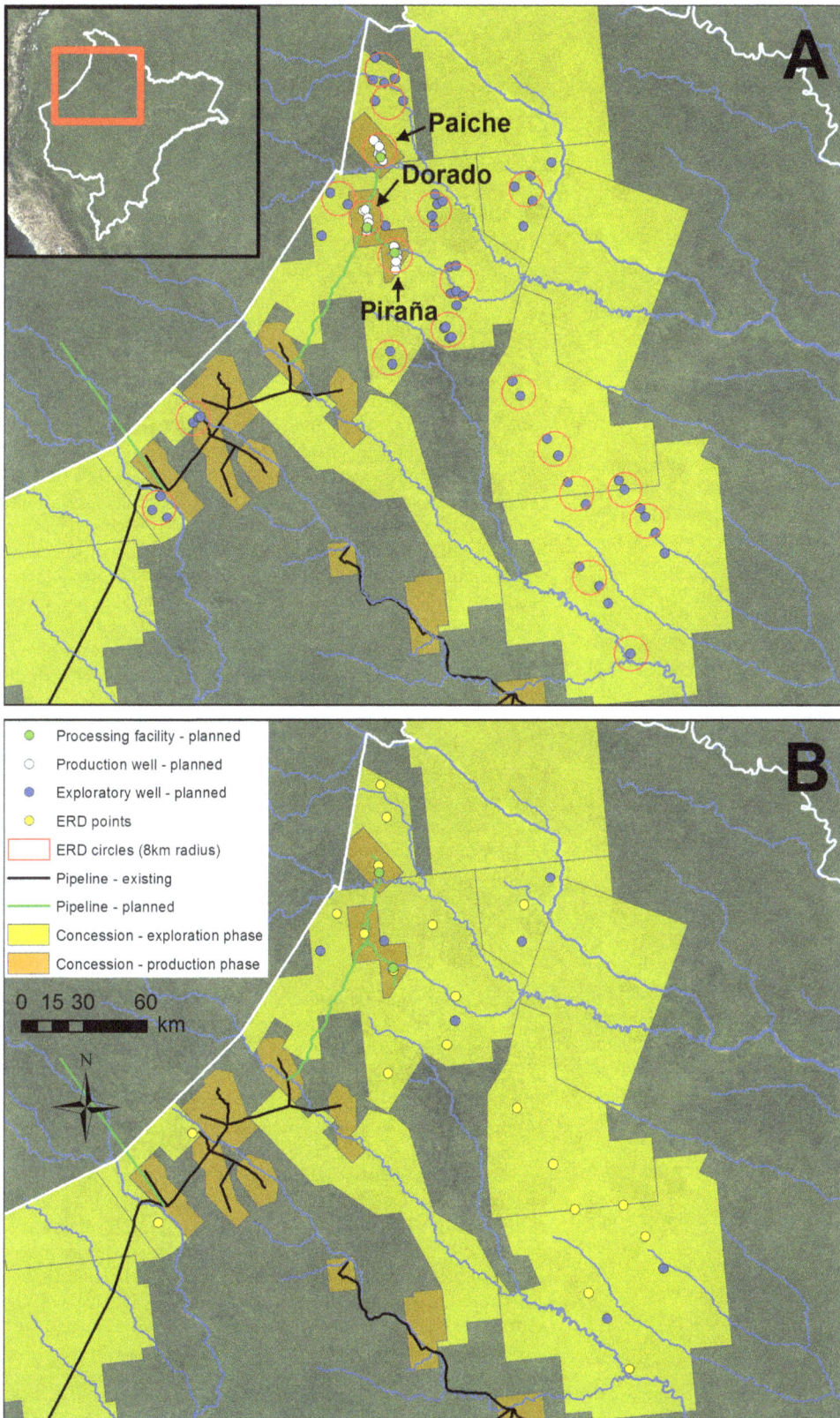

Figure 5. Analysis of planned projects in relation to best practice guidelines in high activity zone of Loreto. (A) Planned exploratory wells, production wells, and processing facilities. Red circles indicate multiple platforms within eight kilometers of a single hypothetical ERD-capable drilling platform. (B) Alternative design based on best practice. Yellow dots indicate where an ERD-capable drilling platform could replace multiple planned platforms. Note that the Dorado processing facility is eliminated due to distance from a navigable river.

Block 64 and seven additional planned exploratory wells in other blocks.

Twelve oil blocks overlap 60% (21,962 km^2) of the proposed reserves for indigenous peoples in voluntary isolation (Figure S4). Note that there is an extremely high level of existing and planned activity within the proposed Napo-Tigre Territorial Reserve (Figure S4). The three recently discovered Block 67 oil deposits, two of the Block 39 oil discoveries, and 48 planned exploratory wells are within the reserve. There are also three planned exploration wells in the proposed Yavari-Tapiche Territorial Reserve.

Twelve blocks overlap white-sand forest patches (Figure S5). Indeed, blocks cover all of the known large patches of white-sand forest outside of Allpahuayo-Mishana National Reserve. Several wells in Blocks 123, 129, and 135 are close to white-sand forests. The Northern Peru Pipeline crosses one of the largest white-sand forest patches.

Finally, fourteen planned exploratory wells in Blocks 123 and 129 are within the Nanay watershed, as are sections of four new blocks included in the new bidding round (Figure S6).

When combining all areas covered by protected areas, indigenous territories, white-sand forests, and the Nanay watershed (Figure S7), we found that nearly half (48%) of the total hydrocarbon block area in Loreto overlaps at least one key ecological or social factor (Figure 6). In addition, 80% of the planned exploratory wells, 100% of the planned production platforms, most of the recent hydrocarbon discoveries, and 59% of the planned pipelines contain such an overlap.

Cost analysis. There is enough data available on costs for the planned Block 67 development project to make a comparison between the proposed conventional project and an alternative project using best practice [11,41]. This cost analysis considered changes due to use of ERD, elimination of the access road network, elimination of one processing facility, and implementation of the green pipeline ROW construction technique.

The average depth of the wells in Block 67 is approximately two kilometers. Therefore, only a well with a horizontal displacement of greater than four kilometers would be considered an ERD well. Assuming a single, central drilling platform in each of the three oil fields, we estimated that one-third of the planned wells would use ERD and the remaining two-thirds would be conventional directional wells. A conventional well costs $3.5 million and the cost of an ERD well increases approximately linearly with its horizontal displacement. Therefore, assuming that an ERD well's average horizontal displacement will be twice that of a conventional well, we estimated an average ERD well cost of $7 million. We calculated that the use of ERD for one-third of the wells would increase costs by about $220 million.

Several other key components of best practice, however, would reduce costs. The elimination of 18 planned drilling platforms due to use of ERD would reduce costs by about $142 million. The elimination of one of the planned processing facilities would reduce costs by about $36.5 million (this estimate includes additional costs for expanding one of the other planned processing facilities to accept more flow).

In terms of transportation costs, the elimination of the access road network would reduce costs by about $45 million. Reliance on extensive jungle road networks and diesel-fueled heavy vehicles, using imported diesel fuel, adds a substantial operational cost. There would be some increase in helicopter flights, though this expense would be offset by the near-elimination of heavy vehicle traffic in the block. In regards to arriving to the site, barges already move on regular schedules from Iquitos to docks of active

concessions throughout Loreto and therefore do not represent a major new expense.

Overall, we found that best practice does not translate to substantially higher costs, and may in fact reduce total expenses. The operating company for Block 67 estimated total costs of $1.339 billion [41]. We estimate that total costs for the best practice alternative is $1.321 billion.

Discussion

Loreto, a vast region larger than Germany or nearly the size of Montana, is one of the most active and dynamic hydrocarbon zones in the Amazon. Forty-eight oil blocks cover over half the department, an affected area of over 215,000 km^2. These blocks cover the full range of project development stages: 4 in production, 25 in various stages of exploration, and the remaining 19 are part of Perupetro's latest international bidding round. Adding to the complexity, 29 companies operate the production and exploration phase blocks, and company turnover is frequent.

Companies have extracted over one billion barrels of oil from Loreto over the past 40 years. However, a major long-term trend of decreasing production has spurred efforts to boost exploration in search of additional deposits. This trend will begin to reverse with the imminent start of production in Block 67, the most recent block to enter into production phase. Two additional recent notable discoveries include heavy oil in Block 39 and light oil in Block 64. The Peruvian Energy Ministry estimates reserves of over 900 million barrels of oil in these three blocks. Together with the remaining reserves in Blocks 1AB and 8, Loreto may have another billion barrels of oil available.

A key wild card is the shale gas discovery in Block 31E. This discovery is significant because of the potentially large size of the shale formation, the novelty of developing this type of gas deposit in Peru, and the possible utilization of shale fracturing techniques [42]. Recent experience in the United States has demonstrated that there are significant and unique risks associated with shale gas production, and that these risks are not yet fully understood [43].

More new discoveries are likely given that exploration activities remain very active. Indeed, 44 of the 48 blocks in Loreto are in either exploration or bidding phase, 13 of which already have finalized environmental studies for seismic testing and exploratory wells. In other words, extensive and widespread amounts of exploration are still to come.

Impacts and the role of best practice

With such a large number of hydrocarbon projects, it is critical to advance best practice as a means of minimizing social and environmental impacts in Loreto. The original design and operations of Blocks 1AB and 8 – characterized by many closely-spaced drilling platforms, dumping toxic production waters directly into local waterways, and extensive access road networks – represent high-impact, 1970s-era technology [11]. In contrast, best practice incorporates a number of technological advances and strategic planning techniques to minimize negative impacts, such as deforestation and contamination.

We demonstrated that the use of technical best practice, in the case of Block 67, would reduce impacts by: 1) reducing the number of drilling platforms from twenty-one to three, 2) eliminating one of the three processing facilities, 3) eliminating virtually the entire access road network, and 4) narrowing the pipeline right-of-way. Furthermore, we estimate that the use of ERD-capable drilling rigs across all exploration blocks in Loreto could eliminate about half of the proposed drilling platforms. In the context of a Strategic Environmental Assessment, this would

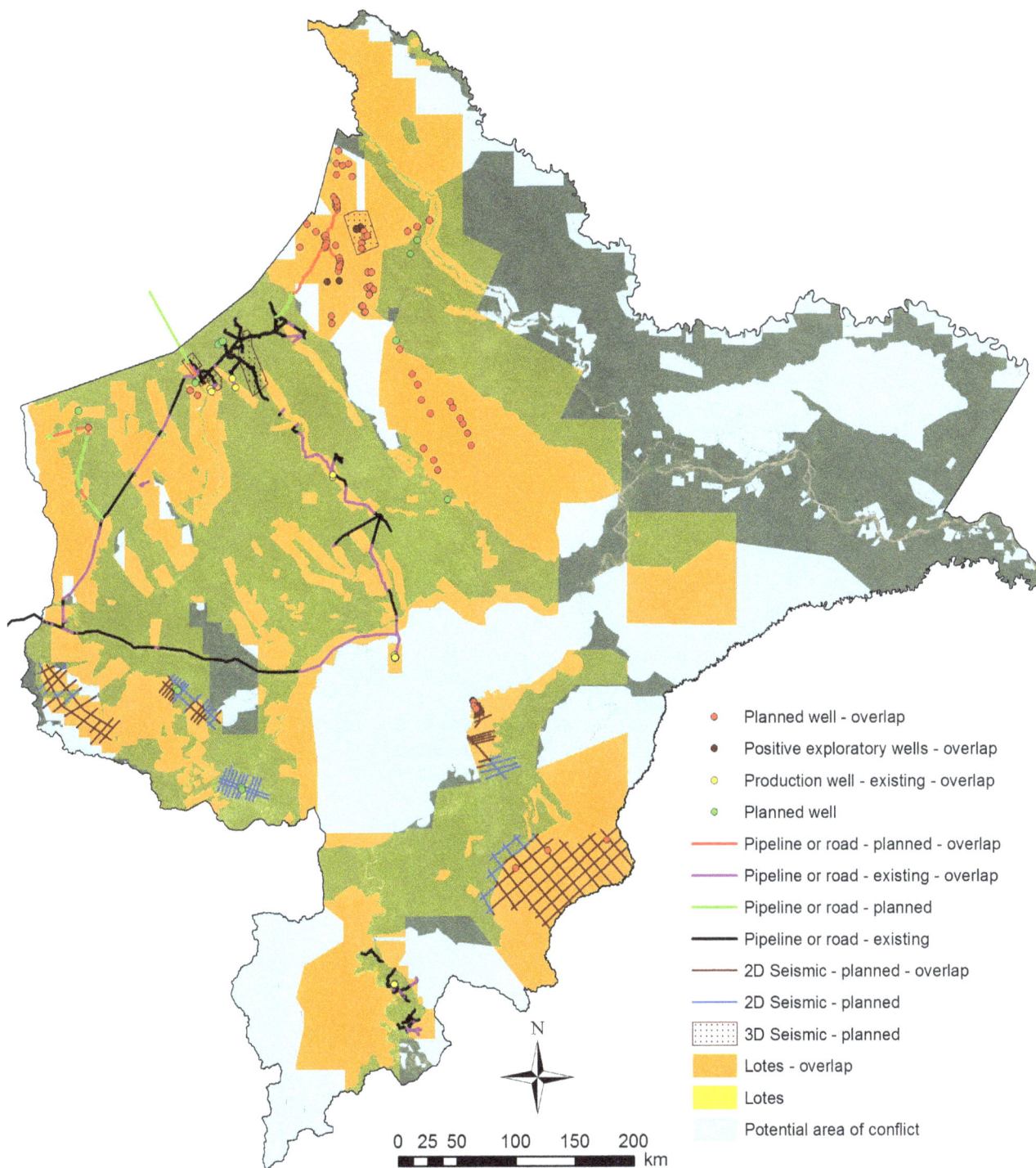

Figure 6. Consideration of key ecological and social factors: overlaps. See Figure S7 for more information on background layer. Light blue indicates an important or sensitive area that is not covered by a hydrocarbon block, while orange indicates an area that is covered by a block. Further, we indicate planned 2D and 3D seismic testing, exploratory and production wells, access roads, and flowlines/pipelines that would overlap with at least one of the key ecological and social factors.

represent a lower-impact, "greener" scenario, in relation to the higher-impact Business-As-Usual scenario.

We further found that this reduction in infrastructure from best practice would directly translate to a reduction in deforestation. In the case of Block 67, forest loss would drop by around 50, 75, and 100% from drilling platforms and processing facilities, the

pipeline, and access roads, respectively. Moreover, the reduction of access roads could prevent substantial secondary deforestation. Fortunately, the isolated existing access roads have not yet triggered significant indirect forest loss from subsequent colonization and logging, as roads have in the neighboring Ecuadorian Amazon. If connected to the rest of Peru's road network, as called for in long-term government plans, indirect deforestation would likely quickly escalate.

The reduction in drilling platforms by employing best practice may also serve to reduce contamination. Blocks 1AB and 8 resulted in nearly four decades of significant contamination through the dumping of toxic production waters into local waterways, until indigenous inhabitants forced an accelerated phase-out of this practice between 2006 and 2009 [10,44]. However, pollution problems continue to plague local communities, as all three current oil producing blocks in Loreto (Blocks 1AB, 8, and 31B) have had major leaks and spills in recent years [45,46]. In addition to the now mandatory practice of reinjecting toxic production waters, best practice serves to reduce contamination by significantly reducing the number of point sources (i.e., drilling platforms) and designing more strategic flowline/pipeline routes.

Our best practice guidelines also aim to minimize the negative impacts from exploration phase seismic testing. Our review of environmental impact studies and post-project reports revealed that traditional seismic projects do cause deforestation, primarily from the need to construct hundreds of helipads, temporary camps, and drop zones. In addition, seismic testing, particularly the more intensive 3D form, results in helicopter noise, an inux of workers, the cutting of hundreds of kilometers of seismic lines through the understory, and the detonation of thousands of underground seismic charges [47]. A recent study found a significant decrease in the group sizes of the endangered white-bellied spider monkey (*Ateles belzebuth*) during 2D seismic testing in Block 39 [48], although these same researchers found no negative impacts on ocelots (*Leopardus pardalis*) [49].

As part of best practice, we contend that the extent of future seismic testing, and therefore its associated impacts, could be greatly reduced by combining existing exploration data with remote sensing data in a state-of-the-art subsurface computer model. The region has already been subject to over 61,000 km of 2D seismic testing, 2,500 km^2 of 3D seismic testing, and 220 exploratory wells. However, companies operating in the region typically do not analyze this existing information in combination with remote sensing data for the purpose of minimizing the amount of new seismic testing. Instead, extensive new seismic testing programs are still the norm, as evidenced by the more than 3,400 km of planned 2D seismic and 1,700 km^2 3D seismic projects. Given the extensive amount of existing exploration data in Loreto, this modeling advance offers a methodology that may greatly minimize the extent of new seismic campaigns.

We also raised the important need to consider ecological and social factors in addition to technical best practice criteria. We found that nearly half of the total block area and the vast majority of planned exploration wells, production platforms, and planned pipeline length overlap sensitive areas in Loreto. For example, oil blocks overlap over one-third of the protected area system, two-thirds of the titled and solicited indigenous territories, nearly all of the large white-sand forest patches, and nearly the entire Nanay watershed. Recognizing and minimizing these types of conflictive overlaps early in the government's concession evaluation process could avoid future conflicts. For example, the current controversy over planned exploratory wells in the Nanay watershed, the source of the capital city's water supply, could have been avoided by excluding this area from concessions in the first place. However, history may be doomed to repeat itself as four of the new bidding round blocks overlap this same watershed.

Identifying overlaps and possible conflicts with indigenous communities is also an important element of the new indigenous consultation law. This law, which entered into force in April 2012,

is debuting in Loreto with the re-leasing of Block 1AB as Block 192 (current contract expires in 2015). Indigenous organizations are demanding a number of important actions, such as the remediation of existing environmental damages, resolution of land-titling disputes, and consultation with affected indigenous communities before the bidding process begins [50]. They are also calling for the elaboration of a Strategic Environmental Assessment for all planned and existing blocks.

Finally, we demonstrated that incorporating best practice does not impose substantially greater costs than a conventional project, and may in fact reduce overall costs. Although costs for ERD wells are around double that of conventional wells, the reduction in costs from elimination of drilling platforms, access roads, and remote processing facilities counterbalance the higher well construction costs.

Large barriers to the widespread implementation of best practice in Loreto and the rest of the Amazon clearly exist. Despite meetings and letters urging Peruvian officials to mandate use of ERD and green pipeline ROW in Block 67, the environmental impact studies were approved without full adoption of these key elements of best practice. Further work is needed to advance the concepts discussed in this paper, ideally in the form of a government-led Strategic Environmental Assessment.

Methods

We obtained all GIS data described below from existing sources, no field work was conducted in this study. However, in some cases we revised the data if obvious differences were observed in satellite imagery.

Analysis of existing and planned activities and infrastructure

We obtained GIS data for hydrocarbon blocks, seismic lines, exploratory wells, and pipelines from Perupetro in November 2011 and October 2012. We acquired GIS data for production wells from Perupetro in July 2012. Additional information on seismic testing, exploratory wells, production wells, oil production, and operating companies is from monthly "Informe Estadístico" and yearly "Anuario Estadístico" reports available on the Ministerio de Energía y Minas website (http://www.minem.gob.pe). We acquired information on whether or not recent exploratory wells encountered hydrocarbon deposits from a Perupetro presentation [51] and press reports. We updated the status of the blocks using the environmental impact studies published on the Ministerio de Energía y Minas website. Data pertaining to the new bidding round blocks are from information included in a Perupetro presentation [52].

For existing pipelines, additional GIS data are from the Loreto Regional Government. We compared the Petroperu pipeline datasets to recent Landsat and higher resolution satellite imagery in Google Earth and ArcGIS basemaps to produce a revised pipeline layer. This revised layer included route corrections for known pipelines and the addition of spurs visible in the satellite imagery but not included in either of the original datasets.

For existing access roads, we obtained two GIS datasets. The first was from the national government via the Ministerio de Transportes y Communicaciones. The second was from the Loreto Regional Government. We compared both datasets to recent Landsat and higher resolution satellite imagery in Google Earth and ArcGIS basemaps to produce a revised data layer. This revised layer included route corrections and the addition of spurs visible in the satellite imagery but not included in either of the original datasets.

Data for planned seismic lines and exploratory wells are from environmental impact studies published on the Ministerio de Energía y Minas website. Information related to the planned production wells in Block 67 is from the relevant environmental impact study [37]. For planned pipelines, we obtained information from the relevant Block 67 environmental impact studies [36,37], a public presentation by a Block 64 operating company representative in Iquitos, Peru (June 2012), and press reports regarding the pipeline extension to Ecuador. For planned access roads, information is from the relevant Block 67 environmental impact study [37] and an operating company report detailing development options for Block 64.

The cut-off date for incorporating new data was March 2013.

Best practice

We analyzed all planned projects in relation to both the engineering guidelines and identified ecological and social factors. For the engineering criteria component, we identified all planned exploratory wells and production platforms that are within eight kilometers of a single central drilling platform. These wells could therefore be drilled from a central drilling platform using an ERD-capable drilling rig. We also identified all river sections with at least 5,000 upstream cells in HydroSHEDS [53], which we used as a proxy for year-round navigability of the river. This data was used to corroborate the feasibility of limiting permanent camps and processing facilities to sites along navigable rivers. For the estimates on avoided deforestation, we collected information on the area required for drilling platforms, processing facilities, and seismic activities from a sampling of current environmental impact studies and post-project reports from Blocks 39, 67, 102, 123, 127, 128, 129, 130 and 135.

For the ecological and social factors component, we analyzed all existing and planned activities and infrastructure in relation to: protected areas, indigenous territories, white-sand forest patches, and the Nanay watershed. Data for protected areas are from SERNANP [54]. Subsequently we digitized three new areas created after the data were obtained from SERNANP. GIS data for indigenous territories are from the Instituto del Bien Común [55]. Data for white sand forest patches are from NatureServe, Field Museum, and published studies [24,56,57]. Analyses were done in ArcGIS 10.1.

For the comparative cost analysis, we used oil industry guidelines on the definition of ERD wells of at least 2:1 ratio of horizontal displacement to vertical depth [58] and the relative cost of an ERD well (proportionate to length of well), industry data on the maximum length of oil and natural gas flowlines [59], and a comparative cost estimate of green pipeline and conventional pipeline ROW construction costs [60]. Specific data for the Block 67 case study came from the actual projected costs estimated by the operating company (Perenco) to fully develop Block 67. These costs were presented in an official environmental impact study response by Perenco [41] and approved by the Energy Ministry in January 2012. This document includes details on the cost of all major Block 67 infrastructure elements, including well development, drilling platforms, processing facilities, permanent camps, roads, docks, and logistical bases.

Supporting Information

Figure S1   Status of hydrocarbon blocks in Loreto. Blocks color-coded to indicate phase of activity within.

Figure S2   Analysis of planned projects in relation to best practice guidelines across Loreto. Red circles indicate multiple platforms within eight kilometers of a single hypothetical ERD-capable drilling platform. Note that of the 66 planned platforms, we estimate that nearly half could be eliminated using ERD.

Figure S3   Hydrocarbon blocks, activities, and infrastructure in relation to protected areas of Loreto.

Figure S4   Hydrocarbon blocks, activities, and infrastructure in relation to indigenous territories of Loreto.

Figure S5   Hydrocarbon blocks, activities, and infrastructure in relation to white-sand forests of Loreto.

Figure S6   Hydrocarbon blocks, activities, and infrastructure in relation to the Nanay watershed. Note that the waters of the Nanay lead to the departmental capital of Iquitos, providing its drinking water.

Figure S7   Consideration of key ecological and social factors: background layer. This background layer indicates important and sensitive areas such as protected areas, indigenous territories, and key ecosystems and watersheds. Details are in Figures S3, S4, S5, S6.

Acknowledgments

We thank Carroll Muffett, Melissa Blue Sky, Amanda Kistler, Enrique Ortiz, and three anonymous reviewers for reviewing the manuscript and providing helpful feedback. We thank Valeria Urbina, Patricia Patron, Asunta Santillan, and Cristina López for valuable comments and assistance in gathering data.

Author Contributions

Conceived and designed the experiments: MF CNJ BP. Performed the experiments: MF CNJ BP. Analyzed the data: MF CNJ BP. Contributed reagents/materials/analysis tools: MF CNJ BP. Wrote the paper: MF CNJ BP.

References

1. Bass MS, Finer M, Jenkins CN, Kreft H, Cisneros-Heredia DF, et al. (2010) Global Conservation Significance of Ecuador's Yasuní National Park. PLoS ONE 5(1): e8767. Available: http://www.plosone.org/article/info:doi/10.1371/journal.pone.0008767. Accessed 2013 Apr 3.

2. Hoorn C, Wesselingh FP, ter Steege H, Bermudez MA, Mora A, et al. (2010) Amazonia through time: Andean uplift, climate change, landscape evolution, and biodiversity. Science 330: 927–931.

3. Red Amazónica de Información Socioambiental Georreferenciada (2012) Mapa Amazonía 2012: Áreas Protegidas y Territorios Indígenas.

4. Killeen TJ (2007) A perfect storm in the Amazon wilderness: Development and conservation in the context of the Initiative for the Integration of the Regional Infrastructure of South America (IIRSA). Arlington, VA: Conservation International.

5. Finer M, Jenkins CN, Pimm SL, Keane B, Ross C (2008) Oil and Gas Projects in the Western Amazon: Threats to Wilderness, Biodiversity, and Indigenous Peoples. PLoS ONE 3(8): e2932. Available: http://www.plosone.org/article/info%3Adoi%2F10.1371%2Fjournal.pone.0002932. Accessed 2013 Apr 3.

6. Finer M, Moncel R, Jenkins CN (2010) Leaving the oil under the Amazon: Ecuador's Yasuní-ITT Initiative. Biotropica 42: 63–66.

7. Haselip J (2011) Transparency, consultation and conflict: Assessing the micro-level risks surrounding the drive to develop Peru's Amazonian oil and gas resources. Nat Resour Forum 35:283–92.

8. Finer M, Orta-Martínez M (2010) Second hydrocarbon boom threatens the Peruvian Amazon: trends, projections, and policy implications. Environ. Res. Lett. 5: 014012. Available: http://iopscience.iop.org/1748-9326/5/1/014012. Accessed 2013 Apr 3.

9. Benavides M (2009) Mapa Amazonia Peruana. Lima: Instituto del Bien Común. Available: http://www.ibcperu.org/mapas/mapa-amazonia-peruana.php. Accessed 2013 Apr 3.

10. Orta Martínez M, Napolitano DA, MacLennan GJ, O'Callaghan C, Ciborowski S, et al. (2007) Impacts of petroleum activities for the Achuar people of the Peruvian Amazon: summary of existing evidence and research gaps. Environ Res Lett 2: 1–10.

11. Powers B (2012) Best Practices: Design of Oil and Gas Projects in Tropical Forests. E-Tech International.

12. República del Perú (2001) Ley del Sistema Nacional de Evaluación del Impacto Ambiental (Ley No. 27446), As modified by Decreto Legislativo No. 1078.

13. República del Perú; Reglamento de la Ley del SEIA (DS 019-2009-MINAM).

14. R. García Consultores SA, ARCAN Ingeniera y Construcciones SA, Centro de Conservación de Energía y del Ambiente (2012). Elaboración de la Nueva Matriz Energética Sostenible y Evaluación Ambiental Estratégica, como Instrumentos de Planificación.

15. Perupetro (2011) Marco general de las actividades de exploración y explotación de hidrocarburos.

16. Schlumberger (2011) Multi-property Earth model building through data integration for improved subsurface imaging. First Break 29.

17. National Petroleum Council (2011) Sustainable drilling of onshore oil and gas wells. Paper #2–23.

18. República del Perú (2006) Reglamento para la Protección Ambiental en las Actividades de Hidrocarburos. Decreto Supremo No. 015-2006-EM.

19. Tollefson J (2011) Fighting for the forest: The roadless warrior. Nature 480: 22–24.

20. Thurber M, Ayarza P (2005) Canopy bridges along a rainforest pipeline in Ecuador. Society of Petroleum Engineers paper SPE-96504-PP.

21. Ministerio de Energía y Minas del Perú (2007) Auditoría integral de los sistemas de transporte de gas natural y líquidos de gas natural del Proyecto Camisea. Reporte Final de Auditoria Integral.

22. The Energy & Biodiversity Initiative (2003) Integrating biodiversity conservation into oil and gas development. Washington DC: Conservation International.

23. Goodland R (2012) Responsible mining: the key to profitable resource development. Institute for Environmental Diplomacy and Security Research Series: A1-2012-4.

24. Fine PVA, Garcia-Villacorta R, Pitman NCA, Mesones I, Kembel SW (2010) A floristic study of the white-sand forests of Peru. Ann. Missouri Bot. Gard. 97: 283–305.

25. International Labour Organisation (1989) Convention No. 169 concerning Indigenous and Tribal Peoples in Independent Countries.

26. United Nations General Assembly (2007) United Nations Declaration on the Rights of Indigenous Peoples.

27. República del Perú (2011) Ley del Derecho a la Consulta Previa a los Pueblos Indígenas u Originarios, reconocido en el Convenio 169 de la Organización Internacional del Trabajo (OIT).

28. Zahniser A (2007) Characterization of greenhouse gas emissions involved in oil and gas exploration and production operations. Review for the California Air Resources Board.

29. Viale C (2011) Renta petrolera y uso del Canon en Loreto.

30. Viale C (2012) Generación, distribución y uso de la renta petrolera en Loreto. Lima: Programa de Vigilancia Ciudadana.

31. Ministerio de Energía y Minas del Perú (2011) Anuario estadístico de hidrocarburos 2011.

32. Petroperú (1981) Informe estadístico de petróleos del Perú 1970–1981.

33. Petroperú (1989) Informe estadístico annual 1989.

34. Ministerio de Energía y Minas del Perú (1999) Anuario estadístico de hidrocarburos 1999.

35. Ministerio de Energía y Minas del Perú (2012) Libro anual de reservas de hidrocarburos 2011.

36. Daimi Peru, Perenco (2010) EIA del proyecto de construcción del oleoducto y línea de diluyente CPF – Andoas.

37. Asamre Perenco (2011) Estudio de impacto ambiental para la fase de desarrollo del Lote 67A Y 67B.

38. Burlington Resources, ConocoPhillips (2012) Plan de abandono del proyecto de prospección sísmica 2D Lote123, Loreto.

39. Burlington Resources, ConocoPhillips (2012) Plan de abandono del proyecto de prospección sísmica 2D Lote129, Loreto.

40. Repsol GEMA (2011) EIA Proyecto de prospección sísmica 3D y perforación de 21 pozos exploratorios – Lote 39.

41. Ministerio de Energía y Minas del Perú (2012) Levantamiento de observaciones del EIA para el Proyecto Fase de Desarrollo de los Lotes 67A y 67B. Informe No. 001–2012-MEM-AAE/MMR.

42. Maple Energy (2012) Oil and gas exploration. Available: http://www.maple-energy.com/Oilmerger1.aspx. Accessed 2013 Apr 3.

43. Kerr RA (2010) Natural Gas From Shale Bursts Onto the Scene. Science 328: 1624–26.

44. Orta Martínez M and Finer M (2010) Oil frontiers and indigenous resistance in the Peruvian Amazon. Ecological Economics 70: 207–218.

45. Pueblos Indígenas de Canaán de Cachiyacu y Nuevo Sucre (2010) Queja al Ombudsman de la CAO relativa a violaciones de derechos humanos y daños ambientales causados por Maple Energy. Available: http://www.accountabilitycounsel.org/wp-content/uploads/2011/12/Demanda-a-la-CAO-Sobre-Maple.pdf. Accessed 2013 Apr 3.

46. Comisión de Pueblos Andinos, Amazónicos y Afroperuanos, Ambiente y Ecología (2012) Grupo de Trabajo sobre la Situación Indígena de las Cuencas de los Ríos Tigre, Pastaza, Corrientes y Marañón.

47. Thomsen JB, Mitchell C, Piland R, Donnaway JR (2001) Monitoring impact of hydrocarbon exploration in sensitive terrestrial ecosystems: perspectives from Block 78 in Peru. In: Bowles IA, Prickett GT, editors. Footprints in the jungle. New York: Oxford University Press.

48. Kolowski JM, Alonso A (2012) Primate abundance in an unhunted region of the northern Peruvian Amazon and the influence of seismic oil exploration. Int J Primatol 33: 958–971.

49. Kolowski JM, Alonso A (2010) Density and activity patterns of ocelots (*Leopardus pardalis*) in northern Peru and the impact of oil exploration activities. Biol Conserv 143: 917–25.

50. Orpio Corpi-Sl, Fediquep Acodecospat, Feconat, et al. (2012) Nuestras condiciones previas para poder iniciar el proceso de consulta previo.

51. Perupetro (2011) Plays and new hydrocarbon potential in the Marañon Basin.

52. Perupetro (2012) Next bidding round 2012: 36 blocks for exploration activities.

53. Lehner B, Verdin K, Jarvis A (2008) New global hydrography derived from spaceborne elevation data. Eos Trans AGU 89: 93–94.

54. SERNANP (Servico Nacional de Áreas Naturales Protegidas por el Estado) (2009) Plan Director de las Áreas Naturales Protegidas (Estrategia Nacional). Ministerio del Ambiente.

55. Instituto del Bien Común (2011) Sistema de Información sobre Comunidades Nativas de la Amazonía Peruana (SICNA), versión 4. Lima: IBC.

56. Vriesendorp C, Pitman N, Rojas Moscoso JI, Pawlak BA, Chávez LR, et al. (2006) Perú: Matsés. Rapid Biological Inventories Report 16. Chicago: The Field Museum.

57. Josse C, Navarro G, Encarnación F, Tovar A, Comer P, et al. (2007) Digital ecological systems map of the Amazon Basin of Peru and Bolivia. Arlington: NatureServe.

58. Mims M (2002) Drilling design and implementation for extended reach and complex wells. KM Technology Group.

59. Lee J (2009) Introduction to offshore pipelines and risers.

60. Amores G (2010) Comparaciones de calidad y costo entre un gasoducto verde y una construcción tradicional – el control de la erosión como medida de protección ambiental. INMAC Perú.

61. Olson DM, Dinerstein E, Wikramanayake ED, Burgess ND, Powell GVN, et al. (2001) Terrestrial ecoregions of the world: a new map of life on Earth. BioScience 51: 933–938.

16

Regulatory Response to Carbon Starvation in *Caulobacter crescentus*

Leticia Britos[1], Eduardo Abeliuk[1,2], Thomas Taverner[3], Mary Lipton[3], Harley McAdams[1], Lucy Shapiro[1]*

1 Department of Developmental Biology, Stanford University School of Medicine, Stanford, California, United States of America, 2 Department of Electrical Engineering, Stanford University, Stanford, California, United States of America, 3 Environmental Molecular Sciences Laboratory, Pacific Northwest National Laboratory, Richland, Washington, United States of America

Abstract

Bacteria adapt to shifts from rapid to slow growth, and have developed strategies for long-term survival during prolonged starvation and stress conditions. We report the regulatory response of *C. crescentus* to carbon starvation, based on combined high-throughput proteome and transcriptome analyses. Our results identify cell cycle changes in gene expression in response to carbon starvation that involve the prominent role of the FixK FNR/CAP family transcription factor and the CtrA cell cycle regulator. Notably, the SigT ECF sigma factor mediates the carbon starvation-induced degradation of CtrA, while activating a core set of general starvation-stress genes that respond to carbon starvation, osmotic stress, and exposure to heavy metals. Comparison of the response of swarmer cells and stalked cells to carbon starvation revealed four groups of genes that exhibit different expression profiles. Also, cell pole morphogenesis and initiation of chromosome replication normally occurring at the swarmer-to-stalked cell transition are uncoupled in carbon-starved cells.

Editor: Roy Martin Roop II, East Carolina University School of Medicine, United States of America

Funding: This work was supported by National Institutes of Health (NIH - http://nih.gov/) grant GM32506 to LS and by Department of Energy (DOE - http://www.energy.gov/) grant DE-FG02ER64136 to LS and HM. Proteomic analyses were performed in the Environmental Molecular Sciences Laboratory, a Department of Energy/Office of Biological and Environmental Research (DOE/BER) national scientific user facility on the Pacific Northwest National Laboratory (PNNL) campus in Richland, Washington. PNNL is a multiprogram national laboratory operated by Battelle for the DOE under Contract DE-AC05-76RL01830. The funders had no role in study design, data collection and analysis, decision to publish, or preparation of the manuscript.

Competing Interests: The authors have declared that no competing interests exist.

* E-mail: shapiro@cmgm.stanford.edu

Introduction

Starvation for nutrient and energy sources are common stresses confronted by bacteria in natural environments. Bacteria have limited energy reserves, so they need robust mechanisms to quickly shift between rapid and slow growth, as well as a strategy for long-term survival during periods of prolonged starvation. The response to starvation comprises an initial stage of scavenging and metabolic adaptation. If the missing essential nutrients are not replenished, there is a second stage of physiological adaptation, which includes the inhibition of growth and cell division, in order to retain viability.

We have used *Caulobacter crescentus*, a gram-negative oligotrophic bacterium [1], as a model system to study the response to carbon deprivation. *C. crescentus* has a dimorphic life cycle. Each asymmetric division yields a chemotactically-competent flagellated cell (swarmer cell) and a sessile cell with a polar stalk (stalked cell). The core genetic network that drives cell cycle progression and cell division in *C. crescentus* is well characterized [2,3,4,5].

In *Escherichia coli* and *Bacillus subtilis*, the RpoS (σ^S) and SigB (σ^B) sigma factors are the master regulators of the general starvation-stress response [6,7]. *C. crescentus* lacks orthologs of the *rpoS* and *sigB* genes, as do all other α-proteobacteria [8,9]. An equivalent master regulator has not been identified. We have recently identified two factors involved in the adaptation of *C. crescentus* to carbon starvation: the SpoT ppGpp synthetase/hydrolase, which contributes to the regulation of the DnaA

protein stability and initiation of DNA replication [10], and the CrfA small noncoding regulatory RNA controlling the mRNA stability of 27 transcripts [11].

A comparison of gene expression profiles of *C. crescentus* growing under carbon- and nitrogen-limited continuous flow cultures was recently reported [12]. In these experiments, slow cell growth was supported by a constant supply of nutrients at low concentration, and the gene expression profiling captured the metabolic adaptations to the relative levels of carbon and nitrogen. This study identified genes differentially induced in carbon- versus nitrogen-limited conditions, among which are those predicted to require the alternative sigma factor RpoN for complete induction [12]. Here, we have studied the cells' response to an abrupt loss of carbon source resulting in the inhibition of mass accumulation and cell cycle progression. We performed a global analysis of the differences between growing and carbon-starved *C. crescentus* cultures, through combined high-throughput proteome and transcriptome assays. By analyzing the response at the protein level, we took into account the fact that *C. crescentus* uses a complex array of regulatory strategies that include targeted proteolytic events [13]. We identified genetic regulatory pathways that mediate the transduction of carbon starvation signals and found that the SigT ECF sigma factor controls a core set of genes that are activated by carbon depletion, osmotic stress and exposure to heavy metals. We identified gene clusters that are differentially expressed upon carbon starvation at specific stages of the cell cycle.

Results

The proteome and transcriptome profile of *C. crescentus* cells starved for carbon

Transfer of *C. crescentus* cultures in exponential growth to media lacking a carbon source results in immediate growth arrest. Under prolonged starvation, there is a pronounced loss in viability [10]. In order to identify proteins that participate in the carbon starvation response, we incubated cultures in minimal M2 medium in the absence of glucose, and the respective non-starved controls in the presence of 0.2% glucose, for 30 and 60 minutes, as described in the Methods section. Samples were analyzed by liquid chromatography coupled to tandem mass spectrometry. We identified 2471 distinct proteins across all conditions tested, accounting for 65.6% of 3767 *C. crescentus* predicted protein-coding genes. This is high coverage for a prokaryotic proteome, lower only than that reported for *Mycoplasma mobile* (88.6%), which has a genome of one-sixth the size of that of *C. crescentus* (approx 0.78 Mb) (Table S1). Previous electrophoresis-based proteome studies of *C. crescentus* identified 81 cell cycle-regulated proteins [14], 39 stalk-specific proteins [15], and 86 membrane-associated proteins [16,17].

The levels of 513 proteins were found to change reproducibly in cells starved for carbon for 30 and/or 60 min (Table S2). Figure 1 shows the upregulated and downregulated proteins by functional categories, according to COG classifications [18]. Those categories with a statistically significant enrichment in either upregulated or downregulated proteins are indicated (see Methods for details on the enrichment analysis). Functional categories comprising transport and metabolism of amino acids (COG E), nucleotides (COG F), carbohydrates (COG G) and cell motility (COG N) were significantly enriched in proteins whose levels decreased upon carbon starvation. On the other hand, the categories comprising energy production and conversion (COG C), inorganic ion transport and metabolism (COG P) and defense mechanisms (COG V) were significantly enriched in proteins whose levels increased upon carbon starvation. Of the 40 COG P proteins that are upregulated upon carbon starvation, 21 belong to the TonB-dependent receptor family, which are highly abundant in the *C. crescentus* genome, including those controlled by the CrfA small noncoding RNA described by Landt *et. al.* [11]. These observations are compatible with metabolic adaptation of the cell to carbon starvation.

The proteome profile of carbon-starved cells also provided insights into the physiological adaptation associated with stasis. Among the proteins that change significantly, there was notable downregulation of three essential cell cycle regulators (CtrA, DnaA and GcrA), and two proteins (PopZ and ParA) that mediate chromosome segregation [13,19,20]. The decrease in the levels of DnaA, an essential protein that serves both as an activator of replication initiation and global transcription factor [21], is consistent with the previously reported starvation-induced ClpP-mediated proteolysis of DnaA [22]. The decrease in DnaA levels prevents the initiation of DNA replication. The CtrA response regulator inhibits DNA replication in the swarmer cell, and acts as a transcription factor regulating the expression of nearly 100 genes. It was previously shown that CtrA is proteolyzed in swarmer cells starved for carbon, but with different kinetics than that observed under normal growth conditions [10,22]. Given the number of CtrA-regulated genes that change upon carbon starvation, it is clear that, at least in part, an important role of CtrA in the starvation response relies on its function as transcription factor. Additionally, the level of the FtsZ cell division protein, that is essential for cytokinesis, is shown to decrease

significantly upon carbon starvation; FtsZ was previously shown to decrease during stationary phase in *C. crescentus* [23].

Total RNA was extracted from samples parallel to those used for the proteome analysis. As described in Methods, cDNA was synthesized, fragmented, labeled and hybridized to the CauloHI1 chip. A minimum two-fold change requirement between the starvation and non-starvation conditions, with a false discovery rate (FDR) below 1%, resulted in upregulation of 607 of the 3767 *C. crescentus* protein coding genes after 30 min of carbon starvation, and 700 genes after 60 min (553 of these were significantly upregulated at both time points); 725 and 618 genes were downregulated after 30 and 60 min of carbon starvation, respectively; 603 genes were equally downregulated for both time points. All genes that changed significantly are listed in Table S3. 16.3% of the 753 genes that were significantly upregulated after either 30 or 60 min of carbon starvation, and 19.9% of the 739 downregulated genes were classified as cell cycle regulated transcripts by Laub *et al.* [24] (Table S3).

Carbon starvation transcriptional regulators

Cell cycle transcription profiles were previously used to identify predicted cell cycle regulons and their conserved promoter motifs. Fourteen conserved promoter motifs were identified, seven of which shared significant similarity with binding motifs for previously characterized regulators [25]. We analyzed these sets of genes to determine if any of them were significantly enriched in genes whose expression was up- or down-regulated upon carbon starvation. As shown in Figure 2A, out of 14 gene sets, each of which shared a distinct promoter motif, six were significantly enriched in genes upregulated upon carbon starvation (sets corresponding to motifs cc_1, cc_2, cc_3, cc_4, cc_7, cc_8), while two were enriched in genes downregulated upon carbon starvation (corresponding to motifs cc_6 and cc_13) (Table S4). Motif cc_6 is the cognate binding motif for the RpoD sigma factor, which drives the expression of biosynthetic and housekeeping genes throughout the cell cycle [26] (Figure 2B); as expected, the great majority of the genes with cc_6 motif (27 out of 34) were down-regulated upon carbon starvation in our experiments.

Carbon starvation FixK regulon. Genes with motifs cc_1 to cc_4 are expressed in the swarmer cell stage (Figure 2B). A subset of these genes are likely activated by the stress suffered by cells during the synchronization process, and not swarmer cell specific genes, as they were activated equally by carbon starvation in stalked cells. Motif cc_3 corresponds to the DNA binding motif of the FixK transcription factor. Microarray analysis of a *C. crescentus* *fixK* knockout strain identified downstream targets whose expression changed in a FixK-dependent manner under hypoxia conditions [27]. We found that five out of eight genes with motif cc_3 were upregulated upon 30 min of carbon starvation. Expression of three of those genes was FixK-dependent under hypoxia conditions [27]. Eleven additional FixK-dependent genes identified by Crosson *et al.*, which lack the cc_3 motif, showed a significant change upon carbon starvation, and are either indirect FixK targets, or are transcribed as part of an operon directly regulated by FixK.

Carbon starvation SigT regulon. Motif cc_7 is the binding motif for the SigT alternative sigma factor, and genes in this cluster have a distinct peak of expression at the swarmer-to-stalked cell transition (Figure 2B). SigT, which belongs to the ECF (extra-cytoplasmic function) sigma factor family, is a regulator of the osmotic and oxidative stress responses in *C. crescentus* [28]. Of 26 genes with motif cc_7, 7 were upregulated after 30 min of carbon starvation, and an additional 7 genes were upregulated after 60 min.

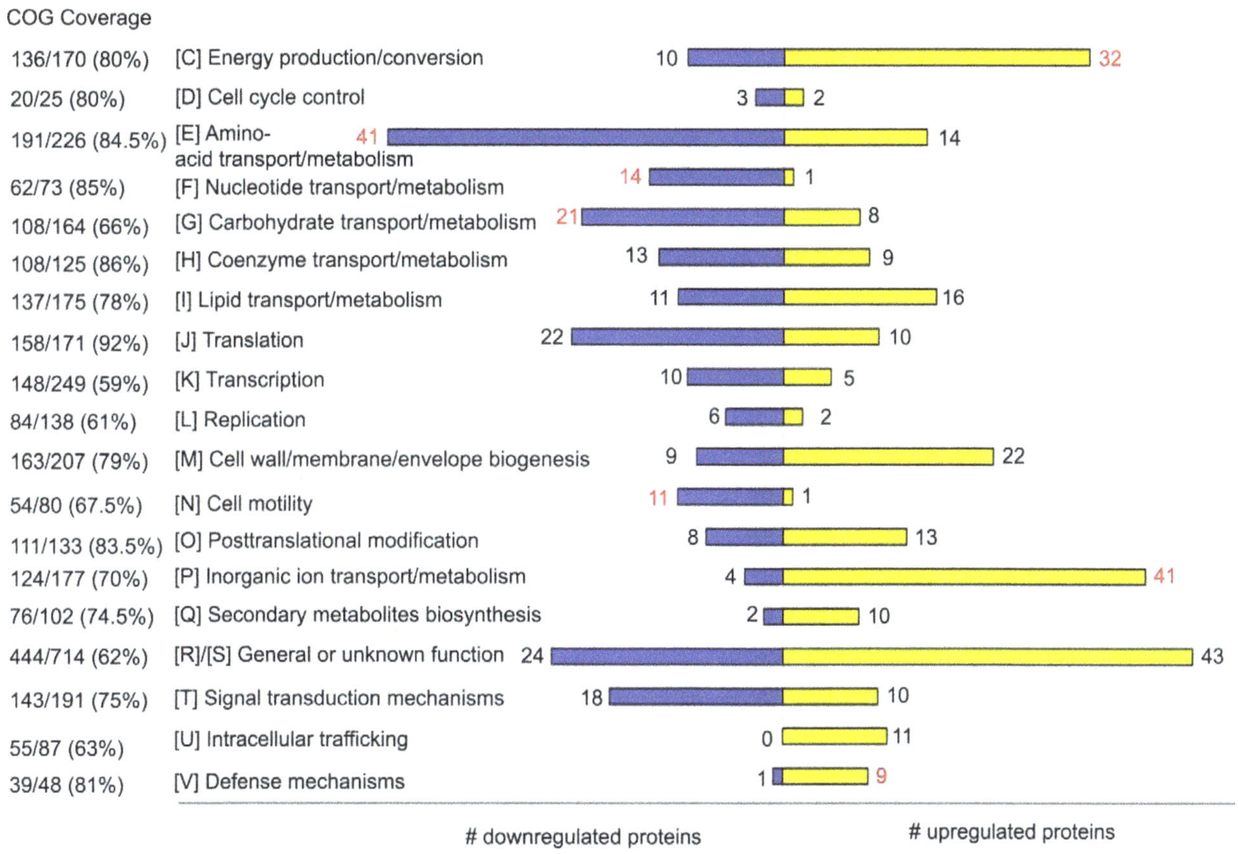
COG Coverage
136/170 (80%) [C] Energy production/conversion 10 32
20/25 (80%) [D] Cell cycle control 3 2
191/226 (84.5%) [E] Amino-acid transport/metabolism 41 14
62/73 (85%) [F] Nucleotide transport/metabolism 14 1
108/164 (66%) [G] Carbohydrate transport/metabolism 21 8
108/125 (86%) [H] Coenzyme transport/metabolism 13 9
137/175 (78%) [I] Lipid transport/metabolism 11 16
158/171 (92%) [J] Translation 22 10
148/249 (59%) [K] Transcription 10 5
84/138 (61%) [L] Replication 6 2
163/207 (79%) [M] Cell wall/membrane/envelope biogenesis 9 22
54/80 (67.5%) [N] Cell motility 11 1
111/133 (83.5%) [O] Posttranslational modification 8 13
124/177 (70%) [P] Inorganic ion transport/metabolism 4 41
76/102 (74.5%) [Q] Secondary metabolites biosynthesis 2 10
444/714 (62%) [R]/[S] General or unknown function 24 43
143/191 (75%) [T] Signal transduction mechanisms 18 10
55/87 (63%) [U] Intracellular trafficking 0 11
39/48 (81%) [V] Defense mechanisms 1 9
downregulated proteins
upregulated proteins

Figure 1. Proteome profile of *C. crescentus* subjected to carbon starvation. Proteins that change significantly upon 30 and/or 60 minutes of carbon starvation classified by NCBI Clusters of Orthologous Genes (COG). Coverage for all categories of the COG scheme (except for A and B for which no proteins were detected), is indicated as the ratio of the number of detected proteins in each category over the total number of proteins assigned to that category (the corresponding percentage value is indicated in parentheses). Yellow bars represent upregulated proteins and blue bars, downregulated proteins in each COG category. Values that meet a statistically significant enrichment, assuming a hypergeometric distribution, are denoted in red. The proteins within each category are listed in Table S2, along with 43 additional proteins that are not included in the COG classification.

In order to characterize the carbon starvation SigT regulon, we performed microarray experiments comparing gene expression profiles of a *C. crescentus* wild-type strain (LS101) and a *sigT* null mutant strain (LS3554), in M2 cultures incubated in the absence of glucose for 15 min. Genes showing a greater than two-fold difference in expression between the mutant and the wild type strain under starvation conditions, and meeting statistical significance as defined in the Methods section, were considered to be SigT-dependent upon carbon starvation (Table 1). Thirteen of the 27 SigT-dependent carbon starvation genes have the previously characterized [25,28] SigT binding motif in their promoters and are presumably direct targets of the regulator. Four additional targets are in close proximity to genes that possess the motif, suggesting they are part of directly regulated operons. Sixteen of the 27 genes whose expression is significantly reduced in the *sigT* mutant, and the two genes with increased expression in the mutant strain, do not show SigT-dependency under osmotic stress, as reported by Alvarez-Martinez *et al.* [28].

The main functional groups represented in the putative SigT carbon starvation regulon are signal transduction and gene regulation (7 genes), and transport and metabolism (5 genes). The gene encoding the SigU ECF sigma factor has a SigT binding motif in its promoter, suggesting it is directly regulated by SigT. SigU is activated by SigT during cell cycle progression [18], upon osmotic stress [25,28] and upon carbon starvation (this work).

In order to determine if any of the SigT-dependent genes lacking the conserved SigT promoter binding motif were regulated by a pathway involving SigU, we used microarrays to compare the carbon starvation-induced expression levels in a wild type strain (LS101) and a *sigU* null mutant strain (LS3547). Only one gene, CC_3466, encoding a hypothetical protein, appeared to be SigU-dependent under carbon starvation. This gene, which was also found to be SigT-dependent, does not have the conserved SigT motif in its promoter, consistent with its SigT dependence being, directly or indirectly through the function of the SigU transcription factor. The CC3466 transcript is cell cycle regulated [24], with a distinct peak at the swarmer-to-stalked cell transition. This pattern was also observed in the cell cycle microarray results of McGrath *et al.* [25]. CC_3466 is not reported to be SigT-dependent upon osmotic stress [28]. On the other hand, its expression is induced by exposure to chromate and dichromate heavy metals [29], incubation in minimal media vs rich media [30], as well as carbon limitation (compared to nitrogen limitation) [12]. The predicted amino acid sequence of CC_3466 (102 amino acids with a predicted molecular weight of 11 KDa) yielded very few low scoring BLASTP hits outside of the *Caulobacter* genus. In

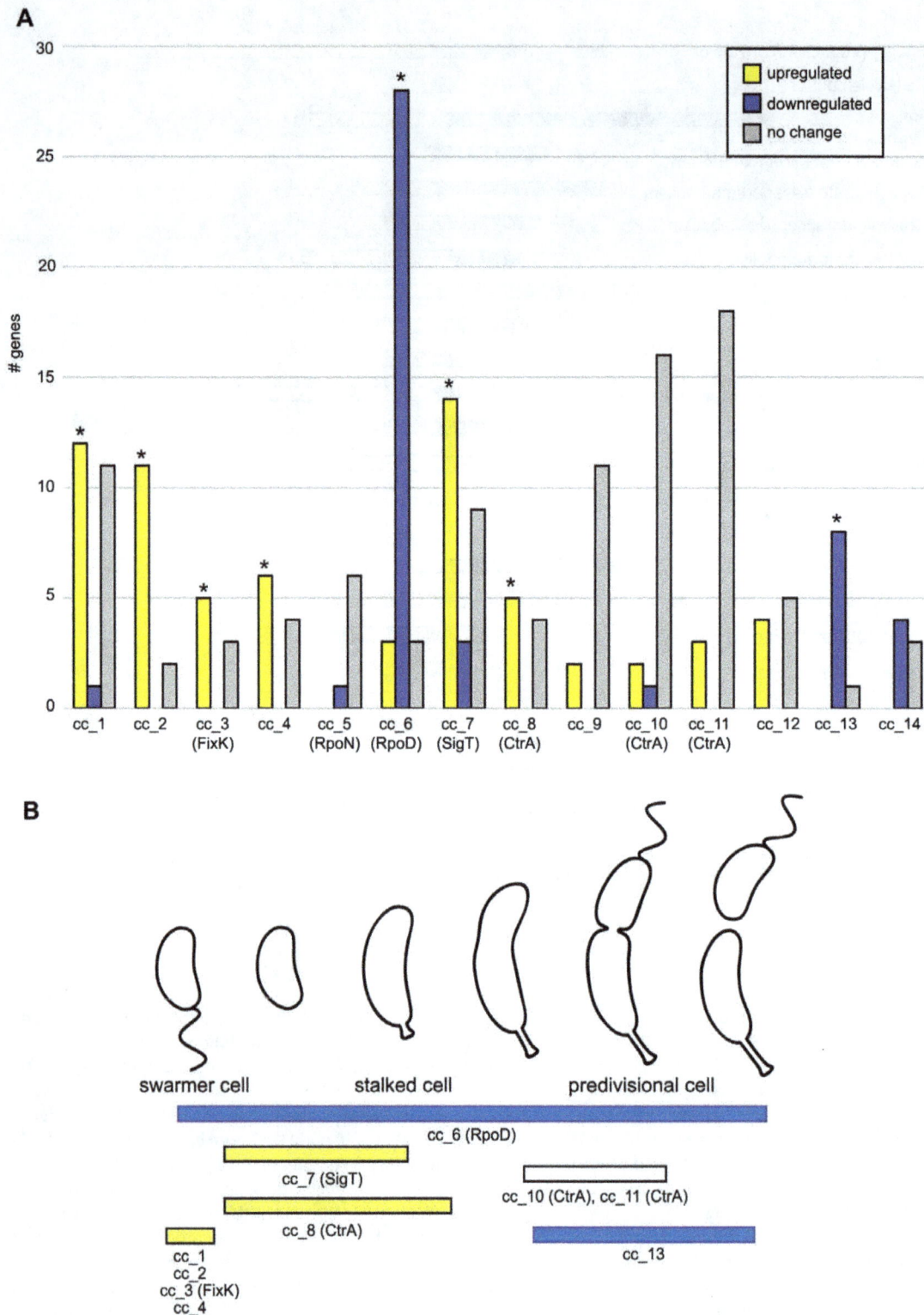

Figure 2. Putative regulators of the carbon starvation response. A. Previously identified clusters of *C. crescentus* genes that share conserved promoter motifs were analyzed for enrichment in genes that change significantly upon carbon starvation. For each gene set, cc_1 through cc_14 (identified by a shared promoter motif), the number of carbon starvation upregulated genes is indicated in yellow, downregulated genes in blue and genes that don't change in gray. Asterisks denote statistically significant enrichment. Gene numbers correspond to 30 min of carbon starvation, except for motif cc_7, for which gene numbers correspond to 60 min of carbon starvation. Identical results were obtained in terms of significant enrichment for the 30 and 60 min carbon starvation data for most gene sets, except for cc_8, which was only enriched in starvation upregulated genes at 30 min, and cc_7, which was only enriched at 60 min. B. Cell cycle patterns of expression, as previously determined [25], of carbon-starvation enriched gene sets. Yellow indicates enrichment in genes up-regulated upon 30 min of carbon starvation, while blue corresponds to enrichment in genes down-regulated upon 30 min of carbon starvation.

Table 1. SigT-dependent carbon starvation regulon.

Gene	Annotation	COG	Fold reduction in $\Delta sigT$ mutant	SigT motif?	Sig-T dependent in osmotic stress?	Up-regulated by exposure to multiple heavy metals?
CC_0284	Two-component receiver domain protein lovR	K	2.65	Yes	Yes	No
CC_1356	Transcriptional regulator	K	3.29	Yes	Yes	No
CC_2883	RNA polymerase ECF-type sigma factor sigU	K	15.93	Yes	Yes	Yes
CC_3477	Hybrid sigma factor/two-component receiver protein phyR	K	3.10	Yes	No	No
CC_0285	Photosensory histidine protein kinase lovK	T	3.94	No (*)	No	No
CC_2330	HTH transcriptional regulator	K	2.25	No	No	No
CC_3258	chemotaxis receiver domain protein cheYIII	K	2.05	No	No	Yes
CC_2549	Putative heme-binding protein	NC	3.97	Yes	Yes	Yes
CC_1048	Acylamino-acid-releasing enzyme	E	0.42	No	No	No
CC_2318	ABC transporter permease protein	Q	2.15	No	No	No
CC_3001	TonB-dependent receptor	P	2.08	No	No	No
CC_3572	Carbonic anhydrase	P	0.43	No	No	No
CC_0280	Conserved cell surface protein	NC	20.99	Yes	Yes	Yes
CC_0673	Hypothetical protein	NC	3.59	Yes	No	Yes
CC_0717	Hypothetical membrane associated protein	S	2.01	Yes	No	No
CC_1179	hypothetical protein	NC	3.14	Yes	Yes	Yes
CC_0163	Conserved hypothetical protein	S	5.60	No (*)	Yes	No
CC_1178	Hypothetical cytosolic protein	R	2.10	No (*)	Yes	Yes
CC_0557	Hypothetical protein	NC	2.78	No	No	No
CC_1532	Conserved hypothetical protein	S	23.41	No	Yes	Yes
CC_3466	Hypothetical protein	NC	7.67	No	No	Yes
CC_3473	Entericidin B homolog	NC	2.57	Yes	No	Yes
CC_1682	Transglycosylase associated protein	S	2.57	Yes	No	Yes
CC_0201	Outer membrane protein	M	2.11	Yes	No	Yes
CC_0428	Methyl-accepting chemotaxis protein	N	2.82	Yes	No	No
CC_0164	Chain length regulator (capsular polysaccharide biosynthesis)	D	2.18	No (*)	No	No
CC_0556	Catalase	NC	2.00	No	No	No

Genes with reduced or increased expression in the $\Delta sigT$ strain upon 15 min of carbon starvation, with respect to the NA1000 wild type strain under the same conditions. SigT-dependency under osmotic stress is according to [28]. Upregulation by exposure to multiple heavy metals corresponds to genes upregulated by two or more metals according to [29]. Genes annotated as No(*) in the 'sigT motif' column correspond to cases where the sigT motif is found upstream of an adjacent gene as part of a putative operon (CC_0284 and CC_0285; CC_1179 and CC_1178; CC_0162, CC_0163 and CC_0164). COG groups: K = Transcription; T = Signal transduction mechanisms; E = Amino acid transport and metabolism; Q = Secondary metabolites biosynthesis; P = Inorganic ion transport and metabolism; S = Function unknown; R = General function prediction only; M = Cell wall/membrane/envelope biogenesis; N = Cell motility; D = Cell cycle control; NC = Not classified.

our proteome analysis, peptide levels for CC_3466 increased significantly after 60 min of carbon starvation.

Carbon starvation CtrA regulon. The gene cluster sharing motif cc_8 was shown to be enriched in genes that were upregulated upon carbon starvation: five out of nine genes with this motif increased significantly upon 30 min of carbon starvation. Motifs cc_08, cc_10 and cc_11 share similarity with the binding motif for the CtrA cell cycle master regulator. While genes with motifs cc_10 and cc_11 showed an expression peak at the predivisional stage, genes with motif cc_8 showed an earlier peak, at the swarmer to stalk cell transition [25]. The CtrA regulon has been characterized by chIP-chip analysis, identifying genomic regions where CtrA binds, and microarray analysis, identifying genes with affected expression in a CtrA temperature sensitive mutant strain [31]. We cross-correlated these results with our carbon starvation microarray dataset and found that the expression of a significant number of both cell-cycle regulated and non cell-cycle regulated CtrA targets was altered upon 30 min of carbon starvation (Table 2). Among these targets was the CC_2644 gene, whose transcript was one of the most strongly upregulated transcripts upon carbon starvation. This gene, which remains uncharacterized in *C. crescentus*, encodes a protein belonging to the PhoH family. The *E. coli phoH* gene, which defines the family, was shown to be induced by phosphate starvation and to have ATPase activity [32].

Table 2. CtrA targets that change significantly upon carbon starvation.

Gene	Annotation	Transcript levels carbon starvation/control	Cell cycle regulated?
CC_2165	division plane positioning ATPase *mipZ*	5.88	Yes
CC_2540	Cell division protein *ftsZ*	2.15	Yes
CC_2063	Flagellar basal-body rod protein *flgF*	2.31	Yes
CC_2316	Transcriptional regulator	2.18	Yes
CC_3219	Two-component sensor histidine kinase	2.54	Yes
CC_2644	*phoH* protein	44.56	Yes
CC_1396	Lactate 2-monooxygenase	6.28	Yes
CC_1101	Protoporphyrinogen oxidase	2.37	Yes
CC_1307	Aspartyl protease *perP*	3.05	Yes
CC_1211	Hypothetical protein	2.01	Yes
CC_3291	Hypothetical protein	9.90	Yes
CC_1072	Ribonuclease H	3.35	Yes
CC_3218	Hypothetical protein	2.12	Yes
CC_1872	peptidoglycan-specific endopeptidase, M23 family	0.30	Yes
CC_0484	peptidyl-tRNA hydrolase	0.41	Yes
CC_1923	SSU ribosomal protein S2P	0.05	Yes
CC_3202	SSU ribosomal protein S12P	0.04	Yes
CC_2166	pantoate-beta-alanine ligase	0.37	Yes
CC_0050	S-adenosylmethionine synthetase	0.02	Yes
CC_0350	hypothetical protein with pentapeptide repeats	0.48	Yes
CC_3454	Acyl-CoA dehydrogenase	4.47	No
CC_0651	Conserved hypothetical protein	2.21	No
CC_0696	GumN superfamily protein	4.72	No
CC_1106	Ice nucleation protein	2.30	No
CC_2869	Hypothetical protein	2.32	No
CC_2966	3-oxoacyl-[acyl-carrier protein] reductase	3.45	No
CC_1892	aspartyl-tRNA synthetase	0.09	No
CC_2451	DNA topoisomerase I	0.11	No
CC_3155	chemotaxis receiver domain protein cheYIII	0.39	No
CC_2241	heat shock protein *hsp33*	0.30	No
CC_1034	GTP-binding protein *lepA*	0.30	No
CC_0249	SCO1/SenC family protein	0.46	No

Genes that change significantly upon 30 min of carbon starvation, and that are directly regulated by CtrA [31], and are either cell cycle regulated and non cell cycle regulated [24].

We performed the same enrichment analysis with the motifs identified by McGrath *et al.* for genes responding to heavy metal stress [25]. Sets sharing motifs m_2 (corresponding to the RpoD binding motif), m_3 (similar to motif cc_13) and m_4 were enriched in genes downregulated upon 30 min of carbon starvation, while sets sharing motifs m_5 (similar to motif cc_1) and m_13 were enriched in genes upregulated upon 30 min of carbon starvation. Since none of the additional sets had a known cognate regulator, this analysis did not identify further pathways related to the carbon starvation response.

Cell-stage specific response to carbon starvation

We explored the response to carbon starvation separately in swarmer and stalked cell populations. We isolated swarmer cells and incubated them in M2 medium lacking glucose for 15 min before collecting samples (Figure 3A). To study the stalked cell response to carbon starvation, a population of swarmer cells was allowed to proceed through the cell cycle for 60 min and transition into stalked cells in complete medium, before removing the carbon source and incubating for 15 min in M2 medium. RNA was extracted from starved swarmer cells and starved stalked cells, and hybridized to the CauloHI1 chip. To account for changes in transcript levels normally occurring as the cell cycle progresses, we used the normalized values corresponding to 15 min and 75 min of the cell cycle microarray profiles obtained from the data of McGrath *et al.* [25], as the non-starved controls for swarmer and stalked cells, respectively. The log-2 values of the expression ratios of the starved and non-starved cells for both stages were used to determine the cell-stage specific responsiveness of each gene, and clustered accordingly (Figure 3B and Table S5). Six clusters were identified. Most of the genes showed a similar expression change in swarmer and stalked cells: out of 667 genes, 165 were upregulated (cluster 3), and 342 genes were downregulated (cluster 6), in both stages. The remaining clusters displayed a stage-differential response to carbon starvation: 27 genes did not change significantly in swarmer cells, but were down-regulated preferentially in stalked cells (cluster 1); 24 genes were upregulated in swarmer cells, but did not change significantly in stalked cells (cluster 2); 71 genes did not change significantly in swarmer cells, but were upregulated in stalked cells (cluster 4); 37 genes were down-regulated in swarmer cells, but did not change significantly in stalked cells (cluster 5). Using MEME motif finder [33], we identified five novel conserved DNA motifs in the upstream regions of genes corresponding to three of the clusters (motif a in cluster 1; b, c and d in cluster 3; and e in cluster 6).

Adaptive changes of swarmer cells subjected to carbon starvation

In *Caulobacter*, the chromosomal replication origin is positioned at the cell pole, and upon initiation of replication, a copy of the origin sequence is moved to the opposite cell pole. Therefore, we can follow the onset of DNA replication by following the cellular position of the origin bound to the ParB segregation factor. In order to connect the stage-specific response to carbon starvation observed at the molecular level with cell cycle and differentiation events, we obtained transmission electron microscope images of swarmer cells immediately following their isolation, and then after subjecting them to carbon starvation. We examined the replication and segregation of the chromosomal origin (*ori*) locus, under the same conditions, using epifluorescence imaging of a strain carrying a fluorescently-tagged version of the ParB centromere-binding protein. 65% of the isolated swarmer cells developed short incipient stalks when incubated in the absence of glucose for 2 hs (Figure 4A). After 8 hs of starvation, these stalks failed to elongate. Under the same conditions, only about a tenth of the population had duplicated the chromosomal *ori* locus (Figure 4B), whereas after 30 min in complete media, 57% of the swarmers showed fully or partially segregated origins (Figure 4C). Thus, for a significant percentage of the population, carbon starvation uncoupled the initiation of stalk biogenesis and the G1-to-S transition, events that occur coincidently when nutrients are sufficient.

SigT-dependent degradation of CtrA in swarmer cells upon carbon starvation

The CtrA cell cycle master regulator is a critical element of the core machinery that regulates cell cycle progression. CtrA binds to the chromosomal origin of replication and blocks replication initiation in the swarmer cell [34]. Upon differentiation of the swarmer cell to the stalked cell, CtrA is cleared from the cell by proteolysis, allowing the initiation of DNA replication. CtrA is re-synthesized following replication initiation. Our proteome studies showed a decrease in the protein levels of the CtrA master regulator in a mixed population of *Caulobacter* cells subjected to carbon starvation: CtrA levels in cells incubated in the absence of glucose for 30 min were half of that of cells incubated in the presence of glucose (Table 2). We observed a similar decrease in the relative levels of CtrA by immunoblot assays of carbon starved mixed cell populations (not shown) and swarmer cells, as previously reported [10]. Swarmer cells incubated for 150 min in the absence of glucose showed low levels of the CtrA protein, while those incubated for 150 min in the presence of glucose had progressed through the cell cycle and reaccumulated CtrA. In contrast, when the same assay was performed with swarmer cells carrying a *sigT* deletion, a less significant decrease in CtrA levels

was observed in the absence of glucose, indicating that SigT contributes to the clearance of CtrA protein in carbon starved swarmer cells (Figure 5A and D). We measured the activity of the CtrA promoter under carbon starvation in a wild type and *sigT* deletion background, using a transcriptional fusion of the CtrA promoter region to *lacZ*. We found only a minimal difference in the transcriptional activity of the reporter in both backgrounds (Figure 5B). Further experiments will be carried out to determine if SigT affects the starvation-induced clearance of CtrA at the post-transcriptional level.

Since SigT is a transcriptional regulator, the simplest explanation for these results is that a downstream target whose expression is SigT-dependent is responsible for the effect on CtrA protein stability. Consequently, we compared the starvation and non-starvation levels of CtrA (as described for Figure 5A) in different deletion mutant backgrounds for selected SigT carbon starvation targets identified in our microarray analysis. We also assayed mutants carrying deletions in genes that are colocated with *sigT* in the chromosome (Figure 5C) and are presumed regulators of its activity. As shown in Figure 5D, deletions in both ECF Sigma factor *sigU* and the conserved hypothetical protein CC_1532 showed levels of CtrA stabilization that are similar to those observed for the *sigT* deletion strain, suggesting they participate in the same pathway that leads to CtrA degradation under starvation. This is also the case for the genes encoding the histidine kinase HK4, and the response regulator PhyR (CC_3473), that was recently postulated to function as an anti-anti-sigma factor for SigT (or homologous sigma factor) in *Methylobacterium extorquens* [35], *Sinorhizobium meliloti* [36] and *C. crescentus* [37]. In this assay, we did not test *nepR*, the postulated anti-sigma factor for SigT [37].

Discussion

When faced with nutrient limitation, bacterial cells must deploy an array of scavenging systems, adapt their metabolic fluxes to compensate for missing compounds, and limit energy-consuming growth and cell division processes. *C. crescentus* is an oligotroph whose physiology is adapted for survival in environments characterized by low and fluctuating nutrient levels [1]. Here we have examined the response of *C. crescentus* to the sudden onset of carbon starvation.

Carbon starvation uncouples swarmer cell differentiation and the G1-to-S phase transition

In the presence of adequate nutrient levels, the swarmer-to-stalked-cell transition involves the loss of the polar flagellum, the biogenesis of a stalk in its place, and the concurrent initiation of chromosome replication. It has been shown however, that these processes can be uncoupled. When the DnaA activator of DNA replication is depleted [21], or when CtrA inactivation is prevented [38], swarmer cells undergo polar morphological changes, but fail to initiate chromosome replication. Meanwhile, swarmer cells carrying a deletion in the gene encoding the PleD response regulator replicate their chromosome in the absence of polar morphogenesis [39]. We show here that wild type swarmer cells subjected to abrupt carbon depletion were capable of initiating, but not completing, stalk development (see Figure 4A), while the replication and segregation of the chromosomal origin was inhibited (see Figure 4B, [10]). Because the incipient stalks could not be detected by light microscopy, it was previously postulated that carbon starvation inhibits swarmer differentiation. Furthermore, in these experiments cell populations were starved for carbon prior to isolation of swarmer cells [10]. However, electron microscope images (see Figure 4A) revealed the presence

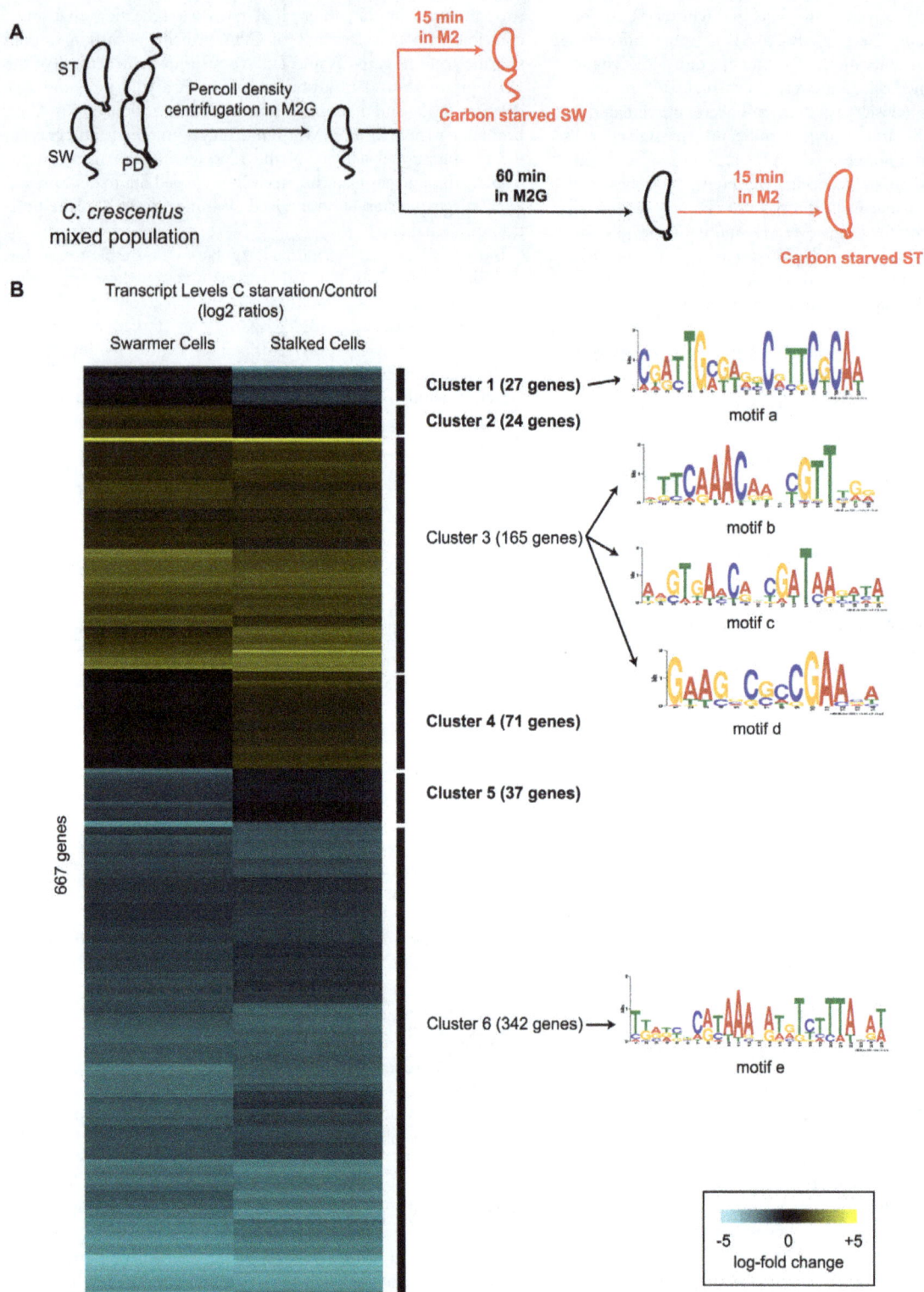

Figure 3. Cell stage-specific carbon starvation response. A. Diagram of the experimental design to explore the cell-stage specific response to carbon starvation. Isolated swarmer (SW) cells were subjected to glucose starvation for 15 min. To assess the response at the stalked cell stage (ST), swarmer cells were allowed to differentiate into stalked cells in complete minimal media for 60 min, and then subjected to glucose starvation for 15 minutes. At the indicated times, cell samples were collected and RNA extracted and transcribed to cDNA to hybridize onto *Caulobacter* microarray chips. PD = predivisional cell. M2 and M2G media are described in the Methods section. **B**. Hierarchical clustering of the transcriptional response to carbon starvation in swarmer cells and stalked cells. The values plotted are the log2-fold change ratios of the cells subjected to 15 minutes of carbon starvation and the non-starved controls. The promoter regions of the genes in each cluster (from −200 to +50 with respect to the translational start site) were used as input in the search for shared motifs using MEME. The five motifs with significant E-values and information content are shown.

A

	0 hs	2 hs	8 hs
% cells with incipient stalks	0.0 % (n= 360 cells)	65.0 % (n= 334 cells)	70.2 % (n= 316 cells)

time post-carbon starvation

B

	0 hs	2 hs	8 hs
% cells with duplicated origins	0.4 % (n= 234 cells)	11.3 % (n= 195 cells)	10.1 % (n= 178 cells)

time post-carbon starvation

C

30 min in M2G: 57.1 % duplicated origins (n=98)

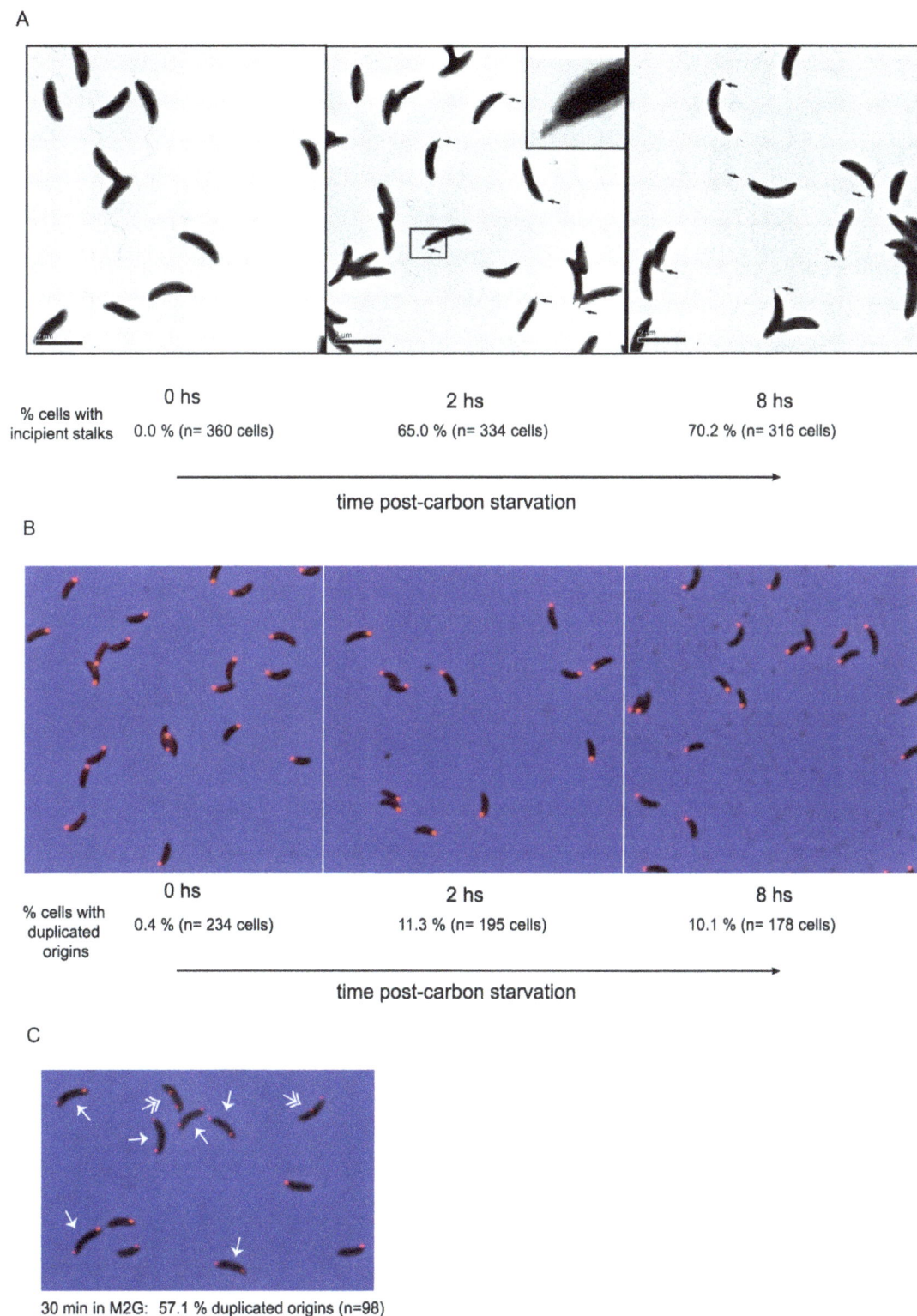

Figure 4. Adaptive changes of swarmer cells upon carbon starvation. A. Polar morphogenesis of wild type swarmer cells subjected to carbon starvation. Swarmer cells were incubated in M2 medium in the absence of glucose for 0, 2 and 8 hs, and visualized by electron microscopy, as described in Methods. The number of cells with incipient stalks (indicated by arrows and detailed in the inset in middle panel) was tallied for several fields and the corresponding percentage is indicated for each time point. **B**. Replication initiation in carbon starved swarmer cells was observed in a strain in which the *parB* gene was replaced with an *ecfp-parB* fusion, treated as described in A. At the indicated times, a sample was removed from the cultures, transferred unto an agarose pad an imaged. The number of cells with duplicated ECFP-ParB foci was counted for each time point and the percentage is indicated. **C**. Swarmer cells isolated for the experiment described in B were incubated in complete M2G media and imaged after 30 min. The percentage of cells with duplicated ECFP-ParB foci is indicated. Arrows indicate cells that have completed origin segregation, as evidence by ECFP-ParB foci in opposite poles, while double arrows indicate cells in the process of segregation.

A

C

B

D

Strain	Ratio CtrA Protein -G/+G
WT	**0.26 ± 0.07**
ΔsigT	0.57 ± 0.18
ΔHK4	0.61 ± 0.09
ΔphyR	0.58 ± 0.18
ΔsigU	0.46 ± 0.01
ΔCC_1532	0.49 ± 0.14

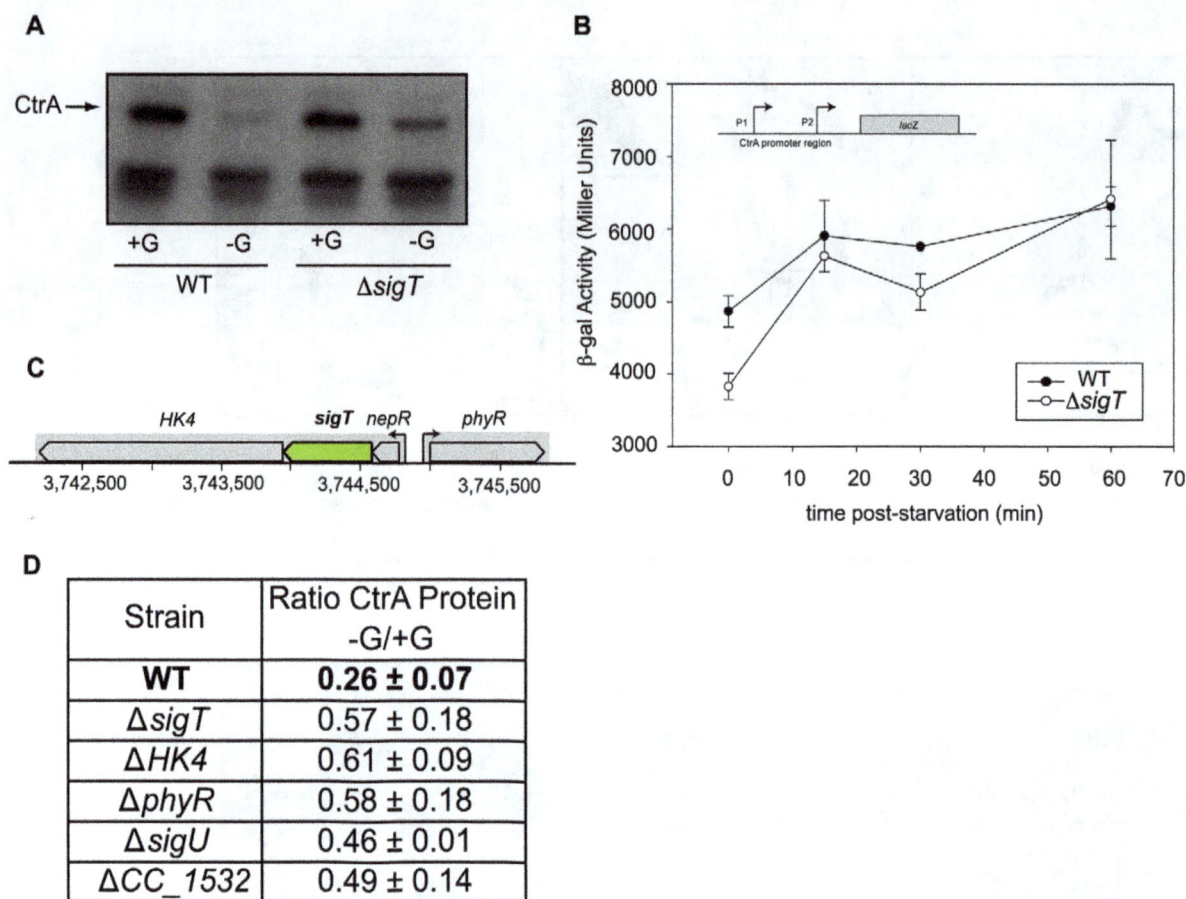

Figure 5. SigT-dependent degradation of CtrA upon carbon starvation. A. Levels of CtrA protein in swarmer cells in the presence and absence of a carbon source, in wild type cells and cells carrying a *sigT* deletion. Isolated swarmer cells from both genetic backgrounds were incubated in M2 medium in the absence or presence of 0.2% glucose. After 150 min, samples from these cultures were subjected to immunoblot analysis with an anti-CtrA polyclonal antibody. The band corresponding to CtrA is indicated with an arrow. **B.** A construct with the complete CtrA promoter region fused to a promoterless *lacZ* reporter in pRKlac290 was introduced into wild-type and *sigT* deletion strains. ß-galactosidase activity was measured in swarmer cells starved for carbon for up to 60 min. **C.** Genomic context of *sigT*. The coordinates correspond to the *C. crescentus* NA1000 genome. CC_3474 (HK = Histidine Kinase) corresponds to CCNA_3588 in NA1000; *sigT* corresponds to CC_3475 (CCNA_3589); *nepR* corresponds to CC_3476 (CCNA_3590); *phyR* corresponds to CC_3477 (CCNA_3591). **D.** Relative levels of the CtrA protein in the presence and absence of carbon, assayed as described in A, are shown for different mutant backgrounds, including deletions of genes in the genomic region of *sigT*, and of SigT-dependent genes (see Table 1). +G corresponds to M2 medium in the presence of 0.2% glucose; -G corresponds to M2 in the absence of glucose. The sampling times are 150 min after starvation, as in A. The error for the ratios corresponds to the standard deviation of the mean for at least two experiments.

of incipient stalks under these conditions. Thus, the point of commitment to initiate stalk morphogenesis appears to precede that of initiation of chromosomal replication. In the population of swarmer cells obtained by the synchronization procedure, the younger cells that are most recently derived from cell division might block both processes, accounting for the 30% of cells that did not develop an incipient stalk by 8 hs of carbon starvation. On the other hand, the majority of swarmer cells in the population that had already committed to polar morphogenesis, blocked the initiation of chromosome replication. Since the stalk is associated with the ability to scavenge for nutrients [15], the initiation of stalk development upon sustained carbon starvation might prime the cell for quick resumption of cell cycle progression once nutrients become available.

Transcript and protein changes in the response to carbon starvation

Transcript profiles as a function of the cell cycle have revealed just-in-time transcriptional activation of distinct functional mod-

ules in *C. crescentus* [24]. However, multiple layers of post-transcriptional regulation are known to be involved in the complex orchestration of cell cycle progression and polar differentiation in this bacterium (see [13,40] for reviews). For this reason, our global analysis included not only the transcriptome but also the proteome profile of *C. crescentus* cells subjected to carbon starvation.

Table 3 shows previously characterized proteins of interest from selected functional groups, that change significantly upon 30 and 60 min of carbon starvation. The levels of FtsZ, essential for cytokinesis in *C. crescentus* [41], and FtsX, a predicted ABC transporter that is needed for cell division in *E. coli* [42], decrease significantly upon carbon starvation. This is consistent with the inhibition of cell division that occurs upon carbon starvation.

Proteases have traditionally been associated with the response to environmental stress, as cells need to re-engineer the cellular landscape, recycle damaged and unwanted proteins, and selectively target key regulators [43]. The levels of two major proteolysis-related factors increased significantly upon carbon starvation, and could be fulfilling this role in *C. crescentus*: the FtsH

Table 3. Proteins of interest in selected functional categories that change upon carbon starvation.

Functional category	Upregulated proteins	Downregulated proteins
Cell wall/membrane/envelope biogenesis (COG M) and cell division	TolA, TolB, TolR, TolQ, BamA, BamB, BamE, BamD, DipM, MreC	FtsZ, FtsX
Cell cycle control		CtrA, DnaA, GcrA
Post-translational modification (COG O)	ClpA, FtsH, SppA	DnaJ, Hsp33
Signal transduction and gene regulation	DivL (HK), FixL (HK), CckA (HK-RR)	FlbD (RR), PleD (RR), DivK (RR), McpA, CheAI, CheYI, CheD
Chromosome structure and dynamics	Smc	PopZ, ParA

The classification is based upon COG (NCBI) and the literature. Abbreviations: HK = histidine kinase; RR = response regulator; HK-RR = hybrid histidine kinase-response regulator.

protease, previously characterized as a component of *C. crescentus'* general stress response [44], and the ClpA chaperone, which normally works in concert with the ClpP protease, and has not been thus far associated with the stress response.

The level of Smc (CC_0373), a nucleoid-associated protein required for chromosome structure maintenance and segregation in *C. crescentus* [45], increased significantly upon carbon starvation. This raises the possibility that the nucleoid of growth-arrested cells adopts a different compaction state than that in exponentially growing cells, possibly contributing to the inhibition of DNA replication and segregation, as well as an increased tolerance to stress-related damaging agents. *E. coli's* nucleoid protein composition has been shown to be growth phase-dependent [46] and a factor in long term survival [47].

Respiratory metabolism generates reactive oxygen species that may damage membranes, DNA and proteins. As long as the environment promotes growth and continued de novo synthesis, the oxidized macromolecules are rapidly diluted, but this is not the case during growth-arrest of metabolically active cells. Correlation of our carbon starvation datasets with microarray analysis of *C. crescentus'* response to metal stress [29], revealed that of 222 genes that were induced by two or more of the four heavy metals tested, 38 were upregulated in our carbon starvation proteomics samples, and 103 were induced after 30 minutes of carbon starvation at the transcript level (Table S6). These results suggest that a subset of the genes that respond to carbon starvation are part of a general stress response. In *E. coli* and *B. subtilis*, growth arrest caused by starvation has also been shown to elicit increased synthesis of proteins normally induced by oxidative stress, and starved cells display cross-resistance to these stresses [7,48,49].

The correlation of the set of transcripts and the set of proteins that change significantly upon starvation revealed interesting insights regarding the levels of regulation that might be operating in the carbon starvation response (Figure 6 and Table S7). Considering the 1364 genes for which proteome data were obtained and there was a significant change upon carbon starvation at either the transcript or protein level, or both, in more than half of the cases, the change was only observed at the transcript level (28.5% of genes were up-regulated and 34.1% were downregulated). Some of these might reflect an inherent greater sensitivity of the microarray technique used for detection of transcript changes, than that of the mass spectrometry method used to probe changes in protein levels. 14.8% of the genes showed changes in the same direction at both the transcript and protein level, consistent with transcriptional regulation. For a similar proportion of genes (13.6%), no change was observed at the transcript level, but the corresponding protein was significantly up-

or down-regulated. This group contains candidates for post-transcriptional regulation, either at the translation or protein stability levels. For 9.0% of the genes, observed changes at transcript and protein levels occurred in opposite directions. Overall, our observations are in line with a complex interplay of regulatory mechanisms operating in bacteria, and the lack of correlation between mRNA and protein levels observed in *E. coli* [50,51].

Regulatory networks controlling *C. crescentus'* response to carbon starvation

Analysis of transcriptional profiles allowed us to identify regulators and regulatory modules involved in the response to carbon starvation and, more specifically, connections with the regulatory network that controls cell cycle progression. The enrichment analysis based on previously characterized clusters of co-expressed cell cycle genes that share conserved promoter motifs [25], indicated that the RpoD housekeeping sigma factor, the FixK transcription factor, the SigT ECF sigma factor, and the CtrA cell cycle master regulator play significant roles in the response of *C. crescentus* to carbon starvation. The diagram shown in Figure 7 integrates our transcriptome and proteome datasets of the response to carbon starvation, with regulatory pathways derived from microarray data published for CtrA [31], FixK [27], and LexA [52] (See Tables 2 and S8). The pathways for SigT and SigU under starvation conditions were inferred from data presented in this paper (See Table 1). The proposed transcriptional regulatory interactions were assumed to be direct when the cognate DNA binding motif for the regulator was located in the promoter region of the target gene.

RpoD. The transcript levels of housekeeping sigma factor RpoD and 30 genes identified as RpoD targets by the presence of its cognate DNA motif (cc_6), changed significantly upon carbon starvation, the great majority of them being downregulated. Five of the proteins predicted to be encoded by these genes also changed significantly upon carbon starvation (Figure 7 and Table S8). A decrease in RpoD protein levels was not observed under these conditions, pointing to an alternative mechanism of regulation of its activity that would explain the down-regulation of its target genes, such as the competition from alternative sigma factors induced upon starvation, in binding to the core RNA polymerase.

FixK. The FixL-FixJ-FixK pathway was shown to be a major component of *C. crescentus'* response to hypoxia by Crosson *et al.* [27]. Our analysis of the response to carbon starvation revealed that the protein levels of sensor histidine kinase FixL increased, while 16 of the genes directly (bearing the motif cc_3 in their

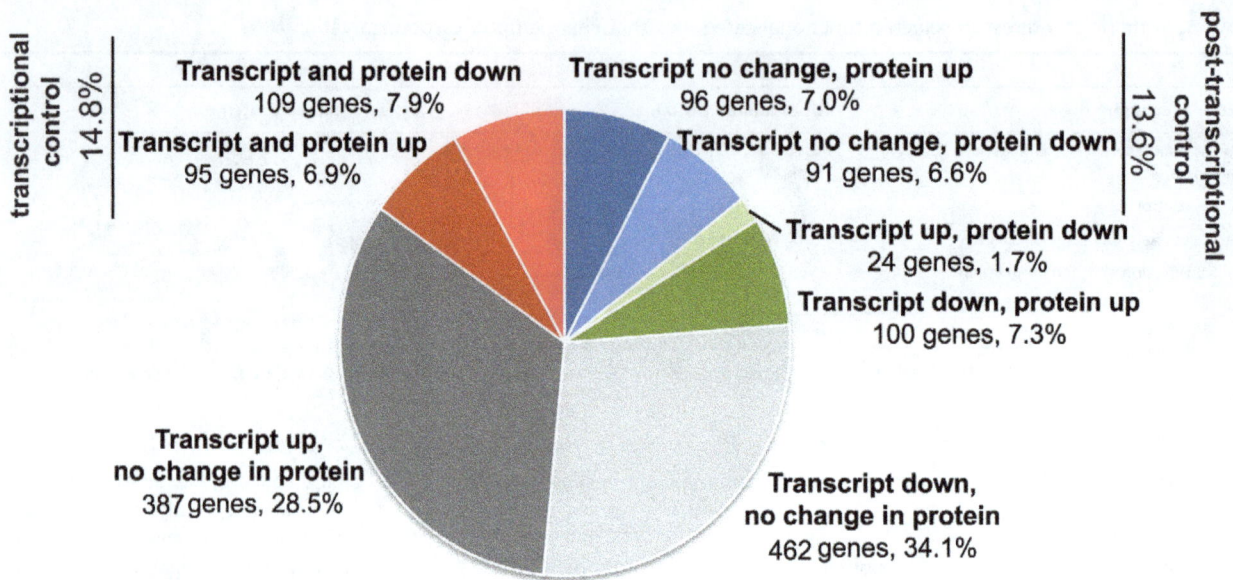

Figure 6. Distribution of carbon starvation-induced changes at the transcript and protein levels. For the 1364 genes for which we had both microarrays and proteomics valid data, and a significant change had been observed for at least one of the datasets, the intersection and exclusion sets are represented. Table S7 lists the genes that belong to each group in the distribution.

promoter regions) or indirectly regulated by transcriptional regulator FixK, changed significantly upon carbon starvation (Figure 7). Two proteins encoded by these FixK-regulated genes were observed to be up-regulated under the same conditions. Interestingly, only half of the genes that have the cc_3 motif were identified as part of the hypoxia-induced FixK regulon by Crosson *et al.*, suggesting that FixK might activate different sets of genes in response to different environmental stimuli or that additional regulatory pathways or factors are involved in these responses.

CtrA. Out of 55 cell cycle-regulated CtrA target genes, we found that 13 were upregulated and 7 were downregulated after 30 minutes of carbon starvation. Out of 32 non-cell cycle-regulated CtrA target genes, 6 were upregulated and 6 were downregulated after 30 minutes of carbon starvation (Table 2 and Figure 7). Our proteome analysis showed a significant decrease in the protein level of CtrA in response to carbon starvation, consistent with previous analyses [10,22]. Of the CtrA-regulated cell cycle transcripts [25] that were upregulated upon carbon starvation (Table 2), 11 out of 13 peaked at the stalked or predivisional stage, while the levels of 10 of these 13 transcripts were reported to increase in a strain carrying a temperature sensitive allele of CtrA (CtrA^{ts}) incubated at restrictive temperature [31]. These observations are compatible with these genes being de-repressed as CtrA protein levels drop in swarmer cells starved for carbon. Conversely, 5 out of the 7 CtrA cell cycle target genes that are down-regulated in response to carbon starvation, show lower transcript levels at the swarmer stage of the cell cycle, with 2 of them displaying increased levels in the CtrA thermosensitive mutant at the restrictive temperature. These genes are most likely deactivated in carbon starved swarmer cells as CtrA levels drop. The remaining 12 CtrA target genes, whose transcript levels remain constant during cell cycle progression in complete media, point to a novel role for CtrA in activating and repressing genes involved in the response to starvation and other stress signals, in addition to those involved in cell cycle progression.

LexA. Upon carbon starvation, the protein levels of LexA –a CtrA target- were observed to decrease to 60% of the non-

starvation levels (although the p-value (0.09) didn't meet the cutoff established for significance (0.05)). The SOS response, the prototypical response to DNA damage in prokaryotes, is controlled by the opposing activities of the LexA repressor and the RecA activator proteins [53]. Genes in the SOS regulon, repressed by LexA under basal conditions, are activated in response to single-stranded DNA regions, often as result of DNA replication inhibition [54]. Activation of the SOS response in *E. coli* leads to blocked FtsZ ring formation and cell division, via the SulA protein [55]. It is possible that a LexA target interacts with FtsZ in *C. crescentus*, to mediate the cell division arrest observed upon starvation [10]. Out of the 44 previously characterized direct LexA target genes [52], two were downregulated and nine were upregulated upon 30 min of carbon starvation, consistent with the decrease in LexA protein levels; one of the upregulated genes, encoding the CC_2589 hypothetical protein, was also upregulated at the protein level.

SigT is a regulator of the starvation-stress response in *C. crescentus* and is involved in the starvation-induced degradation of the CtrA master regulator

Alternative sigma factors play key roles in various stress responses and morphological differentiation in bacteria. Upon activation by environmental or internal signals, alternative sigma factors direct RNA polymerase promoter specificity to activate different regulons. In most gram-negative bacteria, the transcriptional response to environmental stresses is dominated by alternative sigma factor RpoS (see [56] for a review). In gram-positive bacteria, that role falls upon the alternative sigma factor SigB (see [57] for a review). However, the α-proteobacteria lack homologues of either of these sigma factors [8,58].

Three of the 13 *C. crescentus* ECF alternative sigma factors have been implicated in the response to specific stress conditions: SigF mediates the response to oxidative stress in stationary phase [59], SigE mediates the response to cadmium, organic hydroperoxide, singlet oxygen and UV [60], and SigT mediates the response to osmotic and oxidative stress [28]. Our results support a broader

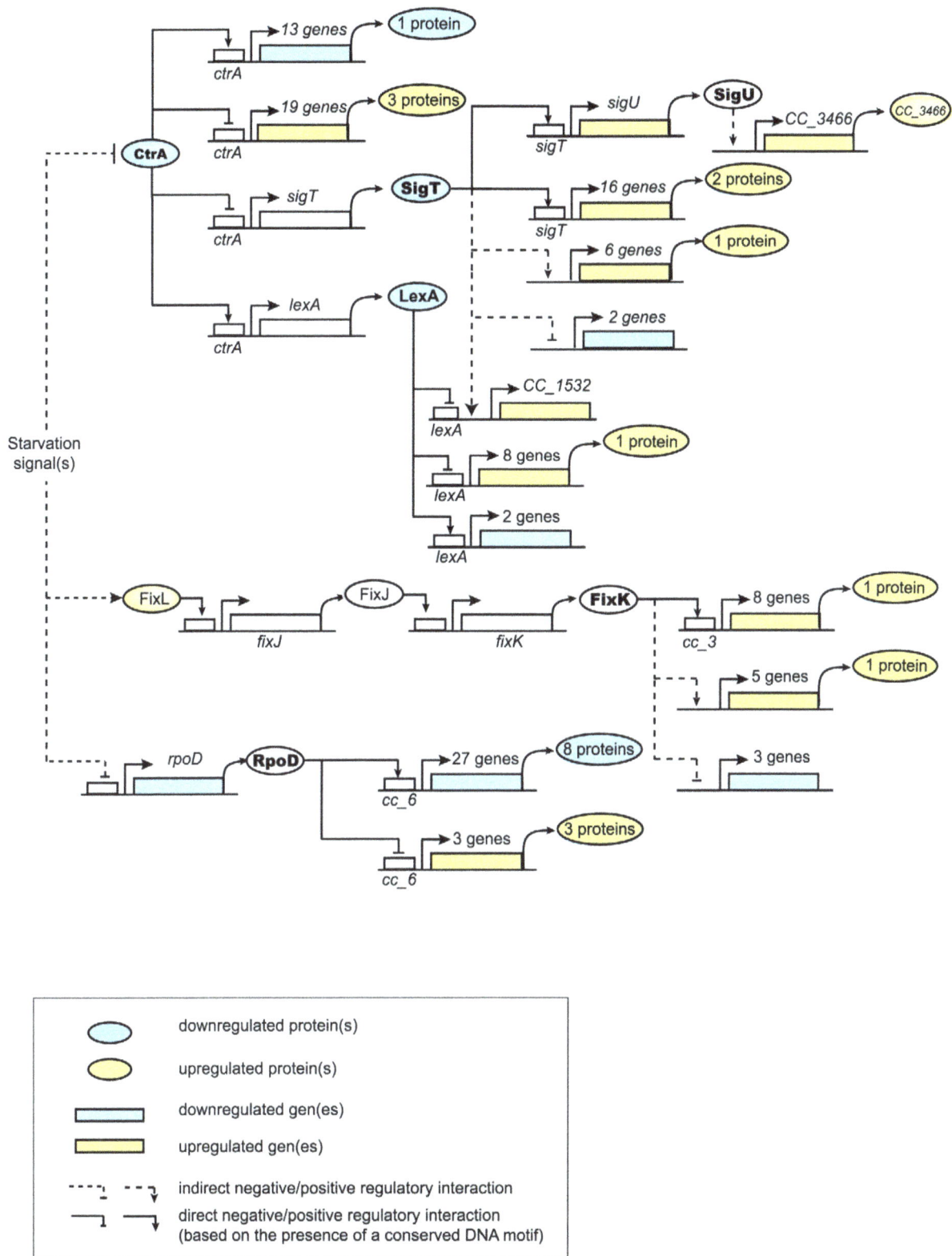

Figure 7. Carbon starvation regulatory pathways. Diagram of regulatory pathways involved in the response to carbon starvation, derived from the analysis of proteomic and gene expression profile changes. Ovals represent proteins and rectangles represent genes. Genes and proteins whose levels increase significantly upon starvation are represented in yellow, and downregulated genes and proteins in blue. For example, we indicate that of the 13 genes controlled by CtrA whose transcripts were downregulated in the absence of carbon (blue rectangle), we were able to detect the down-regulation of one of the corresponding protein products (blue oval). The same schematic representation was used for all genes shown. The genes and proteins that change significantly upon carbon starvation are listed in Table S8. Putative direct regulatory interactions are based on the presence of conserved promoter sequences, while an indirect regulation is proposed if there is no evidence of DNA binding or of the presence of a

conserved promoter element. For the transcript changes, the 30 min time point was considered, while both the 30 and 60 min time points were considered for the protein changes. The CtrA regulon comprises direct CtrA targets, as previously determined [31], that changed upon 30 min of carbon starvation determined by microarray analysis (listed in Table 2), as well as proteins encoded by direct CtrA targets that changed in the proteomic analysis at 30 and/or 60 min of carbon starvation (while most of those proteins were seen to change at the transcript level as well, SigU and LexA were observed to change only at the protein level). CtrA activation or repression was inferred by the direction of the change observed upon starvation: those genes with increased transcript levels upon starvation are inferred to be negatively regulated by CtrA (and consequently derepressed as CtrA protein levels decrease upon starvation); genes with decreased transcript levels were inferred to be positively regulated by CtrA. The SigT regulon includes the genes that showed a significant difference in transcript levels changes upon 15 min of carbon starvation in the sigT deletion strain with respect to wild type (listed in Table 1). Genes with a SigT binding motif in their promoters are represented as direct targets, while those lacking the motif, as indirect targets. The levels of proteins encoded by a subset of these genes changed significantly upon 30 and/or 60 min of carbon starvation. The SigU-dependent gene, CC_3466, showed reduced transcript levels under carbon starvation in a sigU deletion strain with respect to wild type. The LexA (SOS) regulon comprises those genes belonging to the SOS regulon –as previously determined [52] - that changed significantly after 30 min of carbon starvation. The topology of the FixL-FixJ-FixK pathway is as determined by Crosson et al. [27]. The FixK direct targets are those that have the cc_3 motif in their promoter, while the indirect targets were shown to be FixK-dependent by Crosson et al., but lack the cc_3 motif. The RpoD regulon comprises the genes with the cc_6 promoter motif whose transcripts levels were upregulated or downregulated upon 30 min of carbon starvation.

role for SigT. The 27 genes that respond to carbon starvation in a SigT-dependent manner comprise the carbon starvation SigT regulon (Table 1 and Figure 7). Some of these genes were also found to change at the protein level, including CC_3466 (Table 1, Table S2), which appears to be regulated through the SigU sigma factor. Moreover, there is an overlap of the SigT carbon starvation regulon with both the SigT osmotic stress regulon [28] (10 genes in common between the two sets), and the set of genes that are activated by exposure to several heavy metals [29] (12 genes in common). These observations suggest that a core set of SigT regulated genes belongs to a general starvation-stress response. Members of SigT's subfamily of ECF sigma factors from *Sinorhizobium meliloti* and *Methylobacterium extorquens* have been implicated in the regulation of the general stress response [35,61]. Interestingly, some of the transcripts encoding regulatory proteins that were shown to be SigT-dependent upon carbon starvation -namely CC_1356, CC_0284 and CC_1178- required RpoN for induction in carbon versus nitrogen-limited conditions [12]. Two of these genes, CC_1356 and CC_0284 were reported to be SigT-dependent under conditions of osmotic stress [28].

The *sigT* gene belongs to the CtrA cell cycle regulon. CtrA is a negative regulator of *sigT* expression, and *sigT* transcripts peak when CtrA is cleared from the cell at the swarmer to stalked cell transition [31]. The transcript levels of *sigT* were not seen to change significantly upon carbon starvation at the sampled times, in either mixed population or isolated swarmer and stalked cells, while the SigT protein level was found to decrease after 60 minutes of carbon starvation. It is likely that SigT is regulated post-translationally. SigT belongs to the ECF sub-family (ECF15 or EcfG-like) that is characterized by a conserved genomic context, which includes the genes encoding the HK4 histidine kinase and the PhyR response regulator (see Fig. 5C), putative candidates to modulate the activity of the sigma factor [62]. A recent report postulates that *C. crescentus' sigT* is regulated by the anti-sigma factor NepR (CC_3476, cotranscribed with *sigT*), and the gene encoding PhyR, acting as an anti-anti-sigma factor, that is transcribed divergently from *sigT* [37].

We propose that a SigT-dependent pathway is involved in the degradation of CtrA upon carbon starvation, and that this pathway includes the response regulator PhyR and the HK4 histidine kinase. The ability of the anti-anti-sigma factor PhyR to sequester the anti-sigma factor NepR, releasing this SigT to activate transcription, has been shown to be dependent on PhyR phosphorylation state [37]. It is possible that the HK4 histidine kinase, which possesses a predicted signal peptide that would target it to the membrane, is the protein responsible for sensing environmental cues and inducing a response in the CtrA starvation degradation pathway, via PhyR (Figure 8).

Materials and Methods

Bacterial strains and growth conditions

All strains were derived from wild type *C. crescentus* (NCBI Taxonomy ID: 155892) strain CB15N (NA1000) [63]. PYE medium (0.2% Bacto Peptone (Difco), 0.1% yeast extract (Difco), 1 mM $MgSO_4$, and 0.5 mM $CaCl_2$) was used to grow strains for cloning purposes. In all other cases, strains were grown in M2 minimal medium (6.1 mM Na_2HPO_4, 3.9 mM KH_2PO_4, 9.3 mM NH_4Cl, 0.5 mM $MgSO_4$, 10 μM $FeSO_4$ (EDTA chelate; Sigma), 0.5 mM $CaCl2$) with 0.2% glucose as the sole carbon source (referred to as M2G). All *Caulobacter* strains were grown at 28°C and were not allowed to reach an OD_{600} higher than 0.4 at any time, to minimize differences due to physiological adaptation of cultures to stationary phase conditions. *Escherichia coli* strain TOP10 (Invitrogen) was used for cloning following standard procedures.

Construction of gene deletion strains

In frame deletions of *sigT* (CC_3475), *sigU* (CC_2883), *phyR* (CC_3477), *HK4* (CC_3474), CC_3466 and CC_1532 were obtained by a two-step *sacB* counterselection procedure, as previously described [64].

Culture synchronization to obtain isolated swarmer and stalked cells

Swarmer cells were isolated from mixed population cultures using a modified version of the percoll density centrifugation protocol [65]. Isolated colonies from a PYE plate were used to inoculate an M2G culture, grown to an optical density at 600 nm of 0.3. Cultures were cooled on ice, pelleted at 9,000×g at 4°C, and resuspended in 300 μl of ice-cold M2G. After adjusting the volume to 1 ml, one volume of ice-cold Percoll was added and

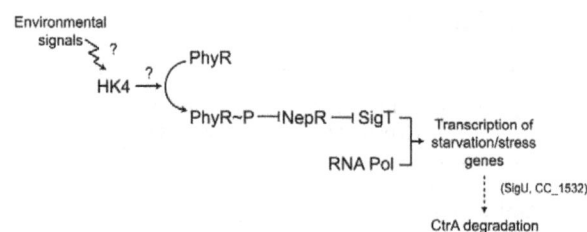

Figure 8. Diagram of the putative HK4-SigT signal transduction pathway. Shown is an inferred pathway based on our results, suggesting that SigT, HK4 (CC_3474), PhyR, SigU and CC_1532 contribute to the degradation of CtrA upon carbon starvation.

mixed thoroughly. The suspension was centrifuged at $11,000 \times g$ for 20 min at 4°C, to separate the swarmer cells (lower band) from the rest of the population (top band). To obtain the stalked cells samples for the experiment in Figure 3, the synchronization procedure was performed as described, swarmers were resuspended in M2G, and allowed to differentiate for 60 min at 28°C. At that point, loss of motility of the majority of the population was confirmed by microscopy.

Carbon starvation of wild-type swarmers and stalked cells, and of mixed populations of wild-type, ΔsigT and ΔsigU strains

Cells (swarmers or mixed populations, as indicated in each case) were washed twice in either ice-cold M2G (control samples) or ice-cold M2 (carbon starvation samples), and resuspended in 2 ml of pre-warmed (28°C) M2G or M2 media. Optical density was adjusted to ~0.3–0.4. After 15 min of incubation at 28°C, cells were pelleted, frozen in liquid nitrogen, and transferred to −80°C. RNA isolation (see below) was always performed within 24 hs of obtaining the samples.

Carbon starvation of C. crescentus mixed population for proteomics and microarrays assays

Cells were grown overnight in small cultures in M2G medium (never reaching $OD_{600\,nm}$ greater than 0.4). These cultures were used to inoculate 1200 ml cultures, which were incubated at 28°C until they had reached an $OD_{600\,nm}$ of 0.3–0.4. Cells were collected by centrifugation (5 min at $9,000 \times g$) and washed twice with a large volume of ice-cold M2 medium, resuspending by gentle pipetting. Control cells were mock-washed with ice-cold M2 with 0.2% glucose. Washed cells were resuspended in pre-warmed M2 or M2G media and incubated at 28°C in a water bath with shaking. Samples were collected at 30 and 60 minutes, frozen in liquid nitrogen and stored at −80°C.

Liquid chromatography-mass spectrometry analysis

Cells were pelleted, frozen in liquid nitrogen and stored at −80°C for no more than 2 days. Lysis and protein treatment was done as described previously [66]. Cell were lysed by bead beating. Proteins were denatured, reduced, alkylated, concentrated by solid-phase extraction, and digested with trypsin. For each injection, 1 µL of each sample of resuspended peptides was injected onto a reversed-phase column using an Isco LC system (Teledyne Isco) and eluted into a Thermo LTQ mass spectrometer (ThermoFinnigan, Inc., San Jose, CA). The mass spectrometer was operated in a data-dependent scan mode as previously described [67]. Raw data were analyzed with the Sequest program with an in silico database obtained from simulated tryptic digestion of the C. crescentus genome. Peptides from this analysis meeting our previously-used scoring standards [67] were used for abundance measurements. Peptide abundances were estimated from ion chromatograms using Viper [68]. Data were extracted and tabulated using SQL queries generated by a custom query from in-house databases. Data analysis, following the general normalization and rollup protocol described previously [69], was performed using the in-house quantitative proteomics tool DAnTE [70] and the statistical software package R (r-project.org). Recorded intensities were converted to natural logarithms throughout. Datasets consisted of the raw peptide intensities, reported as ion counts, for all experiments. All data sets were aligned together as a single batch using MultiAlign, and centered to the most complete data set using the C. crescentus accurate mass tag (AMT) database [68]. For subsequent work, peptide intensities

were linear regressed to the median value of each experimental replicate data set. Normalized peptide intensities were rolled up into effective intensities for each protein in each experimental replicate using the RRollup algorithm [70]. For the binary comparisons of interest, the mean differential expression was calculated for each protein and the statistical significance of the differential expression was established using a two-sided t-test. Significance was established at $p < 0.05$. To take into account very low abundance proteins, the counts of observed peptides under different treatment conditions were also compared using a standard Fisher's Exact Test on the 5 most abundant peptides. Proteins that showed significantly different peptide counts ($p < 0.05$) were pooled together with the proteins that showed a significant change by intensity analysis, to report the final list of significant changes.

Transmission electron microscopy

Cells were collected by centrifugation, fixed for 15 min at room temperature in 4% glutaraldehyde in 100 mM cacodylate-HCl buffer pH 7.4, washed, resuspended in cacodylate buffer, and preserved at 4°C until imaging. Fixed cells were spotted onto glow discharged formvar-carbon coated 300 mesh copper grids (Electron Microscopy Sciences), and allowed to settled for 1 min. Grids were then washed with two drops of mQ water, stained for 15 seconds with 1% uranyl acetate, washed again with two drops of mQ water and air dried. Grids were imaged at 80 kV on a JEOL TEM1230 system. Images were captured with a Gatan 967 slow-scan, cooled CCD camera, using the associated Gatan software. The numbers of cells with or without incipient stalks were counted manually from the exported images.

Light and fluorescent microscopy

Swarmer cells from a strain carrying a ecfp-parB fusion in place of the parB gene in the chromosome (MT190) [71] were isolated, washed as described above to remove glucose, and immobilized onto freshly prepared 1% agarose-M2 pads onto microscopy slides, at the indicated times. Microscopy was performed on a DM6000B upright microscope (Leica) fitted with a 100×1.46 NA HCX Plan APO oil immersion objective (Leica) and a Hamamatsu C9100 EM CCD camera. Phase contrast and fluorescent images (CFP channel, Ex 438/24 nm, Em 483/32 nm) were taken at 40 and 100 ms exposure times, respectively. Images were acquired using KAMS-acquire, a custom software program developed in-house [72], to control the microscope and camera. We used Photoshop CS4 (Adobe) to make false color merges of phase and fluorescent images. In order to tally cells with duplicated and non-duplicated origins, we processed KAMS-acquire files with a Matlab script developed in-house (S. Hong, unpublished), which were then counted manually.

Immunoblot analysis

Cell samples normalized by OD_{600} were lysed by boiling in $2 \times$ sample buffer (4% Sodium Dodecyl Sulphate (SDS), 20% glycerol, 0.01% Bromophenol Blue, 0.125 M Tris-HCl pH 6.8), and loaded in 8–16% Precise polyacrylamide gradient gels (Pierce), followed by electrophoretic transfer to a PVDF membrane (Millipore). Immunoblotting was done using anti-CtrA polyclonal serum (1:10,000), and horseradish-peroxidase conjugated goat anti-rabbit IgG (1:20,000). Signal was detected with chemiluminescence reagent (Perkin-Elmer) and BioMax MR film. The developed film was scanned, processed with Photoshop CS4 (Adobe), and band intensities were determined using ImageQuant (Molecular Dynamics).

β-galactosidase assays

The β-galactosidase activity of a wild type and a ΔsigT strain carrying the plasmid pCtrA290 [73], with the CtrA promoter region fused to lacZ, was assayed after removal of glucose from the cultures in M2 medium, using o-nitrophenyl-β-D-galactoside (ONPG) [74].

RNA Extraction, cDNA synthesis and processing, and microarrays hybridization

Total RNA was extracted from cells using the Purelink Mini Total RNA Purification System (Invitrogen), according to the manufacturer's instructions for on-column DNAse treatment to remove contaminating DNA. A maximum of 2 ml of culture of $OD_{600\ nm}$ between 0.3 and 0.4 was used per ml of Trizol. Integrity of the RNA was confirmed using the RNA 6000 Nano kit (Agilent) on a Agilent 2100 Bioanalyzer, and its concentration was calculated from $OD_{260\ nm}$ measurements on a nanodrop. cDNA was synthesized using Super Script II (Invitrogen) with random hexamers, according to the manufacturer's instructions. After reverse transcription, RNA was removed by 30 min NaOH incubation at 65°C and cDNA was purified using MinElute columns (Qiagen). Purified cDNA was fragmented by a 5 min incubation with DNAseI (Invitrogen; 0.6 U/µg cDNA) in a thermocycler, followed by 15 min enzyme inactivation at 98°C. DNAseI activity was previously titrated, and the same batch of enzyme and thermocycler were used with the same settings across all experiments to obtain reproducible fragmentation. RNA 6000 Nano kit (Agilent) on a Agilent 2100 Bioanalyzer was used to control for homogeneous fragmentation (50–200 bp range) of samples. Fragmented cDNA was biotinylated at the 3' termini with GeneChip DNA Labeling Reagent (Affymetrix) and hybridized onto the CauloHI1 chip (Affymetrix). Hybridization was performed at Stanford's Protein and Nucleic Acid facility, as previously described [25]. All data are MIAME compliant and raw data have been deposited in NCBI Gene Expression Omnibus (www.ncbi.nlm.nih.gov/geo/, with series accession number GSE25999).

Gene annotation

The initial analysis of transcripts and proteins took as a reference the coordinates, ORF prediction and annotation for the C. crescentus reference strain CB15 (GeneBank accession AE005673, RefSeq NC_002696). Upon release of the genomic sequence of the C. crescentus laboratory strain CB15N (also known as NA1000; GeneBank accession CP001340, RefSeq NC_011916), which includes an updated ORF prediction and gene annotation [75], all results were mapped to this new annotation. Supplementary tables include gene IDs corresponding to both genomes nomenclatures. To generate and process diverse gene lists from the microarrays and proteomics results, a Pathways and Genome Database (PGDB) was built using the PathoLogic and Pathway/Genomes Editor software, in the Pathway Tools platform [76]. The C. crescentus NA1000 PGDB was manually curated to incorporate information for diverse published datasets, and is available upon request.

Statistical analysis of microarray data from carbon starved C. crescentus mixed population

The RMA statistical algorithm [77], available under the Bioconductor software package of R, was used for background noise removal, normalization and summarization of microarray data corresponding to two independent experiments for each condition (cells starved for carbon for 30 and 60 min, and the

respective controls). All data are MIAME compliant, and raw and normalized data files were submitted to NCBI Gene Expression Omnibus (accession number GSE25996). A Significance Analysis of Microarrays (SAM) [78] was applied to the dataset, with the following parameters: unpaired, logged, median centered, T statistic. A 2-fold minimum change was selected as cutoff, and a delta value yielding a false discovery rate lower than 1%.

Clustering of transcript changes in swarmer and stalked cells, and search for conserved promoter motifs

Microarray data were normalized and summarized as described in the previous section. All data are MIAME compliant and raw and normalized data has been deposited in NCBI Gene Expression Omnibus (www.ncbi.nlm.nih.gov/geo/, with accession number GSE25997). For each one of the 3767 genes analyzed, we created a profile vector composed of two values: i) log-ratio between expression values measured for swarmer cells after 15 minutes of carbon starvation and the non-starved control, ii) log-ratio between expression values for stalked cells after 15 minutes of carbon starvation and non-starved control. The non-starved controls values were obtained by linearly interpolating the data from [25], at the corresponding time-points (15 minutes into the cell cycle, as a control for swarmer cells, and 75 minutes for stalked cells). We defined a gene as not responding to carbon starvation if the fold-change in expression was less than 2×. Similarly, we defined a gene as responding to carbon starvation if the fold-change in expression was greater than 3×. We only considered genes that, for both carbon starved swarmer and stalked cells, either responded to carbon starvation or did not (no change: $-1 < |log\text{-}ratio| < 1$; change: $|log\text{-}ratio| > 1.585$). We then clustered the profiles of the genes using a bottom-up hierarchical clustering approach (Bioconductor R package), using the Pearson correlation distance and complete linkage for measuring inter-cluster distances. To search for conserved motifs, we extracted 250 bp of sequence (from -200 to $+50$ with respect to the translational start site) for the genes in each cluster, and used MEME [79], with the following parameters: distribution of motif occurrences: zero or one per sequence; minimum motif width: 6; maximum motif width: 25. We selected the motifs with an E-value $< = 0.2$ and average Relative Entropy $> = 1$ bit/bp.

Functional categories enrichment of gene and protein sets

In order to identify functional categories that were significantly over or under-represented within the set of transcripts and proteins that changed upon carbon starvation (30 minutes for transcripts and 30 and/or 60 minutes for proteins), we calculated p-values for each COG-category based on hyper-geometric distributions (p-value $< = 1\%$). Similarly, in order to identify known motifs, associated with cell-cycle and metal stress [25], that were significantly over or under-represented in the upstream regions of the set of genes whose expression changed significantly upon 30 minutes of carbon starvation, we calculated p-values based on hyper-geometric distributions (p-value $< = 5\%$).

Microarray analysis of ΔsigT and ΔsigU strains upon starvation

For each microarray experiment, expression signals for 3767 genes were analyzed. The RMA statistical algorithm [77], available under the Bioconductor software package of R, was used for background noise removal, normalization and summarization of the microarray data. All data are MIAME compliant, and raw and normalized data files were submitted to NCBI Gene

Expression Omnibus (accession number GSE25998). For each gene i, we calculated the difference between sample means and a two-sample t statistic. For each gene and each condition ($\Delta sigT$, $\Delta sigU$ and wild-type strains starved for glucose for 15 minutes), we then calculated a p-value derived from the gene-specific t-statistic, as well as the empirical distribution derived from the t-statistics of all genes. A gene was considered to be differentially expressed, if two conditions were met: the difference between the mutant and WT was greater than two-fold, and the p-value derived from the corresponding t-statistics was smaller than 5%.

Supporting Information

Table S1 Proteome coverage of bacterial species. The highest proteomic coverages for bacterial organisms are shown with the corresponding reference. The proteomic coverage is expressed in the percentage of annotated genes for which the predicted encoded protein has been detected, as reported by the cited publications.

Table S2 List of proteins that change significantly upon carbon starvation. Gene ID (for strains CB15 and CB15N), annotation, COG category (if applicable), fold change and corresponding p-value are shown for the proteins that were found to change reproducibly in cells starved for carbon for 30 and/or 60 min.

Table S3 List of transcripts that change significantly upon carbon starvation. Gene ID (for strains CB15 and CB15N), annotation, COG category (if applicable), average fold change and corresponding standard deviation are shown for the transcripts that were found to change reproducibly in cells starved for carbon for 30 and/or 60 min. For each transcript it is also indicated whether it was previously classified as cell cycle regulated [24].

Table S4 List of genes that change significantly upon carbon starvation in the sets represented in Figure 2. Gene ID (for strains CB15 and CB15N), annotation, and fold change between starved and non-starved cells at 30 min (except for motif cc_7, for which the values correspond to 60 min) are shown.

Table S5 List of genes in the clusters represented in Figure 3B. Gene ID (for strains CB15 and CB15N), log-2 ratio of transcript levels of starved and non-starved swarmer (SW) and stalked (ST) cells after 15 min, annotation, and COG category (if applicable), are shown for all genes clustered in Figure 3B.

Table S6 List of proteins and transcripts that are upregulated upon carbon starvation and heavy metal stress. Genes that were induced by two or more heavy metals [29], and were upregulated in carbon starvation at the level of protein or transcript.

Table S7 Lists of genes represented in Figure 6. This spreadsheet contains eight separate tabs with the lists of genes corresponding to the categories in the chart in Figure 6. Each list was obtained from the intersections of the lists of genes and proteins determined to be up- and down-regulated according to the criteria described in Methods, for those 1378 genes for which there was valid data at both the transcript and the protein level, and a significant change in at least one of the datasets (microarrays or proteomics).

Table S8 Lists of genes represented in Figure 7. Genes in the FixK carbon starvation regulon were obtained from the intersection of up- and down-regulated genes after 30 min of carbon starvation with genes with motif cc_3 according to McGrath et. al. [25], and FixK-dependent genes in hypoxia according to Crosson et. al. [27]. Genes in the LexA carbon starvation regulon were obtained from the intersection of up- (1) and down-regulated (2) genes after 30 min of carbon starvation with genes in the SOS regulon [52]. Genes in the RpoD carbon starvation regulon were obtained from the intersection of up- (1) and down-regulated (2) genes after 30 min of carbon starvation with genes with motif cc_6 according to McGrath et. al. [25]. Genes in the SigT carbon starvation regulon are listed in Table 1. The proteins encoded by SigT targets from Table 1 that are upregulated after 30 or 60 min of carbon starvation are listed here. The CtrA regulon was determined by intersecting the list of CtrA targets [24] (Table 2) with the list of genes up- and down-regulated after 30 min of carbon starvation (Table S3), and the list of proteins that are up- or down-regulated after 30 and/or 60 min of carbon starvation (Table S2). Only the genes for which the transcript and protein changed upon starvation (sets 1 and 2) were included in Figure 7.

Acknowledgments

The authors would like to thank Peter Karp, Alexander Shearer and Suzanne Paley (Bioinformatics Research Group, SRI International, Menlo Park, California, United States), for assistance with development of the *C. crescentus* NA1000 Pathways and Genome Database; John Perrino (Cell Sciences Imaging Facility, Stanford University, Stanford, California, United States), for training in the use of the transmission electron microscope; Elizabeth Zuo and Natalia Kosovilka (Protein and Nucleic Acid Facility, Stanford University, Stanford, California, United States), for microarray samples hybridization; John Coller and Elena Seraia (Stanford Functional Genomics Facility, Stanford University, Stanford, California, United States), for assistance with Bioanalyzer assays; Mike Fero, for development of KAMS image analysis software; Sun-Hae Hong, for image analysis scripts; and Paola Mera (Shapiro Laboratory, Stanford University, Stanford, California, United States), for critical reading of the manuscript.

Author Contributions

Conceived and designed the experiments: LB HHM LS. Performed the experiments: LB TT. Analyzed the data: LB EA TT ML. Wrote the paper: LB LS.

References

1. Poindexter JS (1964) Biological Properties and Classification of the Caulobacter Group. Bacteriol Rev 28: 231–295.
2. Brown PJ, Hardy GG, Trimble MJ, Brun YV (2009) Complex regulatory pathways coordinate cell-cycle progression and development in Caulobacter crescentus. Adv Microb Physiol 54: 1–101.
3. Goley ED, Iniesta AA, Shapiro L (2007) Cell cycle regulation in Caulobacter: location, location, location. J Cell Sci 120: 3501–3507.
4. Holtzendorff J, Reinhardt J, Viollier PH (2006) Cell cycle control by oscillating regulatory proteins in Caulobacter crescentus. Bioessays 28: 355–361.
5. Jacobs-Wagner C (2004) Regulatory proteins with a sense of direction: cell cycle signalling network in Caulobacter. Mol Microbiol 51: 7–13.
6. Lange R, Hengge-Aronis R (1991) Identification of a central regulator of stationary-phase gene expression in Escherichia coli. Mol Microbiol 5: 49–59.

7. Volker U, Maul B, Hecker M (1999) Expression of the sigmaB-dependent general stress regulon confers multiple stress resistance in Bacillus subtilis. J Bacteriol 181: 3942–3948.

8. Chiang SM, Schellhorn HE (2010) Evolution of the RpoS regulon: origin of RpoS and the conservation of RpoS-dependent regulation in bacteria. J Mol Evol 70: 557–571.

9. Hecker M, Pane-Farre J, Volker U (2007) SigB-dependent general stress response in Bacillus subtilis and related gram-positive bacteria. Annu Rev Microbiol 61: 215–236.

10. Lesley JA, Shapiro L (2008) SpoT regulates DnaA stability and initiation of DNA replication in carbon-starved Caulobacter crescentus. J Bacteriol 190: 6867–6880.

11. Landt SG, Lesley JA, Britos L, Shapiro L (2010) CrfA, a small noncoding RNA regulator of adaptation to carbon starvation in Caulobacter crescentus. J Bacteriol 192: 4763–4775.

12. England JC, Perchuk BS, Laub MT, Gober JW (2010) Global regulation of gene expression and cell differentiation in Caulobacter crescentus in response to nutrient availability. J Bacteriol 192: 819–833.

13. McAdams HH, Shapiro L (2009) System-level design of bacterial cell cycle control. FEBS Lett 583: 3984–3991.

14. Grunenfelder B, Rummel G, Vohradsky J, Roder D, Langen H, et al. (2001) Proteomic analysis of the bacterial cell cycle. Proc Natl Acad Sci U S A 98: 4681–4686.

15. Ireland MM, Karty JA, Quardokus EM, Reilly JP, Brun YV (2002) Proteomic analysis of the Caulobacter crescentus stalk indicates competence for nutrient uptake. Mol Microbiol 45: 1029–1041.

16. Molloy MP, Phadke ND, Chen H, Tyldesley R, Garfin DE, et al. (2002) Profiling the alkaline membrane proteome of Caulobacter crescentus with two-dimensional electrophoresis and mass spectrometry. Proteomics 2: 899–910.

17. Phadke ND, Molloy MP, Steinhoff SA, Ulintz PJ, Andrews PC, et al. (2001) Analysis of the outer membrane proteome of Caulobacter crescentus by two-dimensional electrophoresis and mass spectrometry. Proteomics 1: 705–720.

18. Tatusov RL, Natale DA, Garkavtsev IV, Tatusova TA, Shankavaram UT, et al. (2001) The COG database: new developments in phylogenetic classification of proteins from complete genomes. Nucleic Acids Res 29: 22–28.

19. Bowman GR, Comolli LR, Gaietta GM, Fero M, Hong SH, et al. (2010) Caulobacter PopZ forms a polar subdomain dictating sequential changes in pole composition and function. Mol Microbiol 76: 173–189.

20. Figge RM, Easter J, Gober JW (2003) Productive interaction between the chromosome partitioning proteins, ParA and ParB, is required for the progression of the cell cycle in Caulobacter crescentus. Mol Microbiol 47: 1225–1237.

21. Hottes AK, Shapiro L, McAdams HH (2005) DnaA coordinates replication initiation and cell cycle transcription in Caulobacter crescentus. Mol Microbiol 58: 1340–1353.

22. Gorbatyuk B, Marczynski GT (2005) Regulated degradation of chromosome replication proteins DnaA and CtrA in Caulobacter crescentus. Mol Microbiol 55: 1233–1245.

23. Wortinger MA, Quardokus EM, Brun YV (1998) Morphological adaptation and inhibition of cell division during stationary phase in Caulobacter crescentus. Mol Microbiol 29: 963–973.

24. Laub MT, McAdams HH, Feldblyum T, Fraser CM, Shapiro L (2000) Global analysis of the genetic network controlling a bacterial cell cycle. Science 290: 2144–2148.

25. McGrath PT, Lee H, Zhang L, Iniesta AA, Hottes AK, et al. (2007) High-throughput identification of transcription start sites, conserved promoter motifs and predicted regulons. Nat Biotechnol 25: 584–592.

26. Malakooti J, Wang SP, Ely B (1995) A consensus promoter sequence for Caulobacter crescentus genes involved in biosynthetic and housekeeping functions. J Bacteriol 177: 4372–4376.

27. Crosson S, McGrath PT, Stephens C, McAdams HH, Shapiro L (2005) Conserved modular design of an oxygen sensory/signaling network with species-specific output. Proc Natl Acad Sci U S A 102: 8018–8023.

28. Alvarez-Martinez CE, Lourenco RF, Baldini RL, Laub MT, Gomes SL (2007) The ECF sigma factor sigma(T) is involved in osmotic and oxidative stress responses in Caulobacter crescentus. Mol Microbiol 66: 1240–1255.

29. Hu P, Brodie EL, Suzuki Y, McAdams HH, Andersen GL (2005) Whole-genome transcriptional analysis of heavy metal stresses in Caulobacter crescentus. J Bacteriol 187: 8437–8449.

30. Hottes AK, Meewan M, Yang D, Arana N, Romero P, et al. (2004) Transcriptional profiling of Caulobacter crescentus during growth on complex and minimal media. J Bacteriol 186: 1448–1461.

31. Laub MT, Chen SL, Shapiro L, McAdams HH (2002) Genes directly controlled by CtrA, a master regulator of the Caulobacter cell cycle. Proc Natl Acad Sci U S A 99: 4632–4637.

32. Kim SK, Makino K, Amemura M, Shinagawa H, Nakata A (1993) Molecular analysis of the phoH gene, belonging to the phosphate regulon in Escherichia coli. J Bacteriol 175: 1316–1324.

33. Bailey TL, Elkan C (1994) Fitting a mixture model by expectation maximization to discover motifs in biopolymers. Proc Int Conf Intell Syst Mol Biol 2: 28–36.

34. Quon KC, Yang B, Domian IJ, Shapiro L, Marczynski GT (1998) Negative control of bacterial DNA replication by a cell cycle regulatory protein that binds at the chromosome origin. Proc Natl Acad Sci U S A 95: 120–125.

35. Francez-Charlot A, Frunzke J, Reichen C, Ebneter JZ, Gourion B, et al. (2009) Sigma factor mimicry involved in regulation of general stress response. Proc Natl Acad Sci U S A 106: 3467–3472.

36. Bastiat B, Sauviac L, Bruand C (2010) Dual control of Sinorhizobium meliloti RpoE2 sigma factor activity by two PhyR-type two-component response regulators. J Bacteriol 192: 2255–2265.

37. Herrou J, Foreman R, Fiebig A, Crosson S (2010) A structural model of anti-anti-sigma inhibition by a two-component receiver domain: the PhyR stress response regulator. Mol Microbiol 78: 290–304.

38. Hung DY, Shapiro L (2002) A signal transduction protein cues proteolytic events critical to Caulobacter cell cycle progression. Proc Natl Acad Sci U S A 99: 13160–13165.

39. Aldridge P, Jenal U (1999) Cell cycle-dependent degradation of a flagellar motor component requires a novel-type response regulator. Mol Microbiol 32: 379–391.

40. Curtis PD, Brun YV (2010) Getting in the loop: regulation of development in Caulobacter crescentus. Microbiol Mol Biol Rev 74: 13–41.

41. Quardokus E, Din N, Brun YV (1996) Cell cycle regulation and cell type-specific localization of the FtsZ division initiation protein in Caulobacter. Proc Natl Acad Sci U S A 93: 6314–6319.

42. Schmidt KL, Peterson ND, Kustusch RJ, Wissel MC, Graham B, et al. (2004) A predicted ABC transporter, FtsEX, is needed for cell division in Escherichia coli. J Bacteriol 186: 785–793.

43. Gottesman S (2003) Proteolysis in bacterial regulatory circuits. Annu Rev Cell Dev Biol 19: 565–587.

44. Fischer B, Rummel G, Aldridge P, Jenal U (2002) The FtsH protease is involved in development, stress response and heat shock control in Caulobacter crescentus. Mol Microbiol 44: 461–478.

45. Jensen RB, Shapiro L (1999) The Caulobacter crescentus smc gene is required for cell cycle progression and chromosome segregation. Proc Natl Acad Sci U S A 96: 10661–10666.

46. Ali Azam T, Iwata A, Nishimura A, Ueda S, Ishihama A (1999) Growth phase-dependent variation in protein composition of the Escherichia coli nucleoid. J Bacteriol 181: 6361–6370.

47. Claret L, Rouviere-Yaniv J (1997) Variation in HU composition during growth of Escherichia coli: the heterodimer is required for long term survival. J Mol Biol 273: 93–104.

48. Jenkins DE, Chaisson SA, Matin A (1990) Starvation-induced cross protection against osmotic challenge in Escherichia coli. J Bacteriol 172: 2779–2781.

49. Jenkins DE, Auger EA, Matin A (1991) Role of RpoH, a heat shock regulator protein, in Escherichia coli carbon starvation protein synthesis and survival. J Bacteriol 173: 1992–1996.

50. Taniguchi Y, Choi PJ, Li GW, Chen H, Babu M, et al. (2010) Quantifying E. coli proteome and transcriptome with single-molecule sensitivity in single cells. Science 329: 533–538.

51. Lee PS, Shaw LB, Choe LH, Mehra A, Hatzimanikatis V, et al. (2003) Insights into the relation between mrna and protein expression patterns: II. Experimental observations in Escherichia coli. Biotechnol Bioeng 84: 834–841.

52. da Rocha RP, Paquola AC, Marques Mdo V, Menck CF, Galhardo RS (2008) Characterization of the SOS regulon of Caulobacter crescentus. J Bacteriol 190: 1209–1218.

53. Friedberg EC, Walker GC, Siede W (1995) DNA repair and mutagenesis. Washington, D.C.: ASM Press.

54. Sassanfar M, Roberts JW (1990) Nature of the SOS-inducing signal in Escherichia coli. The involvement of DNA replication. J Mol Biol 212: 79–96.

55. Bi E, Lutkenhaus J (1993) Cell division inhibitors SulA and MinCD prevent formation of the FtsZ ring. J Bacteriol 175: 1118–1125.

56. Hengge-Aronis R (2002) Recent insights into the general stress response regulatory network in Escherichia coli. J Mol Microbiol Biotechnol 4: 341–346.

57. Hecker M, Volker U (2001) General stress response of Bacillus subtilis and other bacteria. Adv Microb Physiol 44: 35–91.

58. Mittenhuber G (2002) A phylogenomic study of the general stress response sigma factor sigmaB of Bacillus subtilis and its regulatory proteins. J Mol Microbiol Biotechnol 4: 427–452.

59. Alvarez-Martinez CE, Baldini RL, Gomes SL (2006) A caulobacter crescentus extracytoplasmic function sigma factor mediating the response to oxidative stress in stationary phase. J Bacteriol 188: 1835–1846.

60. Lourenco RF, Gomes SL (2009) The transcriptional response to cadmium, organic hydroperoxide, singlet oxygen and UV-A mediated by the sigmaE-ChrR system in Caulobacter crescentus. Mol Microbiol 72: 1159–1170.

61. Sauviac L, Philippe H, Phok K, Bruand C (2007) An extracytoplasmic function sigma factor acts as a general stress response regulator in Sinorhizobium meliloti. J Bacteriol 189: 4204–4216.

62. Staron A, Sofia HJ, Dietrich S, Ulrich LE, Liesegang H, et al. (2009) The third pillar of bacterial signal transduction: classification of the extracytoplasmic function (ECF) sigma factor protein family. Mol Microbiol 74: 557–581.

63. Evinger M, Agabian N (1977) Envelope-associated nucleoid from Caulobacter crescentus stalked and swarmer cells. J Bacteriol 132: 294–301.

64. Stephens C, Reisenauer A, Wright R, Shapiro L (1996) A cell cycle-regulated bacterial DNA methyltransferase is essential for viability. Proc Natl Acad Sci U S A 93: 1210–1214.

65. Tsai JW, Alley MR (2001) Proteolysis of the Caulobacter McpA chemoreceptor is cell cycle regulated by a ClpX-dependent pathway. J Bacteriol 183: 5001–5007.

66. Shi L, Adkins JN, Coleman JR, Schepmoes AA, Dohnkova A, et al. (2006) Proteomic analysis of Salmonella enterica serovar typhimurium isolated from RAW 264.7 macrophages: identification of a novel protein that contributes to the replication of serovar typhimurium inside macrophages. J Biol Chem 281: 29131–29140.

67. Schutzer SE, Liu T, Natelson BH, Angel TE, Schepmoes AA, et al. (2010) Establishing the proteome of normal human cerebrospinal fluid. PLoS One 5: e10980.

68. Monroe ME, Tolic N, Jaitly N, Shaw JL, Adkins JN, et al. (2007) VIPER: an advanced software package to support high-throughput LC-MS peptide identification. Bioinformatics 23: 2021–2023.

69. Du X, Callister SJ, Manes NP, Adkins JN, Alexandridis RA, et al. (2008) A computational strategy to analyze label-free temporal bottom-up proteomics data. J Proteome Res 7: 2595–2604.

70. Polpitiya AD, Qian WJ, Jaitly N, Petyuk VA, Adkins JN, et al. (2008) DAnTE: a statistical tool for quantitative analysis of -omics data. Bioinformatics 24: 1556–1558.

71. Thanbichler M, Shapiro L (2006) MipZ, a spatial regulator coordinating chromosome segregation with cell division in Caulobacter. Cell 126: 147–162.

72. Christen B, Fero MJ, Hillson NJ, Bowman G, Hong SH, et al. (2010) High-throughput identification of protein localization dependency networks. Proc Natl Acad Sci U S A 107: 4681–4686.

73. Domian IJ, Reisenauer A, Shapiro L (1999) Feedback control of a master bacterial cell-cycle regulator. Proc Natl Acad Sci U S A 96: 6648–6653.

74. Miller JH (1972) Experiments in molecular genetics. Cold Spring Harbor, N.Y.: Cold Spring Harbor Laboratory. xvi, 466 p.

75. Marks ME, Castro-Rojas CM, Teiling C, Du L, Kapatral V, et al. (2010) The genetic basis of laboratory adaptation in Caulobacter crescentus. J Bacteriol 192: 3678–3688.

76. Karp PD, Paley SM, Krummenacker M, Latendresse M, Dale JM, et al. (2010) Pathway Tools version 13.0: integrated software for pathway/genome informatics and systems biology. Brief Bioinform 11: 40–79.

77. Irizarry RA, Hobbs B, Collin F, Beazer-Barclay YD, Antonellis KJ, et al. (2003) Exploration, normalization, and summaries of high density oligonucleotide array probe level data. Biostatistics 4: 249–264.

78. Tusher VG, Tibshirani R, Chu G (2001) Significance analysis of microarrays applied to the ionizing radiation response. Proc Natl Acad Sci U S A 98: 5116–5121.

79. Bailey TL, Williams N, Misleh C, Li WW (2006) MEME: discovering and analyzing DNA and protein sequence motifs. Nucleic Acids Res 34: W369–373.

Spatial Distribution of Soil Organic Carbon and Its Influencing Factors in Desert Grasslands of the Hexi Corridor, Northwest China

Min Wang[1,2]*, Yongzhong Su[1], Xiao Yang[1]

1 Linze Inland River Basin Research Station, Chinese Ecosystem Network Research, Cold and Arid Regions Environmental and Engineering Research Institute, Chinese Academy of Sciences, Lanzhou, Gansu, China, **2** University of Chinese Academy of Sciences, Beijing, China

Abstract

Knowledge of the distribution patterns of soil organic carbon (SOC) and factors that influence these patterns is crucial for understanding the carbon cycle. The objectives of this study were to determine the spatial distribution pattern of soil organic carbon density (SOCD) and the controlling factors in arid desert grasslands of northwest China. The above- and belowground biomass and SOCD in 260 soil profiles from 52 sites over 2.7×10^4 km^2 were investigated. Combined with a satellite-based dataset of an enhanced vegetation index during 2011–2012 and climatic factors at different sites, the relationships between SOCD and biotic and abiotic factors were identified. The results indicated that the mean SOCD was 1.20 (SD:+/− 0.85), 1.73 (SD:+/− 1.20), and 2.69 (SD:+/− 1.91) kg m^{-2} at soil depths of 0–30 cm, 0–50 cm, and 0–100 cm, respectively, which was smaller than other estimates in temperate grassland, steppe, and desert-grassland ecosystems. The spatial distribution of SOCD gradually decreased from the southeast to the northwest, corresponding to the precipitation gradient. SOCD increased significantly with vegetation biomass, annual precipitation, soil moisture, clay and silt content, and decreased with mean annual temperature and sand content. The correlation between BGB and SOCD was closer than the correlation between AGB and SOCD. Variables could together explain about 69.8%, 74.4%, and 78.9% of total variation in SOCD at 0–30 cm, 0–50 cm, and 0–100 cm, respectively. In addition, we found that mean annual temperature is more important than other abiotic factors in determining SOCD in arid desert grasslands in our study area. The information obtained in this study provides a basis for accurately estimating SOC stocks and assessing carbon (C) sequestration potential in the desert grasslands of northwest China.

Editor: Ben Bond-Lamberty, DOE Pacific Northwest National Laboratory, United States of America

Funding: The Strategic Priority Research Program - Climate Change: Carbon Budget and Relevant Issues of the Chinese Academy of Sciences, grant number XDA05050406-3. (2) National Natural Science Foundation of China (91125022). The funders had no role in study design, data collection and analysis, decision to publish, or preparation of the manuscript.

Competing Interests: The authors have declared that no competing interests exist.

* E-mail: wmin85@126.com

Introduction

Soil plays a crucial role in the global carbon cycle by linking carbon transformation with the pedosphere, biosphere, and atmosphere. Therefore, minor changes in the soil carbon pool will greatly affect the alteration of atmospheric CO_2 concentration and have potential feedbacks to climate change [1–4]. SOC storage in temperate grasslands is heavily studied [5–14], while there is little research examining SOC storage in arid regions such as desert-grassland or desert-steppe. Arid regions cover about 47.2% of the earth's land area, with soils containing nearly 241 Pg of soil organic carbon, which is about 40 times more than what was added into the atmosphere through anthropogenic activities [15]. Additionally, soils in these regions are fragile and may experience degradation, desertification, wind erosion, and overgrazing. Small changes in soil conditions can modify the original balance of soil carbon cycle, increase the C loss from soil, and release more greenhouse gases into the atmosphere. Therefore, SOC storage in the desert-grassland ecosystem is a critical component of global C cycle and has a considerable effect on reducing the rate of enrichment of atmospheric CO_2.

In northwest China, desert-grasslands are widely distributed (near 6.5×10^7 hm^2); however, the SOC storage here has not been widely studied. Among the limited estimates of SOC storage in desert-grassland in northwest China [16–18], large differences were found, potentially due to different data sources or approaches. Data from the Second National Soil Survey were usually used for these previous estimates [16–18], but few soil profiles were sampled from the grasslands in northwest China, and these soil profiles lacked data on bulk density and gravel fractions [19]. Regarding different data approaches, previous studies usually calculated SOC stock using average SOC density (SOCD); however, this approach could be constrained by limited soil profiles and large soil heterogeneity. Accordingly, satellite-based approaches will be useful to scale up site-level observations to regional-scale estimates [19].

SOC storage in grasslands is closely correlated with biological, climatic, and edaphic factors. SOC storage exhibits a balance between C inputs from organic material, and C losses through decomposition and mineralization [20–23]. In particular, this balance depends on climatic conditions [18,24–26]. Precipitation and temperature determine the vegetation types, size of plant productivity, and the speed of microbial degradation of soil organic matter. In addition to climate, soil texture plays an important role, which results in an increasing clay content and decreasing C outputs through its stabilizing effect on SOC [5,26,27].

The Hexi Corridor, one characteristically arid area in northwest China, with annual precipitation ranging from about 200 mm in the east to less than 50 mm in the west, represents a desert-grassland ecosystem. Desert-grasslands here always contain transition zones between desert and grassland or between desert and oases, and play a crucial role in maintaining a stable ecological environment and productivity. The desert grassland ecosystem in the Hexi Corridor has unique features, such as limited precipitation, low vegetation cover, coarse soil particles, large gravel content, and highly intensified wind erosion. Accordingly, the vegetation composition varies more than other desert-grasslands. For example, desert steppe in Inner Mongolia is mainly composed of gramineous plants and shrubs with deep roots [28], whereas small shrubs or subshrubs with shallow roots dominate the desert grassland in the Hexi Corridor [29], which results in small biomass productivities and low soil C inputs. These characteristics could lead to greater differences in SOC storage compared with other typical grasslands, but few studies have focused on this desert grassland. The working hypothesis for our study was that the spatial distribution of SOC will show clear relationships with unique climate and vegetation biomass, as well as specific soil conditions. Throughout our field investigations, our research objectives were: (1) to identify SOC density and its spatial distribution characteristics; and (2) to analyze the influence of biotic, climatic, and edaphic factors on SOC density and its distribution in an arid desert grassland. Well water conditions and soil particle composition can promote plant growth and soil organic carbon fixation, and high temperature can accelerate the decomposition of soil organic carbon. For these reasons we hypothesized that SOCD would increase with the increase of moisture and soil clay content, and decrease with the rise of temperature.

Materials and Methods

Ethics Statements

The location of field studies is not privately-owned or protected in any way, so no specific permission was required. All field studies in the desert-grassland were undertaken with support from Linze Inland River Basin Research Station, Cold and Arid Regions Environmental and Engineering Research Institute, Chinese Academy of Sciences. The field studies did not involve endangered or protected species.

Study area

The study area is found in the central region of the Hexi Corridor in Gansu province, northwestern China (spanning from $101°42'36''$ to $97°45'36''$E and $40°31'12''$ to $38°8'26''$N, elevation ranging from 1200 m to 1500 m); the study region has an area of $\sim 2.7 \times 10^4$ km^2. The mean annual precipitation varied from 250 mm to 50 mm from the southeast to the northwest, and 70–80% of the rainfall occurs between June and August. The annual mean temperature varies from 5 to 9°C. The main soil types are Calcic-Orthic Aridosols according to Chinese Soil Taxonomy, which is equivalent to the Aridosols and Entisols of the USDA soil taxonomy classification (Group of Chinese Soil Taxonomy, Institute of Soil Science, Chinese Academy of Sciences, 2001). Soil thickness ranges from 0.2 m to 1.5 m, and most soils contain a large amount of gravel (in 0.5–6 cm),

especially below the 30 cm soil layer. The vegetation population structure in desert grassland is relatively simple, and the main plant species is composed of small shrubs and sub-shrubs including *Asterothamnus centraliasiaticus*, *Reaumuria songarica*, *Salsola passerina*, *Sympegma regelii* and some ephemeral plant species such as *Suaeda glauca*, *Bassia dasyphylla* and *Artemisia scoparia*. The research areas have been subjected to grazing prohibition since 2000.

Sampling sites

A total of 52 sites were selected (Figure 1). Each site contains five plots (the area of a square plot is 1 m×1 m for herbaceous plants and small semi-shrubs or 5 m×5 m for shrubs). A total of 260 soil profiles were sampled (i.e., five profiles at each site) in August of 2011–2012. Based on the pre- investigation of vegetation community types in the study area, we set 52 sampling points. These sampling points basically cover the main vegetation communities of the area (*Reaumuria songarica* community, *Salsola passerine* community, *Sympegma regelii* community, and *Asterothamnus centraliasiaticus* community), and can represent the community characteristics, biomass and soil organic carbon contents and other information of desert grassland. On the other hand, we chose areas with no animal dung, no trace of vegetation were eaten, no trace of trampling, and within the barbed wire enclosure as sampling points. That can ensure selected sample sites without human grazing interference.

Soil sampling, analysis, and biomass survey

At each sampling plot, three soil pits were randomly excavated and mixed into a composite sample at seven depths of 0–5, 5–10, 10–20, 20–30, 30–50, 50–70, and 70–100 cm. Bulk density samples for each depth interval were obtained using a cutting ring (volume of 100 cm^3). Soil moisture (SM) was measured gravimetrically after 24 h desiccation at 105°C. Bulk density was also calculated as the ratio of the oven-dry soil weight to the cutting ring volume. Soil samples were air-dried, hand-picked to remove plant residues, visible soil organisms, and stones, and weighed. The air-dried samples were then sieved through a screen with 2 mm openings, and gravels (>2 mm) were weighed. The weight percent of gravel to soil was obtained. A proportion of samples that passed through the 2 mm sieve were finely ground to pass through a 0.10 mm sieve and analyzed for soil organic carbon

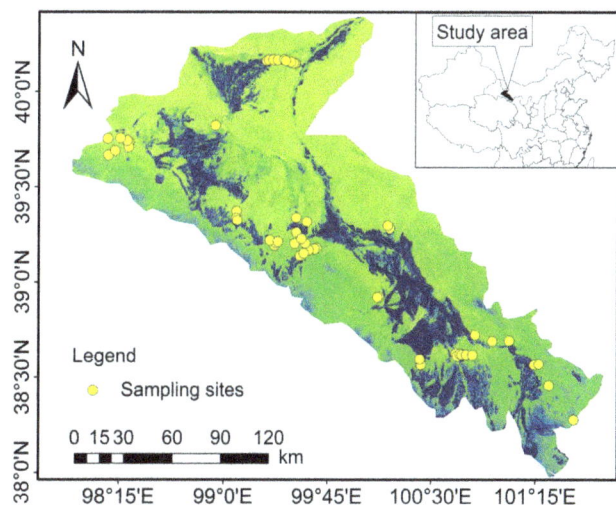

Figure 1. Spatial distribution of sampling sites in the Hexi Corridor.

by the $K_2Cr_2O_7$-H_2SO_4 oxidation method developed by Walkley-Black [30]. A subsample was then analyzed for soil texture by the wet sieve method [31].

Aboveground biomass (AGB) and belowground biomass (BGB) were harvested at 260 plots. At sites with herbaceous plants, which were always present at a low plant density (1 m×1 m), we excavated the total plants to obtain aboveground and below-ground biomass. At sites with shrubs (5 m×5 m), one or several of the dominant plant species in the plot were chosen according to the crown breadth proportion (large: length × width ≥ 50 cm×50 cm; medium: 20 cm×20 cm≤length × width <50 cm×50 cm, and small: length × width <20 cm×20 cm); the whole plants were then dug out in accordance with the crown breadth survey to estimate above and belowground biomass. In the laboratory, the roots samples were soaked in water and cleaned of residual soil using a 0.5 mm sieve. Biomass samples were oven-dried at 65°C to a constant weight and weighed to the nearest 0.01 g.

MODIS data and climate information

The MODIS-EVI data used in this study were obtained from the United States Geological Survey at a spatial resolution of 250 m×250 m and 16-day intervals for the period 2011 to 2012 (http://LPDAAC.usgs.gov). Monthly maximum EVI composites were generated using the Maximum Value Composition method proposed by Holben [32] from 2011 to 2012. The EVI data used were the average of monthly EVI during the growing season from July to August.

Climate data, such as mean annual air temperature (MAT) and annual precipitation (AP), were separated from the climate database of the China monthly ground weather dataset during 2011–2012 (http://cdc.cma.gov.cn). These data were spatially interpolated from the records of 25 climatic stations located throughout the Hexi Corridor.

SOC evaluation

SOC densities for each soil profile at 0–30 cm, 0–50 cm, and 0–100 cm depth intervals were calculated:

$$SOCD = \sum_{h=1}^{n} H_h \times BD_h \times SOC_h \times (1 - C_h)/100 \qquad (1)$$

where SOC density (SOCD) in kg m^{-2}, soil thickness (H_h) in cm, bulk density (BD_h) in g cm^{-3}, SOC (SOC_h) in g kg^{-1}, and volume percentage of the fraction >2 mm (C_h) at layer h were used.

To investigate the spatial distribution of SOCD, we established the relationship between SOCD and MODIS-EVI for three soil depth intervals (Table 1), which was based on the linear relationship between AGB-EVI (Figure 2A) and AGB-SOCD (Figure 2B, and equation for AGB-SOCD$_{0-50\,cm}$ and AGB-SOCD$_{0-100\,cm}$ was SOCD$_{0-50\,cm}$ = 0.014AGB+0.091, R^2 = 0.41, SOCD$_{0-100\,cm}$ = 0.023AGB+0.0039, R^2 = 0.44, respectively).

Using the regression equations (Table 1), each pixel of EVI was converted to SOCD. We then obtained the spatial distribution of SOCD for different soil layers (Figure 3A–C). The spatial distribution of SOCD was performed in ArcGIS, version 9.3 (ESRI, RedLands, California).

Statistical analysis

We used simple linear regression to analyze the relationship between dependent and independent variables (AGB-EVI, SOCD-AGB, SOCD-EVI, SOCD-BGB, and SOCD-TB,

Table 1. Relationships between soil organic carbon density and enhanced vegetation index at three soil depth intervals (0–30 cm, 0–50 cm, and 0–100 cm).

Soil depth	Equation	R^2	P	RMSE	SSE	F-statistic
0–30 cm	SOCD$_{0-30\,cm}$ = 18.87 EVI - 1.05	0.62	<0.001	0.52	13.97	82.77
0–50 cm	SOCD$_{0-50\,cm}$ = 25.64 EVI - 1.33	0.58	<0.001	0.70	30.76	69.98
0–100 cm	SOCD$_{0-30\,cm}$ = 37.12 EVI - 1.75	0.48	<0.001	1.35	95.32	47.69

Notes: RMSE, root-mean-square error; SSE, sum of squares for error.

respectively). Additionally, to evaluate integrative effects of MAT, AP, SM, and soil texture on SOCD, a general linear model (GLM) was employed. Ordinary least squares regression was used to fit these models. Residuals from the linear models were examined for normality (by Shapiro-Wilk test), independence (by Durbin-Watson test), and linearity (by plotting the residuals after fitting linear regression between dependent variable and independent variable) and data met the assumption of normality, independence and linearity. Therefore, linear regressions and GLM models were effective and meaningful for these analyses. Statistical analyses were conducted with R version 3.0.2 package (http://www.R-project.org). We also analyzed the correlations between SOCD and environmental factors (MAT, AP, SM, Silt, Clay, and Sand) using nonlinear regression (by Levenberg-Marquardt and Universal Global Optimization analyses). We constructed exponential equations to describe relationships between SOCD and MAT, AP, SM, soil texture. Nonlinear regression analyses were performed using 1stOpt software, version 1.5 (First Optimization, 7D-Soft High Technology Inc., Xian, China).

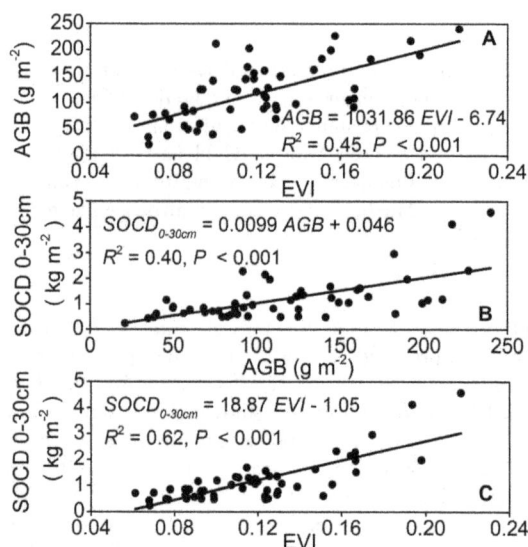

Figure 2. Relationships between above-ground biomass, enhanced vegetation index, and soil organic carbon density at depth interval of 0–30 cm. Notes: AGB, above-ground biomass; EVI, enhanced vegetation index; SOCD, soil organic carbon density.

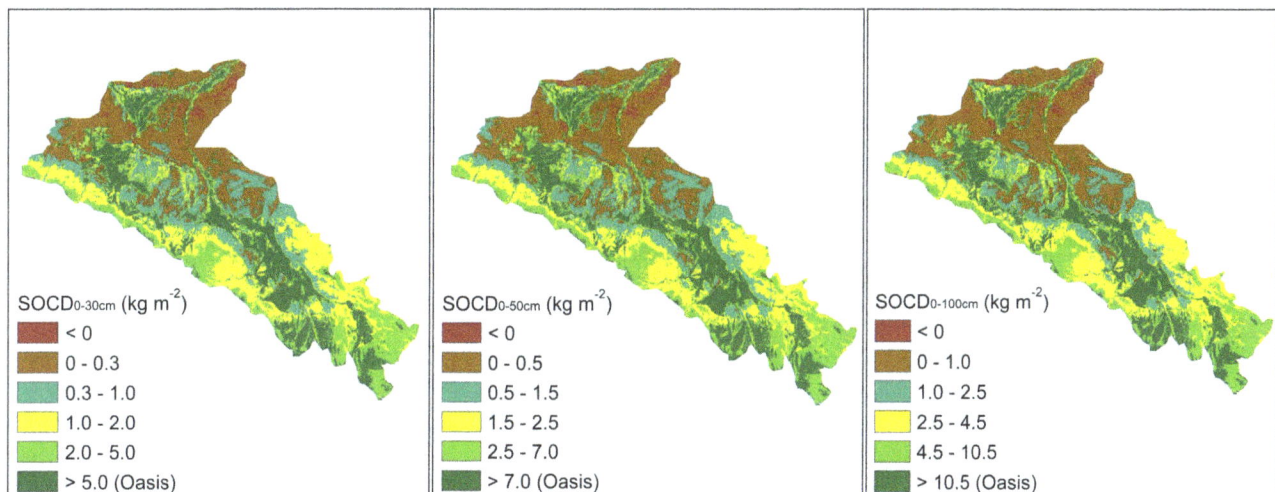

Figure 3. Spatial distributions of soil organic carbon density at different soil layers (0–30 cm, 0–50 cm, and 0–100 cm). *Notes:* SOCD, soil organic carbon density.

Results

SOC stocks and spatial distribution

The statistical description of SOCD across 52 sites at soil depths of 0–30 cm, 0–50 cm, and 0–100 cm is shown in Table 2. As shown, SOCD exhibited large variations among the three soil depths, ranging from 0.24–4.58 kg m^{-2} for 30 cm in depth, 0.35–6.32 kg m^{-2} for 50 cm, and 0.59–9.57 kg m^{-2} for 100 cm, respectively. The corresponding average SOC densities were 1.20, 1.73, and 2.69 kg m^{-2}, respectively. The SOC content in 0–30 cm interval accounted for almost 45% of total SOC in the top 1 m of soil.

The density of SOC decreased from the southeast to the northwest (Figure 3A–C), which corresponds to the precipitation gradient.

Effects of biomass and environmental factors on SOCD

The significant positive relationships between SOCD and AGB, BGB, and TB at different soil depths were characterized by linear functions (Figure 4, $P<0.001$). The R^2 values of regression functions between TB and SOCD were the highest of three biological variables. The R^2 values of regression functions between BGB and SOCD were higher than that between AGB and SOCD and thus showed the closer correlation between BGB and SOCD compared with the correlation between AGB and SOCD (Figure 4, A–C for BGB, D–F for AGB) and exhibited the crucial influence of BGB on SOC distribution (Figure 4G–I). The association of SOC content with BGB and TB was closest in the top soil and decreased at deeper intervals (Figure 4D and G). Nevertheless, the

relationships between AGB and SOCD exhibited only minimal differences at three soil depths (Figure 4A–C).

We established regression relationships between SOCD and various environmental factors, such as MAT, AP, SM, silt, clay, and sand content (Figure 5). In the top 0–30 cm interval, SOCD decreased markedly with MAT (Figure 5A) and sand content (Figure 5P). On the contrary, SOCD increased significantly with an increase SM (Figure 5G). Furthermore, SOCD was positively related with silt (Figure 5J) and clay content (Figure 5M) as well as AP and SM. The regression curves indicated that the closest relationship between SOCD and environmental variables in the top 30 cm soil was with MAT (Figure 5A, $R^2 = 0.79$, $P<0.001$), followed by SM (Figure 5G, $R^2 = 0.67$, $P<0.001$). Additionally, similar relationships between SOCD and environmental factors were also observed in other soil intervals (Figure 5B, E, H, K, N, and Q for 0–50 cm; Figure 5C, F, I, L, O, and R for 0–100 cm). The R^2 value of fitted curves for associations of SOCD with MAT, AP, SM, and sand content declined with soil depth, but an increasing trend was found for association of SOCD with silt and clay content (Figure 5).

According to the above-described relationship between SOCD and environmental factors, we chose five variables (MAT, SM, clay, silt, and sand content, and soil moisture as a measure of water availability) to establish a GLM model. The results suggested that environmental factors explained 69.81%, 74.41%, and 78.87% of the overall variation of SOCD at the soil intervals of 0–30 cm, 0–50 cm, and 0–100 cm, respectively (Table 3).

MAT was the most important parameter for SOCD at 0–30 cm and 0–50 cm (and accounted for 35.53% and 33.58% of

Table 2. Statistics of soil organic carbon density at three soil depth intervals (0–30 cm, 0–50 cm, and 0–100 cm).

SOCD	N	Mean (kg m^{-2})	Std.D. (kg m^{-2})	Min (kg m^{-2})	Median (kg m^{-2})	Max (kg m^{-2})
SOCD$_{0-30\ cm}$	52	1.20	0.85	0.24	0.99	4.58
SOCD$_{0-50\ cm}$	52	1.73	1.20	0.35	1.44	6.32
SOCD$_{0-100\ cm}$	52	2.69	1.91	0.59	2.16	9.57

Notes: SOCD, soil organic carbon density; N, number of samples; Std.D., standard Deviation.

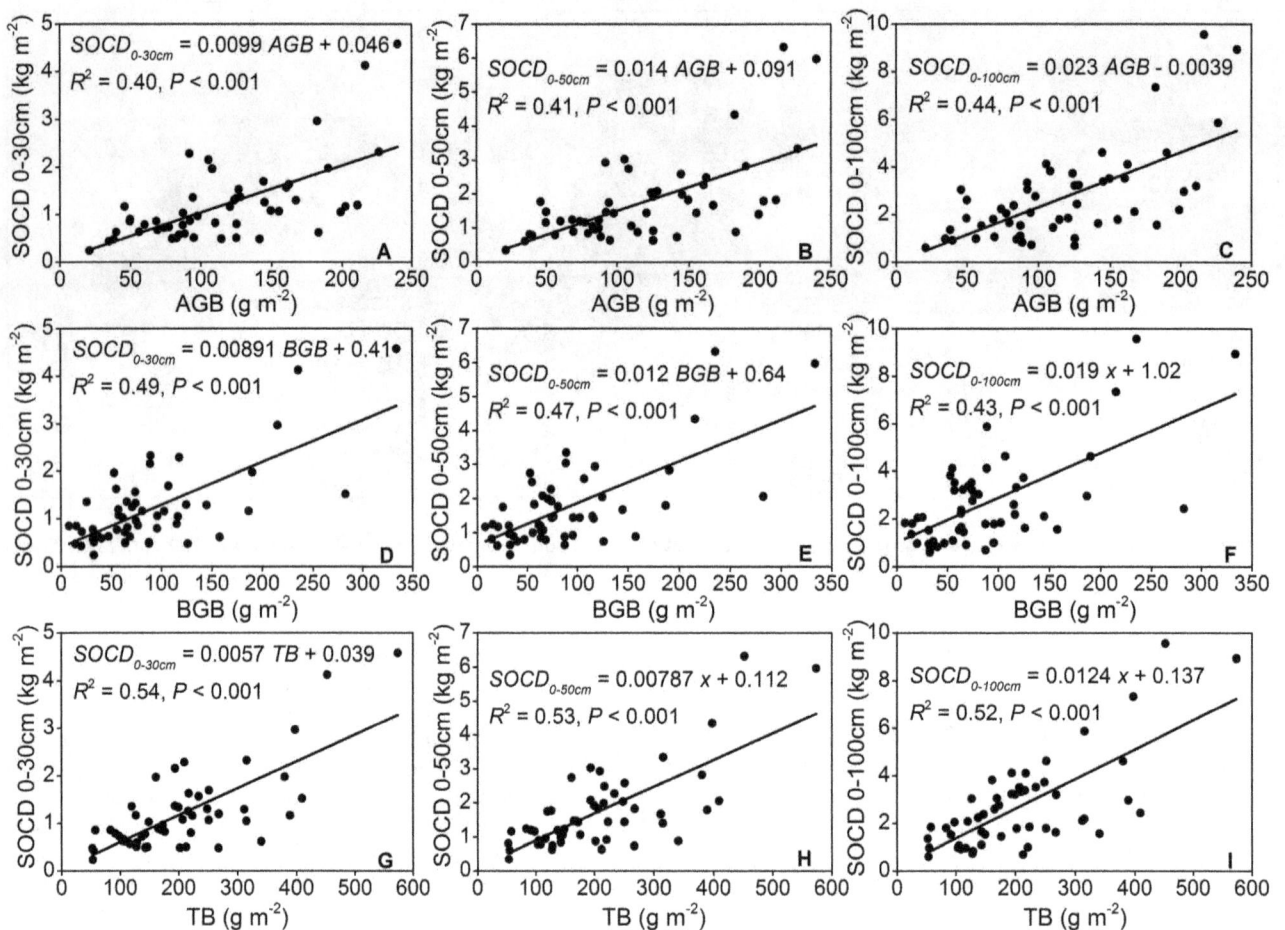

Figure 4. Relationships between soil organic carbon density and biomass at different depth intervals. *Notes*: AGB, above-ground biomass; BGB, below-ground biomass; TB, total biomass; SOCD, soil organic carbon density.

variation), whereas SM was the most important parameter for SOCD at 0–100 cm (where it accounted for 23.61% of variation). The proportion of variances explained by both factors decreased with an increase of soil depth. Soil texture variables (clay, silt, and sand content) explained 7.41%, 15.71%, and 35.95% of the variance at 0–30 cm, 0–50 cm, and 0–100 cm, respectively. In contrast to MAT and SM, the proportion of variance explained by soil texture markedly increased along with soil depth. The proportion of variance explained by sand content increased more rapidly than those explained by the other two variables, and instead silt content become the most important textural variable at soil depth of 0–100 cm (accounted for 13.60% of variation).

Discussion

SOC storage estimation

We summarized previous estimations on SOC storage on different vegetation types at global and regional scales in Table 4. In this study, the mean SOCD of 260 soil profiles at a soil depth of 0–100 cm in the desert grassland of the Hexi Corridor was 2.69 kg m^{-2}. Our results were generally lower than the global mean SOCD (10.8 kg m^{-2}), the average SOCD in China (7.8 kg m^{-2}), and other records based on vegetation types (temperate desert, steppe, and grassland). This difference is most likely due to differences in climate and soil conditions, which play a critical role in determining vegetation types and biomass productivity. The

drier local climate, together with a greater gravel content and thinner soil layer thickness compared to other temperate steppe, grassland, and desert-grassland regions lead to lower vegetation production and limited SOC inputs [29].

Relationship between SOCD and biomass

SOC concentrations are closely linked to biotic processes, such as biomass production, decomposition, and the placement of aboveground litter and root litter in and on the soil [33]. Through regression analyses, we detected a higher coefficient of R^2 for the linear functions between SOCD and BGB compared with the association between SOCD and AGB at the three soil depth intervals. This finding indicated that BGB was most likely the main resource of C inputs and a dominant biological factor on the determination of SOCD. Belowground roots can provide abundant and stable organic material into soil and enhance SOC density. However, due to intense wind erosion in our research area, large amounts of litter fall were blown away by wind, which resulted in only a small amount of aboveground organic matter entering the soil.

Both BGB and TB exhibited the highest correlation in the top 30 cm soil, and this correlation decreased at deeper intervals likely due to high organic matter inputs at surface soil [26,34]. Large gravel content at depths below 30 cm make it hard for roots to extend to deep soil layers, so most of root biomass is concentrated in the upper 0–30 cm soil interval (almost 97%) [29]. According to

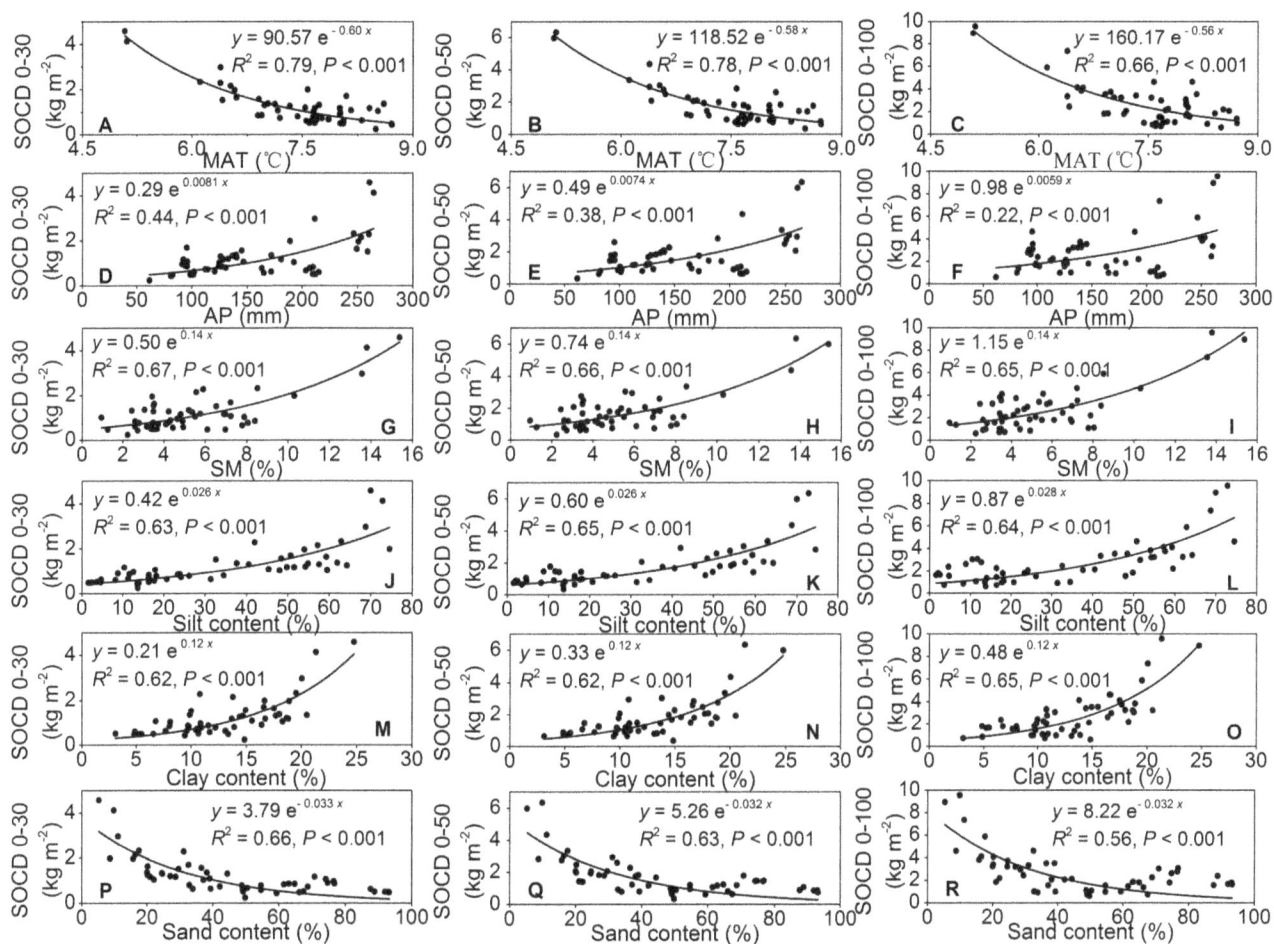

Figure 5. Relationships between soil organic carbon density and environmental factors at different depth intervals. *Notes*: SOCD, soil organic carbon density; MAT, mean annual temperature; AP, annual precipitation; SM, soil moisture.

Table 3. Integrative effects of mean annual temperature, soil moisture and soil texture (clay, silt, and sand) on soil organic carbon density at three soil depth intervals (0–30 cm, 0–50 cm, and 0–100 cm).

Source	0–30 cm				0–50 cm				0–100 cm			
	df	MS	SS%	P	df	MS	SS%	P	df	MS	SS%	P
MAT	1	4.45***	35.53	−0.47	1	8.70***	33.58	−0.66	1	15.13***	19.31	−0.87
SM	1	3.37***	26.87	0.10	1	6.51***	25.12	0.14	1	18.49***	23.61	0.24
Clay	1	0.33*	2.60	0.026	1	1.22**	4.69	0.051	1	7.04***	8.99	0.12
Silt	1	0.52*	4.14	0.013	1	1.89**	7.28	0.024	1	10.46***	13.36	0.057
Sand	1	0.08	0.67	0.004	1	0.97*	3.74	0.014	1	10.65***	13.60	0.047
Residuals	46	0.08	30.19		46	0.14	25.59		46	0.36***	21.13	
Intercept				3.22				3.78				2.25

Notes:
***$P<0.001$;
**$P<0.01$;
*$P<0.05$.
df, degree of freedom; MS, mean squares; SS%, proportion of variances explained by the variable; P, parameters of best-fitted GLM equations; MAT, mean annual temperature; SM, soil moisture.
The results were obtained from general linear model analysis.

Table 4. Comparisons of soil organic carbon density in desert grasslands in the Hexi Corridor with previous estimates.

Research types	Soil organic carbon density (kg m^{-2})				Reference
	0–30 (cm)	0–50 (cm)	0–100 (cm)	Actual depth (cm)	
Global mean	–	–	10.8	–	[23]
Global cool temperate desert	–	–	9.7	–	
Global cool temperate steppe	–	–	13.3	–	
Global temperate grassland	–	–	11.7	–	[26]
Global desert	–	–	6.2	–	
Average of China	3.7	–	7.8	–	[18]
Temperate grassland of Inner Mongolia	4	5.19	6.68	–	[28]
Desert steppe of Inner Mongolia	2.3	3.1	4.01	–	
				–	
Temperate typical steppe	–	–	12.3	–	[40]
Temperate deserted steppe	–	–	8.7	–	
Temperate desert	–	–	6.2	–	
Temperate steppe-desert	–	–	8	–	[10]
Temperate desert-steppe	–	–	8.7	–	
Temperate desert	–	–	6.2	–	
Desert	–	–	4.39	–	[41]
Desert steppe	–	–	7.09	–	
Temperate typical steppe	–	–	–	8	[16]
Temperate deserted steppe	–	–	–	2.8	
Temperate desert	–	–	–	2.2	
Desert grasslands in the Hexi Corridor	1.2	1.73	2.69	–	This study

Notes: "–" mean not measured.

the shallow distribution of roots, the majority of C inputs from roots are concentrated at soil depths of 0–30 cm; additionally, the topsoil is the first layer that directly receives C inputs from aboveground biomass.

Relationship between SOCD and climate and soil environmental factors

Under natural conditions, the distribution of SOC was controlled by climate, vegetation, parent material, and soil texture [3,5,23,26,35]. Our studies observed SOC distribution was positively associated with precipitation and clay content and negatively correlated with temperature and sand content and indicated that the strength of relationships between SOCD and environmental variables decreased with the increase of soil depth.

As shown in Figure 5, SOCD decreased markedly with MAT, likely due to the accelerated mineralization of soil organic matter with increasing temperature [3,24,26]. On the other hand, increasing temperature in arid regions results in a considerable decline in water use efficiency by increasing evapotranspiration can lead to low biomass production and low SOC density [15]. Additionally, the effect of MAT on SOCD in each soil layer was stronger than that for AP and edaphic factors. This result underlines the importance of MAT as a predicator for the SOC density in desert grasslands.

In arid ecosystems, precipitation and soil moisture constrain plant production and decomposition [26]. Our results revealed a substantial increase in the SOCD in desert grasslands with an increase in AP and SM. These results imply that water availability is a powerful parameter for assessment of SOCD. Water is the

limiting factor for plant production in desert grasslands, as a small increase of water could significantly stimulate bio-productivity and thus contribute to the accumulation of SOCD [26,36]. Meanwhile, higher precipitation and soil moisture will affect SOC sequestration mainly through higher soil acidity and lower base saturation at the exchange sites, which would reduce the litter decomposition rate. In addition to the abovementioned results, the relationship between SOCD and SM was stronger than that between SOCD and AP. Precipitation in arid desert-grassland has difficulty entering the soil through infiltration due to low vegetation coverage, less litter content, and coarse soil texture, and consequently, soil moisture can represent the actual water content and play a more important role in model construction between SOCD and environmental factors.

In our study, we observed that the increased SOCD was positively correlated with the accumulation of silt and clay content but negatively correlated with sand content. Finely textured soils with appropriate clay and silt content can increase physical and hydrological protection of SOC by inhibiting decomposition through stabilizing SOC, and increasing residence time to decrease C leaching [3,5,26,37,38]. Moreover, an increase in clay and silt content could enhance the formation of aggregates, which have two advantages for improving SOC content: (1) by improving the soil quality and capacity to efficiently retain water [3,5,39] and (2) by mitigating wind erosion [27]. These two advantages could stimulate plant productivity and thus result in additional C inputs.

The GLM analysis suggested that climate factors were more important in determining the distribution of SOC than soil texture, but the effect decreased gradually with an increase in soil depth. In contrast, the impact of soil texture on SOC distribution in surface soils was not obvious but increased alongside soil depth, which indicates that soil texture plays a critical role in SOC distribution at deeper soil layers. Among all climatic and edaphic parameters, MAT had the highest contribution to explain the distribution of SOC in surface soils (0–30 cm and 0–50 cm). Considering the vast majority of SOC accumulates in the surface soil [26], MAT plays a decisive role in the spatial distribution of SOC in the arid desert- grassland ecosystem in the Hexi Corridor. That may be because biomass production is strictly limited by water in arid desert- grasslands, and the rise of temperature can promote the decomposition of SOC, and conversely promote the accumulation of SOC.

Author Contributions

Conceived and designed the experiments: MW. Performed the experiments: MW YS XY. Analyzed the data: MW YS. Contributed reagents/materials/analysis tools: MW. Wrote the paper: MW.

References

1. Davidson EA, Janssens IA (2006) Temperature sensitivity of soil carbon decomposition and feedbacks to climate change. Nature 440: 165–173.
2. Albaladejo J, Ortiz R, Garcia-Franco N, Navarro AR, Almagro M, et al. (2013) Land use and climate change impacts on soil organic carbon stocks in semi-arid Spain. Journal of Soils and Sediments 13: 265–277.
3. Schimel DS, Braswell BH, Holland EA, McKeown R, Ojima DS, et al. (1994) Climatic, edaphic, and biotic controls over storage and turnover of carbon in soils. Global biogeochemical cycles 8: 279–293.
4. Lal R (2004) Soil carbon sequestration impacts on global climate change and food security. Science 304: 1623–1627.
5. Yang YH, Fang JY, Tang YH, Ji CJ, Zheng CY, et al. (2008) Storage, patterns and controls of soil organic carbon in the Tibetan grasslands. Global Change Biology 14: 1592–1599.
6. Conant R, Paustian K (2002) Spatial variability of soil organic carbon in grasslands: implications for detecting change at different scales. Environmental Pollution 116: S127–S135.
7. Fan JW, Zhong HP, Harris W, Yu GR, Wang SQ, et al. (2008) Carbon storage in the grasslands of China based on field measurements of above-and below-ground biomass. Climatic Change 86: 375–396.
8. Wang GX, Qian J, Cheng GD, Lai YM (2002) Soil organic carbon pool of grassland soils on the Qinghai-Tibetan Plateau and its global implication. Science of the Total Environment 291: 207–217.
9. Han X, He N, Wu L, Wang Y (2008) Storage and Dynamics of Carbon and Nitrogen in Soil after Grazing Exclusion in Grasslands of Northern China. Journal of Environmental Quality 37: 663–668.
10. Ni J (2002) Carbon storage in grasslands of China. Journal of Arid Environments 50: 205–218.
11. Parton W, Scurlock J, Ojima D, Gilmanov T, Scholes R, et al. (1993) Observations and modeling of biomass and soil organic matter dynamics for the grassland biome worldwide. Global biogeochemical cycles 7: 785–809.
12. Piao SL, Fang JY, Zhou LM, Tan K, Tao S (2007) Changes in biomass carbon stocks in China's grasslands between 1982 and 1999. Global biogeochemical cycles 21 Doi: 10.1029/2005gb002634.
13. Qibin W, Linghao L, Xianhua L (1998) Spatial heterogeneity of soil organic carbon and total nitrogen in Xilin River basin grassland, Inner Mongolia. Acta Phytoecologica Sinica 22: 409–414.
14. Yang YH, Fang JY, Guo DL, Ji CJ, Ma WH (2010) Vertical patterns of soil carbon, nitrogen and carbon: nitrogen stoichiometry in Tibetan grasslands. Biogeosciences Discussions 7: 1–24.
15. Lal R (2004) Carbon sequestration in dryland ecosystems. Environ Manage 33: 528–544.
16. Wu HB, Guo ZT, Peng CH (2003) Distribution and storage of soil organic carbon in China. Global biogeochemical cycles 17 Doi: 10.1029/2001gb001844.
17. Wang SQ, Tian HQ, Liu JY, Pan SF (2003) Pattern and change of soil organic carbon storage in China: 1960s–1980s. Tellus B 55: 416–427.
18. Yang YH, Mohammat A, Feng JM, Zhou R, Fang JY (2007) Storage, patterns and environmental controls of soil organic carbon in China. Biogeochemistry 84: 131–141.
19. Fang JY, Yang YH, Ma WH, Mohammat A, Shen HH (2010) Ecosystem carbon stocks and their changes in China's grasslands. Science China Life sciences 53: 757–765.
20. Schlesinger WH, Andrews JA (2000) Soil respiration and the global carbon cycle. Biogeochemistry 48: 7–20.
21. Bastida F, Moreno JL, Hernandez T, Garcia C (2007) The long-term effects of the management of a forest soil on its carbon content, microbial biomass and activity under a semi-arid climate. Applied Soil Ecology 37: 53–62.
22. Johnston CA, Groffman P, Breshears DD, Cardon ZG, Currie W, et al. (2004) Carbon cycling in soil. Frontiers in Ecology and the Environment 2: 522–528.
23. Post WM, Emanuel WR, Zinke PJ, Stangenberger AG (1982) Soil carbon pools and world life zones. Nature 298: 156–159.
24. Burke IC, Yonker CM, Parton WJ, Cole CV, Flach K, et al. (1989) Texture, climate, and cultivation effects on soil organic matter content in US grassland soils. Soil Sci Soc Am J 53: 800–805.
25. Canadell JG, Kirschbaum MUF, Kurz WA, Sanz M-J, Schlamadinger B, et al. (2007) Factoring out natural and indirect human effects on terrestrial carbon sources and sinks. Environmental Science & Policy 10: 370–384.
26. Jobbágy EG, Jackson RB (2000) The vertical distribution of soil organic carbon and its relation to climate and vegetation. Ecological Applications 10: 423–436.
27. Su YZ, Wang XF, Yang R, Lee J (2010) Effects of sandy desertified land rehabilitation on soil carbon sequestration and aggregation in an arid region in China. Journal of Environmental Management 91: 2109–2116.
28. Ma WH (2006) Carbon Storage in the Temperate Grassland of Inner Mongolia. Beijing: Peking University.
29. Wang M, Su YZ, Yang R, Yang X (2013) Allocation patterns of above- and below-ground biomass in desert grassland in the middle reaches of Heihe River, Gansu Province, China. Chinese Journal of Plant Ecology 37: 209–219 (in Chinese).
30. Nelson DW, Sommers LE (1982) Total carbon, organic carbon, and organic matter. In: Methods of soil analysis Part 2 Chemical and microbiological properties: 539–579.
31. Chaudhari SK, Singh R, Kundu DK (2008) Rapid textural analysis for saline and alkaline soils with different physical and chemical properties. Soil Sci Soc Am J 72: 431–441.
32. Holben BN (1986) Characteristics of maximum-value composite images from temporal AVHRR data. International Journal of Remote Sensing 7: 1417–1434.
33. Lal R, Kimble J (2001) Soil erosion and carbon dynamics on grazing land. The potential of US grazing lands to sequester carbon and mitigate the greenhouse effect: Lewis Publishers, Boca Raton, Florida: 231–247.
34. Li Z, Zhao QG (2001) Organic carbon content and distribution in soils under different land uses in tropical and subtropical China. Plant and Soil 231: 175–185.
35. Wang SQ, Huang M, Shao XM, Mickler RA, Li KR, et al. (2004) Vertical distribution of soil organic carbon in China. Environmental Management 33: 200–209.
36. Callesen I, Liski J, Raulund-Rasmussen K, Olsson MT, Tau-Strand L, et al. (2003) Soil carbon stores in Nordic well-drained forest soils - relationships with climate and texture class. Global Change Biology 9: 358–370.
37. Paul EA (1984) Dynamics of organic matter in soils. Plant and Soil 76: 275–285.
38. Wynn JG, Bird MI, Vellen L, Grand-Clement E, Carter J, et al. (2006) Continental-scale measurement of the soil organic carbon pool with climatic, edaphic, and biotic controls. Global biogeochemical cycles 20 Doi: 10.1029/2005gb002576.
39. Schimel DS, Parton WJ (1986) Microclimatic controls of nitrogen mineralization and nitrification in shortgrass steppe soils. Plant and Soil 93: 347–357.
40. Zinke PJ, Stangenberger AG, Post WM, Emanuel WR, Olson JS (1984) Worldwide organic soil carbon and nitrogen data. Oak Ridge National Lab., TN (USA) .
41. Liu WJ, Chen SY, Qin X, Baumann F, Scholten T, et al. (2012) Storage, patterns, and control of soil organic carbon and nitrogen in the northeastern margin of the Qinghai–Tibetan Plateau. Environmental Research Letters 7 Doi: 10.1088/1748-9326/7/3/035401.

Rapid Response of Hydrological Loss of DOC to Water Table Drawdown and Warming in Zoige Peatland: Results from a Mesocosm Experiment

Xue-Dong Lou[1,2], Sheng-Qiang Zhai[1], Bing Kang[2], Ya-Lin Hu[3], Li-Le Hu[1]*

[1] Chinese Research Academy of Environmental Sciences, Beijing, China, [2] College of Life Sciences, Northwest Agriculture & Forestry University, Yangling, Shaanxi, China, [3] Institute of Applied Ecology, Chinese Academy of Sciences, Shenyang, China

Abstract

A large portion of the global carbon pool is stored in peatlands, which are sensitive to a changing environment conditions. The hydrological loss of dissolved organic carbon (DOC) is believed to play a key role in determining the carbon balance in peatlands. Zoige peatland, the largest peat store in China, is experiencing climatic warming and drying as well as experiencing severe artificial drainage. Using a fully crossed factorial design, we experimentally manipulated temperature and controlled the water tables in large mesocosms containing intact peat monoliths. Specifically, we determined the impact of warming and water table position on the hydrological loss of DOC, the exported amounts, concentrations and qualities of DOC, and the discharge volume in Zoige peatland. Our results revealed that of the water table position had a greater impact on DOC export than the warming treatment, which showed no interactive effects with the water table treatment. Both DOC concentration and discharge volume were significantly increased when water table drawdown, while only the DOC concentration was significantly promoted by warming treatment. Annual DOC export was increased by 69% and 102% when the water table, controlled at 0 cm, was experimentally lowered by −10 cm and −20 cm. Increases in colored and aromatic constituents of DOC (measured by $Abs_{254\,nm}$, $SUVA_{254\,nm}$, $Abs_{400\,nm}$, and $SUVA_{400\,nm}$) were observed under the lower water tables and at the higher peat temperature. Our results provide an indication of the potential impacts of climatic change and anthropogenic drainage on the carbon cycle and/or water storage in a peatland and simultaneously imply the likelihood of potential damage to downstream ecosystems. Furthermore, our results highlight the need for local protection and sustainable development, as well as suggest that more research is required to better understand the impacts of climatic change and artificial disturbances on peatland degradation.

Editor: Shiping Wang, Institute of Tibetan Plateau Research, China

Funding: The authors received funding from the Nature Science Foundation of China under Grant Nos. 41103041 and 41271318 to support this study. The funders had no role in study design, data collection and analysis, decision to publish, or preparation of the manuscript.

Competing Interests: All authors have declared that no competing interests exist.

* Email: hulile@craes.org.cn

Introduction

Generally, peat-accumulating wetlands provide waterlogged conditions where carbon accumulation is encouraged [1], and therefore have huge carbon storage potential. However, there is increasing concern that carbon storage in peatlands is unstable and may be susceptible to water table drawdown and higher temperatures over the next two centuries due to projected climatic change [2–6]. Furthermore, the water table in peatlands may also be significantly lowered by drainage resulting from human activities [7,8]. As the largest highland wetland in the world [9] and the largest peat storage area in China, the Zoige alpine wetland serves as a natural barrier and prevents desertification in Northwest China, extending farther toward Southeast China, and is very sensitive to climate change [10]. The peatland in Zoige is also the major water source of the world's largest plateau reserve (i.e., Three-Rivers Source Nature Reserve), supplying water for the three most important rivers in East Asia (i.e. the Yellow, Yangtze, and Lancang rivers) [11]. The Zoige wetland is particularly closely associated with the ecological security of the Yellow River drainage basin [12] because it provides about 40% of the total flow of the Yellow River [13]. Zoige peatland covers an estimated area of 0.5 million hectares and accounts for 47.53% of the total organic carbon reserves in Chinese peatland. Thus, it accounts for the highest organic carbon accumulation of any peatland in China [14].

Unfortunately, due to climate warming, artificial drainage for pastures, and peat exploitation since the 1970s, Zoige peatland has suffered extensive biodiversity loss and ecosystem degradation, including severe peat deterioration [9]. The Zoige wetland has decreased by 30% in the past 30 years due to water table drawdown [15], and artificial drainage has been regarded to be the most important cause of Zoige wetland (including peatland) degradation [9]. Previous studies suggested that the carbon cycle in peatland could change rapidly with climate change [16–18] and that is sensitive to water table [4,19–21]. Therefore, climate warming and a lowered of the water table could potentially create a carbon storage and ecosystem stability crisis in Zoige peatland.

Dissolved organic carbon (DOC) is the most active and sensitive indictor in the carbon cycle [22], and connects the biogeochemical cycle from terrestrial to aquatic ecosystems [23]. Hydrological losses of aquatic carbon can be of significant concern when determining carbon storage in peatlands [24] and may be increasing [25,26]. Among the aquatic constituents of peatlands, DOC is generally considered to have the largest aquatic carbon flux [27,28]. The peculiar water–peat interaction system and strong hydrological connectivity in peatlands ensures that the export of DOC from peatland to downstream plays a key role in the regional redistribution of terrestrial carbon [29] and the carbon balance [30]. Furthermore, the transfer of carbon from terrestrial peatland to fluvial downstream locations has a large influence on the water quality in aquatic ecosystems [31,32]. Previous studies have warned that larger amounts of DOC feeding into downstream locations could increase the levels of aquatic organic acids, decrease the buffering ability of the water, and attenuate the penetration of visible and ultraviolet (UV) light due to changes in the water color [33]. This is likely to cause damage to the sustainable and stable development of aquatic ecosystems, such as their net primary productivity [34] and production of bacteria [2,35]. A large body of literature has reported changes in the color or aromatic components of water in peatlands that has occurred in recent years [11,31,36,37]. $SUVA_{254\ nm}$ was a useful parameter for determining the aromatic characteristic of DOC [38], and absorbance at 400 nm was used as a measure of the color composition [36] and could further indicate changes in DOC composition when combined with specific absorbance [39]. Therefore, DOC is likely to be an important part of the carbon cycle linking peatland and downstream ecosystems, although it is not the only pathway of carbon loss from an upland peatland.

The amount of DOC exported from peatlands is believed to depend on interactions between discharged water through peatland and the production and consumption of DOC within the peatland [4]. However, it has also been reported to increase with a higher discharge [25,40,41] without any effect on DOC concentration. Climate change can regulate the import and export of DOC [42,43], mainly by controlling the most important environmental factors (i.e., temperature and the water table) affecting the peatland carbon cycle. A high water table and low soil temperatures are believed to be major reasons for the low decay rates, which could restrict the production of DOC compounds [44–46]. However, previous observations have indicated that DOC concentrations in peat could be either elevated [47–49] or lowered [4,37,50,51] with a decline in the water level, which could be contributed to the complicated mechanisms and processes involved in the production, consumption, and transport of DOC in peat along with inevitable site-specific characteristics [52]. Similarly, high temperatures can not only improve DOC production through enhanced phenol oxidase activity but also increase the consumption of DOC [25,43]. Thus, it is difficult to determine DOC concentrations in specific regions without performing practical experiments. Moreover, some studies have observed significant changes in water color and aromatic content with shifting water tables and soil temperatures at a range of sites [28,36,49,52–54]. Many previous studies have produced inconsistent results regarding the effects of changes to the water table and/or warming on aquatic DOC release, with both factors able to impact DOC concentrations, the amount of discharge, or both, in a confounding way. Specifically, the response of these variables in peatlands could depend on the length of the observation period. For example, the response of DOC production to drought conditions in the year of drought may differ from that a few years after the drought [55,56], and in a Tibetan alpine meadow

experiment, the response of the aboveground environment to warming treatments in the third year was found to be different from the trend of the first two years [57]. Therefore, our observations in the year immediately after a controlled experiment are helpful for understanding how DOC export might react to climate change and anthropogenic drainage.

Zoige peatland is known to be undergoing a warming and drying climate trend [46], and severe artificial drainage [58]. Several studies in Zoige recently have reported that changes of temperature and/or water table could cause effect on the emissions of CH4 and CO2 [23,26,58,59], and Luo et al. [60] has noticed that DOC could response to experimental warming and grazing. However, there is currently knowledge of the potential response of hydrological DOC loss to the variation of temperature and water table in Zoige peatland. Furthermore, most previous studies on DOC have been conducted countries other than China, particularly in Europe and North America. Therefore, investigating regarding the response of DOC export to warming and water table treatments could provide insight into the impact and mechanisms of climate warming and artificial drainage on the regional carbon budget of Zoige peatland, as well as provide guidance for the local protection and restoration of this deteriorating natural environment. Thus, we undertook a mesocosm experiment to investigate how the hydrological loss of DOC would respond to climatic warming and artificial drainage. The specific objectives were to determine whether the export quantity and concentration, as well as the qualities of DOC and the discharge of flow water, could respond significantly to water table and temperature manipulations. In terms of potential changes of DOC export, the study provides evidence of possible changes to the carbon cycle and storage under the impact of climate change or artificial disturbance and provides evidence for the need to protect and further restore the Zoige peatland.

Methods

Field Site

The peat columns used for mesocosms were collected from Zoige peatland in Hongyuan County, Sichuan Province, on the northeastern margin of the Qinghai–Tibet Plateau (32.76°N, 102.5°E), with a mean altitude of about 3,500 m. Peat was extracted in the area for energy production until 2003, which has left a peat layer of approximately 2 m deep and created severe long-term water shortages [61]. The vegetation community mainly consists of *Carex muliensis* (relative coverage of 41%) and *Kobresia setchwanensis* (39%), as well as a small number of scattered *Potentilla anserina* (15%) and *Plantago depressa* (11%). The topography and vegetation characteristics of the study area are shown in Figure S1. During the period 2002–2011, the site experienced a mean annual temperature and precipitation level of 2.27°C and 700 mm year^{-1}, respectively. During that period, the mean temperature and precipitation from May to October were 8.26°C and 596.34 mm, respectively (data obtained from the China Meteorological Data Sharing Service System at http://cdc.cma.gov.cn/home.do). This study was conducted from May to October in 2012, when the mean temperature and precipitation were 8.65°C and 808 mm, respectively. Therefore, the site experienced higher rainfall and higher temperatures than the average of the previous 10 years.

The study was carried out on the private land of Mr. Jiang in Hongyuan County. Please contact the author first if further information is required. No further permits were required for the locations/activities in the study, and our work did not involve any endangered or protected species.

Mesocosm Experiment

All peat columns were extracted intact from the source plot in December 2011, when the peat was totally frozen and easy to move and reset. Frozen peat cores (cuboid-shaped, with intact vegetation and peat structure) with a surface area of 1 m^2 and a depth of 50–66 cm were carefully placed into stainless-steel barrels with only an open top. We used perforated stainless steel (diameter 9.0 cm) as a pocket sand filter (gravel particle size <4 mm), passing water through its inlet to maintain a near-natural infiltration rate. The perforated stainless-steel filter was buried into the peat column and connected by a drainage system to 5-L tanks in the closed bottom used to store the discharge [62]. The drainage system was connected to a manostat device with a similar pocket sand filter in the interface to lessen the peat outflow. Eighteen mesocosms were constructed for the manipulation of temperature and water table levels (three water table levels, two temperature, and three replicates, n = 18) in a crossed factorial experiment that commenced in May 2012. Positions of the water table level were controlled by hanger loops of the drainage system and set to 0 cm (W0), −10 cm (W1), and −20 cm (W2). They were calibrated using engraved rulers placed adjacent to the bottom of the steel barrels (i.e., the height of the water table was equal to the depth of the peat column plus the observed value). Warming treatment was achieved by using open top chambers (OTCs) during the snow-free period following Walker et al. [63], with 0.43-m-high polycarbonate solar panels placed outside of the mesocosms instead of infrared lamps. Actually, OTCs realize warming mainly through reducing both wind-speed and air convection and increasing incoming solar radiation [64,65]. It can be confirmed by results of previous studies [63,65,66]. During the first growing season, we observed an overall temperature increase of 1.35°C (on an annual basis) was observed for the peat with a −10 cm water table in the warming mesocosms (T1) compared to the ambient mesocosms (T0) (Figure S2). The details on the experiment design in the study were shown in Figure S4.

To closely monitor the output–input water budget in the mesocosms, water discharged from the mesocosms, rainfall, and recharge water were measured using a gauge at least once a week, and more frequently for the first two measurements when rainfall occurred. Water in each mesocosm was mainly supplied by natural precipitation and supplemented by water pumped from a nearby drainage ditch to maintain the preset water table level when necessary. As the drainage ditch extended from the same continuous *C. muliensis* peatland, thus this supplementary water had a similar attributes to the water at the field site. We buried four HOBOPro data-loggers to record peat temperature at −10 cm depth in the mesocosms: two in warming mesocosms with a water table level of 0 cm and two in control mesocosms with a water table level of −20 cm. Monthly weather data from the Hongyuan County weather station were collected for reference.

Sample Analysis

During the study period, discharged water was collected every month for DOC analysis during the growing season (May–October) in 2012. Water samples collected from the manostat tanks were mixed well before sampling, stored in sterile containers (volume 100 ml), and then filtrated through a syringe microfilter (0.45 μm) as preparation for further testing. The DOC concentration was equivalent to total carbon (TC) minus dissolved inorganic carbon (DIC), and both were determined directly using a TOC/TN analyzer (Multi N/C3100TOC/TN; Analytik Jena, Germany). TC was measured by wet combustion, and DIC was measured after sample acidification by 10% H_3PO_4 as proposed

by Guo et al. [67]. The water budget data and the measured DOC concentrations were used to estimate DOC export by Method 3 proposed by Walling and Webb [68]. The UV absorbances of filtered water samples at wavelengths of 254 nm and 400 nm were determined using a UV-visible spectrophotometer (UV-2600; Shimadzu, Kyoto, Japan). UV absorption characteristics of DOC are generally measured to obtain information regarding changes in the composition of DOC compounds. We thus determined the characteristics of DOC composition by means of four related measurements of specific- and UV absorption (i.e., $Abs_{254 nm}$, $SUVA_{254 nm}$, $Abs_{400 nm}$, and $SUVA_{400 nm}$).

Statistical Analysis

Statistical analysis was done using a three-way ANOVA, including the effect of interactions between the time variable (month) and the two treatments on the monthly changes of DOC. Then a sequential full model of two-way repeated-measures ANOVA and main effect analysis and a Sidak post hoc comparison of means test were successively conducted to determine the effects of two treatments. All of these analyses were conducted after testing for essential homogeneity of variance ($p > 0.05$, meaning that variances were homogenous; see Table 1). Further correlation and regression analyses were conducted to determine the relationships of the monthly mean DOC concentrations and DOC export values with the corresponding peat temperature and precipitation. We also conducted linear regression analysis to determine what proportion of the two treatments and discharged volumes respectively. All statistical analyses were performed using SPSS 17.0 (SPSS Inc., Chicago, IL, USA).

Results

Microclimate in Mesocosms and Its Correlation with DOC

Soil Temperature in Peat and Precipitation. The mean monthly temperatures in the mesocosms showed that peat temperature (from May to October) was significantly higher on average in warmed (11.95 ± 4.03°C) than in the controlled mesocosms (10.60 ± 4.25°C, $p = 0.003$; n = 164; Figure S2). The mean monthly precipitation measured with the rain gauge in the mesocosms was 139.14 mm (139.14 L m^{-2}) during the growing season, which was very similar to the value of 134.92 mm recorded by the meteorological station in Hongyuan County. In general, precipitation in all mesocosms was nearly identical regardless of the possible changes in discharge and evapotranspiration among mesocosms located at the same site. A mean number of 17.4 rainy days per month occurred from May to September. Specifically, October only had five rainy days, and June and July each had twenty-one rainy days. The average discharge in the mesocosms was 281.19 L m^{-2} $year^{-1}$, which accounted for 32% of the water input (precipitation and water recharge) and 34% of the rainfall. In contrast, the annual precipitation of temperate biomes (e.g. temperate forest) in China ranges from 400 to 650 mm, i.e., less than the average precipitation (834.84 mm for 6 months) at Zoige. This suggests an abundance of precipitation in Zoige peatland, which is necessary to maintain its year-round spongy condition. It was found that the effect of warming on discharge was nonsignificant in the experiment ($p > 0.05$; Table 1). Meanwhile, the interactive effect of warming and water table on discharge was also insignificant ($p > 0.05$; Table 1).

Correlation of Peat Temperature and Precipitation with DOC. One year study provides limited perspective on the interannual patterns of DOC hydrological export. However, we obtained extra information by the correlation analysis between the two main microclimatic factors and the mean monthly concen-

Table 1. *P*-values of a two-way ANOVA and Levene's test for the effects of the water table level, temperature, and their interactions on the amount of annual DOC export, DOC concentration, absorbance and specific absorbance, and water discharge.

Treatment	DOC		Discharge	Absorbances and specific absorbances			
	Export	Concentration	Discharge	Abs$_{254 nm}$	SUVA$_{254}$	Abs$_{400 nm}$	SUVA$_{400}$
Water table	<0.001**	<0.001**	0.037*	0.001**	0.005**	0.003**	0.008**
Temperature	0.075	0.012**	0.764	0.007**	0.008**	0.010**	0.018**
Water table × Temperature	0.734	0.735	0.553	0.689	0.077	0.439	0.431
Levene's Test	0.179	0.318	0.235	0.520	0.318	0.616	0.110

*indicates a significant difference (*p*<0.05, n = 18)
**indicates a highly significant difference (*p*<0.01, n = 18).

tration and export of DOC, see Figure S3. Monthly DOC export was positively correlated with peat temperature ($R^2 = 0.4984$, $p <$ 0.01; n = 24) and precipitation ($R^2 = 0.8982$, $p<0.01$; n = 6), while the mean monthly DOC concentration was significantly correlated with peat temperature ($R^2 = 0.4025$, $p<0.01$; n = 24), but had a nonsignificant correlation with precipitation ($R^2 = 0.3046$, $p = 0.128$; n = 6).

Annual Export of DOC

Neither the water table × warming nor the month × two controlling-factor interactions were statistically significant. Thus, we examined the single effect of water table and temperature manipulation on the measured variables in the study (i.e. DOC export, concentration, and quality, and the water budget).

We found a difference between the experimental warming treatment and water table treatment in terms of the annual amount of DOC exported. During the period of the study, the manipulation of water table depth in the mesocosms significantly influenced DOC export ($p<0.001$; Figure 1). The export of DOC displayed an upward trend with decreasing water levels. DOC export was 5.76 ± 0.63 g C m^{-2} year^{-1} when the water table was at 0 cm, significantly lower than the levels of 9.75 ± 0.84 g C m^{-2} year^{-1} (-10 cm water-level; $p = 0.004$) and 11.65 ± 1.68 g C m^{-2} year^{-1} (-20 cm water-level; $p<0.001$) respectively. Although all values were within the range (5–40 C m^{-2} year^{-1}) found in the natural peatland [3], the results indicated that DOC export would increase by 69% and 102% annually if the water table at 0 cm was lowered by 10 cm and 20 cm, respectively. In contrast, no significant effect on DOC export was observed in the warming treatments when the peat temperature was raised by 1.35°C (9.81 ± 3.32 g C m^{-2} year^{-1} vs. 8.30 ± 1.82 g C m^{-2} year^{-1}, $p = 0.059$) throughout the growing season. Previous studies have demonstrated that DOC loss during the nongrowing season is similar between controlled and manipulated sites [41]. Therefore, our estimation of DOC export within the growing season did not suggest an alteration in its overall tendency throughout the year.

Eighty-seven percent of the variability in DOC annual export was explained by the combination of the water table level, temperature, and discharge. Among these variables, the level of the water table was the most important predictor, explaining more than 68% of the variation in DOC annual export. Furthermore, discharge was found to be significantly affected only by the water table treatment ($p = 0.028$, n = 18; Figure 2). The discharge

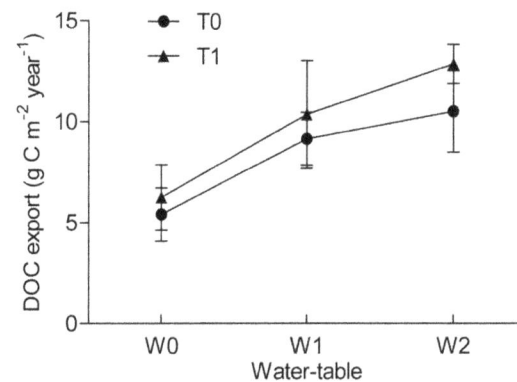

Figure 1. Effect of water table levels and temperature on DOC annual export. Data are means ± standard error. T0 and T1 correspond to ambient temperature and warming temperature, respectively, and W0, W1, and W2 indicate water table depths of 0 cm, −10 cm, and −20 cm, respectively.

volume was 265.08 ± 1.88 L year^{-1} when the water table was at 0 cm, slightly lower than $293.58.04 \pm 6.71$ L year^{-1} (-10 cm water-level; $p = 0.063$) and significantly lower than 294.92 ± 10.84 L year^{-1} (-20 cm water-level; $p = 0.050$). In contrast, the discharge in the warming treatment was nonsignificantly smaller (283.11 ± 17.66 L vs. 279.28 ± 29.28 L). Correlation analysis also showed that water table levels were significantly negatively correlated with discharge ($R^2 = 0.31$, $p = 0.008$; n = 18), while experimental warming was almost irrelevant to discharge ($R^2 = 0.004$, $p = 0.398$; n = 18).

DOC Concentrations

DOC concentrations in discharged water varied significantly between the warmed and ambient temperature treatments as well as among the three water table treatments, despite the differences in their effects on DOC export. As shown in Figure 2, the DOC concentration was lower (23.18 ± 5.76 mg L^{-1}) when the position of the water table level was at 0 cm, significantly lower than 32.81 ± 2.88 mg L^{-1} (-10 cm water-level; $p = 0.028$) and 38.92 ± 4.98 mg L^{-1} (-20 cm water-level; $p = 0.001$). Similarly, DOC concentration was higher in the warmed mesocosms (34.90 ± 5.25 mg L^{-1}, $p = 0.005$) than in the ambient temperature mesocosms (28.48 ± 8.13 mg L^{-1}). In addition, almost 78% of the variation in DOC concentrations was explained by water table and temperature treatments, and 61% of these could be contributed to the water table treatment alone.

Qualities of DOC

There were similar trends for the effects of experimental warming and water table level on absorbance (at wavelengths of 254 nm and 400 nm) and specific absorbance (SUVA$_{254 \text{ nm}}$ and SUVA$_{400 \text{ nm}}$) of DOC in the filtered discharge. These four measures of the quality of DOC were all significantly higher under the warming treatment ($p = 0.004$, $p = 0.016$, $p = 0.004$, and $p = 0.015$, respectively; Table 1, Figure 3) than in the control. Similarly, values of the four measures for a lower water table were significantly higher than those observed at higher water table level ($p<0.001$, $p = 0.010$, $p<0.001$, $p = 0.006$, respectively; Table 1). We assessed the impact of the three positions of the water table level on the four DOC quality measures using a multiple comparison analysis as shown in Figure 3. Therefore, the results above showed that the water table and warming treatments clearly led to several changes in both the quality and absolute DOC concentrations of DOC, indicating a higher aromatic content and changes in the color of downstream water.

Discussion

Effect of Water Table Treatment on Export, Concentration, and Qualities of DOC

Water table manipulation had significant effects on the annual amount of DOC exported, DOC concentration, and the water discharge. Lower water tables were often accompanied by higher DOC exports and concentrations, and a larger discharge volume. These effects were significant when the water table at 0 cm was lowered to -10 cm and -20 cm. This supports the observed variation in the export of DOC, possibly due to site-specific characteristics, and is mainly derived from both fluctuations in the DOC concentration in runoff water and the quantity of water discharged, which disagrees with several previous studies [4,26,69]. Peatland has its own specific features such as plant community construction [70] and hydro-topographical characteristics [67]. Meanwhile, the export of DOC varies with catchment properties and hydrogeologic setting [71], such as precipitation, evapotranspiration [72] and annual runoff [73]. Therefore, it may have disparate performances in DOC production and export which may result in difficulty to draw a universal conclusion [67]. Actually, most reported studies on the response of DOC to water table were done in Europe [74] or North America [4,69]. In this study, when the water table at 0 cm was artificially lowered to -10 cm and -20 cm, DOC export increased by 69% and 102%, respectively. This suggests that an estimated extra 18.4×10^9 g C and 27.2×10^9 g C, respectively, would be transported downstream during the growing season in Zoige. Our ranges of estimated DOC exports with a water table level of -10 cm and -20 cm ($9.75–11.65$ g C m^{-2} year^{-1}) were similar to those ($8.4–11.3$ C m^{-2} year^{-1}) observed in Quebec, Canada, in two growing seasons following a water table drawdown [41]. This could result in a substantial DOC loading into downstream ecosystems, potentially altering the physical and chemical characteristics of aquatic ecosystems, such as acidity, light penetration, and metal and nutrient availability [75]. Such indirect effects on aquatic ecosystems may be the most severe consequence of the elevated DOC export from peatlands. Increases in DOC export may also indicate shifts in the carbon budget, suggesting either a decrease in carbon uptake and storage, or an increase in the turnover of organic carbon [39].

As expected, DOC concentrations were elevated when the water level declined, which agreed with the results of several previous studies [47–49], but disagreed with some other published observations [4,37,50,51]. These inconsistencies reflect the complicated mechanisms and processes involved in the production, consumption, and transport of DOC in peat [52]. Initially, water

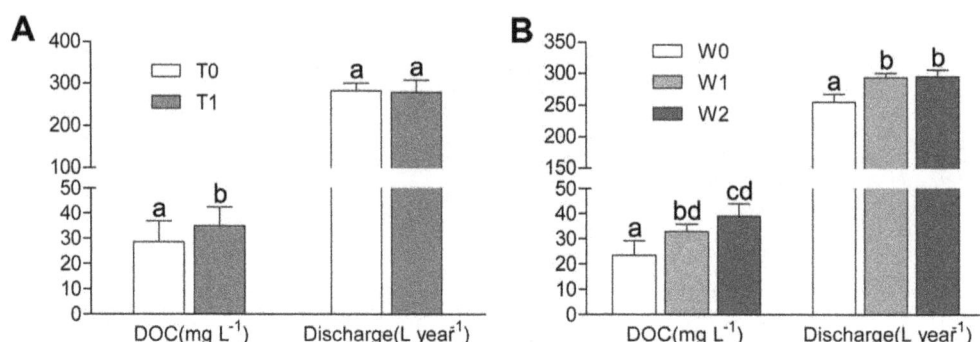

Figure 2. Variations in DOC concentrations and discharge volumes under different treatments. Data are means ± standard error. Same letter superscripts denote insignificant differences among the three water table levels from post hoc tests.

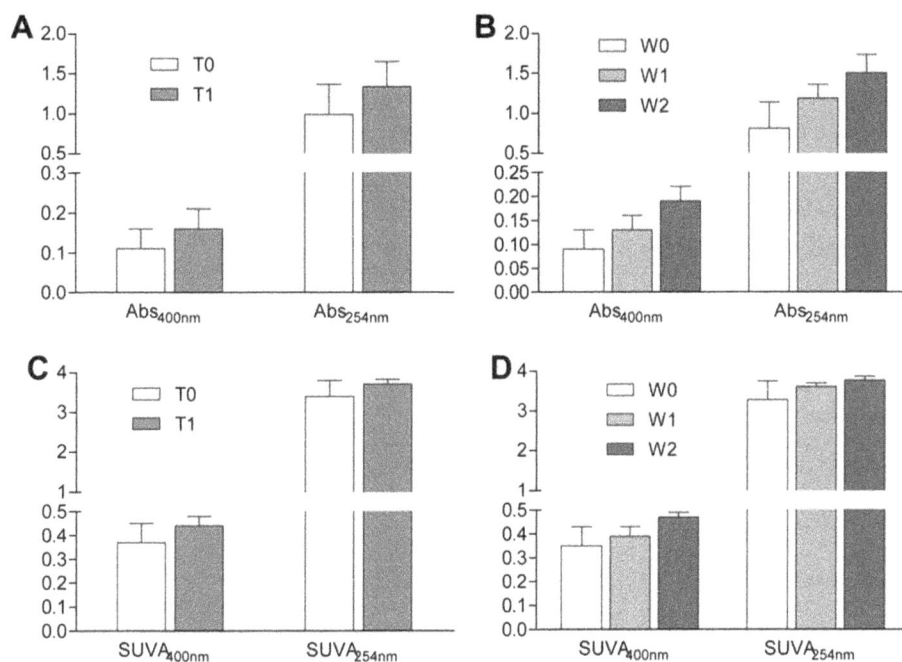

Figure 3. Effects of water table levels and temperature treatments on $Abs_{254\ nm}$, $Abs_{400\ nm}$, $SUVA_{254\ nm}$, and $SUVA_{400\ nm}$. Data are means $\pm$ standard error. Same letter superscripts denote nonsignificant differences among the three water table levels from post hoc tests.

table drawdown, which is closely associated with peat moisture content, can promote aerobic respiration [76,77] and release the activity of degrading hydrolase enzymes in peat, which is supported by reported changes in the specific absorbance of DOC [78]. Biomass is often related to water table, larger biomass being associated with lower water table [79]. Besides, water table could also affect decomposition indirectly through changes in plant community composition [80] or reduce productivity and even cause death in wetland species as the water table is lowered too far [79], thus it can't be determined water table effects on decomposition through changes in plant community or microbial activity within the results of our study. These two mechanisms described above lead to an increase in DOC production, which can then be flushed out from stagnant peat horizons [41] during rainfall. Furthermore, studies have shown that this could lead to peat subsidence and a lower porosity when the water level declines [42,81], likely resulting in slower interflow and a longer residence time for water moving through peat. This may also contribute to more DOC compounds from peat transferring into flowing water. There were 17.4 rainy days every month from May to September 2012 in Zoige, with a maximum of 21 in each of June and July. According to Harrison et al. [82], this high rainfall rate may have potential effect on DOC concentration in water by promoting DOC release from peat soil. Moreover, the water discharge also increased when the level of the water table was lowered in the mesocosm experiment. This suggests a decreased capacity for water storage, which might exacerbate severe water loss and peat erosion if artificial water drainage continues in Zoige peatland. Higher DOC concentrations could also create problems for the stable and sustainable development of peatland ecosystems because it would alter aquatic habitats through its effect on pH and various biological activities (e.g., transportation of nutrients). The weak correlation between precipitation and DOC concentration also indicates that precipitation could promote the export of DOC with a limited enhancement of the DOC concentration. It

also indicates that the elevated DOC in runoff water was the main contributor to the high discharge rather than the high concentration observed under higher rainfall.

The changes in DOC concentrations resulting from the water table treatment were accompanied by changes in UV absorbance characteristics. Our results showed that the aromatic and colored components of DOC increased when the water table was low, suggesting a possible increase in peat degradation. The colored components of DOC, as measured by absorbance at 400 nm and $SUVA_{400\ nm}$, increased sharply when the water table at 0 cm level was lowered to -20 cm, but showed only a slight decline when it was lowered to -10 cm. The increase in DOC aromaticity, measured by absorbance at 254 nm and $SUVA_{254\ nm}$, implied that more aromatic DOC substances should occur in discharge water following a decline in the water table level. Recent studies of Zoige peatland have demonstrated that aromaticity has substantially increased in sites that have experienced a long period of aerobic oxidation and water loss [11]. This is in accordance with our observation of a higher aromatic content when the water table was lower because more peat would be exposed to the air, resulting in increased aerobic respiration [76]. This would also support a reduction in the colored components and aromatic content when the water table rises, which has also been observed elsewhere [36,49,53]. Consequently, DOC in runoff consists of more colored and aromatic components when the water table was drawdown, making it less accessible to microbes within the fluvial ecosystem [54]. Finally, as our results suggested, it would lead to carrying more DOC compounds downstream.

Effect of Warming on Export, Concentration, and Qualities of DOC

Experimental warming at a rate of $1.35 °C \cdot year^{-1}$ in peat had a significant effect on DOC concentration but limited effects on DOC export and discharge. DOC concentrations in warming mesocosms were higher than in normal mesocosms. However, the observed variability of DOC concentrations generally had a limited impact on the hydrological export of DOC which is explained by the relevance of the discharge volume to DOC exports and displayed a nonsignificant correlation with warming in our experiment. Some studies have observed high DOC concentrations at higher temperatures that increased the amount of DOC exported for as long as 12 years under natural warming conditions [25], whereas others have reported a decrease in DOC exports due to the lower discharge following a temperature increase of 1.6–4.1°C [4]. In the study, warming mesocosms have significantly higher DOC concentration but with insignificant lower discharge, which may result from its specific feature (such as evapotranspiration) [72] in Zoige that differs from anywhere else. Therefore, we assumed that temperature variation may have a complex influence on DOC export that is probably associated with both the rate of warming and the temporal scale. It is important to consider both present and future climate change when investigating the effects of experimental warming on peatland carbon turnover.

Temperature is the main factor influencing bacterial metabolism and the rate of decomposition of organic materials, and it also affects DOC dynamics in ecosystems [83]. High temperatures can not increase DOC production through enhanced phenol oxidase activity, but it also can increase the consumption of DOC [25,43]. Thus, determining the DOC concentration in specific regions is difficult without conducting practical experiments. Warming can decrease plant species richness but increase aboveground net primary production [84], thus it may influence inputs of carbon into peatland and lead to unstable DOC concentrations. Our correlation analysis in our experiment indicated that the monthly mean DOC concentration in the discharge increased with temperature. This suggests that the higher temperature increased the DOC concentration because the enhanced decomposition exceeded the gain in DOC consumption in the first year following a rise in temperature. Walker et al. [63] suggested that warming in OTC experiments increased the height and cover of deciduous shrubs and graminoids and decreased species diversity and evenness, implying that the increased DOC concentration we observed might also be attributable to the changes in plant primary productivity. However, temperature had a weak effect on the water discharge during the whole growing season, indicating a relatively stable discharge volume independent of the warming treatment, and differing slightly from the observations of Bridgham et al. [62]. However, this is understandable given the relatively plentiful precipitation at the study site as well as the relatively small change in peat temperature.

Similarly, all of the colored components displayed an upward trend with warming in the study, which was contrary to the results obtained in a laboratory experiment by Tang et al. [52]. But one recent study showed that warming could cause a shift in the composition of bacterial communities in the surface (1–3 cm) and middle layers (9–11 cm) of peat [85], which supported our observations of greater colored and aromatic content under warmer conditions. Therefore, our results indicated that the rising temperature could influence the composition of peat (especially color and aromatic content). Furthermore, peat degradation might arise following climatic warming according to the results, probably due to potential shifts in the function and structure of microbial communities in the peatland, a hypothesis that requires further investigation.

Conclusion

We investigated the response of the hydrological export of DOC in Zoige peatland to changes in the water table level and temperature. Our one-year study provides a basis for understanding the rapid response of the carbon budget in Zoige peatland to climate change and/or artificial drainage, as well as the potential damage to downstream ecosystems, particularly in the Yellow River.

The differences between the water table and peat-temperature treatments implied that future short-duration water table drawdown events could have a greater impact than rising temperature on the export of DOC. In this study, water table effects DOC concentration and export as well as discharge, while temperature treatment only causes obvious effect on DOC concentration. It probably derives from that water table drawdown influenced temperature patterns in the decomposing litter [80] in the crossed factorial experiment. Meanwhile, the experimental warming may also not be high enough for changing the amount of DOC export, as warming could also increase evapotranspiration and therefore decrease discharge [62], which can be known from the relatively smaller discharge in warming mesocosm in the study. Thus, it supported the view that the influence of the water table or water content on peatland ecosystems (such as DOC loss in the study) is stronger than in variations in other environmental conditions, such as temperature [79]. The two experimental water table positions (−10 m and −20 m) resulted in increases in the annual export of DOC by 69% and 102%, respectively, through both a higher DOC concentration and larger discharge volume, indicating the potential release of both carbon and water from peatland after the level of the water table is lowered. The temperature treatment resulted in clear changes in the DOC concentration but had a limited effect on DOC export, probably indicating a shift in the turnover rate of organic carbon in peat because temperature is the main factor affecting bacterial metabolism and the rate of decomposition of organic materials. The nonsignificant effect of warming on DOC export and the notably positive relationship between mean monthly DOC export and peat temperature, which resulted from the shortage of recorded data for all mesocosms, appear contradictory, and suggest that further studies should be careful to consider these issues. Variable water levels and temperatures changed the absorbances of DOC in the year immediately after this experiment. This result suggests the varied nature or qualities of DOC that might influence fluvial systems, and also warns that using absorbance records as a proxy for DOC concentrations when studied in peatland should be done with caution.

Therefore, our observations in the first year immediately after the controlled experiment were helpful for understanding how the carbon budget might react to climate change and anthropogenic interference (i.e., drainage). This mesocosm experiment also provides useful information for local protection and sustainable development in Zoige peatland. Additional experiments and observations are required to achieve a comprehensive understanding of the carbon cycle (such as DOC fluxes) in peatlands facing changing environmental conditions.

Supporting Information

Figure S1 Topography and vegetation characteristics of the study area. (A) Drainage ditch. (B) Vegetation community growing in shallow water. (C) Scattered vegetation surrounded by surface water. All photographs were taken during May 2012 in Hongyuan County located in Zoige peatland.

Figure S2 Peat temperature at −10 cm depth and precipitation during the growing season in 2012.

Figure S3 Correlation analysis of temperature and precipitation with DOC. (A) Results of the correlation analysis between peat temperature recorded in four mesocosms and corresponding mean monthly export and concentration of DOC (DOC concentration: $y = 1.4573x + 9.9416$, $R^2 = 0.4025$, $p < 0.01$, n = 24; DOC export: $y = 0.2043x - 0.9177$, $R^2 = 0.4984$, $p < 0.01$, n = 24). (B) Results of the correlation analysis between mean monthly precipitation and mean monthly export and concentration of DOC in all mesocosms (DOC concentration: $y = 0.0379x + 26.409$, $R^2 = 0.3046$, $p = 0.128$, n = 6; DOC export: $y = 0.013x - 0.2987$, $R^2 = 0.8982$, $p < 0.01$, n = 6).

Figure S4 The schematic drawing of the mesocosm in the study. The references of the number are shown as below: 1. Polycarbonate solar panels; 2. Water intake system; 3. Drainage system; 4. Water storage barrel; 201. Observation tube of water-level; 202. Sand filter pocket of inlet; 301. Sand filter pocket of outlet; 302. High pressure valves; 303. Water pipe; 304. Hanger loop; 305. Observation rule.

Acknowledgments

The authors wish to acknowledge the assistance of Prof. Zhang Fen Chun at CRAES (Chinese Research Academy of Environmental Sciences) for his expert revision of the manuscript. The authors are also grateful to two anonymous reviewers for their valuable comments on the manuscript.

Author Contributions

Conceived and designed the experiments: LLH BK. Performed the experiments: XDL SQZ. Analyzed the data: XDL BK YLH. Contributed reagents/materials/analysis tools: XDL SQZ BK LLH. Wrote the paper: XDL LLH.

References

1. Aselmann I, Crutzen PJ (1989) Global distribution of natural fresh-water wetlands and rice paddies, their net primary productivity seasonality and possible methane emissions. Journal of Atmospheric Chemistry 8: 307–358.
2. Hobbie JE (1992) Microbial control of dissolved organic-carbon in lakes - research for the Future. Hydrobiologia 229: 169–180.
3. Moore TR, Roulet NT, Waddington JM (1998) Uncertainty in predicting the effect of climatic change on the carbon cycling of Canadian peatlands. Climatic Change 40: 229–245.
4. Pastor J, Solin J, Bridgham SD, Updegraff K, Harth C, et al. (2003) Global warming and the export of dissolved organic carbon from boreal peatlands. Oikos 100: 380–386.
5. Turunen J (2008) Development of Finnish peatland area and carbon storage 1950–2000. Helsinski, FINLANDE: Finnish Environment Institute. 16 p.
6. Yu Z, Beilman DW, Jones MC (2013) Sensitivity of northern peatland carbon dynamics to Holocene climate change. Carbon Cycling in Northern Peatlands: American Geophysical Union. pp. 55–69.
7. Moore TR, Dalva M (1993) The influence of temperature and water table position on carbon dioxide and methane emissions from laboratory columns of peatland soils. Journal of Soil Science 44: 651–664.
8. Price JS (2003) Role and character of seasonal peat soil deformation on the hydrology of undisturbed and cutover peatlands. Water Resources Research 39: 1241.
9. Xiang S, Guo R, Wu N, Sun S (2009) Current status and future prospects of Zoige Marsh in Eastern Qinghai-Tibet Plateau. Ecological Engineering 35: 553–562.
10. Shi C-c, Tu J (2009) Remote Sensing Monitory Study on Land Desertification in Ruoergai Plateau of Sichuan Province during 40 Years. Southwest China Journal of Agricultural Sciences 6: 035 (in Chinese).
11. Guo X, Du W, Wang X, Yang Z (2013) Degradation and structure change of humic acids corresponding to water decline in Zoige peatland, Qinghai-Tibet Plateau. Sci Total Environ 445–446: 231–236.
12. Zhang X, Lv X, Gu H (2005) To analyze threats, to describe present conservation situation and to provide management advices of the Ruoergai marshes. Wetland Sci 3: 292–297(in Chinese).
13. SAFS (Sichuan Academy of Forest Science) (2006) Scientific investigation report on Zoige marsh. Chengdu: Sichuan Science and Technology Press(in Chinese).
14. Wang M, Liu Z, Ma X, Wang G (2012) Division of organic carbon reserves of peatlands in China. Wetland Sci 10: 156–163(in Chinese).
15. Gao J (2006) Degradation factor analysis and solutions of Ruoergai Wetland in Sichuan. Sichuan Environ 25: 48–53(in Chinese).
16. Gorham E (1991) Northern peatlands- role in the carbon-cycle and probable responses to climatic warming Ecological Applications 1: 182–195.
17. Trettin CC, Laiho R, Minkkinen K, Laine J (2006) Influence of climate change factors on carbon dynamics in northern forested peatlands. Canadian Journal of Soil Science 86: 269–280.
18. Bridgham SD, Pastor J, Dewey B, Weltzin JF, Updegraff K (2008) Rapid carbon response of peatlands to climate change. Ecology 89: 3041–3048.
19. Hogg EH, Lieffers VJ, Wein RW (1992) Potential carbon losses from peat profiles - effects of temperature, drought cycles, and fire. Ecological Applications 2: 298–306.
20. Freeman C, Lock MA, Reynolds B (1993) Fluxes of CO2, CH4 and N2O from a Welsh peatland following simulation of water-table draw-down - potential feedback to climatic-change. Biogeochemistry 19: 51–60.
21. Bohrer G, Chen H, Wu N, Wang Y, Zhu D, et al. (2013) Inter-Annual Variations of Methane Emission from an Open Fen on the Qinghai-Tibetan Plateau: A Three-Year Study. PLoS ONE 8: e53878.
22. Evans CD, Chapman PJ, Clark JM, Monteith DT, Cresser MS (2006) Alternative explanations for rising dissolved organic carbon export from organic soils. Global Change Biology 12: 2044–2053.
23. Zhang G, Tian J, Jiang NA, Guo X, Wang Y, et al. (2008) Methanogen community in Zoige wetland of Tibetan plateau and phenotypic characterization of a dominant uncultured methanogen cluster ZC-I. Environmental microbiology 10: 1850–1860.
24. Billett MF, Palmer SM, Hope D, Deacon C, Storeton-West R, et al. (2004) Linking land-atmosphere-stream carbon fluxes in a lowland peatland system. Global Biogeochemical Cycles 18: GB1024.
25. Freeman C, Evans CD, Monteith DT, Reynolds B, Fenner N (2001) Export of organic carbon from peat soils. Nature 412: 785–785.
26. Chen H, Yao S, Wu N, Wang Y, Luo P, et al. (2008) Determinants influencing seasonal variations of methane emissions from alpine wetlands in Zoige Plateau and their implications. Journal of Geophysical Research: Atmospheres 113: D12303.
27. Limpens J, Berendse F, Blodau C, Canadell JG, Freeman C, et al. (2008) Peatlands and the carbon cycle: from local processes to global implications – a synthesis. Biogeosciences 5: 1475–1491.
28. Dinsmore KJ, Billett MF, Dyson KE (2013) Temperature and precipitation drive temporal variability in aquatic carbon and GHG concentrations and fluxes in a peatland catchment. Global Change Biology 19: 2133–2148.
29. Aerts R, De Caluwe H (1999) Nitrogen deposition effects on carbon dioxide and methane emissions from temperate peatland soils. Oikos 84: 44–54.
30. Arnosti C, Holmer M (2003) Carbon cycling in a continental margin sediment: contrasts between organic matter characteristics and remineralization rates and pathways. Estuarine Coastal and Shelf Science 58: 197–208.
31. Wallage ZE, Holden J (2010) Spatial and temporal variability in the relationship between water colour and dissolved organic carbon in blanket peat pore waters. Science of The Total Environment 408: 6235–6242.
32. Chin W-C, Lennon JT, Hamilton SK, Muscarella ME, Grandy AS, et al. (2013) A Source of Terrestrial Organic Carbon to Investigate the Browning of Aquatic Ecosystems. PLoS ONE 8: e75771.
33. Evans CD, Monteith DT, Cooper DM (2005) Long-term increases in surface water dissolved organic carbon: Observations, possible causes and environmental impacts. Environmental Pollution 137: 55–71.
34. Carpenter SR, Pace ML (1997) Dystrophy and eutrophy in lake ecosystems: Implications of fluctuating inputs. Oikos 78: 3–14.
35. Wetzel RG (1992) Gradient-dominated ecosystems - sources and regulatory functions of dissolved organic-matter in fresh-water ecosystems. Hydrobiologia 229: 181–198.
36. Wallage ZE, Holden J, McDonald AT (2006) Drain blocking: An effective treatment for reducing dissolved organic carbon loss and water discolouration in a drained peatland. Science of the Total Environment 367: 811–821.
37. Grayson R, Holden J (2012) Continuous measurement of spectrophotometric absorbance in peatland streamwater in northern England: implications for understanding fluvial carbon fluxes. Hydrological Processes 26: 27–39.
38. Weishaar JL, Aiken GR, Bergamaschi BA, Fram MS, Fujii R, et al. (2003) Evaluation of specific ultraviolet absorbance as an indicator of the chemical composition and reactivity of dissolved organic carbon. Environmental Science & Technology 37: 4702–4708.
39. Worrall F, Armstrong A, Adamson JK (2007) The effects of burning and sheep-grazing on water table depth and soil water quality in a upland peat. Journal of Hydrology 339: 1–14.
40. Fraser CJD, Roulet NT, Moore TR (2001) Hydrology and dissolved organic carbon biogeochemistry in an ombrotrophic bog. Hydrological Processes 15: 3151–3166.
41. Strack M, Waddington JM, Bourbonniere RA, Buckton EL, Shaw K, et al. (2008) Effect of water table drawdown on peatland dissolved organic carbon export and dynamics. Hydrological Processes 22: 3373–3385.

42. Sommer M (2006) Influence of soil pattern on matter transport in and from terrestrial biogeosystems—A new concept for landscape pedology. Geoderma 133: 107–123.

43. Briggs J, Large DJ, Snape C, Drage T, Whittles D, et al. (2007) Influence of climate and hydrology on carbon in an early Miocene peatland. Earth and Planetary Science Letters 253: 445–454.

44. Strack M, Waddington JM, Tuittila ES (2004) Effect of water table drawdown on northern peatland methane dynamics: Implications for climate change. Global Biogeochemical Cycles 18: GB4003.

45. Scanlon D, Moore T (2000) Carbon dioxide production from peatland soil profiles: The influence of temperature, oxic/anoxic conditions and substrate. Soil Science 165: 153–160.

46. Morris PJ, Belyea LR, Baird AJ (2011) Ecohydrological feedbacks in peatland development: A theoretical modelling study. Journal of Ecology 99: 1190–1201.

47. Dai Y, Luo Y, Wang C, Shen Y, Ma Z, et al. (2010) Climate variation and abrupt change in wetland of Zoig Plateau during 1961 and 2008. Journal of Glaciology and Geocryology 32: 35–42(in Chinese).

48. Jager DF, Wilmking M, Kukkonen JVK (2009) The influence of summer seasonal extremes on dissolved organic carbon export from a boreal peatland catchment: Evidence from one dry and one wet growing season. Science of The Total Environment 407: 1373–1382.

49. Blodau C, Siems M (2012) Drainage-induced forest growth alters belowground carbon biogeochemistry in the Mer Bleue bog, Canada. Biogeochemistry 107: 107–123.

50. Sapek A, Sapek B, Chrzanowski S, Urbaniak M (2009) Nutrient mobilisation and losses related to the groundwater level in low peat soils. International Journal of Environment and Pollution 37: 398–408.

51. Ellis T, Hill PW, Fenner N, Williams GG, Godbold D, et al. (2009) The interactive effects of elevated carbon dioxide and water table draw-down on carbon cycling in a Welsh ombrotrophic bog. Ecological Engineering 35: 978–986.

52. Tang R, Clark JM, Bond T, Graham N, Hughes D, et al. (2013) Assessment of potential climate change impacts on peatland dissolved organic carbon release and drinking water treatment from laboratory experiments. Environmental Pollution 173: 270–277.

53. Watts CD, Naden PS, Machell J, Banks J (2001) Long term variation in water colour from Yorkshire catchments. Science of The Total Environment 278: 57–72.

54. Wilson L, Wilson J, Holden J, Johnstone I, Armstrong A, et al. (2011) Ditch blocking, water chemistry and organic carbon flux: Evidence that blanket bog restoration reduces erosion and fluvial carbon loss. Science of the Total Environment 409: 2010–2018.

55. Mitchell GN (1990) Natural discoloration of freshwater: Chemical composition and environmental genesis. Progress in Physical Geography 14: 317–334.

56. Mitchell G, McDonald AT (1992) Discolouration of water by peat following induced drought and rainfall simulation. Water Research 26: 321–326.

57. Li G, Liu Y, Frelich LE, Sun S (2011) Experimental warming induces degradation of a Tibetan alpine meadow through trophic interactions. Journal of Applied Ecology 48: 659–667.

58. Chen H, Wu N, Wang Y, Zhu D, Zhu Qa, et al. (2013) Inter-Annual Variations of Methane Emission from an Open Fen on the Qinghai-Tibetan Plateau: A Three-Year Study. PLoS ONE 8: e53878.

59. Yanbin H, Yanfen W, Xurong M, Xiangzhong H, Xiaoyong C, et al. (2008) CO2H2O and energy exchange of an Inner Mongolia steppe ecosystem during a dry and wet year. Acta Oecologica-international Journal Of Ecology 33: 133–143.

60. Luo C, Xu G, Wang Y, Wang S, Lin X, et al. (2009) Effects of grazing and experimental warming on DOC concentrations in the soil solution on the Qinghai-Tibet plateau. Soil Biology and Biochemistry 41: 2493–2500.

61. Zhang XH, Liu HY, Baker C, Graham S (2012) Restoration approaches used for degraded peatlands in Ruoergai (Zoige), Tibetan Plateau, China, for sustainable land management. Ecological Engineering 38: 86–92.

62. Bridgham SD, Pastor J, Updegraff K, Malterer TJ, Johnson K, et al. (1999) Ecosystem control over temperature and energy flux in Northern peatlands. Ecological Applications 9: 1345–1358.

63. Walker MD, Wahren CH, Hollister RD, Henry GH, Ahlquist LE, et al. (2006) Plant community responses to experimental warming across the tundra biome. Proc Natl Acad Sci U S A 103: 1342–1346.

64. Debevec EM, MacLean JrSF (1993) Design of greenhouses for the manipulation of temperature in tundra plant communities. Arctic and Alpine Research: 56–62.

65. Turetsky M, Treat C, Waldrop M, Waddington J, Harden J, et al. (2008) Short-term response of methane fluxes and methanogen activity to water table and soil warming manipulations in an Alaskan peatland. Journal of Geophysical Research: Biogeosciences (2005–2012) 113.

66. Chivers M, Turetsky M, Waddington J, Harden J, McGuire A (2009) Effects of experimental water table and temperature manipulations on ecosystem CO2 fluxes in an Alaskan rich fen. Ecosystems 12: 1329–1342.

67. Guo Y, Wan Z, Liu D (2010) Dynamics of dissolved organic carbon in the mires in the Sanjiang Plain, Northeast China. Journal of Environmental Sciences 22: 84–90.

68. Walling DE, Webb BW (1981) The reliability of suspended sediment load data: IAHS Publication.

69. Clair TA, Arp P, Moore TR, Dalva M, Meng FR (2002) Gaseous carbon dioxide and methane, as well as dissolved organic carbon losses from a small temperate wetland under a changing climate. Environmental Pollution 116, Supplement 1: S143–S148.

70. Weltzin JF, Pastor J, Harth C, Bridgham SD, Updegraff K, et al. (2000) Response of bog and fen plant communities to warming and water-table manipulations. Ecology 81: 3464–3478.

71. Fraser C, Roulet N, Moore T (2001) Hydrology and dissolved organic carbon biogeochemistry in an ombrotrophic bog. Hydrological Processes 15: 3151–3166.

72. Moore T (1989) Dynamics of dissolved organic carbon in forested and disturbed catchments, Westland, New Zealand: 1. Maimai. Water Resources Research 25: 1321–1330.

73. Urban N, Bayley S, Eisenreich S (1989) Export of dissolved organic carbon and acidity from peatlands. Water Resources Research 25: 1619–1628.

74. Worrall F, Reed M, Warburton J, Burt T (2003) Carbon budget for a British upland peat catchment. Science of The Total Environment 312: 133–146.

75. Steinberg (2003) Ecology of humic substances in freshwaters: determinants from geochemistry to ecological niches. Berlin: Springer.

76. Clymo RS (1984) The limits to peat bog growth. Philosophical Transactions of the Royal Society of London B, Biological Sciences 303: 605–654.

77. Mars H, Wassen MJ, Peeters WHM (1996) The effect of drainage and management on peat chemistry and nutrient deficiency in the former Jegrznia-floodplain (NE-Poland). Vegetatio 126: 59–72.

78. Freeman C, Ostle N, Kang H (2001) An enzymic 'latch' on a global carbon store. Nature 409: 149–149.

79. Moore PD (2002) The future of cool temperate bogs. Environmental Conservation 29: 3–20.

80. Strakova P, Penttilä T, Laine J, Laiho R (2012) Disentangling direct and indirect effects of water table drawdown on above-and belowground plant litter decomposition: consequences for accumulation of organic matter in boreal peatlands. Global Change Biology 18: 322–335.

81. Whittington PN, Price JS (2006) The effects of water table draw-down (as a surrogate for climate change) on the hydrology of a fen peatland, Canada. Hydrological Processes 20: 3589–3600.

82. Harrison AF, Taylor K, Scott A, Poskitt J, Benham D, et al. (2008) Potential effects of climate change on DOC release from three different soil types on the Northern Pennines UK: examination using field manipulation experiments. Global Change Biology 14: 687–702.

83. Froberg M, Berggren D, Bergkvist B, Bryant C, Mulder J (2006) Concentration and fluxes of dissolved organic carbon (DOC) in three norway spruce stands along a climatic gradient in sweden. Biogeochemistry 77: 1–23.

84. Lin X, Zhang Z, Wang S, Hu Y, Xu G, et al. (2011) Response of ecosystem respiration to warming and grazing during the growing seasons in the alpine meadow on the Tibetan plateau. Agricultural and Forest Meteorology 151: 792–802.

85. Kim SY, Freeman C, Fenner N, Kang H (2012) Functional and structural responses of bacterial and methanogen communities to 3-year warming incubation in different depths of peat mire. Applied Soil Ecology 57: 23–30.

Carbon Footprint of Telemedicine Solutions - Unexplored Opportunity for Reducing Carbon Emissions in the Health Sector

Åsa Holmner[1]*, Kristie L. Ebi[2,4], Lutfan Lazuardi[3], Maria Nilsson[4]

1 Department of Radiation Sciences, Umeå University, Umeå, Sweden, 2 ClimAdapt, LLC, Seattle, Washington, United States of America, 3 Department of Public Health, Faculty of Medicine, Gadjah Mada University, Yogyakarta, Indonesia, 4 Department of public health and clinical medicine, epidemiology and global health, Umeå University, Umeå, Sweden

Abstract

Background: The healthcare sector is a significant contributor to global carbon emissions, in part due to extensive travelling by patients and health workers.

Objectives: To evaluate the potential of telemedicine services based on videoconferencing technology to reduce travelling and thus carbon emissions in the healthcare sector.

Methods: A life cycle inventory was performed to evaluate the carbon reduction potential of telemedicine activities beyond a reduction in travel related emissions. The study included two rehabilitation units at Umeå University Hospital in Sweden. Carbon emissions generated during telemedicine appointments were compared with care-as-usual scenarios. Upper and lower bound emissions scenarios were created based on different teleconferencing solutions and thresholds for when telemedicine becomes favorable were estimated. Sensitivity analyses were performed to pinpoint the most important contributors to emissions for different set-ups and use cases.

Results: Replacing physical visits with telemedicine appointments resulted in a significant 40–70 times decrease in carbon emissions. Factors such as meeting duration, bandwidth and use rates influence emissions to various extents. According to the lower bound scenario, telemedicine becomes a greener choice at a distance of a few kilometers when the alternative is transport by car.

Conclusions: Telemedicine is a potent carbon reduction strategy in the health sector. But to contribute significantly to climate change mitigation, a paradigm shift might be required where telemedicine is regarded as an essential component of ordinary health care activities and not only considered to be a service to the few who lack access to care due to geography, isolation or other constraints.

Editor: Igor Linkov, US Army Engineer Research and Development Center, United States of America

Funding: This work was partly undertaken within the Umeå Centre for Global Health Research, with support from FAS, the Swedish Council for Working Life and Social Research (grant no. 2006-1512). The work was also supported in part by funding from the Swedish International Development Cooperation Agency (SIDA). The funders had no role in study design, data collection and analysis, decision to publish, or preparation of the manuscript.

Competing Interests: ClimAdapt, LLC is a sole proprietorship consulting company working on health and global change issues. Clients are predominantly international organizations, such as WHO and UNEP, and national institutions, such as Health Canada. Dr. Kristie Ebi conducts research on the impacts of and adaptation to climate change, including on extreme events, thermal stress, foodborne safety and security, and vectorborne diseases. Her work focuses on understanding sources of vulnerability and designing adaptation policies and measures to reduce the risks of climate change in a multi-stressor environment.

* Email: Asa.Holmner-Rocklov@vll.se

Introduction

The health care sector is facing a series of major challenges. In developed countries, the health care sector is committed to improving the provision of care, while at the same reducing costs. In developing countries, the health sector is facing the challenge of meeting the fundamental right of all citizens to adequate health according to the WHO and UN goals of universal health coverage and healthy life expectancy. Challenges range from aging populations with an increasing prevalence of lifestyle related disease, to changing disease patterns because of development patterns and global environmental changes [1,2]. At the same time, the health care sector contributes significantly to climate change [3,4], which means that if the sector does not introduce climate friendly policies and practices, it paradoxically will continue to contribute directly and indirectly to negative health impacts through its emissions of CO_2 and other greenhouse gases. To meet the growing demand of health care resources without furthering climate change, future health services must be built on sustainable and low-carbon systems and work models.

Information and communication technology (ICT) has been suggested as one solution for reducing the carbon footprint of many sectors, including the health care sector [5]. Telemedicine, which is defined as the use of ICT to provide health services across distance, time, or other barriers, has been suggested to be a potent tool to reduce emissions from travel that, according to the UK National Health Services (NHS), represent as much as 18% of the total carbon footprint of the UK health sector [4]. Telemedicine covers a broad range of technologies and activities that have been applied to many different clinical disciplines, such as radiology, pathology, dermatology, rehabilitation, and chronic disease management. Large-scale clinical studies are often lacking, but in general telemedicine studies show high patient satisfaction and acceptance [6,7], although evidence on cost effectiveness is still relatively weak [8]. Regarding clinical outcomes, some telemedicine applications, such as specialist rehabilitation using videoconferencing, have shown to be comparable to or even more efficient than traditional interventions [9]. In our experience, this can be explained by the increased access to specialists that makes it possible to commit to a more intense rehabilitation regime that would have been feasible if the patient had been required to travel to the clinic for each appointment.

Attempts have been made to evaluate telemedicine programs from a climate mitigation perspective by estimating the potential reduction in tailpipe emissions [10–14]. However, to disregard the impacts from the technology used to facilitate telemedicine can be seriously misleading and it is important to take into account the cost of manufacturing, using and discarding the equipment, to evaluate the actual impact of different virtual meeting solutions [15,16].

To assess the carbon costs and benefits of telemedicine, we performed a simplified life cycle inventory for telemedicine activities within two rehabilitation units at the Umeå University Hospital in Northern Sweden. These two units have provided evidence that rehabilitation using telemedicine, i.e., telerehabilitation, is a cost efficient and well-accepted alternative to traditional care in Northern Sweden (personal communication). Net emissions for the telemedicine appointments in these two cases were compared to care-as-usual scenarios constructed using the patient's place of residence. Thresholds for when the telemedicine work model becomes favorable were estimated and sensitivity analyses performed to provide insight into (1) what factors contribute most to carbon emission in the telemedicine work model, (2) what influence technological set-ups and different use scenarios have on net carbon emission and (3) what would be the impact of greener transport options on the carbon reduction potential. These are the first such assessments of telemedicine activities and have the potential to contribute important information for the development of telemedicine guidelines in sustainable health care practice.

Methods

Ethics statement

This study is based on data from two clinical units within the University Hospital of Umeå. The study is based on aggregated data and patient information used in the study was anonymized and de-identified prior to analysis. It was thus judged by the Regional Ethical Review Board in Umeå that no ethical approval was needed.

Data capture

To assess the carbon footprint of telerehabilitation at the University Hospital of Northern Sweden we analyzed data from two units offering specialist rehabilitation using telemedicine. These are the rehabilitation unit of the hand and plastic surgery section, which is part of the centre for reconstructive surgery, and the speech therapy clinic, which is a county clinic belonging to the ear, nose and throat clinic. The reason for including these two clinics in our study was twofold; telemedicine has become a well integrated part of the clinical activities in these units and there are extensive records of their telemedicine appointments available for analysis. Telemedicine appointments were compared with care-as-usual scenarios that require the patient travel to the hospital for a face-to-face visit. Patients enrolled in the telemedicine programs are primarily inhabitants of Västerbotten County, but the hospital also provides specialized health care to the northern care region (approximately the northern half of Sweden), as well as to some patients from other counties. Travel distances are estimated as the distance from the town closest to the patient's place of residence to Umeå where the university hospital is located. One-way distances vary from less than one km to 700 km although the hand rehabilitation unit only registers travel distances above 2.5 km (one way). Car is the transport option assessed in our scenarios as most patients, particularly in the speech therapy clinic, utilize car or subsidized taxi services to reach the hospital. Patients from other counties may travel by other means, such as airplane or bus, but this is not accounted for in our study. All calculations were performed using Microsoft Excel.

Hand- and plastic surgery section

The study included all patients involved in the telerehabilitation program from January to December 2012. Appointments included follow-ups, interventions, consultations, and assessments of various conditions, such as amputations of one or more fingers, osteoarthritis, flexor tendon injuries, radius fractures, finger fractures, and ligament injuries. There were 238 telemedicine appointments during this period that thus avoided travel to Umea. Of these, 81 were conducted in the patient's home using a PC or tablet computer and 157 at the closest primary health centre using standard videoconferencing equipment. Travel to appointments made at the primary health centre is not accounted for. Based on the patient's places of residence, an accumulated travel distance of 82,310 km was avoided during the study. Data on the exact length of individual appointments are lacking, but they typically varied from 10 to 50 minutes with an average of 25 minutes (verbal communication).

Speech therapy unit

The study included data for all patients involved in a telemedicine project conducted in 2005–2006 at the speech therapy unit. Today, telemedicine is well integrated into the clinical activities; however, readily accessible statistics on the patients' place of residence (used to calculate travel-related emissions for comparison) was only available from the time of the project. Telemedicine treatment was delivered to patients of all ages for conditions including aphasia, dysarthria, and dyslexia. 481 therapy sessions using telemedicine were performed either in the patients' home or at the closest primary health centre, although details about exact location were not available. Data on the length of individual appointments were not available but varied from 30 to 40 minutes, with an average of 35 minutes (verbal communication). Based on the patient's place of residence, an accumulated travel distance of 154,842 km was avoided during the study.

Life cycle assessment

A Life Cycle Assessment (LCA) is defined as a complete ecological assessment of all energy, material, and waste flows of a product, and their impact on the environment. This "cradle to grave" evaluation begins with the design of the product and ends with the disposal or recycling of material and components (end-of-life). Typically, LCAs are performed in several steps, starting with a life cycle inventory to determine the raw materials and energy used and the emissions that occur during the life cycle of a product. This inventory is followed by a life cycle impact assessment to estimate the impacts of these emissions and raw material depletions on, for example, human health or certain ecosystems. If the focus of the study is on a single waste product it can be specified as a stream lined LCA.

This study adopts the form of a simplified, stream lined life cycle inventory, with the aim to evaluate the most important aspects of telemedicine with respect to CO_2 emissions. We chose one hour as the functional unit in the telemedicine scenario. This is a typical meeting duration and is, in addition, a convenient entity to work with considering that the units W and kWh are applied in our calculations.

The study builds on results from life cycle inventories on hardware and software required to connect the patient and the specialist, called mediated meeting solutions or videoconferencing solutions. We largely adopted the strategy described in Ong et al. [17] that takes into account end-point devices, such as computers [18], monitors, cameras, local area network (LAN) components [19] and video codecs used to compress and decompress digital video signals, as well as the costs for Internet traffic [19], although technically, data is transmitted using the dedicated hospital network, Sjunet. The method was modified to fit our device set-ups (see Figure 1 and Table 1) and assumptions on life length and use rates of the equipment. To make the protocol more generic, life cycle and operating carbon costs are given in kWh to account for the carbon footprint of energy production itself, which may differ significantly between countries and different stages of the product life cycle.

The total carbon cost of a virtual meeting was estimated by adding i) the cumulative emissions generated by all equipment during the use-phase (from energy consumption) and ii) emissions generated during design, manufacturing, disposal and recycling of the equipment, whenever these data were available. To obtain an hourly carbon cost of a mediated meeting (functional unit), the cost of operating the equipment for one hour was added to the carbon emissions generated throughout all other life cycle phases amortized over the whole life length of the equipment. In energy terms this can be expressed as;

$$E_h = \sum_i (E_o + E_{emb}/T_{life}) \qquad (1)$$

where E_h is the total hourly energy cost in kWh (functional unit) for a videoconferencing set-up consisting of a specific combination of end-point devices or set of devices (i). E_o is the hourly energy consumption of the end-point device during operation, E_{emb} is the accumulated embodied energy for all other life cycle phases available (see Table 1) and T_{life} is the expected number of hours in use. A generalized conversion factor of 0.6 kCO_2e/kWh was used to convert these results to carbon dioxide equivalents, based on the methods developed by Ong and Malmodin [17,20], because the processes of designing, manufacturing, using, and disposing of the technology may differ across locations and countries. It is thus likely that we overestimate the carbon cost of the use phase because northern Sweden has access to environmentally friendly electricity in the form of hydropower and wind power. Life cycle data on computers, monitors, and other end-point devices are summarized in Table 1. Data on the Internet operating expenditures (opex) and embodied energy for different bandwidths are summarized in Table 2.

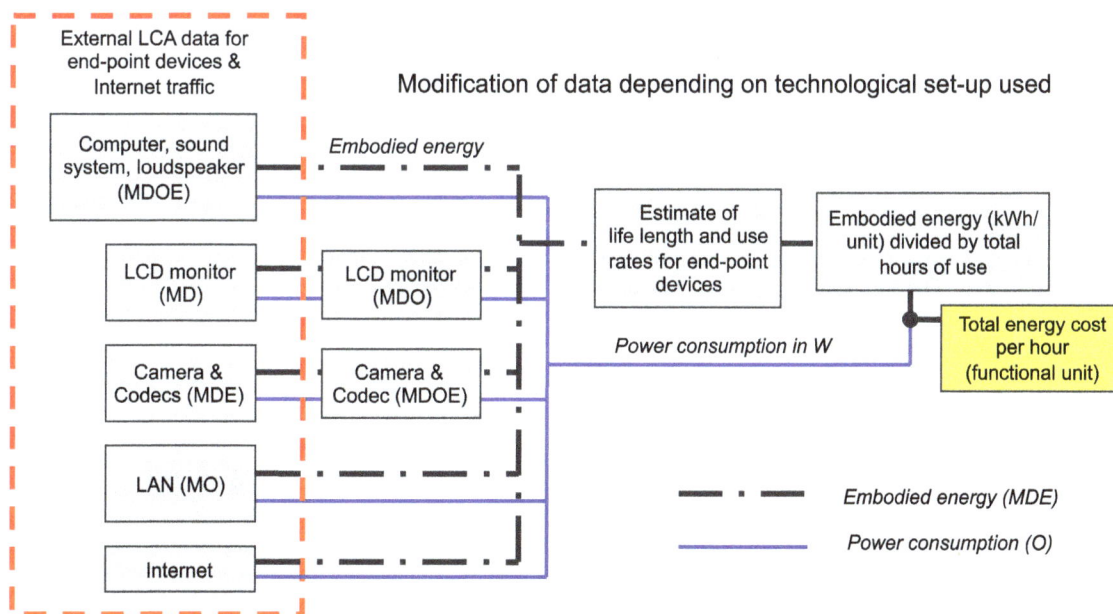

Figure 1. Methods summary. The study is based on the method used by Ong et al. that builds on existing LCA data for end-point devices used in videoconferencing and emission estimates for Internet traffic [17]. Completeness of the LCA data varies between devices, but includes energy costs and/or carbon emissions generated during manufacturing (M), distribution (D), operation (O) and end-of-life stages (E). Emissions data for MDE are provided in, or has been converted to, energy equivalents (kWh/unit) and is called embodied energy. Data for the videoconferencing solution (monitors, camera and video codecs) were modified to better fit our technological set-up. To obtain the hourly carbon cost of telerehabilitation in $kgCO_2e$, we divided the tembodied energy with estimates of the life length and use rates of all equipment, and applied a conversion factor of 0.6 kg CO_2e/kWh [17,20]. See also Equation 1.

Table 1. Life cycle and operating costs of end point devices.

	Power consumption (W)	Embodied energy (kWh/unit)^	LCA phases included*
PC			
Desktop	150	583	M D O E
Laptop	40	378	M D O E
Camera	9.5	33	M D O E
Sound system	4.1	104	M O
Microphone	2.5	52	M O
Monitor			
46" NEC LCD	188	145[□]	M D O
Videoconference			
Camera + codec (Cisco SX20)	40	134[#]	M D O E
Local Area Network (LAN) end-points	20	278	M O

* M = manufacturing, D = distribution, O = operation, E = end-of-life (disposal & recycle)
^In the original reference, some values were converted from CO_2e to MJ using a conversion factor of 0.6 kg CO_2e/kWh [17,20]. Data were further converted from MJ to kWh to better fit our calculations.
[#]Embodied energy of the Cisco SX20 camera and codec was estimated from the embodied energy of camera and entry level video codec used in Ong et al. [17].
[□]Data on the embodied energy of the NEC LCD monitor was based on an active screen size of 1018×572.7 mm and calculated from data provided by Ong et al. [17].

To enable comparison with emissions generated during a traditional physical meeting, we accounted for the life cycle carbon costs of travelling. The estimates were primarily based on the research by Lenzen et al. [21] that addresses tailpipe emissions as well as energy consumption and/or carbon emissions generated during other life cycle phases of transport, such as manufacturing of cars, fuels, and road infrastructure. According to Lenzen, the total carbon cost of a car is 0.86 kgCO_2e/km. To account for improved energy efficiency in more modern set-ups, we assessed more recent data by Leduc et al. [22] that accounts for life cycle phases related to automobiles and fuels, exclusively. This study generated a total of 0.25–0.27 kgCO_2e/km for the most commonly purchased petrol and diesel cars in Europe in 2010; we utilized the value of 0.26 kgCO_2e/km in our study. The EU emissions target of 0.130 kg CO_2 for automobiles in 2015 is also addressed in the sensitivity analysis.

System identification

Telemedicine appointments were conducted with different technological set-ups depending on location of the treatment; at

home or at the closest primary health centre. For home treatments, the patient and the specialist are both assumed to be using a standard desktop PC with additional web camera, microphone, and loud speakers. The patient's computer is assumed to be used for 780 hours in total before disposal, based on a use rate of 5 hours per week for three years. This is a very rough estimate because the users vary significantly in age and computer habit; therefore, use rates are further addressed in the sensitivity analysis. The PC of the specialist is used for 7,300 hours before disposal based on an average use rate of 5 hours per day and a life expectancy of four years (data from the investment database). The bandwidth is set to 512 kbps, which is considered to be an acceptable lower limit for treatment provided in the home.

For treatment provided in the primary health centre, the patient is assumed to be utilizing a Cisco TelePresence SX20 with a 46" LCD monitor from NEC (or equivalent) that is used for 360 hours before disposal, based on an average use of 60 hours per year in six years. This estimate is derived from the average use-rate of videoconferencing equipment in the county council, which is

Table 2. Estimates of the Internet opex and embodied energy.

	Maximum estimates *	
Bandwidth	**Networking energy opex (kWh)**	**Embodied energy cost (kW)**
20 Mbit/s	32	29.3
10 Mbit/s	16	14.6
4 Mbit/s	6.4	5.6
2 Mbit/s	3.2	2.9
1 Mbit/s	1.6	2.2
768 kbit/s	1.2	1.5
512 kbit/s	0.8	1.1

*For simplicity and to avoid underestimating the cost of Internet traffic, the network operating expenditure (opex) and embodied energy (in kWh) are based on the maximum estimate of the operating energy intensity (3.61 kWh/GB) and embodied energy (3.33 kWh/GB) of the Internet. To calculate the opex in kWh, the bandwidth was converted to bytes per hour (or MB/h).

120 hours per year, with the number decreased to better fit the rural primary health centers. The specialist is using the same Cisco system but the use rate is assumed to be 240 hours per year for six years on average, i.e., 1440 hours. This estimate is based on the average use rate of videoconferencing equipment, with the number increased to better fit the two clinics as they are more frequent users of telemedicine than the average department. The LAN is expected to be operating for 14,600 hours derived from 10 hours per day for four years. The data transfer rate is estimated as 4Mbps, slightly exceeding the bandwidth typically used for telerehabilitation. The power consumption and embodied energy of computers [18], monitors, and audio/video peripherals are based on literature values for equivalent products, including the LAN end-point devices.

Access to information on the technological set-ups used for individual appointments was only available for the hand and plastic surgery unit. Therefore, we estimated a realistic upper and lower emission limit based on the following scenarios. These scenarios were applied to both clinics.

Upper bound scenario: The upper bound calculations were based on the scenario described above for treatment in the primary health centre, but with the addition of a second 46" screen in the videoconference room of the specialist.

Lower bound scenario: The lower bound calculations were based on the scenario described above for treatments provided in the patients' home; two standard desktop PCs with use rates of 780 hours (patient) and 7,300 hours (specialist) and a bandwidth of 512 kbps. Regarding carbon costs of Internet traffic, the same assumptions apply as for the upper bound scenario to avoid underestimating the carbon costs. The patients home LAN is likely to be of significantly lower complexity than the hospital LAN and both power consumption and embodied energy costs are likely to be significantly lower. Nonetheless, we chose to apply the same estimates as the higher bound scenario, with the only exception that the use rate applied to the patient scenario is 780 hours, similar to the other end-point devices.

Results

Carbon cost comparisons

For the hand and plastic surgery clinic, the carbon cost of the 238 telemedicine appointments during 2012 was 602 $kgCO_2e$ based on our baseline assumptions. This corresponds to an average of 1.4–2.8% of the carbon costs of travelling to/from the clinic by car or subsidized taxi services, based on a total avoided travel distance of 82,310 km for study patients. Based on the upper and lower bound scenarios, a telerehabilitation visit in the hand and plastic surgery clinic generated 0.4–0.9% and 3.2–6.4% of the expected carbon costs for a traditional face-to-face appointment, respectively. Similar numbers were obtained in the speech therapy clinic. In summary, the telerehabilitation activities of the two clinics resulted in a cut in carbon emissions by 15–250 times for the telemedicine work model compared to traditional care. Data on the carbon costs of the different scenarios are summarized in Table 3 and in Figure 2.

Based on the upper and lower bound scenarios defined in this paper, a one hour telemedicine appointment was estimated to generate 1.86 and 8.43 $kgCO_2e$, respectively. Consequently, telerehabilitation is carbon cost-effective once there is a need for the patient to travel at least 3.6 km by car for a one-hour appointment using the Lenzen estimate [21] and 7.2 km based on the Leduc estimate [22]. Corresponding values for the upper bound videoconference scenario are 16 km and 32 km, respectively. In reality, appointments are often shorter than one hour.

For the care model described in this paper, these distances may well be reduced by 50%.

Sensitivity analysis

Sensitivity analyses were performed to gain a better understanding of how different choices in technological set-up and usage of videoconference equipment affect the magnitude of carbon emission for telerehabilitation. Sensitivity analyses were also performed to address LCA data on passenger transport. Results from the sensitivity analyses are summarized below and in Figure 3.

It is clear from the upper and lower bound scenarios that the choice of hardware affects net emission rates. Based on the upper and lower bound scenarios, a one-hour meeting emits 8.43 $kgCO_2e$ and 1.85 $kgCO_2e$, respectively. Thus, a standard desktop solution produces approximately 20% of the carbon equivalents of the Cisco SX20 based on our assumptions. The difference would be even more significant for dedicated desktop videoconferencing solutions, such as the Cisco EX60 that has a significantly lower power consumption in the use phase, or when compared to more complex tele-presence systems or multipart meeting solutions. Meeting duration is another of the dominant factors influencing the carbon emissions of telerehabilitation. A two-hour meeting would essentially produce twice the emission of a one-hour meeting if the use rates are kept constant and within reasonable limits. Another important factor is bandwidth given that data transfer is one of the most energy consuming processes of videoconferencing. For the upper bound scenario, data transfer contributes 87% of the total emissions and the corresponding emissions for the lower bound scenario are approximately 50%. Because the energy consumption of the Internet is calculated in kWh/GB, there is a linear increase in carbon emission with an increase in bandwidth, at least theoretically. Hence, for a one-hour meeting using the otherwise fixed lower bound set-up, the net carbon emissions would increase by more than 200% when increasing the bandwidth from 4 to 10 Mbps. A decrease to 1 Mbps would yield a corresponding 4-times reduction in energy consumption for data transfer and a 3-times reduction in total carbon emissions.

The use rates and expected life of the equipment also will influence the hourly energy and carbon cost of telerehabilitation given that the embodied energy is amortized over the lifetime of the equipment. The use rate of the specialist applied to the lower bound scenario is estimated at 5 hours per day for four years on average, i.e., 7,300 hours in total. A ten-time reduction in use rate would increase the net carbon emission by 20% for a one-hour meeting. However, if the use rates were to increase ten times the corresponding reduction in carbon emission would only be about 2%. Thus, the contribution from the embodied energy to the hourly emission is most influential on equipment used infrequently with the current lower bound estimates.

Finally, the Lenzen [21] reference used to assess the carbon emissions from patient transport is relatively old (1999) and using the newer reference [22] is unrealistic because all cars used are unlikely to be newer models. Moreover, this reference did not account for additional emissions generated by building and maintaining the road infrastructure, for example. Hence, sensitivity analysis was used to address the impact of lower tailpipe emissions and other life cycle contributions to emissions due to utilization of more modern automobiles. If we account for an annual 5% reduction of all life cycle phases of a private car based on an average emission of 0.86 kgCO2e/km according to Lenzen et al., the net emissions of all face-to-face meetings in the hand rehabilitation clinic in 2012 would be 22,950 $kgCO_2e$, which still

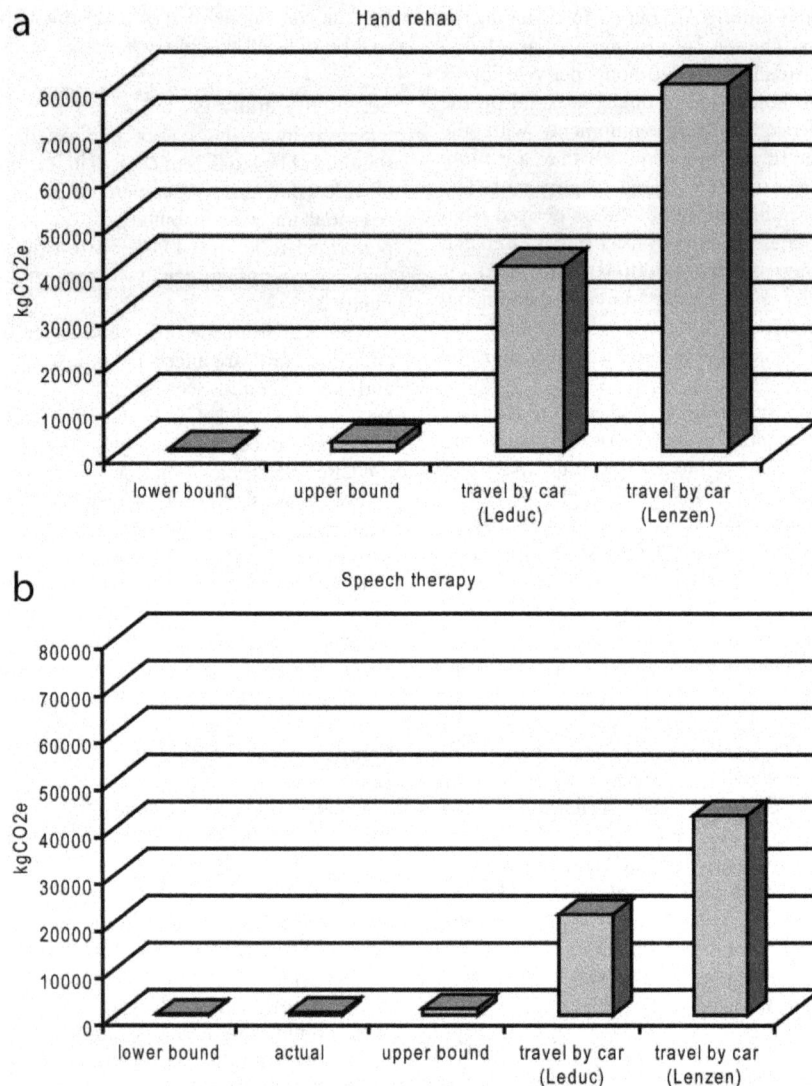

Figure 2. Summary of results. Net emissions for the two scenarios. Hand and plastic surgery section (a), and speech therapy unit (b). Information on the technological set-up used for individual appointments was only available in the hand and plastic surgery section (actual). Therefore, we applied the upper and lower bound scenarios in the speech therapy clinic.

corresponds to 38 times the emissions of the telerehabilitation scenario. If we perform the same calculations using the EU target for automobiles in 2015 of 0.130 $kgCO_2e$/km, emissions are reduced to 6,419 $kgCO_2e$, which are ten times the carbon emissions with respect to telerehabilitation in the hand rehabilitation clinic. These calculations do not take into account corresponding reductions in life cycle costs of the videoconferencing equipments.

Discussion

The carbon footprint of telemedicine services has been assessed by estimating the reduction in carbon emission due to reduced need for transportation [10–14]. However, the environmental impact of ICT is complex and to look at travel savings alone is misleading[5]. By performing a simplified LCA we accounted for the carbon cost of telemedicine and not just the potential to reduce tailpipe emissions. This is a new contribution to the scientific literature. Our analysis reveals that factors such as choice of teleconferencing solutions, duration of the appointment, capacity

of the Internet connection, and use rates of the technology contribute to emissions to various degrees. These results are highly policy relevant. The outcome stresses the benefits of using telemedicine for short meetings and implies that the choice of bandwidth should be based on the clinical need rather than the access to highest possible Internet capacity. At higher bandwidths, data transfer is the main contributor to emissions of telerehabilitation, reaching 87% of total emissions in our upper bound scenario. It is also clear that traditional videoconferencing solutions emit more carbon than desktop solutions per hour of usage, and that use rates should be considered, particularly for equipment used infrequently. Careful planning is thus needed on the local level to make the best use of a videoconference system, which could be a very poor investment if wrongly placed or when in low demand due to the rapid growth of high-quality desktop and mobile videoconferencing solutions.

Based on our results, the magnitude of the carbon reduction *per appointment* is extensive and clearly indicates that up-scaling the use of telemedicine could have a large impact on the over-all

Table 3. Accumulated life cycle carbon costs of telemedicine versus face-to-face meetings.

	Speech therapy (481 visits)		Hand rehabilitation (238 visits)	
	kWh	kgCO$_2$e*	kWh	kgCO$_2$e
Telemedicine				
Authentic conditions^			1004	602
Lower bound	741	409	305	183
Upper bound	3307	1984	818	1364
Face-to-face visit				
Travel by car (Lenzen)		79 909		42 472
Travel by car (Leduc)		40 258		21 400
Emissions with respect to car travel (Lenzen/Leduc)				
Authentic conditions			1.4%/2.8%	
Lower bound	0.6%/1,0%		0.4%/0.9%	
Upper bound	2.5%/4,9%		3.2%/6.4%	

*Based on a conversion factor of 0.6 CO2e/kWh
^Based on 157 connections to primary health centres using videoconferencing solutions and 81 connections to the patients home using desktop solutions. Such detailed data were not available in the speech therapy section.

carbon footprint of the health sector. The hand- and plastic surgery clinic reduced the carbon emissions per appointment by more than 70 times without having to make major financial investments. The yearly monetary cost of a videoconferencing equipment of this standard is approximately 1100 EUR or 1500 USD. Further, when taking into account trends towards greener transports, telerehabilitation is the most climate-smart work model based on our sensitivity analyses.

These results are from a rural perspective and it is reasonable to expect that the carbon reduction potential of this work model could be significantly smaller in urban environments where patients have shorter distances to travel to receive specialist rehabilitation. However, carbon emissions from a one-hour meeting using a desktop solution are exceeded by the emissions from a car driving as little as a few kilometers based on our baseline assumptions and car queues and traffic jams will significantly increase the emission rates by any type of motor vehicle [23]. Thus, telemedicine also might be promising in cities, particularly those in regions that suffer from poor air quality and experience the majority of all traffic accidents. This new way of thinking about telemedicine and virtual meetings in general could thus be very relevant for city planning and future megacities; reducing risks while saving time and being climate friendly.

When evaluating the potential of telemedicine as climate mitigation strategy, it is important to consider its full potential rather than only assessing the effects from single visits or individual telemedicine programs. Traditional telemedicine activities reduce the carbon costs by only a small fraction in relation to the over all burden of travel, according to a study in the Grampaign region of UK [13]. In this study, telemedicine was estimated to reduce net travel by as little as 0.1%. This brings us to the question whether telemedicine should only be used to serve the needs of those who lack access to health care by virtue of geography, isolation, or other constraints, or if telemedicine can be accepted as an essential component of any ordinary health care activities? For the sake of the environment, we strongly support the latter. To gain true insight into the potential impact of telemedicine on health systems and environment, we anticipate that larger scale studies are needed from the viewpoint of telemedicine as a fully integrated part of any medical or health care activity. Only when becoming

part of mainstream medical care and health care can telemedicine reach its full potential as climate mitigation strategy. Some steps in this direction have been taken by the Västerbotten County, which is a geographically widespread and sparsely populated County (population of 260,000) in northern Sweden. Virtual meetings have been highly prioritized for more than a decade and telemedicine is already an integrated part of many clinical units, supporting activities as diverse as surgical planning and follow-ups, specialist rehabilitation, tele-pathology, radiation treatment planning, collaborative care planning and chronic disease management. Consequently, the number of logged videoconferencing hours increased on average 30% per year for the past five years, reaching almost 20,000 hours in 2012 according to the county council statistics. For sake of debate, this corresponds to a reduction of several tons of CO_2e had all these meetings been applied to clientele and activities similar to the ones addressed in our study. When applying this work model in new clinical contexts it is, however, crucial to take into account possible trade-offs between environmental benefits and clinical outcomes. Intuitively, a clinician is unlikely to adapt a more environmentally friendly solution when there is the potential for any negative impact on the patient or on the quality of services. However, intuitive choices might not accurately reflect actual outcomes. Further, promoting sustainability means the adjustments to current practice should be part of standard evaluation. To support this complex decision process, we suggest continuing this work by performing life cycle impact assessments to estimate the impact on climate change and potential health co-benefits from reductions in CO_2 emissions.

The method applied in this paper has its limitations; there are very few studies available on this topic and the LCA data on videoconferencing peripherals and transport are sometimes rough estimates and not always completely up to date. It is thus important to keep in mind that the aim of our study was to describe the central aspects of emissions generated by the system under investigation to guide future studies and policy development

There are a lot of global initiatives focusing on reducing the impacts from travel, for example by implementing regulations to reduce tailpipe greenhouse gas emissions, improving vehicle technology, and introducing lower-carbon fuels. At the same time, the rapid development of the ICT sector is leading to more

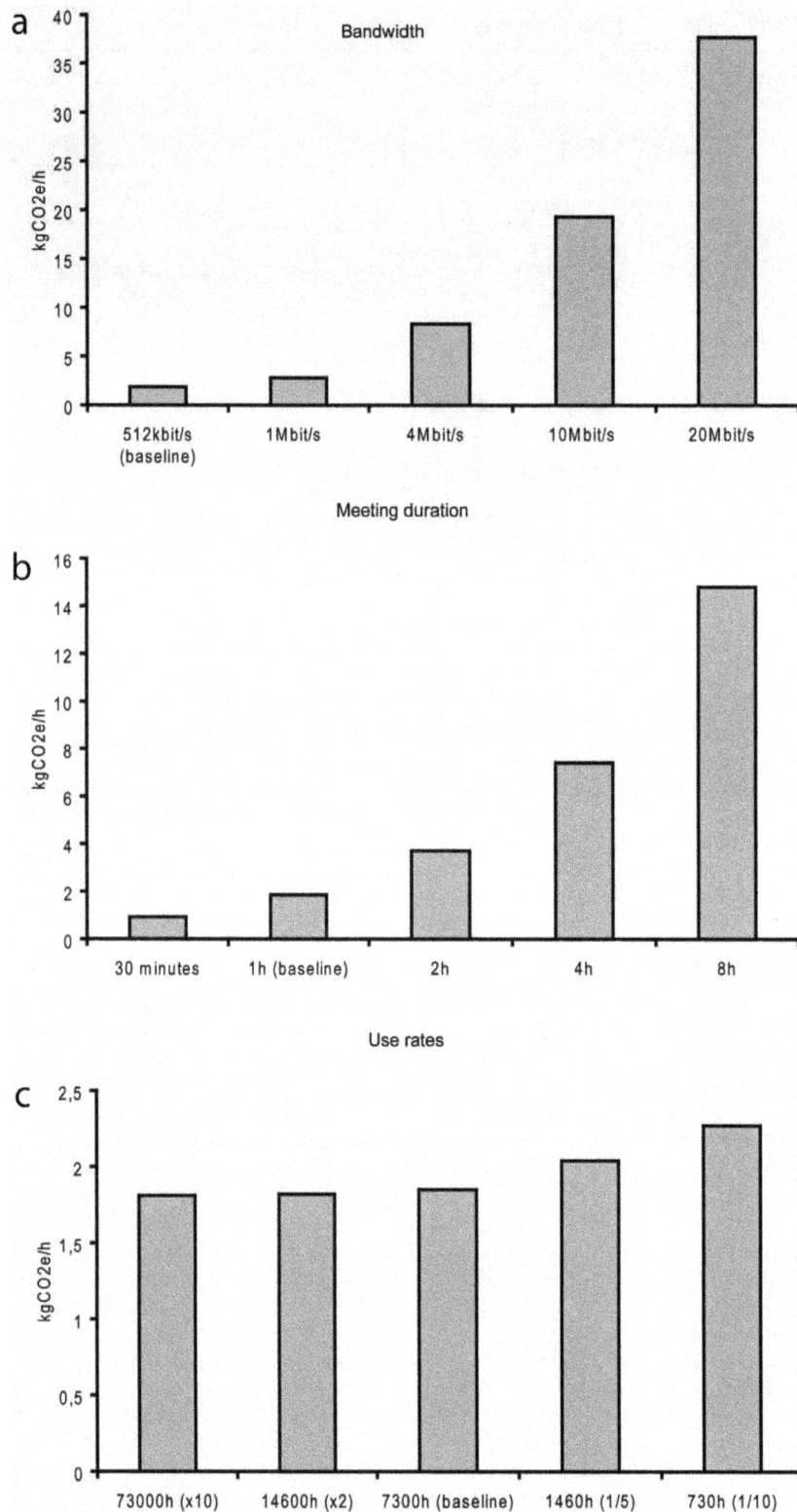

Figure 3. Sensitivity analyses. The analyses are based on the lower bound emissions scenario that emits 1.85 kgCO$_2$e during a one hour meeting. All other factors were kept constant while changing the bandwidth (a), meeting duration (b) or use rates (c), respectively.

integrated and energy efficient local area networks and less energy demanding PCs and videoconferencing equipments. In contrast to this positive development, the automobile industry is growing steadily, and there are tendencies towards people investing in larger screen sizes and more advanced videoconference systems. These initiatives have the potential to significantly shift the anticipated net reduction in emissions from choosing virtual meeting solutions over face-to-face visits, in either direction. A way to address this in future studies is to adapt methods for calculating time-corrected life cycle emissions intensity for all scenarios by taking into account future development in all relevant sectors, such as ICT and transport [24,25].

It is a common perception that telemedicine can cut costs, extend health services to remote areas, maintain or improve clinical outcomes, and save time for patients and health workers. Some scientific evidence challenges these perceptions, particularly with respect to cost efficiency [8] but for many telemedicine services, these expectations are unquestionably met [9,26]. Nonetheless, there is a striking delay in the up scaling and implementation of successful telemedicine work models [27,28]. There is a lack of studies investigating the impacts of such work models on routine care [29] and a lack of clear organization and economic policies and guidelines [30,31]; such policies require empirical evidence and dissemination of results to relevant stakeholders and decision makers. It is our hope that the empirical evidence of the environmental benefits of telemedicine generates additional momentum and accelerates the implementation of successful telemedicine systems and work models in health systems globally, for the benefit of patients, health care providers, and the planet.

Conclusions

This study shows that telemedicine generates far fewer carbon emissions than do traditional care models where patients and health workers are expected to travel by car to appointments. Telemedicine is thus judged to be a potent climate change mitigation strategy, not just in rural areas but also in urban environments where additional co-benefits might be even greater if few people use public or active transport. Although previous research indicates that current telemedicine programs reduce travel-related emissions to a limited extent [14], we conclude that implementing telemedicine more broadly could make a significant contribution to reducing the greenhouse gas emissions driving climate change. This requires a paradigm shift in health care and proactive efforts from health care decision makers. This also calls for empirical evidence on the clinical, economical and environmental benefits of telemedicine. Research is needed on a larger scale to evaluate the current and future impact of different telemedicine solutions on carbon emissions, from the viewpoint of telemedicine as a well-accepted and fully integrated part of any health care activity.

Acknowledgments

The authors wish to thank Agneta Carlsson and Käte Alrutz for supplying the data and for generous support throughout the study.

Author Contributions

Conceived and designed the experiments: ÅH. Performed the experiments: ÅH. Analyzed the data: ÅH KE LL MN. Wrote the paper: ÅH KE LL MN.

References

1. Woodward A, Smith KR, Campbell-Lendrum D, Chadee DD, Honda Y, et al. (2014) Climate change and health: on the latest IPCC report. Lancet 383: 1185–1189.
2. McMichael AJ (2013) Globalization, Climate Change, and Human Health. New England Journal of Medicine 368: 1335–1343.
3. Pencheon D, Rissel CE, Hadfield G, Madden DL (2009) Health sector leadership in mitigating climate change: experience from the UK and NSW. N S W Public Health Bull 20: 173–176.
4. NHS (2009) Saving Carbon, Improving Health, NHS carbon reduction strategy for England. Available: http://www.sduhealth.org.uk/documents/publications/1237308334_qylG_saving_carbon,_improving_health_nhs_carbon_reducti.pdf. Accessed 2013 September 29.
5. Holmner A, Rocklov J, Ng N, Nilsson M (2012) Climate change and eHealth: a promising strategy for health sector mitigation and adaptation. Glob Health Action 5.
6. Johansson T, Wild C (2011) Telerehabilitation in stroke care - a systematic review. Journal of Telemedicine and Telecare 17: 1–6.
7. Rozenblum R, Donze J, Hockey PM, Guzdar E, Labuzetta MA, et al. (2013) The impact of medical informatics on patient satisfaction: A USA-based literature review. International Journal of Medical Informatics 82: 141–158.
8. Mistry H (2012) Systematic review of studies of the cost-effectiveness of telemedicine and telecare. Changes in the economic evidence over twenty years. Journal of Telemedicine and Telecare 18: 1–6.
9. Kairy D, Lehoux P, Vincent C, Visintin M (2009) A systematic review of clinical outcomes, clinical process, healthcare utilization and costs associated with telerehabilitation. Disability and Rehabilitation 31: 427–447.
10. Yellowlees PM, Chorba K, Parish MB, Wynn-Jones H, Nafiz N (2010) Telemedicine Can Make Healthcare Greener. Telemedicine Journal and E-Health 16: 230–233.
11. Connor A, Lillywhite R, Cooke MW (2011) The carbon footprints of home and in-center maintenance hemodialysis in the United Kingdom. Hemodialysis International 15: 39–51.
12. Lewis D, Tranter G, Axford AT (2009) Use of videoconferencing in Wales to reduce carbon dioxide emissions, travel costs and time. Journal of Telemedicine and Telecare 15: 137–138.
13. Wootton R, Tait A, Croft A (2010) Environmental aspects of health care in the Grampian NHS region and the place of telehealth. Journal of Telemedicine and Telecare 16: 215–220.
14. Wootton R, Bahaadinbeigy K, Hailey D (2011) Estimating travel reduction associated with the use of telemedicine by patients and healthcare professionals: proposal for quantitative synthesis in a systematic review. Bmc Health Services Research 11: 185.
15. Borggren C, Moberg A, Rasanen M, Finnveden G (2013) Business meetings at a distance - decreasing greenhouse gas emissions and cumulative energy demand? Journal of Cleaner Production 41: 126–139.
16. Arnfalk P, Kogg B (2003) Service transformation - managing a shift from business travel to virtual meetings. Journal of Cleaner Production 11: 859–872.
17. Ong D, Moors T, Sivaraman V (2012) Complete life-cycle assessment of the energy/CO2 costs of videoconferencing vs face-to-face meetings. 2012 Ieee Online Conference on Green Communications, p50–55.
18. Fujitsu (2011) White paper - Life Cycle Assessment and Product Carbon Footprint - Fujitsu ESPRIMO E9900 Desktop PC. Available: http://globalsp.ts.fujitsu.com/dmsp/Publications/public/wp-LCA-PCF-ESPRIMO-E9900.pdf. Accessed 2013 September 25.
19. Raghavan B, Ma J (2011) The energy and emergy of the internet. Proceedings of the 10th ACM Workshop on Hot Topics in Networks. Cambridge, Massachusetts: ACM. 1–6.
20. Malmodin J, Moberg A, Lunden D, Finnveden G, Lovehagen N (2010) Greenhouse Gas Emissions and Operational Electricity Use in the ICT and Entertainment & Media Sectors. Journal of Industrial Ecology 14: 770–779.
21. Lenzen M (1999) Total requirements of energy and greenhouse gases for Australian transport. Transportation Research Part D-Transport and Environment 4: 265–290.
22. Leduc G, Mongelli I, Uihlein A, Nemry F (2010) How can our cars become less polluting? An assessment of the environmental improvement potential of cars. Transport Policy 17: 409–419.
23. Barkenbus JN (2010) Eco-driving: An overlooked climate change initiative. Energy Policy 38: 762–769.
24. Kendall A (2012) Time-adjusted global warming potentials for LCA and carbon footprints. International Journal of Life Cycle Assessment 17: 1042–1049.
25. Kendall A, Price L (2012) Incorporating Time-Corrected Life Cycle Greenhouse Gas Emissions in Vehicle Regulations. Environmental Science & Technology 46: 2557–2563.
26. Hilty DM, Ferrer DC, Parish MB, Johnston B, Callahan EJ, et al. (2013) The Effectiveness of Telemental Health: A 2013 Review. Telemedicine and E-Health 19: 444–454.
27. Zapka J, Simpson K, Hiott L, Langston L, Fakhry S, et al. (2013) A mixed methods descriptive investigation of readiness to change in rural hospitals participating in a tele-critical care intervention. Bmc Health Services Research 13:33.
28. Christensen MC, Remler D (2009) Information and Communications Technology in US Health Care: Why Is Adoption So Slow and Is Slower Better? Journal of Health Politics Policy and Law 34: 1011–1034.

29. Schmidt S, Grimm A (2009) Health service research of telemedicine applications. Bundesgesundheitsblatt-Gesundheitsforschung-Gesundheitsschutz 52: 270–278.

30. Koch S (2006) Home telehealth - Current state and future trends. International Journal of Medical Informatics 75: 565–576.

31. Broens TH, Huis in't Veld RM, Vollenbroek-Hutten MM, Hermens HJ, van Halteren AT, et al. (2007) Determinants of successful telemedicine implementations: a literature study. J Telemed Telecare 13: 303–309.

Combining XCO$_2$ Measurements Derived from SCIAMACHY and GOSAT for Potentially Generating Global CO$_2$ Maps with High Spatiotemporal Resolution

Tianxing Wang*, Jiancheng Shi, Yingying Jing, Tianjie Zhao, Dabin Ji, Chuan Xiong

State Key Laboratory of Remote Sensing Science, Institute of Remote Sensing and Digital Earth, Chinese Academy of Sciences. Beijing, China

Abstract

Global warming induced by atmospheric CO$_2$ has attracted increasing attention of researchers all over the world. Although space-based technology provides the ability to map atmospheric CO$_2$ globally, the number of valid CO$_2$ measurements is generally limited for certain instruments owing to the presence of clouds, which in turn constrain the studies of global CO$_2$ sources and sinks. Thus, it is a potentially promising work to combine the currently available CO$_2$ measurements. In this study, a strategy for fusing SCIAMACHY and GOSAT CO$_2$ measurements is proposed by fully considering the CO$_2$ global bias, averaging kernel, and spatiotemporal variations as well as the CO$_2$ retrieval errors. Based on this method, a global CO$_2$ map with certain UTC time can also be generated by employing the pattern of the CO$_2$ daily cycle reflected by Carbon Tracker (CT) data. The results reveal that relative to GOSAT, the global spatial coverage of the combined CO$_2$ map increased by 41.3% and 47.7% on a daily and monthly scale, respectively, and even higher when compared with that relative to SCIAMACHY. The findings in this paper prove the effectiveness of the combination method in supporting the generation of global full-coverage XCO$_2$ maps with higher temporal and spatial sampling by jointly using these two space-based XCO$_2$ datasets.

Editor: Juan A. Añel, University of Oxford, United Kingdom

Funding: The work described in this paper has been jointly supported by project of "Climate Change: Carbon Budget and Related Issues" (Grant nr. XDA05040402) from Chinese Academy of Sciences (CAS), the CAS/SAFEA International Partnership Program for Creative Research Teams (Grant nr. KZZD-EW-TZ-09) and National Natural Science foundation of China (Grant nr. 41301177). The funders had no role in study design, data collection and analysis, decision to publish, or preparation of the manuscript.

Competing Interests: The authors have declared that no competing interests exist.

* Email: wangtx@radi.ac.cn

Introduction

In recent years, global warming caused by emission of CO$_2$ has attracted considerable attention from the public. During the past decade, although tremendous efforts have been made toward improving the understandings of the mechanism between CO$_2$ increase in the atmosphere and global warming, some uncertainties still exist in the spatiotemporal characteristics of CO$_2$ sinks/sources on regional and global scales due to the lack of high-density measurements of such variables with good accuracy [1,2]. To date, the estimates of CO$_2$ flux from inverse methods rely mainly on ground-based measurements [3,4]. Although providing highly accurate atmospheric CO$_2$ records, the traditional ground-based networks intrinsically suffer from sparse spatial coverage [2,5]. Satellite-based measurements with various spatial and temporal resolutions provide a unique opportunity to accurately map atmospheric CO$_2$ in both daytime and nighttime over large areas, thus having the potential to bridge this gap. As a result, various satellite-based platforms have been equipped in recent years for deriving the CO$_2$ concentrations.

Generally, methods for retrieving CO$_2$ from space can be grouped into two categories: (1) inferring CO$_2$ concentrations by measuring shortwave infrared (SWIR) reflected solar radiation around 1.6 and 2.0 μm with sufficient spectral resolution. This includes the Greenhouse gases Observing SATellite (GOSAT), operating since 2009 [6], the Scanning Imaging Absorption spectrometer for Atmospheric CartograpHY (SCIAMACHY), in orbit since 2002 [7], and the second Orbiting Carbon Observatory (OCO-2), which, as a rebuild of OCO [8,9], is planned to be launched in July 2014. In addition, CarbonSat will also be scheduled to be launched in 2018 (http://www.iup.uni-bremen.de/carbonsat/). These measurements have a nearly uniform sensitivity to CO$_2$ from the surface up through the middle troposphere, and thus are frequently used to derive the column-average dry air mole fraction of atmosphere CO$_2$ (XCO$_2$) during the daytime; (2) retrieving CO$_2$ concentrations by interpreting the recorded spectra of the Earth-atmosphere system in thermal infrared (TIR) bands (around 15 μm). Instruments that work in such a way include AIRS [10,11], IASI [12,13], and FTS (Band 4) of GOSAT [6]. These measurements bring the advantage that they can detect CO$_2$ during both day and night time, while the lack of sensitivity in the lower troposphere makes them inappropriate to estimate CO$_2$ near the surface where the largest signals of CO$_2$ sources and sinks occur [1]. The complementarities of these platforms allow us to combine the SWIR and TIR measurements for obtaining enhanced understanding of CO$_2$ spatiotemporal variations globally. Since XCO$_2$ is much less affected by vertical transport of CO$_2$, it is particularly useful for investigation of CO$_2$ sources and sinks using inversion modeling [14,15]. On the other hand, the spatial and temporal variations in XCO$_2$ are even smaller than that in the surface CO_2; therefore, unprecedented measurement precision and accuracy are highly required for such column measurements [16–19]. SCIAMACHY (operation stopped in April 2012) and GOSAT are two typical instruments that can be used to derive XCO_2 from space, and a variety of retrieval algorithms have been developed for SCIAMACHY [1,20–27] and GOSAT [2,4,5,28–30] with eyes on improving XCO_2 retrieval accuracy to a great extent. At present, a number of XCO_2 products have been released. These will definitely enhance our understanding of the global carbon cycle.

Unfortunately, almost all typical instruments currently used to derive atmospheric CO_2 concentration are working in the infrared spectral range (less than 16 µm). Thus, except for the instrument's observation mode (for example, GOSAT observes in lattice points), the spatial coverage of the derived CO_2 is severely restricted by the presence of clouds. In addition, the lower signal-to-noise level over ice/snow covered surfaces and ocean for SWIR instruments (e.g., SCIAMACHY) also contributes to the CO_2 sparse coverage. For instance, it has been pointed out that only about 10% of GOSAT data can be used for retrieval of XCO_2 due to the cloud contaminations [4]. The amount of CO_2 measurements will be even smaller if additional screening criteria such as quality of spectral fit, aerosol loadings, etc. are further applied. Although the amount of remaining CO_2 measurements from certain space-based instruments may largely surpass that of ground-based sites, it is still not sufficient enough for accurately quantifying the spatiotemporal distribution of CO_2 over the global scale. As a result, it is greatly desired to jointly use these available CO_2 measurements derived from various space-based data. Recently, a novel method has been proposed for combining CO_2 values from seven different algorithms, and a new Level-2 CO_2 database (EMMA) from one algorithm is composed according to the median of monthly average of seven CO_2 products in each $10° \times 10°$ latitude/longitude grid box [31]. In fact, this method cannot increase the number of CO_2 observations but chooses a product with moderate oscillation among the available products. Despite the usefulness of the XCO_2 measurements (Level 2) in their own right, further spatiotemporal analysis for interpreting their scientific merit is essentially necessary due to the retrieval uncertainties and sparse coverage of such Level-2 observations [32]. For this point, many works have attempted to generate global full-coverage (i.e., Level 3) maps from XCO_2 values derived from single satellite observations using a geospatial statistics approach [32–34]. However, as reflected in these studies (for instance, Fig. 1 in the work of [33]), a compromise has to be made between the interpolated accuracy and the spatiotemporal resolution of Level-3 product because of the limited amount of Level-2 XCO_2 observations being used. For this point, instead of using Level-2 XCO_2 from a single dataset (e.g., GOSAT or OCO-2) as performed in the existing literature, we attempt to explore the potential of combining two CO_2 datasets (GOSAT and SCIAMACHY) in assisting in global Level-3 generation, aiming to: (1) propose a general strategy for combining (fusing) various CO_2 datasets with different instruments, algorithms, averaging kernels, etc.; and 2) increase the number of daily CO_2 points (utilized in Level-3 map interpolations) through the combination of two datasets, so that potentially improved Level-3 maps with higher accuracy and shorter time scale can be generated. The better the interpretation of the satellite-based CO_2 observations one can make, the higher the resolution (both temporal and spatial) of the generated global CO_2 maps.

Datasets

For GOSAT, the Fourier transform spectrometer (FTS) on GOSAT is the fundamental unit to retrieve atmospheric CO_2 and CH_4. It observes sunlight reflected from the earth's surface, and light emitted from the atmosphere and the surface. It is composed of three narrow bands in the SWIR region (0.76, 1.6, and 2.0 µm) and a wide TIR band (5.5–14.3 µm) at a spectral and spatial resolution of 0.2 cm^{-1} and 10.5 km, respectively [35]. Specifically, four CO_2 products from GOSAT have currently been released to the public: University of Leicester product [9,36], the RemoTeC product [28], NIES GOSAT product [35] and the product generated by NASA's Atmospheric CO_2 Observations from Space (ACOS) team (hereafter called ACOS product) [2,30]. The difference between some of the above mentioned products with various versions have been investigated in a recent study [37]. In the present paper, the ACOS product of 2009–2010 with version v2.9 has been employed.

SCIAMACHY was successfully launched on board Environmental Satellite (ENVISAT) in 2002 (unfortunately ceased in April 2012), which is a detector elements satellite spectrometer covering the spectral range 0.24–2.38 µm with a moderate spectral resolution of about 0.2–1.6 nm, and spatial resolution at nadir of 60×30 km [7]. It has eight spectral channels, with 1024 individual detector diodes for each band, observing the spectral regions 0.24–1.75 µm (band 1–6), 1.94–2.04 µm (band 7), and 2.26–2.38 µm (band 8) simultaneously in nadir and limb and solar and lunar occultation viewing geometries [22]. As mentioned in Section 1, till today, a number of CO_2-retrieval algorithms have been developed for SCIAMACHY. The IUP/IFE of University of Bremen has released two XCO_2 products, i.e., WFM-DOAS product [21,22] and the Bremen Optimal Estimation DOAS (BESD) product [1,26]. In this study, the BESD product with the versions of v02.00.08 for 2009–2010 is used.

In addition, CO_2 profiles of CT [38] are also collected here to allow the data mentioned above to be properly fused. CT is a NOAA data assimilation system, which provides the 3D profiles of CO_2 mole fractions in the atmosphere over the globe. For this study, CT data with version CT2011 is collected. This dataset provides global CO_2 profiles with $3° \times 2°$ latitude/longitude grid and 3 hours temporal resolution (a total 8 times from 01 to 22 in UTC) spanning the time period from January 2000 to December 2010. The CT dataset is used here mainly to assist in adjusting and time-shifting of the two CO_2 products being combined.

Methodologies

For combining the different space-based CO_2 measurements, three steps are adapted in this study. First, taking the global ground measurements of CO_2 as reference, remove the bias of the individual CO_2 retrievals for ensuring the accuracy of the fused CO_2 product; then make some adjustment for both the ACOS and BESD products, so that they can be physically comparable and thus combined; finally fuse the ACOS and BESD CO_2 products considering their retrieval uncertainties, spatial scales, differences in averaging kernels and overpass times, etc.

3.1 Global bias corrections

Removal of any global bias of the retrieved CO_2 when compared with the ground *in situ* measurements is essential before performing joint use. Many researches [4,25] frequently pointed out that CO_2 retrievals from GOSAT are low biased with different levels due to the uncertainties in pressure, radiometric calibration, line shape model, cloud and aerosol scattering, etc.

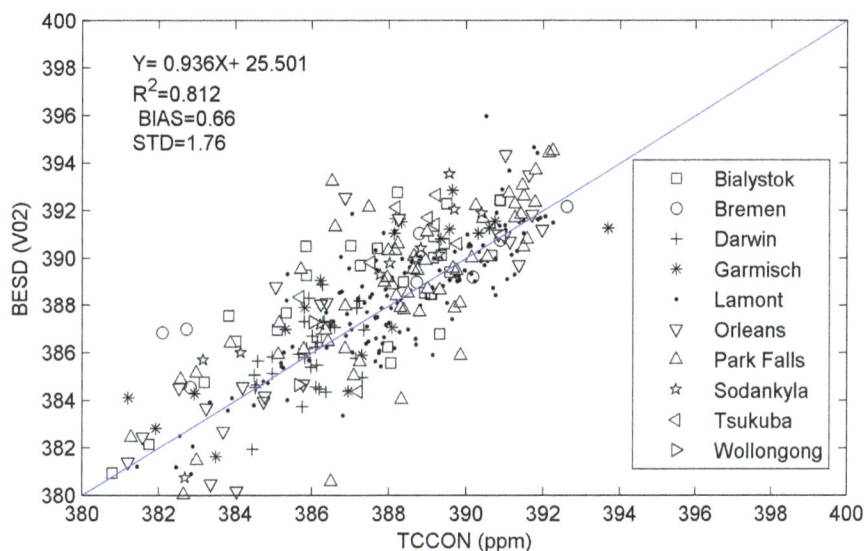

Figure 1. Validation of the BESD products against *in situ* TCCON CO$_2$ measurements over globe for 2009–2010.

Fortunately, a recent study has proposed a method for evaluating systematic errors in CO$_2$ and showed that the new version of ACOS product (v2.9) has a low global bias (<0.5 ppm) [39]. Thus, there is no global bias correction for the ACOS product being conducted here, but only the ACOS retrievals that pass the filter of table B1 in the work of [38] and marked as "good" in the quality flag are used. For the BESD product, we select Total Carbon Column Observing Network (TCCON) [15] measurements for 2009–2010 as the ground truth to determine its global bias. Specifically, BESD retrievals within ±2.5° and ±2.5° latitude/longitude box centered at each TCCON site and the mean FTS value (within ±1 h time window of satellite overpass time) are

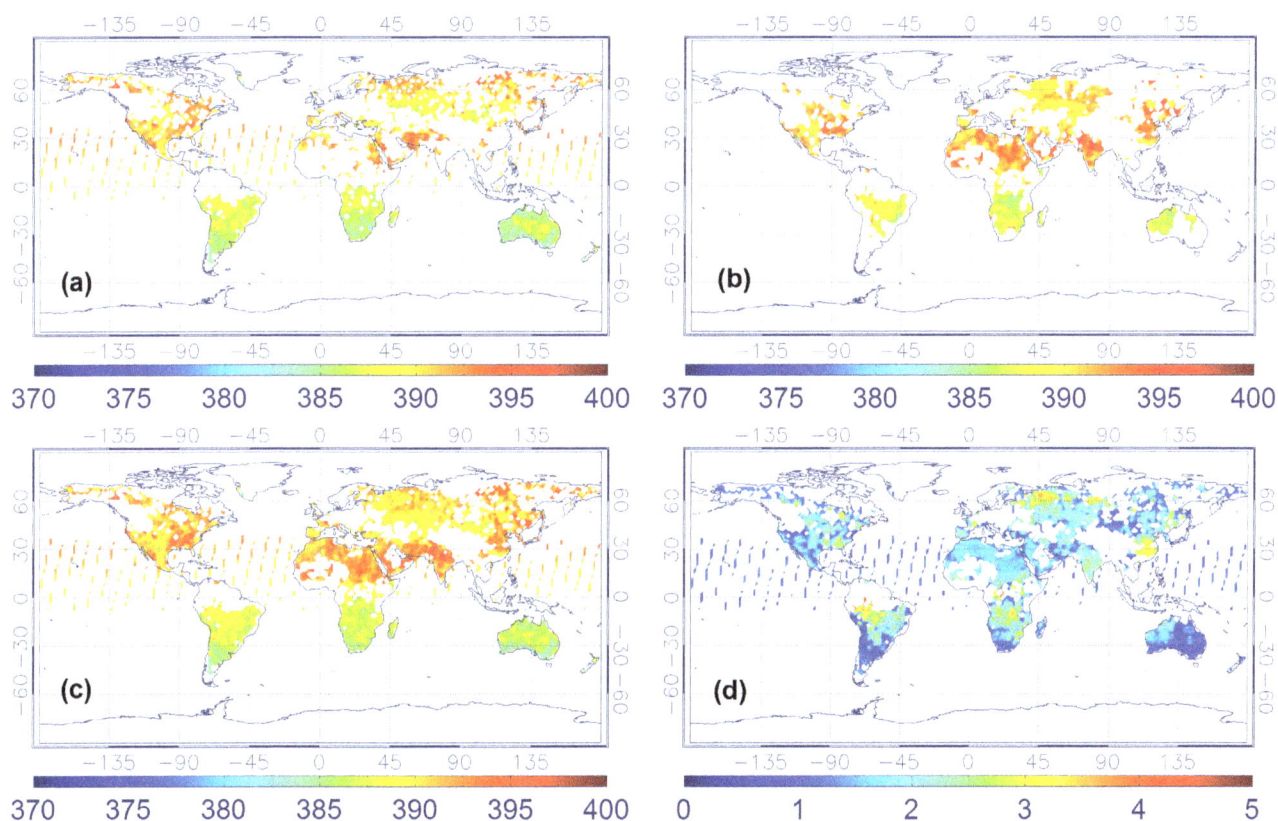

Figure 2. XCO$_2$ monthly mean maps in May of 2010. ((a) ACOS XCO$_2$, (b) BESD XCO$_2$, (c) combined product, and (d) XCO$_2$ uncertainties of the combined product).

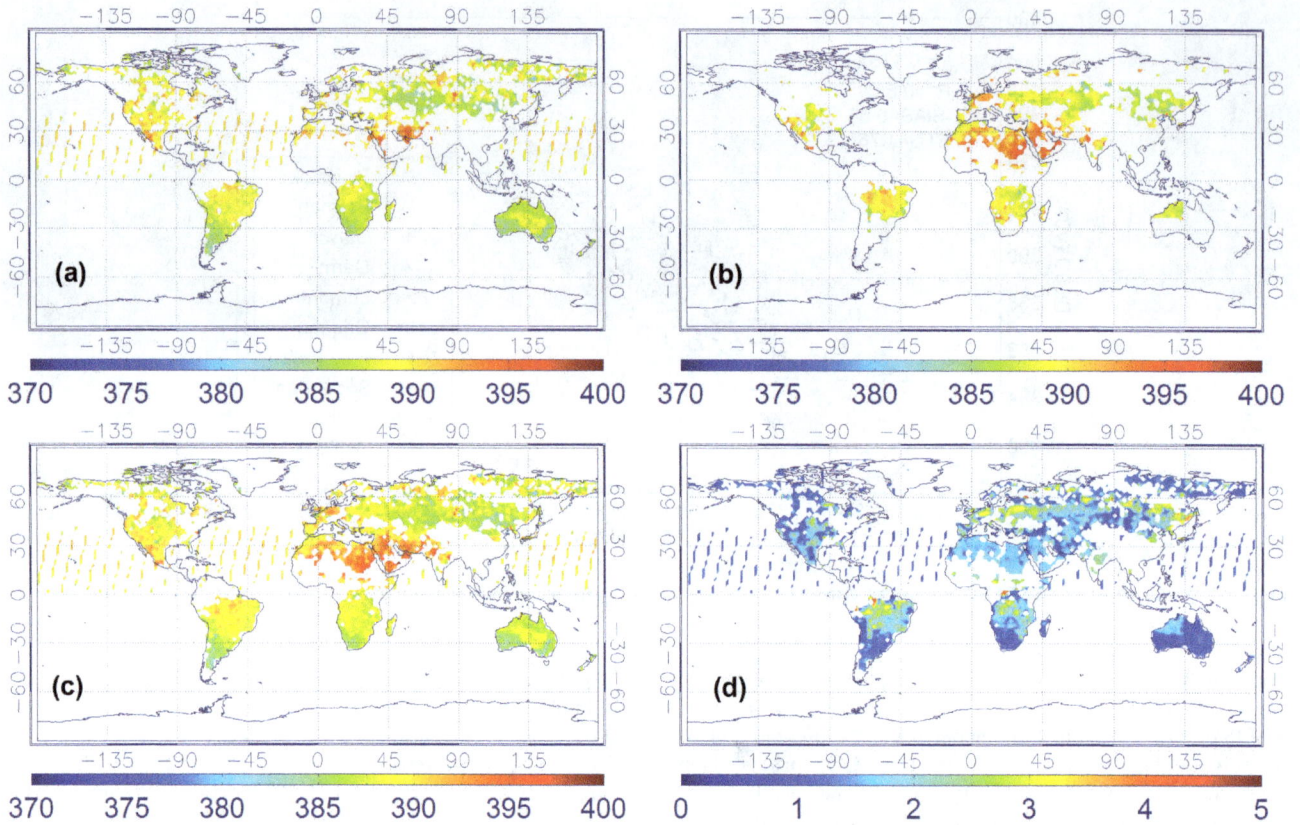

Figure 3. XCO$_2$ monthly mean maps in June of 2010. ((a) ACOS XCO$_2$, (b) BESD XCO$_2$, (c) combined product, and (d) XCO$_2$ uncertainties of the combined product).

extracted and compared (totally ten TCCON sites are utilized). The coincidence criteria mentioned above ultimately yield a total of 338 pairs of CO$_2$ measurements. The comparison result is shown in Fig. 1.

3.2 Retrieval adjustments

As pointed out by most researchers, it is not reasonable to directly compare or use two XCO$_2$ measurements. A suitable way to do that is to take the a priori profiles and variations in averaging kernel into account during the comparison [26,40]. To tackle the a priori issue, after correcting their global biases, both BESD and ACOS products are adjusted for a common a priori profile, which we assume to be the CT profile interpolated at the middle of the two overpass times (Equation (1)). Specifically, the a priori CO$_2$ profile of both the ACOS and BESD are first interpolated or extrapolated to the level of the CT CO$_2$ profile according to their pressure layers. After interpolation, the a priori profiles for both ACOS and BESD have the same dimension as the CT profile. Here the reason we take the CT profile at the middle of the two overpass times is that the time difference for GOSAT (1:00 pm) and SCIAMACHY (10:00 am) is relative large (3 hours), if we take one satellite time as reference, the induced error would be large for the other satellite measurements considering the CO$_2$ natural diurnal variation. So a middle time between these two satellite overpass times is selected for minimizing the CO$_2$ uncertainties during the adjustment.

$$XCO_2\_adj = XCO_2\_ret + (h^T I - a)(xCT - xa) \qquad (1)$$

Here, XCO_2_adj is the adjusted XCO$_2$ for ACOS or BESD; XCO_2_ret corresponds to retrieved XCO$_2$ of ACOS or BESD; a is the column-averaging kernel (row vector) of ACOS or BESD; h is pressure-weighting function (column vector); I is an identity matrix; xCT and xa (column vectors) are the common CT CO$_2$ profile and the corresponding a priori CO$_2$ profile for ACOS or BESD, respectively.

While it is not trivial to accurately consider the smoothing error without an estimate of the true atmospheric variability which is generally not readily available for most cases [39]. Fortunately, some works revealed that the smoothing error is generally small [26,39]. Consequently, for the remainder of this paper, only the adjustment in Equation (1) is applied for both the ACOS and BESD CO$_2$ products (after bias corrections).

3.3 Combination and time shifting

Based on the processes described above, the world is divided into a number of 0.5°×0.5° latitude/longitude grid box (totally 720×360). For each grid cell, Equation (2) is used to combine the corresponding CO$_2$ measurements within that grid.

$$XCO_2\_Fued = \sum_{i=1}^{m} \left(XCO_2\_i \times \frac{1 - Uncert\_ratio^i}{\sum_{i=1}^{m}(1 - Uncert\_ratio^i)} \right) \qquad (2)$$

where XCO_2_Fued is the combined XCO$_2$; m is the total number of space-based CO$_2$ retrievals (ACOS and/or BESD) within a certain grid; XCO_2_i is the ith XCO$_2$ retrieval in a grid for which

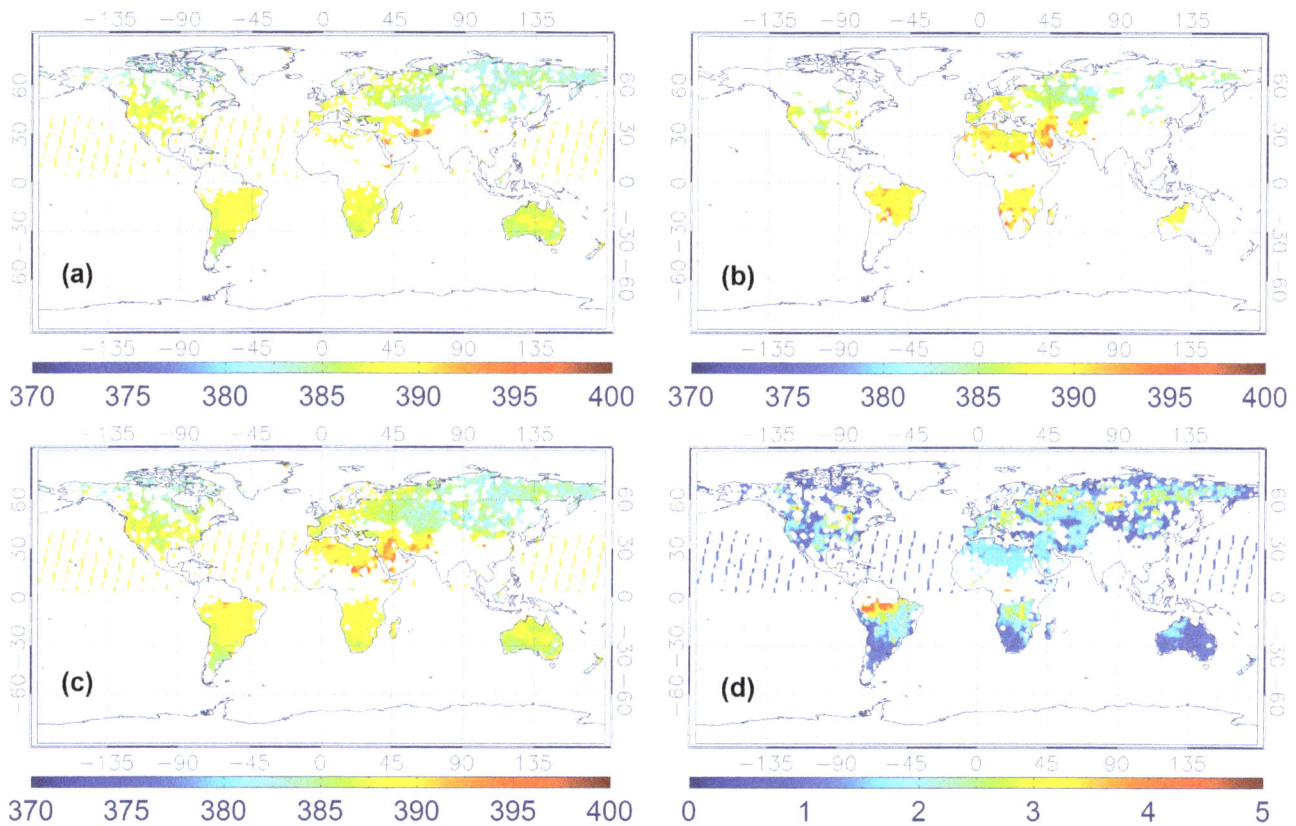

Figure 4. XCO$_2$ monthly mean maps in July of 2010. ((a) ACOS XCO$_2$, (b) BESD XCO$_2$, (c) combined product, and (d) XCO$_2$ uncertainties of the combined product).

the global bias and Equation (1) are supposed to be applied; *Uncert_ratio^i* is the ratio of uncertainty of the ith XCO$_2$ retrieval to its XCO$_2$ value.

Please note that since different CO$_2$ retrievals have distinct overpass times, it is necessary to unify them to avoid uncertainties induced from the time discrepancy before fusion. To this end, a method for considering the CO$_2$ shifting along time has been developed (Equation (3)). First, designate a specific time or select one overpass time as reference, then transfer CO$_2$ measurements at various overpass times to that of the reference time by interpolating the CT CO$_2$ at temporal scale. Here, it should be pointed out that despite the CO$_2$ absolute values of CT not being accurate enough, the daily cycle pattern of atmospheric CO$_2$ it reflects is assumed to be correct.

$$XCO_2\_ref = \frac{\omega^T X\_ref^{CT}}{\omega^T X\_t^{CT}} \times XCO_2\_t \qquad (3)$$

Here, XCO_2_ref is the transformed XCO$_2$ (ACOS or BESD) at the reference time; XCO_2_t is the retrieved XCO$_2$ from ACOS or BESD at overpass time t; X_ref^{CT} and X_t^{CT} are CO$_2$ profiles of CT at times of reference and t, respectively; ω is the pressure-weighting vector (column vector).

Based on the time-shifting strategy proposed here, a global CO$_2$ map at any specific time can be theoretically produced by employing the pattern of the CO$_2$ daily cycle reflected by CT data. For instance, we can unify all XCO$_2$ retrievals being combined with various overpass times to that of UTC = 1.

Results

Evaluation analysis showed that the global bias for the BESD product is generally small. In this study, the bias of the BESD product is corrected by subtracting 0.6 ppm from all XCO$_2$ values according to the results in Fig. 1. Although the systematic bias of the XCO$_2$ retrievals is removed, it is supposed that the error characteristics (random error) within the data are still unchanged. The bias-corrected XCO$_2$ retrievals of both ACOS and BEDS are used as fundamental data for the combination algorithm.

By applying the series of processes shown in Section 3, daily, weekly, as well as monthly maps of combined XCO$_2$ for 2009 and 2010 are generated. Here, as an example, only four maps (from May to August) of monthly mean XCO$_2$ of 2010 are shown here (Fig. 2–Fig. 5). In addition, the total XCO$_2$ uncertainties of the combined product which mainly depend on the uncertainties of the original ACOS or BESD XCO$_2$ retrievals are also illustrated.

From Fig. 2–Fig. 5, it is not difficult to observe that the combined data realize the physical complementary of the two products in terms of spatial coverage. The number of valid CO$_2$ measurements in the fused product is the union of the CO$_2$ data from both the ACOS and BESD at the same geographical location. In addition, the combined XCO$_2$ demonstrates similar spatiotemporal characteristics with that of ACOS and BESD over the globe, which implies that all processes associated with the combination do not distort the essential information of the original XCO$_2$ products (ACOS or BESD). Similar findings can also be observed in the daily mean and weekly mean XCO$_2$ maps. To quantitatively investigate the improvement of fused XCO$_2$ in spatial coverage, the fractional coverage of all three variables

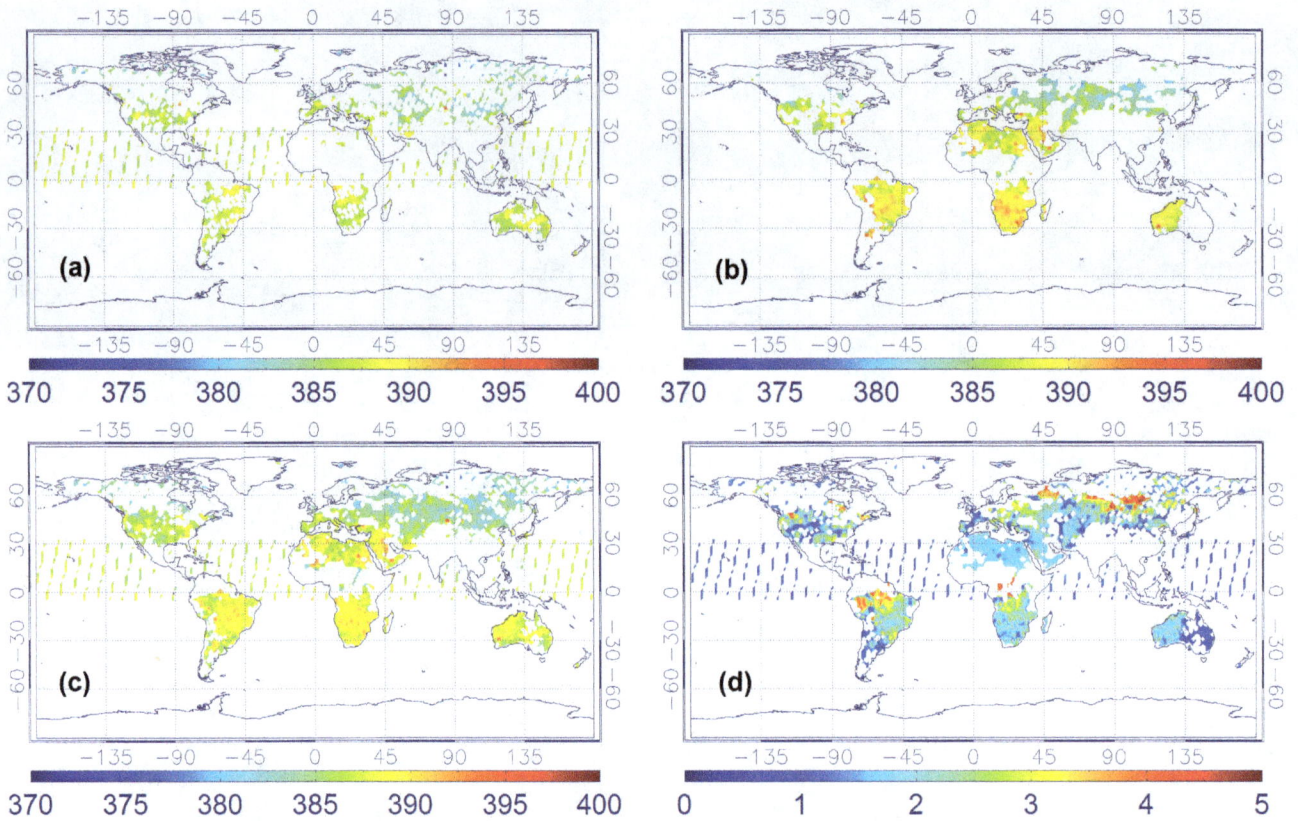

Figure 5. XCO$_2$ monthly mean maps in August of 2010. ((a) ACOS XCO$_2$, (b) BESD XCO$_2$, (c) combined product, and (d) XCO$_2$ uncertainties of the combined product).

(ACOS, BESD, and combined XCO$_2$) on both daily and monthly scales is calculated (Fig. 6). From Fig. 6, it can be seen that the average global coverage of ACOS and BESD is around 0.46% and 0.21%, respectively, on a daily scale. The monthly mean coverage of such products accounts for about 5.70% and 3.75%, respectively. While spatial coverage of combined XCO$_2$ can reach up to 0.65% and 8.42% on daily and monthly scales, respectively, it accounts for increments of 41.3% and 47.7% on the daily and

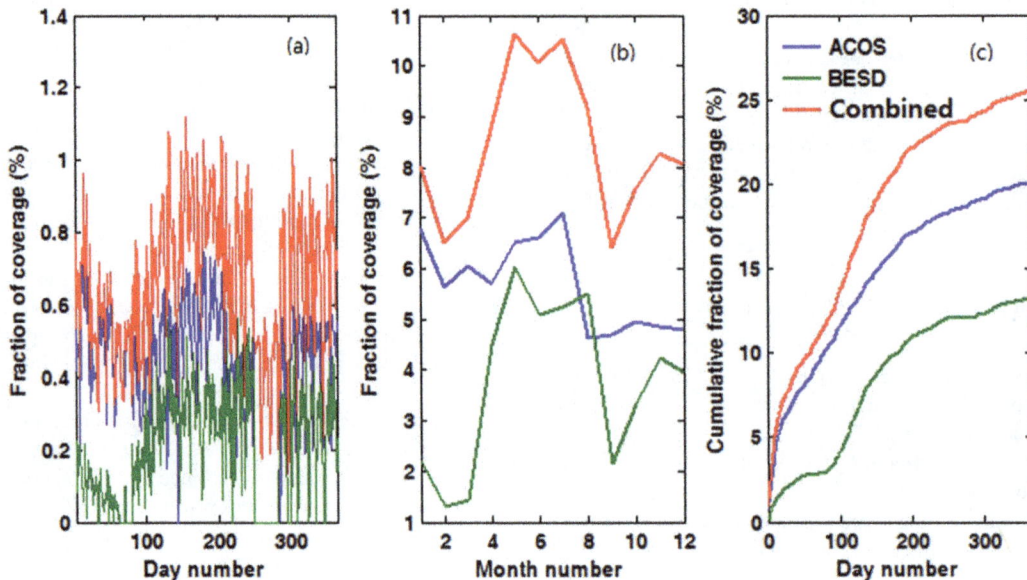

Figure 6. XCO$_2$ fraction of coverage of ACOS, BESD, and combined products. (a) Daily coverage. (b) Monthly coverage. (c) Cumulative coverage.

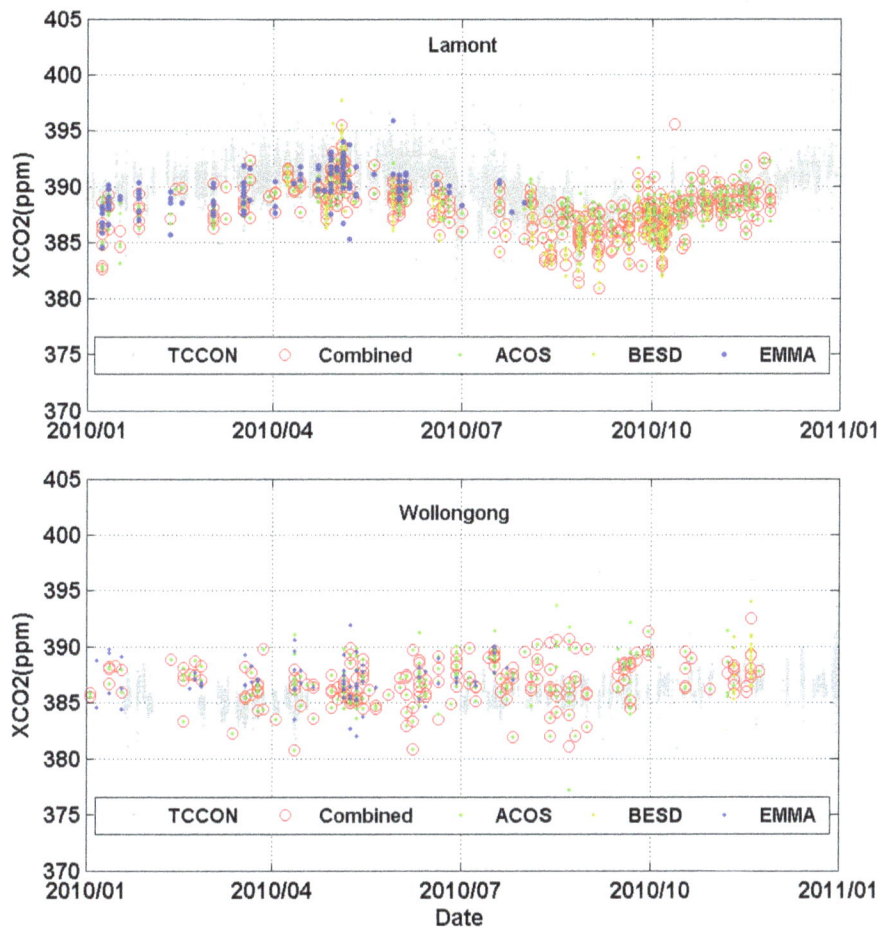

Figure 7. Comparison of XCO$_2$ measurements from TCCON, ACOS, BESD, EMMA, and our new combination method over Wollongong and Lamont sites (distance<0.25 degree, temporal difference<1 hour).

monthly scales with respect to that of GOSAT and it is even higher relative to the coverage of SCIAMACHY. Likewise, the cumulative fraction of coverage of the combined XCO$_2$ has risen to 25% when compared with 20% and 13% for ACOS and BESD, respectively. The increase in the XCO$_2$ spatial coverage indicates the potential advantage of the combined XCO$_2$ observations in generating global Level-3 XCO$_2$ maps when compared with any single dataset by providing more satellite-based XCO$_2$ retrievals used for optimal interpolating.

For evaluating the performance of our combination strategy, the combined XCO$_2$ values are compared with that retrieved from ACOS and BESD as well as XCO$_2$ in the EMMA database at two TCCON sites (Fig. 7). The results reveal that the XCO$_2$ values from the combination method show generally consistent variation in time with TCCON measurements except for a small overall bias (especially for the Lamont site). On the whole, the new combined XCO$_2$ product shows good consistency with the EMMA data, and they are comparable in terms of CO$_2$ magnitude, while the combined XCO$_2$ are shown with a longer time period, which is in line with the satellite observations, and possess more data points even over the same period.

Discussions and Conclusions

Despite the fact that space-based measurements can provide a unique opportunity to map atmospheric CO$_2$ over large areas, the number of valid CO$_2$ measurements from a single space-based instrument is generally limited for a certain day over a specific region due to the presence of clouds. In addition, although these Level-2 XCO$_2$ retrievals themselves are very important for inversion modeling of surface carbon sources/sinks, further comprehensive analysis by investigating the spatiotemporal full-coverage XCO$_2$ (Level 3) distribution is needed for interpreting their significant scientific merit [32]. While the limited satellite observations restrict the generation of Level-3 XCO$_2$ maps with high spatial and temporal resolutions when only a single satellite-based XCO$_2$ dataset is considered. This is our main motivation in this paper.

In this study, a strategy for combining SCIAMACHY and GOSAT CO$_2$ measurements has been proposed by fully accounting for the CO$_2$ global bias, differences in averaging kernels and overpass times, and the Level-2 retrieval errors of the CO$_2$ measurements being used. The results indicated that the average global coverage of both ACOS and BESD is less than 0.5% on a daily scale, and less than 6% on a monthly scale. While spatial coverage of combined XCO$_2$ can reach up to 0.65% and 8.42% on daily and monthly scales, respectively, the comparison analysis reveals that the combined XCO$_2$ product is consistent with TCCON and EMMA in both temporal variation and magnitude except for a small bias when compared with the TCCON measurements. All these findings herein prove the effectiveness of the combination method in supporting generation global full-coverage XCO_2 maps with higher temporal and spatial sampling by jointly using two space-based XCO_2 datasets. Similar to the existing studies (e.g. [32–34]), although these combined XCO_2 are not intended to be used in inverse modeling studies, they deliver a key complement for such research, and can be deemed as an independent dataset for comparison with model predictions. Similar to the existing study [31], an improved fusion approach (based on multiple XCO_2 datasets) to create Level-2 XCO_2 measurements that can be directly used for inverse modeling is also attempted and will be presented in another paper.

A last point that needs to be addressed is that although we employed CO_2 data of GOSAT and SCIAMACHY in this study, the proposed strategies are not restricted to such data. As a general strategy, it can be refined and adapted to further combine other XCO_2 products, such as OCO-2, CarbonSat, etc. in the future,

and even to be applied to the fusion of other trace gases, such as O_3, CH_4.

Acknowledgments

The authors would like to thank the SCIAMACHY team at University of Bremen IUP/IFE as well as the ACOS scientific teams for providing us the CO_2 products. The authors also thank the anonymous reviewers for their helpful and valuable comments to improve this work.

Author Contributions

Conceived and designed the experiments: TW JS YJ TZ DJ CX. Performed the experiments: TW JS YJ TZ DJ CX. Analyzed the data: TW JS YJ TZ DJ CX. Contributed reagents/materials/analysis tools: TW JS YJ TZ DJ CX. Wrote the paper: TW JS YJ TZ DJ CX.

References

1. Reuter M, Bovensmann H, Buchwitz M, Burrows JP (2010) A method for improved SCIAMACHY CO_2 retrieval in the presence of optically thin clouds. Atmos. Meas. Tech 3: 209–232.

2. O'Dell CW, Connor B, Bösch H, O'Brien D, Frankenberg C, et al. (2012) The ACOS CO_2 retrieval algorithm – Part 1: Description and validation against synthetic observations. Atmos Meas Tech 5(1): 99–121.

3. Baker DF, Law RM, Gurney KR, Rayner P, Peylin P, et al (2006) TransCom 3 inversion intercomparison: Impact of transport model errors on the interannual variability of regional CO_2 fluxes, 1988–2003. Global Biogeochem Cy 20: GB1002, doi:10.1029/2004GB002439.

4. Morino I, Uchino O, Inoue M, Yoshida Y, Yokota T, et al. (2011) Preliminary validation of column-averaged volume mixing ratios of carbon dioxide and methane retrieved from GOSAT short-wavelength infrared spectra. Atmos Meas Tech 4: 1061–1076.

5. Butz A, Hasekamp OP, Frankenberg C, Aben I (2009) Retrievals of atmospheric CO_2 from simulated space-borne measurements of backscattered near-infrared sunlight: accounting for aerosol effects. Appl Opt 18: 3322–3336.

6. Kuze A, Suto H, Nakajima M, Hamazaki T (2009) Thermal and near infrared sensor for carbon observation Fourier-transform spectrometer on the Greenhouse Gases Observing Satellite for greenhouse gases monitoring, Appl Opt 35: 6716–6733.

7. Bovensmann H, Burrows JP, Buchwitz M, Frerick J, Noel S, et al (1999) SCIAMACHY –Mission objectives and measurement modes. J Atmos Sci 56: 127–150.

8. Crisp D, Atlas RM, Breon FM, Brown LR, Burrows JP, et al (2004) The Orbiting Carbon Observatory (OCO) mission. Adv Space Res 34: 700–709.

9. Boesch H, Baker D, Connor B, Crisp D, Miller C (2011) Global Characterization of CO_2 Column Retrievals from Shortwave-Infrared Satellite Observations of the Orbiting Carbon Observatory-2 Mission. Remote Sens 3: 270–304.

10. Aumann HH, Chahine MT, Gautier C, Goldberg MD, Kalnay E, et al. (2003) AIRS/AMSU/HSB on the Aqua Mission: Design, science objectives, data products, and processing systems. IEEE Trans Geosci Remote Sens 41: 253–264.

11. Chahine M, Barnet C, Olsen ET, Chen L, Maddy E (2005) On the determination of atmospheric minor gases by the method of vanishing partial derivatives with application to CO_2. Geophys Res Lett 32: L22803, doi:10.1029/2005GL024165.

12. Phulpin T, Cayla F, Chalon G, Diebel D, Schlussel P (2002) IASI on board Metop: Project status and scientific preparation. Proceedings of the 12th International TOVS Study Conference. Lorne, Vi ctoria, Australia.

13. Turquety S, Hadji-Lazaro J, Clerbaux C, Hauglustaine DA, Clough SA, et al. (2004) Operational trace gas retrieval algorithm for the Infrared Atmospheric Sounding Interferometer. J Geophys Res 109: D21301, doi:10.1029/2004JD004821.

14. Yang Z, Washenfelder RA, Keppel-Aleks G, Krakauer NY, Randerson JT, et al. (2007) New constraints on Northern Hemisphere growing season net flux. Geophys Res Lett 34: L12807. doi:10.1029/2007GL029742.

15. Wunch D, Toon GC, Blavier JFL, Washenfelder R, Notholt J, et al. (2011) The Total Carbon Column Observing Network. Philos T Roy Soc A 369: 2087–2112.

16. Rayner PJ, O'Brien DM (2011) The utility of remotely sensed CO_2 concentration data in surface source inversions. Geophys Res Lett 28: 175–178.

17. Houweling S, Breon FM, Aben I, Rodenbeck C, Gloor M, et al. (2004) Inverse modeling of CO_2 sources and sinks using satellite data: a synthetic inter-comparison of measurement techniques and their performance as a function of space and time. Atmos Chem Phys 4: 523–538.

18. Olsen SC, Randerson JT (2004) Differences between surface and column atmospheric CO_2 and implications for carbon cycle research. J Geophys Res 109; D02301. doi:10.1029/2003JD003968.

19. Miller CE, Crisp D, DeCola PL, Olsen SC, Randerson JT, et al. (2007) Precision requirements for space-based XCO_2 data. J Geophys Res 112: D10314, doi:10.1029/2006JD007659.

20. Buchwitz M, Beek R, Burrows JP, Bovensmann H, Warneke T, et al. (2005) Atmospheric methane and carbon dioxide from SCIAMACHY satellite data: initial comparison with chemistry and transport models, Atmos Chem Phys 5: 941–962.

21. Buchwitz M, Beek R, Nol S, Burrows JP, Bovensmann H, et al. (2005) Carbon monoxide, methane and carbon dioxide columns retrieved from SCIAMACHY by WFM-DOAS: year 2003 initial data set. Atmos Chem Phys 5: 3313–3329.

22. Buchwitz M, Rozanov VV, Burrows JP (2000) A near-infrared optimized DOAS method for the fast global retrieval of atmospheric CH_4, CO, CO_2, H_2O, and N_2O total column amounts from SCIAMACHY Envisat-1 nadir radiances. J Geophys Res 105: 15231–15245, doi:10.1029/2000JD900191.

23. Houweling S, Hartmann W, Aben I, Schrijver H, Skidmore J, et al. (2005) Evidence of systematic errors in SCIAMACHY-observed CO_2 due to aerosols. Atmos Chem Phys 5: 3003–3013.

24. Barkley MP, Frieβ U, Monks PS (2006) Measuring atmospheric CO_2 from space using Full Spectral Initiation (FSI) WFM-DOAS, Atmos Chem Phys 6: 3517–3534.

25. Schneising O, Buchwitz M, Burrows JP, Bovensmann H, Reuter M, et al. (2008) Three years of greenhouse gas column-averaged dry air mole fractions retrieved from satellite – Part 1: Carbon dioxide. Atmos Chem Phys 8: 3827–3853.

26. Reuter M, Bovensmann H, Buchwitz M, Burrows JP, Connor BJ, et al. (2011) Retrieval of atmospheric CO_2 with enhanced accuracy and precision from SCIAMACHY: Validation with FTS measurements and com-parison with model results. J Geophys Res 116: D04301, doi:10.1029/2010JD015047.

27. Bösch H, Toon GC, Sen B, Washenfelder RA, Wennberg PO, et al. (2006) Space-based near-infrared CO_2 measurements: Testing the Orbiting Carbon Observatory retrieval algorithm and validation concept using SCIAMACHY observations over Park Falls, Wisconsin. J Geophys Res 111: D23302, doi:10.1029/2006JD007080.

28. Butz A, Guerlet S, Hasekamp O, Schepers D, Galli A, et al. (2011) Toward accurate CO_2 and CH_4 observations from GOSAT. Geophys Res Lett 14: L14812. DOI:10.1029/2011GL047888.

29. Yoshida Y, Ota Y, Eguchi N, Kikuchi N, Nobuta K, et al. (2011) Retrieval algorithm for CO_2 and CH_4 column abundances from short-wavelength infrared spectral observations by the Greenhouse Gases Observing Satellite. Atmos Meas Tech 4: 717–734.

30. Crisp D, Fisher B, O'Dell C, Frankenberg C, Basilio R, et al. (2012)The ACOS CO_2 retrieval algorithm - Part II: Global XCO_2 data characterization. Atmos Meas Tech 5: 687–707.

31. Reuter M, Bösch H, Bovensmann H, Bril A, Buchwitz M, et al. (2013) A joint effort to deliver satellite retrieved atmospheric CO_2 concentrations for surface flux inversions: the ensemble median algorithm EMMA. Atmos Chem Phys 13: 1771–1780.

32. Hammerling DM, Michalak AM, O'Dell C, Kawa SR (2012) Global CO_2 distributions over land from the Greenhouse Gases Observing Satellite(GO-SAT). Geophys Res Lett 39: L08804, doi:10.1029/2012GL051203.

33. Hammerling DM, Michalak AM, Kawa SR (2012) Mapping of CO_2 at high spatiotemporal resolution using satellite observations: Global distributions from OCO-2. J Geophys Res 117: D06306, doi:10.1029/2011JD017015.

34. Zeng Z, Lei L, Hou S, Ru F, Guan X, et al. (2014) A Regional Gap-Filling Method Based on Spatiotemporal Variogram Model of CO_2 Columns. IEEE Trans Geosci Remote Sens 6: 3594–3603.

35. Yokota T, Yoshida Y, Eguchi N, Ota Y, Tanaka T, et al. (2009) Global Concentrations of CO_2 and CH_4 Retrieved from GOSAT: First Preliminary Results. SOLA 5: 160–163.

36. Parker R, Boesch H, Cogan A, Fraser A, Feng L, et al. (2011) Methane observations from the Greenhouse Gases Observing SATellite: Comparison to

ground-based TCCON data and model calculations. Geophys Res Lett 15: L15807, doi:10.1029/2011GL047871.

37. Wang TX, Shi JC, Jing YY, Xie YH (2013) Investigation of the consistency of atmospheric CO_2 retrievals from different space-based sensors: Intercomparison and spatio-temporal analysis, Chin Sci Bull, 33: 4161–4170.

38. Peters W, Jacobson AR, Sweeney C, Andrews AE, Conway TJ, et al. (2007) An atmospheric perspective on North American carbon dioxide exchange: Carbon Tracker. Proc Natl Acad Sci U.S.A. 48: 18925–18930.

39. Wunch D, Wennberg PO, Toon GC, Connor BJ, Fisher B, et al. (2011) A method for evaluating bias in global measurements of CO_2 total columns from space. Atmos Chem Phys 11: 12317–12337.

40. Rodgers CD (2000) Inverse Methods for Atmospheric Sounding: Theory and Practice. World Scientific Publishing Co. Ltd.193 p.

Permissions

All chapters in this book were first published in PLOS ONE, by The Public Library of Science; hereby published with permission under the Creative Commons Attribution License or equivalent. Every chapter published in this book has been scrutinized by our experts. Their significance has been extensively debated. The topics covered herein carry significant findings which will fuel the growth of the discipline. They may even be implemented as practical applications or may be referred to as a beginning point for another development.

The contributors of this book come from diverse backgrounds, making this book a truly international effort. This book will bring forth new frontiers with its revolutionizing research information and detailed analysis of the nascent developments around the world.

We would like to thank all the contributing authors for lending their expertise to make the book truly unique. They have played a crucial role in the development of this book. Without their invaluable contributions this book wouldn't have been possible. They have made vital efforts to compile up to date information on the varied aspects of this subject to make this book a valuable addition to the collection of many professionals and students.

This book was conceptualized with the vision of imparting up-to-date information and advanced data in this field. To ensure the same, a matchless editorial board was set up. Every individual on the board went through rigorous rounds of assessment to prove their worth. After which they invested a large part of their time researching and compiling the most relevant data for our readers.

The editorial board has been involved in producing this book since its inception. They have spent rigorous hours researching and exploring the diverse topics which have resulted in the successful publishing of this book. They have passed on their knowledge of decades through this book. To expedite this challenging task, the publisher supported the team at every step. A small team of assistant editors was also appointed to further simplify the editing procedure and attain best results for the readers.

Apart from the editorial board, the designing team has also invested a significant amount of their time in understanding the subject and creating the most relevant covers. They scrutinized every image to scout for the most suitable representation of the subject and create an appropriate cover for the book.

The publishing team has been an ardent support to the editorial, designing and production team. Their endless efforts to recruit the best for this project, has resulted in the accomplishment of this book. They are a veteran in the field of academics and their pool of knowledge is as vast as their experience in printing. Their expertise and guidance has proved useful at every step. Their uncompromising quality standards have made this book an exceptional effort. Their encouragement from time to time has been an inspiration for everyone.

The publisher and the editorial board hope that this book will prove to be a valuable piece of knowledge for researchers, students, practitioners and scholars across the globe.

List of Contributors

Junjun Zhi, Shengpan Lin, Cao Zhang and Qiankun Liu
College of Environmental and Resource Sciences, Zhejiang University, Hangzhou, China

Changwei Jing and Jiaping Wu
Ocean College, Zhejiang University, Hangzhou, China

Stephen D. DeGloria
Department of Crop and Soil Sciences, Cornell University, Ithaca, New York, United States of America

Nie Xiaojun
Key Laboratory of Mountain Surface Processes and Ecological Regulation, Chinese Academy of Sciences, Institute of Mountain Hazards and Environment, Chinese Academy of Sciences and Ministry of Water Conservancy, Chengdu, China
School of Surveying and Land Information Engineering, Henan Polytechnic University, Jiaozuo, China

Zhang Jianhui and Su Zhengan
Key Laboratory of Mountain Surface Processes and Ecological Regulation, Chinese Academy of Sciences, Institute of Mountain Hazards and Environment, Chinese Academy of Sciences and Ministry of Water Conservancy, Chengdu, China

Carol Arnosti, Andrew D. Steen, Kai Ziervogel and Sherif Ghobrial
Department of Marine Sciences, University of North Carolina, Chapel Hill, North Carolina, United States of America

Wade H. Jeffrey
Center for Environmental Diagnostics and Bioremediation, University of West Florida, Pensacola, Florida, United States of America

Dennis G. A. B. Oonincx
Department of Plant Sciences, Wageningen University, Wageningen, The Netherlands

Imke J. M. de Boer
Animal Department of Animal Sciences, Wageningen University, Wageningen, The Netherlands

Jiri Jablonsky and Olaf Wolkenhauer
Department of Systems Biology and Bioinformatics, University of Rostock, Rostock, Germany

Martin Hagemann and Doreen Schwarz
Department of Plant Physiology, University of Rostock, Rostock, Germany

Edward W. Maibach and Connie Roser-Renouf
Center for Climate Change Communication, George Mason University, Fairfax, Virginia, United States of America

Anthony Leiserowitz
Yale Project on Climate Change, Yale University, New Haven, Connecticut, United States of America

C. K. Mertz
Decision Research, Eugene, Oregon, United States of America

Brian D. Lutz
Department of Biology, Kent State University, Kent, Ohio, United States of America

Emily S. Bernhardt
Department of Biology, Duke University, Durham, North Carolina, United States of America

William H. Schlesinger
Cary Institute of Ecosystem Studies, Millbrook, New York, United States of America

Douglas A. Schaefer and Kai-Jie Dai
Key Laboratory of Tropical Forest Ecology, Xishuangbanna Tropical Botanical Garden, Chinese Academy of Sciences, Mengla, Yunnan, China

Xishuangbanna Tropical Botanical Garden, Chinese Academy of Sciences, Kunming, China

Yi-Ping Zhang, Li-Qing Sha, Yun Deng and Xiao-Bao Deng
Key Laboratory of Tropical Forest Ecology, Xishuangbanna Tropical Botanical Garden, Chinese Academy of Sciences, Mengla, Yunnan, China
Xishuangbanna Tropical Botanical Garden, Chinese Academy of Sciences, Kunming, China Xishuangbanna Station for Tropical Rain Forest Ecosystem Studies, Chinese Ecosystem Research Net, Mengla, Yunnan, China

Wen-Jun Zhou
Key Laboratory of Tropical Forest Ecology, Xishuangbanna Tropical Botanical Garden, Chinese Academy of Sciences, Mengla, Yunnan, China
Xishuangbanna Tropical Botanical Garden, Chinese Academy of Sciences, Kunming, China Xishuangbanna Station for Tropical Rain Forest Ecosystem Studies, Chinese Ecosystem Research Net, Mengla, Yunnan, China

University of Chinese Academy of Sciences, Beijing, China

Jianbo Wu, Xuyang Lu, Jihui Fan, Yanjiang Cai and Xiaodan Wang
The Key Laboratory of Mountain Environment Evolution and Its Regulation, Institute of Mountain Hazard and Environment, CAS, Chengdu, China

Jiangtao Hong and Jian Sun
The Key Laboratory of Mountain Environment Evolution and Its Regulation, Institute of Mountain Hazard and Environment, CAS, Chengdu, China
University of Chinese Academy of Sciences, Beijing, China

Xiujun Wang
State Key Laboratory of Desert and Oasis Ecology, Xinjiang Institute of Ecology and Geography, Chinese Academy of Sciences, Urumqi, Xinjiang, China
Earth System Science Interdisciplinary Center, University of Maryland, College Park, Maryland, United States of America

Jiaping Wang and Juan Zhang
State Key Laboratory of Desert and Oasis Ecology, Xinjiang Institute of Ecology and Geography, Chinese Academy of Sciences, Urumqi, Xinjiang, China
Graduate University of Chinese Academy of Sciences, Beijing, China

Philip Nuss
Center for Industrial Ecology, School of Forestry and Environmental Studies, Yale University, New Haven, Connecticut, United States of America

Matthew J. Eckelman
Department of Civil and Environmental Engineering, Northeastern University, Boston, Massachusetts, United States of America

Agustín J. Sánchez-Medina, Leonardo Romero-Quintero
Department of Economics and Management, University of Las Palmas de Gran Canaria, The Canary Islands, Spain
University Institute of Cybernetic Science and Technology, University of Las Palmas de Gran Canaria, The Canary Islands, Spain

Silvia Sosa-Cabrera
Department of Economics and Management, University of Las Palmas de Gran Canaria, The Canary Islands, Spain

Ariell Friedman, Oscar Pizarro, Stefan B. Williams and Matthew Johnson-Roberson
Australian Centre for Field Robotics, University of Sydney, Australia

Lele Hu
Institute of Systems Biology, Shanghai University, Shanghai, China
Department of Chemistry, College of Sciences, Shanghai University, Shanghai, China

Tao Huang
Key Laboratory of Systems Biology, Shanghai Institutes for Biological Sciences, Chinese Academy of Sciences, Shanghai, China
Shanghai Center for Bioinformation Technology, Shanghai, China

Xiao-Jun Liu
College of Animal Science and Technology, Shihezi University, Shihezi City, Xinjiang, China

Yu-Dong Cai
Institute of Systems Biology, Shanghai University, Shanghai, China
Centre for Computational Systems Biology, Fudan University, Shanghai, China

Matt Finer
Biodiversity Program, Center for International Environmental Law, Washington D.C., United States of America

Clinton N. Jenkins
Department of Biology, North Carolina State University, Raleigh, North Carolina, United States of America

Bill Powers
E-Tech International, Santa Fe, New Mexico, United States of America

Leticia Britos, Harley McAdams and Lucy Shapiro
Department of Developmental Biology, Stanford University School of Medicine, Stanford, California, United States of America

Eduardo Abeliuk
Department of Developmental Biology, Stanford University School of Medicine, Stanford, California, United States of America
Department of Electrical Engineering, Stanford University, Stanford, California, United States of America

Thomas Taverner and Mary Lipton
Environmental Molecular Sciences Laboratory, Pacific Northwest National Laboratory, Richland, Washington, United States of America

Min Wang
Linze Inland River Basin Research Station, Chinese Ecosystem Network Research, Cold and Arid Regions Environmental and Engineering Research Institute, Chinese Academy of Sciences, Lanzhou, Gansu, China

University of Chinese Academy of Sciences, Beijing, China

Yongzhong Su and Xiao Yang
Linze Inland River Basin Research Station, Chinese Ecosystem Network Research, Cold and Arid Regions Environmental and Engineering Research Institute, Chinese Academy of Sciences, Lanzhou, Gansu, China

Xue-Dong Lou
Chinese Research Academy of Environmental Sciences, Beijing, China
College of Life Sciences, Northwest Agriculture & Forestry University, Yangling, Shaanxi, China

Sheng-Qiang Zhai and Li-Le Hu
Chinese Research Academy of Environmental Sciences, Beijing, China

Bing Kan
College of Life Sciences, Northwest Agriculture & Forestry University, Yangling, Shaanxi, China

Ya-Lin Hu
Institute of Applied Ecology, Chinese Academy of Sciences, Shenyang, China

Åsa Holmner
Department of Radiation Sciences, Umeå University, Umeå, Sweden

Kristie L. Ebi
ClimAdapt, LLC, Seattle, Washington, United States of America

Department of public health and clinical medicine, epidemiology and global health, Umeå University, Umeå, Sweden

Lutfan Lazuardi
Department of Public Health, Faculty of Medicine, Gadjah Mada University, Yogyakarta, Indonesia

Maria Nilsson
Department of public health and clinical medicine, epidemiology and global health, Umeå University, Umeå, Sweden

Tianxing Wang, Jiancheng Shi, Yingying Jing, Tianjie Zhao, Dabin Ji and Chuan Xiong
State Key Laboratory of Remote Sensing Science, Institute of Remote Sensing and Digital Earth, Chinese Academy of Sciences. Beijing, China

Index

A

Alpine Steppe, 57-60, 62-63

Animal Protein, 22, 25

Arid Land, 65-66

Autonomous Underwater Vehicles, 95-96

B

Bacteria, 16-21, 27, 29, 32, 131, 141-142, 148, 159

Beef, 22, 24-25

Biodiversity, 26, 47, 95, 117, 120, 129-130, 158

C

Calcareous Soils, 65-66, 68-70

Calcimeter, 65-66, 68-70

Carbon Balance, 26, 48, 55, 158-159

Carbon Cycle, 1, 8-9, 15-16, 20-21, 54-55, 57, 63, 65-66, 70, 150, 158-159, 164-165, 178, 184

Carbon Density, 4, 8, 150, 152-156

Carbon Dynamics, 8, 10, 15, 48, 55, 70, 157, 165

Carbon Sequestration, 9, 43, 47, 157

Carbon Starvation, 131-148

Carbon-starved Cells, 131-132

Caulobacter Crescentus, 131, 147-149

Cell Cycle, 110-111, 113-115, 131-137, 140-142, 144, 146-148

Cells, 27-29, 129, 131-132, 136-142, 144-148

Chicken, 22, 24-25

Chromosome, 131-132, 137, 140-141, 145, 148-149

Coal, 43-47

D

Data Acquisition, 96

Deforestation, 117, 119, 121, 124, 126-129

Degradation, 16, 19-20, 22, 47, 131, 137, 140, 142, 144, 148, 150-151, 158, 163-166

Dry Season, 48, 51-53

E

Economic Activity, 83

Ecosystem, 8, 43, 45-48, 53-55, 57, 60, 63-64, 71-72, 74-75, 77, 120, 150-151, 157-158, 163, 166

Emissions, 22-26, 34, 37-38, 41, 71, 74-76, 78-79, 82, 121, 130, 159, 165, 167-175

Energy, 22-26, 34, 37-41, 47, 71-72, 74-76, 78-80, 82, 84, 87, 120-121, 126, 129-132, 137, 159, 166, 168-171, 174-175

Enzyme, 15-21, 26-27, 29-31, 116, 135, 146

Eukaryotic Cells, 27

F

Feed Conversion Ratio, 23-24

Fine-scale Bathymetric, 95-96

G

Gas, 22-24, 26, 34-35, 38, 41, 71, 75-76, 80, 117-121, 124, 126, 129-130, 173, 175, 184

Gaseous, 48, 53, 55, 166

Genes, 18, 21, 31, 33, 109-110, 113, 115-116, 131-138, 140-144, 146-148

Genetics, 20, 109, 113, 116, 149

Global Gradient, 16

Global Warming, 8, 19, 23-25, 34-43, 47, 71-72, 75, 78, 175, 177

Globalization, 83, 175

Glycolysis, 27-32

Glycolytic Enzyme, 27

Grassland, 15, 45-46, 57-58, 63-64, 66, 150-151, 154, 156-157

Greenhouse Gas Emissions, 22, 26, 34, 71, 75-76, 121, 130, 173, 175

H

Hydrocarbon, 117-124, 126-130

I

In-situ, 95-97, 99, 102, 107

Isometric Allocation Hypothesis, 57-58, 60, 63

L

Latitudinal Gradients, 16-17, 19

Life Cycle, 22, 26, 71-72, 74, 77, 79-80, 82, 119, 131, 167-173, 175

Loss-on-ignition, 65-66, 70

M

Marine, 16-21, 70, 95-96, 104, 107-108

Mealworm, 22-25

Metals, 71-82, 131, 133, 135, 141, 144, 147

Microbial Biomass Carbon, 9, 15

Microbial Communities, 16-21

Milk, 22, 24-25

Mining, 43-47, 71-76, 78-79, 81, 109-110, 130

Morphogenesis, 116, 131, 137, 139-140

O

Oil, 117-124, 126, 128-130, 145

One-isoenzyme, 27, 30

Organic Carbon Stocks, 1, 4, 8, 157

Organisms, 16, 18-19, 57, 92, 109-110, 114-116, 147, 151

P

Particulate Inorganic C, 48

Phenotypes, 109-116

Phosphoglycerate-mutase, 27

Plant Growth, 57, 60, 151

Pork, 22, 24-25

Production Systems, 71-72, 74, 84

Professional Knowledge Based, 1, 3, 5-6

Prokaryotic Cells, 27

Proteome, 110, 116, 131-133, 135, 137, 140-142, 147-149

Public Administration, 83

Purification, 71, 73-75, 79, 146

R

Rainforest, 48, 50-55, 130

Rainy Season, 48, 51-53

Refining, 71-74, 79, 82

Remotely Operated Vehicles, 95

Rivers, 3, 44, 47-48, 54-56, 120, 123, 129, 158

Root-to-shoot Ratio, 57, 59, 62

Rugosity, 95-99, 102, 104, 106-108

S

Simple Linear Regression, 44, 152

Slope, 9-15, 44-46, 59, 68, 70, 95-96, 99, 102, 104, 106-108, 112-113

Soil, 1-15, 43-45, 48, 52-55, 57-60, 63-70, 74, 120, 150-157, 159-160, 163, 165-166

Soil Profile Statistics, 1, 3

Soil Quality, 1, 15, 156

Standardized Major Axis, 57, 59, 63

Stream Impairment, 43-46

Stream Water Temperature, 48, 51, 54

Survey Method, 40

T

Tillage Erosion, 9-15

Total Organic C, 48

Transcriptome, 131-132, 140-141, 148

Tropical Seasonal Rainforest, 48, 50, 52-55

W

Walkley-black Method, 65-70

Water Erosion, 9-12, 14-15

Watersheds, 10, 44, 53, 55, 117, 119-120, 129

Y

Yeast, 23, 144